STUDY GUIDE

Chemistry: The Science in Context

Second Edition

STUDY GUIDE

Chemistry: The Science in Context

Second Edition

Thomas R. Gilbert, Rein V. Kirss, Natalie Foster, Geoffrey Davies

Stephanie Myers

AUGUSTA STATE UNIVERSITY

Richard Lavrich

COLLEGE OF CHARLESTON

W • W • NORTON & COMPANY • NEW YORK • LONDON

Printed in the United States of America

ISBN 13: **978-0-393-93202-7** (pbk.)

W. W. Norton & Company, Inc., 500 Fifth Avenue, New York, NY 10110
www.wwnorton.com

W. W. Norton & Company, Ltd., Castle House, 75/76 Wells Street, London W1T 3QT

1 2 3 4 5 6 7 8 9 0

CONTENTS

CHAPTER 1 Matter, Energy, and the Origins of the Universe

OUTLINE

REVIEW

Chapter Overview

The study of **chemistry** is the scientific study of matter. **Matter** is defined as any substance that has mass and takes up space. In the study of matter, chemists also study the changes that matter undergoes and the energy that accompanies those changes.

One of the first steps in studying something is to categorize it. One of the common ways that chemists use to categorize matter is by its composition. All matter can be classified as either a mixture or a pure substance. **Pure substances** can be categorized as either compounds or elements. An **element** is the simplest form of matter; therefore, it cannot be separated into simpler substances by any chemical means. The reason for this is that it is made up of **atoms**, the smallest fraction of an element that retains its identity, which are all the same type. A **compound** is made of two or more types of atoms that are connected to each other. This connection is called a **chemical bond**. Because these connections are very specific, the amount of each element in a compound is constant. This is stated in the **law of constant composition**: every sample of a particular compound always contains the same elements combined in the same proportions. A compound has very different properties than the elements from which it is composed. It requires a **chemical separation** to break the bonds and turn a compound into its component elements. The smallest fraction of a compound that still retains its identity as a compound is a **molecule**. A symbol that expresses the number of atoms of each type in a compound is its **chemical formula**.

In **mixtures**, combinations of pure substances in variable proportions, substances do not chemically bond to each other. Thus each pure substance retains its identity. When substances mix in such a way that boundaries between the different substances are not observed and every part of the mixture has the same amount of each substance, it is called a **homogeneous mixture** or a **solution**. If the boundaries between substances are visible, it is classified as a **heterogeneous mixture**. In both cases, the amount of each substance in a mixture may vary. For example, sugar and water can be mixed to make a homogeneous solution, but that mixture may be only slightly sweet if a small amount of sugar is added, or very sweet when much more sugar is added. Similarly, when oil and water are mixed, they form a heterogeneous mixture, but the layer of oil can be thin or thick. Mixtures can also be **physically separated**. That is, the components may be separated without changing their identities. This can be done as simply as skimming the oil from the oil and water mixture or using more specific techniques such as **filtration**. In filtration the mixture is passed through a porous substance (filter) where the holes are small enough that part of the mixture passes through and the other part is retained by the filter. Another method of physical separation is **distillation**. In this process, the mixture is heated so that part of the substance is turned into a gas; the gas is collected separate from the remaining liquid and reliquefied.

It is also useful to categorize the properties of matter. One classification depends on whether or not the property depends on the amount of substance. **Extensive properties**, such as mass and volume, do depend on the amount of substance. **Intensive properties**, such as color and hardness, do not depend on the amount of substance. One important intensive property is density. **Density** is the ratio of mass to volume of a substance.

Another way to categorize the properties of matter is as a physical or chemical property. **Physical properties** can be observed without a change in the identity of the substance. Density and mass are both examples of physical properties. **Chemical properties** are observed when a substance does change its identity. For example, a chemical property of water is that it can be decomposed into hydrogen and oxygen.

A third way to categorize matter is by its physical state. **Solids** have a definite shape and volume. This is because the atoms or molecules of the solid are adjacent to each other and not changing positions. **Liquids** have a definite volume, but a variable shape. This is because although the atoms or molecules are adjacent, their positions are not fixed and they can rearrange themselves to the shape of the container. **Gases** have neither definite shape nor volume, but will fill their container. The molecules of gases are moving and separate from each other. Thus the molecules continue in whatever direction they are traveling until they encounter another molecule or the side of the container. Changes in temperature can change one physical state to another. However, this does not change the identity of the substance, thus physical states are physical properties and changes in physical state are physical changes. Normally, with an increase in temperature, a solid will change to a liquid (melting) and then from a liquid to a gas (vaporization). However, sometimes a solid may change directly to a gas. This process is called **sublimation**. Its reverse, gas directly to solid, is **deposition**.

To study matter in a scientific manner is a specific type of process called the **scientific method**. This is an approach to acquiring knowledge based on careful observation of phenomena, development of a simple, testable explanation or **hypothesis**, and then conducting additional experiments to test the validity of the hypothesis. When the hypothesis has been well-tested and accepted by the scientific community it may then be called a **theory**.

Because the scientific method depends so heavily on careful observations and measurements, the quality of the measurements must be expressed in a scientific communication. Two aspects of the quality of the measurement are its accuracy and its precision. **Accuracy** is the closeness of a measurement to its true value. This requires comparing the measurement to a value that is known to be correct. **Precision** is the closeness of answers to each other and requires repeated measurements.

One way of expressing the precision of a measurement is with significant figures. Each measurement will have some numbers that are always the same and some values that vary slightly between measurements. **Significant figures** are all the digits that are certain (do not vary) and one digit that is estimated (has some variability). When many measurements are combined to determine a value, the significant figures in this value will depend on the least precise measure that is used. There are very specific rules for determining how to round the answer of a calculation so that it has the appropriate number of significant figures. The rules depend on the type of calculation. For addition and subtraction, the answer must have the same number of decimal places as the measurement with the fewest decimal places. For multiplication and division, the answer must have the same number of significant figures as the measurement with the fewest significant figures. Numbers that come from definitions or are counted do not have any variability in their values. These numbers are exact and therefore, have an infinite number of significant figures.

Numbers themselves are not a sufficient description of a measurement; all measurements should also include units. Units express the type of measurement made (length, mass, time) and general information on the magnitude of the measurement (miles are longer than inches). In this text, the Système International d'Unités (**SI units**) are used. SI units have a base value for each common type of unit. These base values can be modified for magnitude by using a prefix that changes the value by some factor of 10.

It is often necessary to convert one unit to another. This is sometimes done using an equation. Temperature is one example where unit conversion is done by equation. Because most scientific measurements are made in Celsius (°C) and most scientific calculations use the SI unit **Kelvin (K)**, the conversion

$$K = {}^\circ C + 273.15$$

is used. Kelvin is a temperature scale designed so that **absolute zero** (the coldest temperature that can theoretically exist) has a value of zero. Thus, the Kelvin scale has no negative values for temperature.

Another common way to convert one unit to another is using **dimensional analysis**. This method uses conversion factors, which have equal numerators and denominators but are expressed in different units. The starting unit is multiplied by one or more conversion factors canceling the unwanted units to end up with the desired unit. Because this process is effectively "multiplying by one," it is a change in the way the measurement is expressed but not its actual value.

Regardless of the type of chemistry problem, a useful strategy for approaching the problem uses the acronym COAST. In this method, the information from the problem is *Collected and Organized* so that pertinent information is extracted and supplemental information is obtained. The problem is then *Analyzed*, where connections between the information given and the answer are determined. Next, the information from the first two steps are put together so that the problem can be *Solved*. The last step is to *Think* about the answer, considering whether or not it makes sense and is in the proper format, such as having appropriate significant figures and units.

Worked Examples

Mixtures and How to Separate Them

Example 1.
Classify the following as heterogeneous mixture, homogeneous mixture, compound, or element.

a. 10-k gold b. neon c. fog d. steel e. salt

Collect and Organize We are given five materials and are asked to determine what category of matter they belong to.

Analyze The first step in classifying a given substance is to determine whether it is a mixture or a pure substance. The variable composition of mixtures is usually the best indication that it is a mixture rather than a pure substance. Mixtures can be physically separated. Methods of physical separation particular to chemistry include filtration and distillation. The ratio of substances in a mixture can have many values. Pure substances are often harder to identify because they are usually defined by what they are not. They are not mixtures. They cannot be physically separated.

Having identified the substance as a mixture the next step is to determine whether it is heterogeneous or homogeneous. Heterogeneous mixtures are not uniform across the sample. In other words, the substance will have different characteristics in different locations. A homogeneous mixture should be the same no matter where you sample it.

Once you have identified a pure substance, you can determine whether it is an element or a compound. A compound can be chemically separated. A clue that it is a chemical rather than a physical separation is that the ratio of components will always be the same (composition is *not* variable). Usually, it is easier to see the elements, since every element is listed in the periodic table. If it is named and that name is on the list, it is an element.

Solve

a. 10-k gold is a homogeneous mixture; 24-k gold is known as "pure gold," but is too soft to be used for most purposes. Jewelry commonly comes as 10-, 14-, or 18-karat gold (variable composition).

b. Neon is an element. It is number 10 on the periodic table.

c. Fog is a mixture of water and air. Whether it is homogeneous or heterogeneous depends on how closely you look at it. When caught in a fog bank you might observe that everything looks the same; therefore, it is a homogeneous mixture. When you look closely, you may actually be able to differentiate between the air and water droplets, making it a heterogeneous mixture. If you can get a wider view you might notice that it is denser in some places than in others; thus it is a heterogeneous mixture. Thus its classification depends on the context in which the classification is used.

d. Steel is a homogeneous mixture. There are many types of steel that are easily modified for different properties by changing the ratios of the components.

e. Salt is a compound. It can be separated into sodium and chlorine (elements), but always in the same ratio. You may have heard salt referred to as NaCl, which are the symbols for sodium and chlorine. Compounds can always be shown as formulas of their component elements. (So if you can write a molecular formula for it and the formula has more than one element, it is a compound.)

Think About It Pure substances are the most basic components of nature. They can be either elements or compounds. Elements are the simplest form of matter and are listed in the periodic table. Compounds are combinations of elements that occur in definite proportions, such as H_2O or CH_4. Mixtures are composed of two or more pure substances.

Properties of Matter

Example 1.
Classify the following as physical or chemical properties of water.

a. density = 1.00 g/mL

b. forms ice at 0°C

c. makes hydrogen gas when combined with sodium metal

d. is colorless

e. is formed when wood is burned

Collect and Organize We are given five properties of water and are asked to classify them as either chemical or physical.

Analyze Chemical properties are characterized by a change in identity, whereas physical properties are not. Any property that can be seen or measured is probably a physical property. Any property that describes changes in the substance is a chemical property. Changes in physical state (liquid, solid, or gas) involve physical properties, not chemical ones. Water has different names, depending on its physical state (ice is solid water, steam is gaseous water), but it is still the same substance.

Solve

a. This is a physical property; density can be measured without changing a substance to anything else.

b. This is a physical property; freezing is just changing the physical state, not the identity of the substance.

c. This is a chemical property; hydrogen gas is something different from water. Another tip-off is that another substance was required to achieve the change. Another substance is often, but not always, present in a chemical change.

d. This is a physical property; it is observed without altering the substance.

e. This is a chemical property; that it was not present to begin with suggests an identity change.

Think About It An important distinction between a physical and chemical property is the appearance of a new substance. Physical properties describe the characteristics of a single substance. Chemical properties on the other hand are a description of how the substance is changed into a different substance.

Making Measurements and Expressing the Results

Example 1.
Convert the following values in decimal notation to scientific notation.

a. 0.0000000002 b. 66,700,000 c. 0.000507 d. 7600

Collect and Organize We are asked to express the numbers given in decimal notation in scientific notation.

Analyze Many numbers in chemistry are especially large or especially small. Scientific notation is a way to express these values conveniently and the precision to which the values are measured. Scientific notation takes the form of $N \times 10^x$, where N is a value between 1 and 10, which can be either positive or negative, and x is an integer. The N part of the value expresses the precision of the measurement. The x part of the value expresses the magnitude or position of the decimal place. One convenient way to remember the direction associated with x is that positive values of x represent large numbers and negative values represent fractions.

Solve

a. It is a fraction, so x is negative. For N to be between 1 and 10, the decimal must be moved behind the 2. That is 10 digits, so the value in scientific notation is 2×10^{-10}.

b. It is a large number, so x will be positive. For N to be between 1 and 10, the decimal must move between the two 6's—seven places. So the value is 6.67×10^7.

c. 5.07×10^{-4}

d. 7.6×10^3

Think About It Scientific notation provides a convenient way to express very large or very small numbers in compact form while retaining the essential information concerning the precision to which the value was measured.

Example 2.
Convert the following values in scientific notation to decimal notation.

a. 7×10^9 b. 2.4×10^{-5} c. 7.27×10^{-8} d. 8.4×10^3

Collect and Organize In this example we are asked to convert the numbers given in scientific notation to decimal notation.

Analyze Many numbers in chemistry are especially large or especially small. Scientific notation is a way to express these values conveniently and the precision to which the values are measured. Scientific notation takes the form of $N \times 10^x$, where N is a value between 1 and 10, which can be either positive or negative, and x is an integer. The N part of the value expresses the precision of the measurement. The x part of the value expresses the magnitude or position of the decimal place. One convenient way to remember the direction associated with x is that positive values of x represent large numbers and negative values represent fractions.

Solve

a. a large number, so 7,000,000,000

b. a decimal, so 0.000024

c. 0.0000000727

d. 8400

Think About It Scientific notation takes the form $N \times 10^x$. Positive values of x denote numbers that are greater than one while negative values of x denote numbers that are less than one.

EXAMPLE 3.
Perform the following calculations.

a. $(1.48 \times 10^{-2})(2.6 \times 10^{-4}) =$

b. $\dfrac{4.97 \times 10^{-5}}{8 \times 10^{8}} =$

c. $5.73 \times 10^{8} + 6.1 \times 10^{7} =$

d. $\dfrac{(5.09 \times 10^{5} - 4.9 \times 10^{4})}{3.78 \times 10^{10}} =$

COLLECT AND ORGANIZE We are given several arithmetic operations involving numbers expressed in scientific notation and are asked to solve.

ANALYZE Report the value displayed by your calculator for the above operations.

SOLVE

a. 3.848×10^{-6} or 0.000003848 (They are the same number!)

b. 6.2125×10^{-14}

c. 6.39×10^{8}

d. $5.09 \times 10^{5} - 4.9 \times 10^{4} = 4.6 \times 10^{5}$

$$\frac{4.6 \times 10^{5}}{3.78 \times 10^{10}} = 1.2169312 \times 10^{-5}$$

THINK ABOUT IT Calculators do not take into account the precision to which the values used their calculations are measured. As a result, not all of the numbers produced in the above mathematical operations will be significant.

EXAMPLE 4.
How many significant figures are in the following numbers?

a. 0.0058 b. 310.00 c. 503 d. 0.090 e. 4.670×10^{-6}

COLLECT AND ORGANIZE We are given four numbers and are asked to determine the number of significant figures in each.

ANALYZE To determine the number of significant figures in a number, use the following rules to determine which digits are significant and add.

- All nonzero digits are significant.
- All zeros before the first nonzero digit are *not* significant.
- All zeros between nonzero digits are significant.
- All zeros after the last nonzero digit *and* to the right of the decimal are significant.
- Zeros to the right of the last nonzero digit and before the decimal are significant if the decimal is shown. If the decimal is not shown they are ambiguous (you can't tell).
- If the number is expressed in scientific notation, all digits in the *N* part are significant.

SOLVE

a. Two. Leading zeros are not significant.

b. Five. Zeros after the decimal and all in between are significant.

c. Three. Zeros between nonzero digits are significant.

d. Two. Leading zeros are not significant; trailing zeros are.

e. Four. All digits in the *N* part of scientific notation are significant.

THINK ABOUT IT Some of the leading and trailing zeros in a number may not be significant. In other words, they serve as placeholders. The above rules provide a quick and easy way to determine which zeros are significant.

EXAMPLE 5.
Report answers to the correct number of decimal places.

a. 0.742 + 0.0259 =

b. 4.00 – 3.5 =

c. $1.1 \times 10^{7} + 9.11 \times 10^{6} =$

COLLECT AND ORGANIZE In this problem we are asked to perform three arithmetic operations and to express the result to the correct number of decimal places.

ANALYZE After doing the arithmetic operation we must report the result using the correct number of decimal places. It is important (and necessary) to do the calculations using all digits given. Only then should the result be evaluated for the number of decimal places to be reported. Correctly reporting the result of the calculation requires two steps. First we must determine the correct number of decimal places to use for the particular arithmetic operation under consideration and then round the last decimal place to the correct value. For addition and subtraction, the result will have the same number of decimal places as the value with the fewest decimal places. Once the correct number of decimal places has been determined the following rules are used to correctly round the last decimal place: (1) If the value of the number directly after the last decimal place we will keep is less than five than keep the last digit used the same. (2) If the value of the number directly after the last decimal place we will keep is greater than five than round the last digit we will keep up. (3) If the value of the number directly after the last decimal place is equal to five (and no nonzero digits follow the number), keep the last digit used the same if it is even, otherwise round up.

SOLVE

a. 0.7679 = 0.768. The first value only had three decimal places, whereas the second had four. The last digit, 9, rounds up.

b. 0.5. There is only one decimal place in the second answer. In subtraction it is common to lose significant figures.

c. To determine decimal places it would be better to convert the values to decimal notation. 11,000,000 + 9,110,000 = 20,110,000. In the scientific notation of the first value, only the 1s are significant, whereas the second value has significant digits farther to the right. Consequently, the answer can only go to the millions decimal place and the answer is 20,000,000. So that the reader can determine how many of these digits are significant, the answer should be reported in scientific notation = 2.0×10^7.

THINK ABOUT IT Even nonzero numbers in a value may not be significant.

EXAMPLE 6.

Perform the calculation and report the answer with appropriate significant figures.

a. (0.969)(0.0078) =

b. $\dfrac{8.950}{0.040600} =$

c. $(6.0 \times 10^5)(6.195 \times 10^{-7}) =$

d. $\dfrac{(4.80 \times 10^{13})}{(8.00 \times 10^4)} =$

COLLECT AND ORGANIZE We are given several multiplication and division problems and are asked to evaluate, ensuring that the result is expressed with the correct number of significant figures.

ANALYZE For multiplication and division the answer should have the same number of significant figures as the value with the fewest significant figures. As with addition and subtraction, the value should be calculated first and then rounded. The same rules for rounding are used.

SOLVE

a. 0.969 has three significant figures; 0.0078 has two significant figures. The answer, 0.0075582, should have two significant figures. Therefore, the answer is 0.0076.

b. 8.950 has four significant figures; 0.040600 has five significant figures. The answer should then have four significant figures = 220.4.

c. 6.0×10^5 has two significant figures; 6.195×10^{-7} has four significant figures. The answer is 0.37.

d. Both values have three significant figures. Therefore, the answer should have three significant figures. The answer is 6.00×10^8.

THINK ABOUT IT The rules for determination of significant figures are different for addition/subtraction and multiplication/division.

EXAMPLE 7.

Perform the following operations and report the answer to the appropriate significant figures.

a. $\dfrac{(0.428 + 0.0804)}{0.009800} =$

b. $(31.6 \times 24.78) + (0.569 \times 6.64) =$

COLLECT AND ORGANIZE We are given problems involving mixed operations, problems involving both addition/subtraction and multiplication/division, and are asked to express the result with the correct number of significant figures.

ANALYZE For mixed operations (both addition and multiplication), use each rule in the same order as the mathematical operations.

SOLVE

a. The addition is first: 0.428 + 0.0804 = 0.5084. Only 0.508 is significant, but rounding is a last step, so calculate $\dfrac{0.5084}{0.009800} = 51.87755$. The sum/numerator had three significant figures and the denominator had four, so the answer should have three: 51.9.

b. In this example the multiplication comes first $(31.6 \times 24.78) = 783.048$. Three digits are significant or it is significant to the decimal. $(0.569 \times 6.64) = 3.77816$. Three digits are significant. The sum of the preceding answers is 786.82616. Since the first value is only significant to the decimal, the answer should be reported only that far and the answer is 787.

THINK ABOUT IT Mixed operations involve a series of individual operations. These individual operations can consist of combinations of addition/subtraction and multiplication/division. The number of significant figures for the mixed operation can easily be determined by following the set of rules provided.

EXAMPLE 8.

Report the answers to Example 3 in the scientific notation section to the appropriate significant figures.

COLLECT AND ORGANIZE From Example 3 we have the following results:

a. $(1.48 \times 10^{-2})(2.6 \times 10^{-4}) = 3.848 \times 10^{-6}$ or 0.000003848 (They are the same number!)

b $\dfrac{4.97 \times 10^{-5}}{8 \times 10^{8}} = 6.2125 \times 10^{-14}$

c. $5.73 \times 10^{8} + 6.1 \times 10^{7} = 6.39 \times 10^{8}$

d. $\dfrac{(5.09 \times 10^{5} - 4.9 \times 10^{4})}{3.78 \times 10^{10}} = 1.2169312 \times 10^{-5}$

ANALYZE Use the rules for the number of significant figures for addition/subtraction and multiplication/division in order to correctly determine the number of significant figures for the result of these mixed operation problems.

SOLVE

a. 3.8×10^{-6}

b. 6×10^{-14}

c. 6.39×10^{8}

d. $509{,}000 - 49{,}000 = \dfrac{4.60 \times 10^{5}}{33.78 \times 10^{10}} = 1.22 \times 10^{-5}$

THINK ABOUT IT Significant figures are an essential part of measurements in science. They provide essential information about the precision of the value being reported.

EXAMPLE 9.

$$1.234 \text{ m} + 35.4 \text{ cm} =$$

COLLECT AND ORGANIZE We are asked to add two values given in different units.

ANALYZE Before being able to perform the operation we must first convert the values to the same unit.

SOLVE In addition and subtraction, the units must be the same. Since 1.234 m = 123.4 cm, the solution is 123.4 cm + 35.4 cm = 158.8 cm. (Note that both values, once in the appropriate units, go to the tenth decimal place, so the answer is also to the tenth decimal place, according to the significant-figure rules.) Alternatively, since 35.4 cm = 0.354 m, the solution is 1.234 m + 0.354 m = 1.588 m. Since 1.588 m = 158.8 cm, both answers are the same and both answers are equally correct!

THINK ABOUT IT Because the metric system is based on powers of ten, conversion between units is simply a matter of moving a decimal point.

EXAMPLE 10.

a. (0.316 g/mL)(21 mL) =

b. (1.8646 cm)(1.28 cm) =

COLLECT AND ORGANIZE We are asked to perform two arithmetic operations and to express each result using the correct number of significant figures and units.

ANALYZE Use the rules of significant figures to correctly express the result. Units are treated like algebraic variables in multiplication and division.

SOLVE

a. 6.6 g. Two significant figures; the milliliters cancel.

b. 2.39 cm^2. Three significant figures; cm × cm = cm^2.

THINK ABOUT IT The same unit in both the numerator and the denominator will cancel. When two units are multiplied, the unit is the squared.

UNIT CONVERSIONS AND DIMENSIONAL ANALYSIS

EXAMPLE 1.
How many centimeters are in 7.51 m?

COLLECT AND ORGANIZE We are asked to convert meters into centimeters.

ANALYZE Unit conversions are done using a technique called *dimensional analysis*. This method consists of canceling units by multiplying by 1. This works because multiplying by 1 does not change a value. It also works because a value over the same value is equal to 1. Some of these equivalences are the same as those indicated by the SI prefix relationships.

SOLVE There are 100 cm in 1 m, so 100 cm = 1 m. Our starting value is 7.51 m, so the meter part of the relationship needs to go on the bottom to cancel out.

$$7.51 \text{ m} \times \left(\frac{100 \text{ cm}}{1 \text{ m}}\right) = 751 \text{ cm}$$

There are three significant figures. Since metric prefixes are defined, the relationship has infinite significant figures.

THINK ABOUT IT Dimensional analysis can be used as a check as to whether a problem in chemistry has been solved correctly. One sure-fire way to determine that a mistake has been made in a calculation is to examine the units resulting from the dimensional analysis. If they do not correspond to what is expected, an error has been made.

EXAMPLE 2.
How many centimeters are in 208.0 μm?

COLLECT AND ORGANIZE We are asked to convert 208.0 μm into centimeters.

ANALYZE It doesn't matter how many times the value is multiplied by 1, so you don't need a relationship for every conversion but can use a series of relationships to a central unit. In this case, both use SI prefixes. Therefore, these relationships can be used.

SOLVE There are 100 cm = 1 m and 1,000,000 μm = 1 m, so

$$208.0\ \mu\text{m} \times \left(\frac{1\ \text{m}}{1{,}000{,}000\ \mu\text{m}}\right) \times \left(\frac{100\ \text{cm}}{1\ \text{m}}\right) = 0.02080\ \text{cm}$$

Hint: When doing a long series of conversions on your scientific calculator, ignore all values of 1 (it's just a chance to hit the wrong button), use the "×" button before each number that appears on the top of a conversion and the "÷" button before each number on the bottom. With this method it is unnecessary to use your parenthesis buttons. You would enter the preceding problem as "208.0 ÷ 1,000,000 × 100 =."

THINK ABOUT IT This technique is sometimes referred to as the *factor-unit* method, or the *factor-label* method.

EXAMPLE 3.
What is 9.5 min^{-1} in Hz?

COLLECT AND ORGANIZE We are asked to convert a frequency in units of min^{-1} to units of Hz.

ANALYZE Most of the time, the starting unit is on the top. Occasionally, this is not the case. When the unit is on the bottom of a fraction, it will be represented as u^{-1} or 1/u, or use the term *per* before the unit. That is the case in this problem. Recall that Hz is the same as s^{-1}.

SOLVE

$$9.5/\text{min} \times \left(\frac{1\ \text{min}}{60\ \text{s}}\right) = 0.16\ \text{s}^{-1} = 0.16\ \text{Hz}$$

It is also possible to need to change units on both the top and the bottom. In this case, treat each unit separately. It doesn't matter whether you do the top or the bottom unit first.

THINK ABOUT IT Sometimes units are named for a person. Such is the case for the unit Hz named for the German physicist Heinrich Hertz.

EXAMPLE 4.
What is the speed in km/s of a car going 65.0 mph?

COLLECT AND ORGANIZE Given the speed of a car in mph we are asked to convert to the SI unit of km/s.

ANALYZE Unit conversions are done using a technique called *dimensional analysis*. This method consists of canceling units by multiplying by 1. This works because multiplying by 1 does not change a value. It also works because a value over the same value is equal to 1. Some of these equivalences are the same as those indicated by the SI prefix relationships.

SOLVE Miles per hour is abbreviated as mph or mi/hr. The needed conversions are 1.6 km = 1 mi, 60 s = 1 min, and 60 min = 1 hr. Units will have to be changed on both the top and bottom of the relationship.

$$65.0\ \text{mi/hr} \times \left(\frac{1.6\ \text{km}}{1\ \text{mi}}\right) \times \left(\frac{1\ \text{hr}}{60\ \text{min}}\right) \times \left(\frac{1\ \text{min}}{60\ \text{s}}\right) = 0.029\ \text{km/s}$$

There are only two significant figures. The "weak link" was the conversion to km from miles. Although SI-to-SI conversions are exact, when the unit system changes, the values are measured or approximate. The interpretation of "1.6 km = 1 mi" is that a measured 1.6 km is exactly 1 mile. The time conversions are exact.

THINK ABOUT IT You can convert either unit first. It is a property of multiplication that the calculation can be done in any order. For this problem, you could enter this conversion in your calculator as

65.0 × 1.6 ÷ 60 ÷ 60 =

EXAMPLE 5.
How many cents will it cost to buy 48.0 oz of meat at \$1.89/lb?

COLLECT AND ORGANIZE We are asked to calculate the cost in cents for 48.0 oz of meat that costs \$1.89 per pound.

ANALYZE Since the price of the meat is given in dollars per pound it is first necessary to convert the amount of meat purchased in ounces to pounds. We then calculate the cost in dollars given the price per pound. Finally we convert the price in dollars to cents. Dimensional analysis can be used to solve each step of the problem.

SOLVE The problem tells you that 1 lb = \$1.89. You also need to know that 16 oz = 1 lb (exactly, the unit system is the same) and that 100 cents = \$1.

$$48.0\ \text{oz} \times \left(\frac{1\ \text{lb}}{16\ \text{oz}}\right) \times \left(\frac{\$1.89}{1\ \text{lb}}\right) \times \left(\frac{100\ \text{cents}}{\$1}\right) = 567\ \text{cents}$$

In this problem the answer has three significant figures because of the price per pound and the starting value. It is also fair to use common sense and realize that this is a situation that would not recognize a fraction of a cent.

THINK ABOUT IT Sometimes unit conversions are given as part of the problem.

EXAMPLE 6.
What is the volume in m^3 of a 0.711-in^3 box?

COLLECT AND ORGANIZE Given the volume of a box as 0.711 in^3, we are asked to convert this value to m^3.

ANALYZE We must use the relationship between inches and meters to perform the requested conversion. Care must be taken to account for the fact that the problem deals with cubic meters (m^3) and cubic inches (in^3).

SOLVE Since 1 in = 2.54 cm,

$$(1 \text{ in})^3 = (2.54)^3 \text{ cm}^3$$

$$1 \text{ in}^3 = 16.387 \text{ cm}^3$$

and since 100 cm = 1 m,

$$(100)^3 \text{ cm}^3 = (1)^3 \text{ m}^3$$

$$1{,}000{,}000 \text{ cm}^3 = 1 \text{ m}^3$$

$$0.711 \text{ in}^3 \times \left(\frac{16.387 \text{ cm}^3}{1 \text{ in}^3}\right) \times \left(\frac{1 \text{ m}^3}{1{,}000{,}000 \text{ cm}^3}\right) = 1.17 \times 10^{-5} \text{ m}^3$$

THINK ABOUT IT Derived units, such as cm^2 or m^3, can also have derived conversions. For example, since 1 m = 100 cm, 1 m = 100 cm, and 1 m^2 = 10,000 cm^2.

EXAMPLE 7.
How much does 95.4 mL of mercury, which has a density of 13.6 g/mL, weigh in kilograms?

COLLECT AND ORGANIZE We are given a volume of mercury (95.4 mL) and are asked to determine its weight (in kg) given its density (13.6 g/mL).

ANALYZE The density of a substance is a measure of the mass of the substance per unit volume (1 mL) and is given by the expression;

$$d = \frac{\text{mass}}{\text{volume}}$$

where the symbol d denotes the density. Using the above expression we will obtain a mass in units of grams. We must then convert to kilograms.

SOLVE

$$\text{density} = \frac{\text{mass}}{\text{volume}}$$

$$13.6 \text{ g/mL} = \frac{\text{mass}}{95.4 \text{ mL}}$$

$$1297.44 \text{ g} = \text{mass}$$

$$1297.44 \text{ g} \times \left(\frac{1 \text{ kg}}{1000 \text{ g}}\right) = 1.29744 \text{ kg} = 1.30 \text{ kg}$$

Density is a measured value and does have significant figures. Both the volume and the density have three significant figures. The kg/g relationship is exact (infinite significant figures), so the final answer has three significant figures.

THINK ABOUT IT Density is a physical property of a substance. It describes the amount (grams) of the substance is contained in a specified volume (mL). A substance with a greater density than another substance will contain a greater amount of material in the specified volume than the other.

EXAMPLE 8.
What is the temperature in Celsius at 84°F?

COLLECT AND ORGANIZE We are given a temperature in Fahrenheit and asked to find the corresponding temperature in Celsius.

ANALYZE The relationship between temperatures in the Fahrenheit (F) and Celsius (C) scales is $C = \left(\frac{5}{9}\right)(F - 32)$.

In order to convert from Fahrenheit to Celsius we subtract 32 from the temperature given in Fahrenheit, multiply the result by 5 and then divide by 9.

SOLVE

$$F = 84$$

$$C = \frac{5}{9}(F - 32) = \frac{5}{9}(84 - 32) = \frac{5}{9}(52) = 28.8889 =$$

29°C (two significant figures)

THINK ABOUT IT Without being given the scale in which the temperature was measured, one would expect very different weather conditions. A very warm day in Canada (a country that uses the Celsius scale) would be 29 degrees. In the United States, in which the Fahrenheit scale is used, a temperature of 29 degrees would be quite cold.

EXAMPLE 9.
What is the temperature in Fahrenheit at 25.0°C?

COLLECT AND ORGANIZE In this problem we are asked to convert a temperature given in Celsius to a temperature given in Fahrenheit.

ANALYZE The conversion of temperature in Celsius (C) to a temperature in Fahrenheit (F) is $F = \left(\frac{9}{5}\right)(C) + 32$. In this case we first multiply the given temperature in Celsius degrees by 9, divide by 5, and then add 32 to the result.

SOLVE

$$C = 25$$

$$25 = \frac{5}{9}(F - 32)$$

$$25\left(\frac{9}{5}\right) = F - 32$$

$$25.0\left(\frac{9}{5}\right) + 32 = 77.0°F$$

THINK ABOUT IT A comfortable summer day has a temperature of about 25.0°C (77.0°F).

EXAMPLE 10.
What is the temperature, in Kelvin, at 25°C?

COLLECT AND ORGANIZE We are asked to convert a temperature given in Celsius to a temperature given in Kelvin.

ANALYZE Temperatures from the Celsius scale are easily and directly converted to the Kelvin scale by simply adding 273.15 to the Celsius temperature.

SOLVE

$$25 + 273.15 = 298.15 = 298 \text{ K}$$

THINK ABOUT IT The zero of the Kelvin scale is referred to as absolute zero, a temperature in which all molecular motion nearly stops. Scientists have been able to come very close but have never achieved the absolute zero of temperature.

Self-Test

KEY VOCABULARY COMPLETION QUESTIONS

__________ 1. Having only one kind of atom

__________ 2. The smallest fraction of an element

__________ 3. Can be separated by distillation

__________ 4. The study of material things

__________ 5. Density is an example of this type of property

__________ 6. Two or more substances not equally distributed

__________ 7. Tentative explanation

__________ 8. Basic SI unit of mass

__________ 9. One-one thousandth of a meter

__________ 10. It's low for a ton of feathers and high for a ton of bricks

__________ 11. Temperature scale without negative values

__________ 12. Consistent

__________ 13. Digits including and to the right of the first nonzero digit

__________ 14. Having the true value

__________ 15. Another name for a homogeneous mixture

__________ 16. Observed when a substance changes identity

__________ 17. Changing from a solid to a gas without becoming a liquid

__________ 18. Having definite volume but variable shape

__________ 19. Its molecules are far apart and moving

__________ 20. The way atoms are attached to each other in a molecule

__________ 21. Every sample of this substance has the same proportion of its elements

__________ 22. Explanation accepted by the scientific community

__________ 23. Has mass and occupies space

__________ 24. Prefix increasing the magnitude of a unit by a factor of one thousand

__________ 25. The same in every location

MULTIPLE-CHOICE QUESTIONS

1. Which is the longest?
   a. 1 m
   b. 1 μm
   c. 1 g
   d. 1 kg

2. Coffee is an example of a(n)
   a. heterogeneous mixture
   b. homogeneous mixture
   c. element
   d. compound

3. What should you wear when the temperature is 303 K?
   a. winter coat
   b. light sweater
   c. swimsuit
   d. butter, because you're a turkey basting in the oven

4. Which of the following is the lightest?
   a. 1 m
   b. 1 μm
   c. 1 g
   d. 1 kg

5. Which is the smallest value?
   a. 1.3×10^5
   b. 1.3×10^7
   c. 5.0×10^{-4}
   d. 5.0×10^{-7}

6. How many milliliters are in 1 liter?
   a. 10
   b. 0.001
   c. 1×10^{-6}
   d. 1000

7. 1.771 kg + 213.453 g =
   a. 1984.453 g
   b. 215.2 g
   c. 215.224 g
   d. 1984 g

8. Which is not true of a scientific theory?
   a. It has been tested.
   b. It is repeatable.
   c. It explains observations.
   d. It is known to be true.
   e. All of the above are true.

9. Which measurement, if the true value is exactly 3, is the most precise?
   a. 3.1 ± 0.1
   b. 3.10 ± 0.01
   c. 3 ± 1
   d. All values are the same.

10. Which measurement, if the true value is exactly 3, is the most accurate?
    a. 3.1 ± 0.1
    b. 3.10 ± 0.01
    c. 3 ± 1
    d. All values are the same.

11. How many significant figures does the number of centimeters in exactly one meter have?
    a. 1
    b. 2
    c. 3
    d. infinite

12. How many millimeters in a kilometer?
    a. 10^3 mm
    b. 10^6 mm
    c. 10^{-6} mm
    d. 10^{-9} mm
    e. 1 mm

13. How many kilometers can be traveled in 1.5 hr at 1.0×10^2 m/s?
    a. 0.0019 km
    b. 540 km
    c. 42 km
    d. 54,000 km

14. What is the volume in cubic centimeters of an object 3.35 m × 0.405 m × 0.619 m?
    a. 0.840 cm^3
    b. 84.0 cm^3
    c. 840 cm^3
    d. 8.40×10^5 cm^3

15. What is the mass of 27.6 cm^3 of nickel, which has a density of 8.90 g/cm^3?
    a. 246 g
    b. 0.322 g
    c. 3.10 g
    d. 27.6 g

16. Wood produces water and carbon dioxide when burned. This is an example of a
    a. heterogeneous mixture
    b. homogeneous mixture
    c. physical property
    d. chemical property

17. A substance that is composed of atoms that all have the same identity is classified as
    a. an element
    b. a compound
    c. a homogeneous mixture
    d. a heterogeneous mixture

18. A material can be consistently decomposed to 7.8% carbon and 92.2% chlorine. This material is best classified as
    a. an element
    b. a compound
    c. a homogeneous mixture
    d. a heterogeneous mixture

19. A substance where all the molecules are adjacent and in fixed positions is best described as a
    a. gas
    b. liquid
    c. solid
    d. mixture

20. A substance that has a very low density and is easily compressed is best described as a
    a. gas
    b. liquid
    c. solid
    d. mixture

21. $(3.800 \times 10^{-7})(0.00021) =$
    a. 7.98×10^{-10}
    b. 8.0×10^{-11}
    c. 7.980×10^{-11}
    d. 8×10^{-10}

22. The answer to the following is 7.880 + 8.1 =
    a. 15.980
    b. 15.98
    c. 16.0
    d. 16

Additional Practice Problems

1. Express the answer with appropriate significant figures and units.
   a. 0.057543 g – 0.00065 g =
   b. (0.057543 g)(0.00065 g) =
   c. $(3.2 \times 10^8$ g/mL$)(8.943 \times 10^{-3}$ mL$)$ =
   d. $\dfrac{(0.0057 \text{ m} + 0.00065 \text{ m})}{(6.3899 \times 10^3 \text{ s})} =$
   e. $\dfrac{(9 \times 10^6 \text{ J})}{(1.81 \times 10^9 \text{ J})} =$
2. Express the answer with appropriate significant figures and units.
   a. (0.141 g)(0.00026000 g) =
   b. $(1.0099 \times 10^3$ g$) + (1.2 \times 10^{-2}$ g$)$ =
   c. 1.162 mg + 34 μg =
   d. $(6.96 \times 10^{-15}$ m$)(9.8 \times 10^{-6}$ m$)(0.0000690$ m$)$ =
   e. $(4.4 \times 10^5$ mL $+ 7.1 \times 10^4$ mL$)(4.000 \times 10^{-4}$ mL$)$ =
3. Classify the following as a heterogeneous mixture, homogeneous mixture, element, or compound.
   a. pizza
   b. a substance that always decomposes to make three times as much (by mass) carbon as hydrogen
   c. filtered seawater
   d. chlorine
4. Identify the following descriptions about this substance as chemical or physical.
   a. It is attracted to a magnet.
   b. It combines with oxygen.
   c. It cannot be chemically separated.
   d. It has luster.
   e. It melts at 1538°C.
   f. It has a density of 7.87 g/cm^3.
5. Convert the following values to the requested units.
   a. 72°F to °C and K
   b. the speed of light, 2.998×10^8 m/s, to miles per hour (*Hint:* 1.61 km = 1 mile)
   c. 12 oz to mL (*Hint:* 29.57 mL = 1 oz)
6. What is the appropriate SI unit to measure
   a. a marathon
   b. the mass of a fly
   c. a glass of water
7. Mercury has a density of 13.6 g/cm^3. What is the volume, in liters, of 1.211 kg of mercury?
8. Based on the following pictures of molecules, describe each substance as an element, compound, or mixture.
   a.
   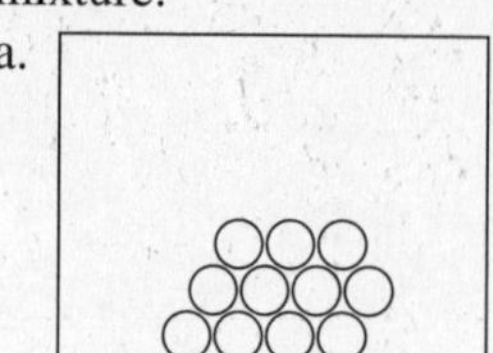
   b.
   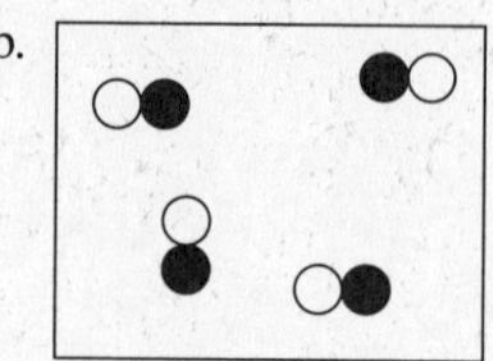
   c.
   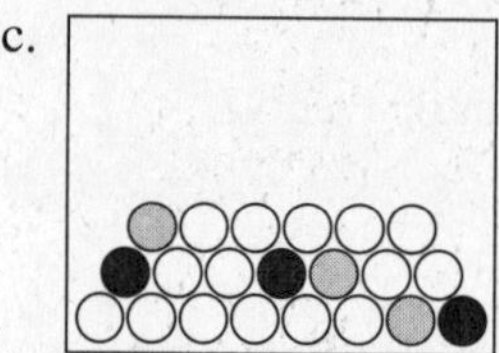
9. For the same three diagrams shown in question 8, identify each as a solid, liquid, or gas.
10. For each pair, identify the largest quantity.
    a. 1.0 μL or 1.0 mL
    b. 1.0×10^3 g or 10 kg
    c. 1 cm^3 or 1 mL
    d. 10 mJ or 1.0 MJ

Self-Test Answer Key

Key Vocabulary Completion Answers

1. element
2. atom
3. mixture
4. chemistry
5. intensive
6. heterogeneous mixture
7. hypothesis
8. gram
9. millimeter
10. density
11. Kelvin
12. precise
13. significant figures

14. accurate
15. solution
16. chemical property
17. sublimation
18. liquid
19. gas
20. bond
21. compound
22. theory
23. matter
24. kilo
25. homogeneous

Multiple-Choice Answers

For further information about the topics these questions address, you should consult the text. The appropriate section of the text is listed in parentheses after the explanation of the answer.

1. a. Grams are not a unit of length, so those two values can be eliminated. A micrometer (μm) is 1/1,000,000 of a meter, so it is smaller, not larger, than a meter. (1.8)
2. b. Coffee has a variable composition, it can be weak, strong, roasted. . . . Therefore, it is a mixture. Once made, however, it is the same throughout the whole pot; therefore, it is homogeneous. (1.3)
3. c. 303 K is 30°C or 86°F, swimsuit weather. (1.10)
4. c. Grams are the unit of mass. Kilograms are 1000 times heavier than a gram; therefore, grams are the lightest. (1.8)
5. d. Negative values on the exponent represent fractions; positive values represent large numbers. The more negative the exponent, the smaller the fraction. (1.8)
6. d. A "milli" is 1/1000 of the base unit. (1.8)
7. d. To add values, the units must be the same. If kilograms are changed to grams, it is 1771 g, then added to 213.453 = 1984.453 g. Then the value must be rounded (based on decimal places, since it is addition) to appropriate significant figures. (1.8 and 1.9)
8. d. Scientific theories are always open to modification and change. (1.7)
9. b. Precision reflects the repeatability of a measurement. More significant figures show the point at which a measurement stops becoming repeatable. Therefore, more significant figures reflect more precise values. (1.8)
10. d. Accuracy reflects the correct value. All these values could easily reflect a measurement of the same object, but with measuring devices of different precision. (1.8)
11. d. The relationship between centimeters and meters is a defined, and therefore, an exact number with infinite significant figures. (1.8)
12. b. $1\text{ km} \times \left(\frac{10^3\text{ m}}{1\text{ km}}\right) \times \left(\frac{10^3\text{ mm}}{1\text{ m}}\right) = 10^6\text{ mm}$ (1.9)
13. b. $1.5\text{ hr} \times \left(\frac{60\text{ min}}{1\text{ hr}}\right) \times \left(\frac{60\text{ s}}{1\text{ min}}\right) \times \left(\frac{1.0 \times 10^2\text{ m}}{1\text{ s}}\right) \times \left(\frac{1\text{ km}}{1000\text{ m}}\right) = 540\text{ km}$ (1.9)
14. d. Since 100 cm = 1 m, $(100\text{ cm})^3 = (1\text{ m})^3$ and $10^6\text{ cm}^3 = 1\text{ m}^3$

    $3.35 \times 0.405 \times 0.619 = 0.840\text{ m}^3 \times \left(\frac{10^6\text{ cm}^3}{1\text{ m}^3}\right) = 8.40 \times 10^5\text{ cm}^3$ (1.9)
15. a. Density = mass/volume. Therefore, mass = (density)(volume) = $(8.90\text{ g/cm}^3)(27.6\text{ cm}^3)$ = 246 g. (1.4)
16. d. Chemical properties involve a change in the identity of the substance, like wood changing to water and carbon dioxide. (1.4)
17. a. In an atomic view, an element is composed of atoms of the same type. (1.2)
18. b. The law of constant composition states that a compound always has the same proportion of elements by mass. (1.1)
19. c. A solid maintains its shape and volume because its component molecules are in fixed positions. It is dense because the molecules are close together. (1.6)
20. a. Gases are composed of molecules that are far apart from each other. Therefore, gases are easily compressed. In addition, this large distance means that gases take up a lot of volume without having many molecules to contribute to the mass. Therefore, gases have a low density. (1.6)
21. b. The calculator, used correctly, gives the answer of 7.98×10^{-11}. However, the second value has only two significant figures, so the answer should have only two significant figures and the answer expressed with two significant figures is 8.0×10^{-11}.

22. c. When adding (or subtracting), the significant figures are counted by whichever value has the fewest spaces on the right. 8.1 goes only to the tenth decimal place, so the final answer should also go to the tenth decimal place. Significant figures, as such, are not counted. The answer is actually the same in all choices.

Additional Practice Problem Answers

1. a. 0.057543 g – 0.00065 g = 0.056893 g = 0.05689 g

   b. (0.057543 g)(0.00065 g) = 3.740295×10^{-5} g^2 = 3.7×10^{-5} g^2

   c. (3.2×10^8 g/mL)(8.943×10^{-3} mL) = 2861760 g = 2.9×10^6 g

   d. $\dfrac{(0.0057 \text{ m} + 0.00065 \text{ m})}{(6.3899 \times 10^3 \text{ s})} = \dfrac{0.00635 \text{ m}}{6.3899} \times 10^3 \text{ s} =$

   9.93756×10^{-7} m/s = 9.9×10^{-7} m/s (*Note:* The last digit of 0.00635 is not significant.)

   e. $\dfrac{(9 \times 10^9 \text{ J})}{(1.81 \times 10^9 \text{ J})} = 4.9723757 \times 10^{-3} = 5 \times 10^{-3}$
   (no unit; it cancelled)

2. a. (0.141 g)(0.00026000 g) = 3.666×10^{-5} g^2 = 3.67×10^{-5} g^2

   b. (1.0099×10^3 g) + (1.2×10^{-2} g) = 1009.9 + 0.012 = 1009.912 g = 1009.9 g

   c. 1.162 mg + 34 μg = 1162 μg + 34 μg = 1196 μg or 1.162 mg + 0.034 mg = 1.196 mg

   d. (6.96×10^{-15} m) (9.8×10^{-6} m)(0.0000690 m) = 47.06352×10^{-24} m^3 = 4.7×10^{-24} m^3

   e. The problem must be solved according to the order of operation of the problem. First: 4.4×10^5 mL + 7.1×10^4 mL = 440,000 + 71,000 = 511,000 and only the first two digits are significant. To finish the problem:

   (511,000)(4.000×10^{-4}) = 204.4 mL^2 = 2.0×10^2 mL^2

   Even though we used all the digits in the calculation, the sum from the first part of the calculation had only two significant figures; therefore, the answer can have only two significant figures.

3. a. Pizza is a heterogeneous mixture with crust on the bottom and cheese on top.

   b. This is a compound, since it decomposes into elements in a fixed ratio.

   c. Seawater is a homogeneous mixture. The sample is the same everywhere, but proportions are variable, even within the sea.

   d. Chlorine is element number 17.

4. a. Physical property. The identity of the substance is unchanged.

   b. Chemical property. By combining with an element, the substance in changed.

   c. Chemical property. Consider the description.

   d. Physical property. Observed by looking.

   e. Physical property. As are all melting and boiling points, since they describe changes in physical state.

   f. Physical property. Observed by measuring without changing the substance.

5. a. $C = \frac{5}{9}(F - 32) = \frac{5}{9}(72 - 32) = \frac{5}{9}(40) =$

   22.22 = 22°C

   C + 273.15 = K

   22.22 + 273.15 = 295.37 = 295 K

   b. $2.998 \times 10^8 \text{ m/s} \times \left(\dfrac{1 \text{ km}}{1000 \text{ m}}\right) \times \left(\dfrac{1 \text{ mi}}{1.61 \text{ km}}\right) \times$

   $\left(\dfrac{60 \text{ s}}{1 \text{ min}}\right) \times \left(\dfrac{60 \text{ min}}{1 \text{ hr}}\right) = 6703602848 \text{ mi/hr} =$

   6.70×10^8 mi/hr

   c. $12 \text{ oz} \times \left(\dfrac{29.57 \text{ mL}}{1 \text{ oz}}\right) = 354.84 \text{ mL} = 3.5 \times 10^2 \text{ mL}$

6. For the answer to be correct, the base unit *must* be what is given in the answers below. The prefix can vary but is generally chosen so that the numbers do not require scientific notation.

   a. km

   b. mg

   c. mL

7. $1.211 \text{ kg} \times \left(\dfrac{1000 \text{ g}}{1 \text{ kg}}\right) = 1211 \text{ g}$

   13.6 g/mL = 1211 g/V; therefore, V = 89.0 mL

   Density is a measured value and has only three significant figures:

   $89.04 \text{ mL} \times \left(\dfrac{1 \text{ L}}{1000 \text{ mL}}\right) = 0.0890 \text{ L}$

8. a. Element, all atoms are the same type.

   b. Compound, there are two types of atoms, but they are connected and in the same ratio every time.

   c. Mixture, there are different pieces randomly scattered through another substance.

9. a. Solid. Atoms are adjacent, in specific positions, and do not fill the container.

   b. Gas. Molecules are spread apart and found in all parts of the container.

   c. Liquid. The substance fills the bottom of the container. Molecules and atoms are adjacent.

10. a. 1.0 μL is smaller. The "μ" represents a smaller quantity than "m."

    b. 10 kg. 1.0×10^3 g is equal to 1.0 kg.

    c. These are two different ways of expressing the same quantity. They are the same.

    d. 1.0 MJ. The uppercase "M" designates a much higher quantity than the lowercase "m."

CHAPTER 2 Atoms, Ions, and Compounds

OUTLINE

REVIEW

Chapter Overview

Although originally atoms were believed to be indivisible, around 1900 this was determined not to be true. Nevertheless, atomic theory accounts for much observed behavior, such as the **law of multiple proportions**, which states that if the same two elements (call them X and Y) form more than one compound, then the ratio of the masses of Y that react with a given mass of X to form any two compounds is the ratio of two small whole numbers. Chemists primarily see the world from the atomic point of view, but they also find it useful to explore the structure of the atom itself.

The first subatomic particle was discovered by **J. J. Thompson** who used magnetic and electric fields to deflect the beam of a cathode ray tube. From these experiments, he determined that all materials emitted the same type of negatively charged particle. He determined the mass-to-charge ratio of this particle, which was named the **electron**. Studies of the electron were furthered by **Robert Millikan** who, with his famous oil drop experiment, was able to determine the charge of an electron. Then, using that value and Thompson's mass-to-charge ratio, Millikan was able to determine the mass of the electron.

Since atoms are known to be neutral, positive particles must exist as well. The discovery of radioactivity, particularly β particles and α particles, allowed more information about the arrangement of these subatomic particles (atomic structure) to be determined. **Ernest Rutherford** is credited with discovering the nuclear structure of the atom. This was done by accelerating the large, postively charged α particles toward a thin sheet of gold. Unexpectedly, a number of the particles were deflected back toward the source. Using this information, Rutherford was able to determine that only one ten-thousandth of the atom was a positive region. This region was called the **nucleus**, which contains all the positive charge and most of the mass of the atom. Later experiments actually isolated the particles in the nucleus, which are called **nucleons**. These consist of positively charged particles called **protons** and neutral particles (that account for the rest of the mass of the atom) called **neutrons**. The effect of the negatively charged electrons is cancelled by an equal number of positively charged protons, making an atom that is neutral overall.

Because the mass of each of these particles is so small, a mass unit called an **atomic mass unit (amu)** was defined as one-twelfth of a carbon atom containing 6 protons and 6 neutrons. Thus, each proton and neutron has a mass of approximately 1 amu and exists in the nucleus of the atom. Since the electron has a mass of about 1/2000 amu, its mass is usually estimated as 0 amu. It exists outside the nucleus.

The identity of an atom is determined by its **atomic number (Z)**, which is its number of protons. The mass of an atom is primary determined by the sum of the protons and neutrons called its **mass number** (*A*). However, atoms may have the same identity (number of protons) but different masses (number of neutrons). Such atoms are called **isotopes**. Isotopic notation is used to differentiate isotopes. In **isotopic notation**, the mass number (*A*) is indicated by a superscript to the left of the element symbol (*E*): ${}^{A}E$. The atomic number may also be included as a subscript to the left of the symbol: ${}^{A}_{Z}E$. Isotopes can also be designated by using the name of the atom followed by the mass number: element-A. For example, the carbon atom with 6 protons and 6 neutrons can be designated as ${}^{12}C$, ${}^{12}_{6}C$, or carbon-12.

Chemists usually work with billions of atoms, which are mixtures of their various isotopes. Therefore, the most useful value of a mass is a **weighted average** of the mass of the isotopes, which takes into account the natural abundance of each isotope. The **natural abundance** is the fraction of the isotope that exists in a natural sample of the element. This is the value reported on the periodic table and generally used as **atomic weight**.

The **periodic table** is a useful organization of all the known elements. It was first developed by **Dimitri Mendeleev**, who organized the elements two ways: first by similar chemical and physical properties, keeping such elements together in columns called **groups** or **families**; then by atomic weight, creating various rows called **periods**. The groups are numbered from left to right across the table. Group 1 is called the **alkali metals**, group 2 contains the **alkaline earths**, group 17 consists of the **halogens**, and group 18 is made up of the **noble gases**. The periods (rows) can also be numbered from the top to the bottom. The two rows that are offset at the bottom of the periodic table actually belong in periods 6 and 7, respectively, so that the sequential atomic numbers are maintained.

Another useful grouping within the periodic table is the staircase-like division that runs between aluminum (Al) and silicon (Si) and continues between germanium (Ge) and arsenic (As) to the bottom of the table. Elements to the left of the staircase can be classified as **metals**. Properties of metals include a metallic luster (they are shiny), malleability (they can be dented), and ductility (they can be pulled into a wire). Elements to the right of the staircase are **nonmetals**, which are neither malleable nor ductile. Elements that actually touch the staircase are **metalloids** or **semimetals**. Metalloids tend to have the physical properties of metals but the chemical properties of nonmetals.

Groups 1, 2, and 13–18 are classified as the **representative** or **main-group elements**. Groups 3–12 are the **transition metals**. The first row—offset at the bottom (elements 58–71)—are the **lanthanides**; the second offset row (elements 90–103) are the **actinides**.

When different elements combine, they form compounds. When the elements that combine are both nonmetals, the atoms are held together by shared electrons called a **covalent bond**, and the combination is called a **molecular compound**. However, when a nonmetal combines with a metal, it usually forms an **ionic compound**. In this case the metal donates one or more of its electrons to the nonmetal. Since the metal now has fewer electrons than protons, it becomes positively charged. Positively charged atoms or groups of atoms are called **cations**. Since the nonmetal now has more electrons than protons, it has a net negative charge. Negatively charged atoms or groups of atoms are called **anions**. The attraction between the cation and anion is called an **ionic bond**, and it is what holds an ionic compound together. An ionic compound has anions and cations in a ratio that has a net charge of 0.

One of the properties that characterize a group is the type of ion it forms.

- Alkali metals (group 1) lose one electron, to form cations with a +1 charge.
- Alkaline earths (group 2) lose two electrons, to form cations with a +2 charge.
- Halogens (group 17) normally gain one electron, to form anions with a –1 charge.
- Hydrogen is a special case that can either lose or gain one electron depending on the type of element it combines with.

The formulas of ionic compounds are always expressed as the simplest ratio of their component ions that allows for electrical neutrality, called a **formula unit**. The cation is always listed first: for example, Na^+ and Cl^- combine to form NaCl rather than ClNa. For group 1 and 2 metals, the name of the cation is simply the name of the metal it came from. Thus, Na^+ is sodium ion. For other metals, the name of the metal is followed by its charge in parentheses. The charge is written as a roman numeral: for example, Fe^{3+} is iron(III) ion. Monatomic (one-atom) anions change the end of the element name to *ide:* for example, Cl^- is chloride ion. For polyatomic (multi-atom) ions, the system is complex; so it is generally easier to memorize the names of the common ions than to decipher the system. The names of ionic compounds are simply the name of the cation followed by the name of the anion. The number of cations and anions is not part of the name of ionic compounds. Since the net charge of the cations must exactly cancel that of the anions in the simplest ratio, that ratio can be deduced from the charges. Therefore, Na^+ and Cl^- combine to create NaCl called sodium chloride and Fe^{3+} and SO_4^{2-} combine to make $Fe_2(SO_4)_3$ called iron(III) sulfate.

The simplest whole number ratio of elements in a compound is its **empirical formula**. However, for molecular compounds, the simplest ratio does not always express the way the atoms are connected. Therefore, a prefix on the name of the element is used to express the number of that type of atom. If there is only one of that type of atom, no prefix is used. In addition, the second element has its ending changed to *ide:* for example, when one carbon combines with two oxygens, its name is carbon dioxide.

Acids produce hydrogen cations in water and normally have hydrogen as the first element in the formula. The naming system depends on the type of acid. Acids made from oxoanions (anions containing oxygen) change the suffix of the oxoanion and add the word *acid.* If the name of the oxoanion ends in *ate,* the acid suffix becomes *ic*: for example, the acid formed from nitrate ion (NO_3^-), HNO_3, is nitric acid. If the name of the oxoanion ends in *ite,* the acid suffix becomes *ous*: for example, if the acid was formed from nitrite ion (NO_2^-), HNO_2, it is called nitrous acid. Binary acids combine hydrogen with one other element. These are named as *hydro[element root]ic acid.* For example, HCl is hydrochloric acid. Organic acids are acids that come from compounds based on carbon. The organic naming system is not addressed here, but organic acids can be identified by the COOH at the end of the formula: for example CH_3COOH is acetic acid. While there are likely to be other hydrogens in the formula, only the hydrogen that is part of the COOH is released into water to become H^+.

Elements can be created through nuclear reactions. Unlike chemical reactions, as its nucleus changes, so might the identity of that element. Isotopic notation can be used to describe nuclear reactions. When used this way, the Z (of ${}^{A}_{Z}E$) actually expresses the charge on the particle. Since the charge on the nucleus of an atom is determined by the number of protons, this does not change the way atomic isotopes are written. To balance a nuclear reaction, the sum of the mass numbers (A) on each side of a nuclear equation must be equal. In addition, the sum of the charges (Z) on each side of the nuclear equation must also be equal. There are various types of nuclear reactions, including:

- **Fusion**, in which atoms or subatomic particles combine to create larger atoms or particles.
- **Neutron-capture**, in which an atomic nucleus captures a neutron.
- **Beta-decay**, in which a beta particle is emitted by an unstable nucleus.

Various particles participate in nuclear reactions, including:

- Protons (${}^{1}_{1}p$), which are the same as hydrogen atoms (${}^{1}_{1}H$).
- Various isotopes of hydrogen, which include **deuterium** (${}^{2}_{1}H$) and **tritium** (${}^{3}_{1}H$).
- **Beta particles** (${}^{0}_{-1}\beta$), which are actually high-energy electrons (${}^{0}_{-1}e$).
- Neutrons (${}^{1}_{0}n$).
- **Alpha particles** (${}^{4}_{2}\alpha$), which are equivalent to a helium nucleus (${}^{4}_{2}He$).

Worked Examples

Atomic Structure

Example 1.
How many protons, neutrons, and electrons are in the following isotopes?

a. ${}^{3}H$ b. ${}^{112}Cd$ c. ${}^{37}Cl$ d. ${}^{16}O$

Collect and Organize Given four isotopes, we are asked to determine the number of protons, neutrons, and electrons in each.

Analyze The number of protons of an element is the same value as its atomic number. The atomic number is listed on the periodic table. In atoms, which are electrically neutral, the number of electrons is the same as the number of protons. For ions, the number of protons and the number of electrons are not the same. None of these are ions.

The number of neutrons depends on the type of isotope. Because the periodic table gives only the average mass, it cannot be used to determine the number of neutrons. However, the mass number, which is the sum of the protons and neutrons, is included as a superscript before the chemical symbol. If the mass number is not given, a reasonable guess for the mass number of the most abundant isotope is to round the average atomic mass to the nearest whole number.

For an atomic isotope of ${}^{A}E$, where A is the mass number and E is the symbol of the element, the number of protons (p) and the number of electrons (e) are equal to the atomic number, Z, and the number of neutrons (n) is $A - Z$.

Solve

a. Hydrogen has an atomic number of 1. So $p = 1$, $e = 1$, and $n = 3 - 1 = 2$.

b. Cadmium is atomic number 48. So $p = 48$, $e = 48$, and $n = 112 - 48 = 64$.

c. Chlorine is atomic number 17. So $p = 17$, $e = 17$, and $n = 37 - 17 = 20$.

d. Oxygen is atomic number 8. So $p = 8$, $e = 8$, and $n = 16 - 8 = 8$.

Think About It The periodic table is an invaluable tool for determining the characteristics of an atom. Each block of the periodic table contains, at a minimum, the symbol representing the identity of the element, the atomic number of the element, and the average element's isotopes mass of the element.

EXAMPLE 2.
What is the average atomic mass of lead if its isotopic abundance is 1.4% ^{204}Pb, 24.1% ^{206}Pb, 52.4% ^{208}Pb, and 22.1% ^{207}Pb?

COLLECT AND ORGANIZE We are given the naturally occurring isotopes of lead and their corresponding percentages and are asked to determine its average atomic mass.

ANALYZE The average atomic mass takes into account both the mass of each naturally occurring isotope and the percentage of that isotope in a large sample. Because these values are experimentally determined, they must be given as part of the problem. To determine average atomic mass, change the percentage to the corresponding fraction and multiply by the mass of that isotope; add all these products together. The mass number is the number of protons plus neutrons; it is not the actual mass of the atom. However, if that is all the information you have, it is a good estimate.

SOLVE Assume the mass number is the mass of the atom (at least until the first decimal place). The fraction of the average mass due to that isotope is that fraction times the mass number. Therefore

$$\text{average mass} = 0.014(204) + 0.241(206) + 0.524(208) + 0.221(207)$$
$$= 2.856 + 49.646 + 108.992 + 45.747$$
$$= 207.241 \text{ amu} = 207 \text{ amu}$$

THINK ABOUT IT Isotopes are any of the several different forms of an element, each having different atomic mass (mass number). Isotopes of an element have nuclei with the same number of protons (the same atomic number) but different numbers of neutrons. Therefore, isotopes have different mass numbers, which give the total number of nucleons—the number of protons plus neutrons. Because the different types of isotopes are not equally distributed, the average atomic mass must account for the abundance of each isotope as well as each mass.

EXAMPLE 3.
What is the average atomic mass of gallium, which has two isotopes, 60.0% ^{69}Ga (actual mass = 68.9256 amu) and 40.0% ^{71}Ga (actual mass = 70.9247 amu)?

COLLECT AND ORGANIZE We are given the naturally occurring isotopes of gallium and their corresponding percentages and are asked to determine the average atomic mass.

ANALYZE The average atomic mass takes into account both the mass of each naturally occurring isotope and the percentage of that isotope in a large sample. Because these values are experimentally determined, they must be given as part of the problem. To determine average atomic mass, change the percentage to the corresponding fraction and multiply by the mass of that isotope; add all these products together. The mass number is the number of protons plus neutrons; it is not the actual mass of the atom. However,in this case, actual masses are given and therefore should be used.

SOLVE

$$\text{average mass} = 0.600(68.9256) + 0.400(70.9247)$$
$$= 41.35536 + 28.3699$$
$$= 69.72524 \text{ amu} = 69.7 \text{ amu}$$

THINK ABOUT IT Elements do not necessarily have the same number of isotopes. Gallium has only two naturally occurring isotopes, whereas lead has four.

NAVIGATING THE PERIODIC TABLE

EXAMPLE 1.
What are the group and period of the following elements?

a. Re   b. Cu   c. Ca   d. Pb
e. Ga   f. Ce   g. F

COLLECT AND ORGANIZE We are asked to determine the group and period of the provided elements.

ANALYZE Once you locate the element in the periodic table, you can read its group and period from the table. The group refers to the number of the column. The period refers to the number of the row.

SOLVE

a. Group 7, period 6
b. Group 11, period 4
c. Group 2, period 4
d. Group 14, period 6
e. Group 13, period 4
f. Lanthanide (doesn't have a group number), period 6. (The lanthanide series is displaced from its proper position in the periodic table; however, if you look at the atomic numbers, you can see where it belongs.)
g. Group 17, period 2

THINK ABOUT IT The elements in the periodic table are divided into different groups and periods. Members of the same group share similar behaviors that systematically vary with period, allowing for trends in different physical and chemical properties to be quickly determined.

EXAMPLE 2.
Classify the following as metals, nonmetals, or metalloids.

a. A shiny substance that tends to become an anion
b. Calcium (Ca)
c. Sulfur (S)
d. Neon (Ne)
e. Arsenic (As)

COLLECT AND ORGANIZE Given five elements, we are asked to classify them as a metal, nonmetal, or metalloids.

ANALYZE One way to categorize the element as a metal, nonmetal, or metalloid is by its properties. Physical properties of metals include luster, malleability, and electrical conductivity. A chemical property of metals is that they form cations. A nonmetal is not malleable or conductive, nor does it have a metallic shine. If it forms any type of ion, a nonmetal is more likely to be an anion. Metalloids tend to have physical properties similar to metals but chemical properties similar to nonmetals.

The easiest way to categorize an element is to use its position on the periodic table. A staircase-like line starts to the left of boron (B) and extends down between polonium (Po) and astatine (At). Elements to the left of the staircase are metals, those to the right are nonmetals, and those touching it are metalloids.

SOLVE

a. *Metalloid:* Shininess is a physical property of metals, making anions a chemical property of nonmetals.

b. *Metal:* It is in the second column of the periodic table.

c. *Nonmetal:* It is in the third row, toward the right side.

d. *Nonmetal:* It is a gas on the last column of the periodic table.

e. *Metalloid:* It touches the staircase.

THINK ABOUT IT Metals, nonmetals, and metalloids are subcategories of elements. Although mixtures and compounds contain elements, these subcategories apply only to the elements themselves.

NAMING CHEMICAL COMPOUNDS

EXAMPLE 1.
Name the following binary molecular compounds.

a. NI_3 b. N_2O c. ClF_5 d. CCl_4 e. P_4O_{10}

COLLECT AND ORGANIZE
We are given several molecular formulas and are asked to name them. We can tell they are molecular rather than ionic compounds because each compound contains only nonmetals.

ANALYZE To name binary molecular compounds, add a prefix to the name of each element to express the number of that type of atom in the compound. If there is only one atom of the first element, no prefix is used. The name of the first element is unchanged. The name of the second element has its ending changed to *ide.*

SOLVE

a. There is only one nitrogen; so the first element does not have a prefix. The root of *iodine* is *iod* and the prefix for 3 is *tri:* nitrogen triiodide

b. There are two nitrogen atoms this time; so the prefix *di* will be used. The root of oxygen is *ox.* There is only one; so the prefix *mono* is optional. When the prefix *mono* is used with *oxide,* it is acceptable to drop the last *o* of *mono.* However, it is not required that you do so. There are three acceptable answers for this question: (1) dinitrogen oxide, (2) dinitrogen monoxide, and (3) dinitrogen monooxide.

c. There is no prefix for the first element. The prefix for 5 is *penta*; the root of *flourine* is *flour:* chlorine pentafluoride.

d. There is no prefix for the first element. The prefix for 4 is *tetra;* the root of *chlorine* is *chlor:* carbon tetrachloride.

e. tetraphosphorus decaoxide.

THINK ABOUT IT The name of the compound must provide sufficient information to identify the number and type of each atom in a molecule. Therefore, systems for naming have been developed. The type of system depends on the type of compound.

EXAMPLE 2.
What is the formula of each of the following binary molecular compounds?

a. oxygen difluoride

b. tetranitrogen decaoxide

c. arsenic tribromide

d. boron trichloride

e. iodine heptafluoride

COLLECT AND ORGANIZE We are asked to determine the molecular formula of several binary molecular compounds from their names. The use of prefixes, as well as the nonmetal components of the compound, indicate that these are molecular rather than ionic compounds.

ANALYZE The prefix attached to each element determines the stoichiometric amount in the molecular formula.

SOLVE

a. *Di* refers to two. No prefix on the first element means there is only one oxygen; so OF_2.

b. *Tetra* refers to 4 and *deca*, to 10; so N_4O_{10}.

c. There is no prefix on arsenic and *tri* is three, so $AsBr_3$.

d. There is no prefix on boron and *tri* is still 3; so BCl_3.

e. There is no prefix on iodine and *hepta* is 7; so IF_7.

THINK ABOUT IT Molecular formulas can easily be constructed from their prefix and their names after one knows the elements' names and symbols.

EXAMPLE 3.
What are the names of the following cations?

a. Mn^{2+} b. K^+ c. Cr^{3+} d. Zn^{2+} e. Cu^+

COLLECT AND ORGANIZE We are asked to determine the names of five cations based on their formula, consisting of both the atomic symbol and the charge.

ANALYZE For elements not contained in groups 1 and 2, the name of the cation is formed by writing the full name of the element followed by its charge given as a roman numeral in parentheses. Cations formed from elements in groups 1 and 2 can form only one type of cation, +1 and +2, respectively; so the roman numeral notation is unnecessary. Aluminum always has a +3 charge; zinc always has a +2 charge; and silver always has a +1 charge; therefore, like groups 1 and 2, roman numerals are not used for these cations.

SOLVE

a. It is not group 1 or 2 or one of the three exceptions; so manganese(II) ion.

b. It is group 1; so there is no roman numeral: potassium ion.

c. It is not group 1, 2, or an exception: chromium(III) ion.

d. It is one of the exceptions so there is no roman numeral: zinc ion.

e. It is not one of the exceptions; so copper(I) ion. An equally acceptable answer is cuprous ion.

THINK ABOUT IT Elements that can form more than one cation require the use of roman numerals.

EXAMPLE 4.
What is the formula of each of the following ions?

a. lithium ion

b. magnesium ion

c. bismuth(V) ion

d. ferric ion

e. silver ion

COLLECT AND ORGANIZE We are asked to write the atomic formulas given the names of several cations.

ANALYZE The charge on the cation is determined either by the roman numeral following the name or, for names without the roman numeral notation, by the element's location in the periodic table.

SOLVE

a. It's a group 1 ion; so Li^+.

b. It's a group 2 ion; so Mg^{2+}.

c. It tells you the charge; so Bi^{5+}.

d. This is why it's good to know the old system. You don't have to name things with it, but you must understand when others do: Fe^{3+}.

e. This is an exception that needs to be memorized: Ag^+.

THINK ABOUT IT Quickly learning the names of common elements will help in writing chemical formulas from names. Other than the three exceptions, you do not have to memorize charges. However, you must remember that group 1 metals will always have a +1 charge and group 2 metals will always have a +2 charge. The ones do not appear in the formulas of the charge.

EXAMPLE 5.
What are the names of each of the the following anions?

a. Cl^- b. P^{3-} c. NO_2^- d. AsO_4^{3-} e. HCO_3^-

COLLECT AND ORGANIZE We are asked to determine the names of several anions given as molecular formulas.

ANALYZE Monatomic anions are named with the root element name and the suffix *ide.* Although there are some systems for naming polyatomic ions, it is often easier just to memorize the common ones.

SOLVE

a. *Monatomic:* So just change suffix and it's chloride ion.

b. *Monatomic:* The root element is *phosph;* so it's phosphide (or alternate spelling phosfide) ion.

c. This is a polyatomic ion. There are two common combinations of nitrogen and oxygen. This has fewer oxygens than the other choice; so it has an *ite* ending: nitrite ion.

b. Aluminum is Al^{3+} and sulfate is SO_4^{2-}; so the formula is $Al_2(SO_4)_3$. Notice that the charges, without the + or – sign, are the subscripts on the other ion. Because there is more than one sulfate ion, the sulfate ion is in parentheses and the subscript tells how many sulfate ions there are.

c. The cation is Pb^{2+}. You can tell the charge from the roman numeral. The formula for sulfite ion is SO_3^{2-}. So the compound is $PbSO_3$. The charges are the same; so the simplest ratio is 1:1, not 2:2, which is suggested by the charge-to-subscript method.

d. The ferrous ion is Fe^{2+} and hydroxide is OH^-. Therefore, the formula is $Fe(OH)_2$.

e. Ammonium ion is NH_4^+ and phosphate is PO_4^{3-}; so the formula is $(NH_4)_3PO_4$. Since there is only one phosphate, it does not have parentheses. However, because there is more than one ammonium, it does have parentheses.

f. SO_2. Notice the prefixes? This was a molecular compound. Don't forget that system too!

THINK ABOUT IT The name of a compound provides all the information needed to properly write its chemical formula.

EXAMPLE 9.
Fill in the following table.

Formula	Cation	Anion	Name
CaO	----------	----------	----------
----------	Pb^{4+}	F^-	----------
----------	----------	----------	sulfur dichloride
----------	----------	----------	copper(II) hydroxide
XeF_4	----------	----------	----------

COLLECT AND ORGANIZE We are given an incomplete table of molecular formulas, cations, anions, and names, and we are are asked to fill in the missing entries.

ANALYZE The missing entries can be found by considering the entries that are given. The type of compound, ionic or molecular, must be determined first so that the correct naming system is used.

SOLVE
Row 1: Calcium is a metal and oxygen is a nonmetal; so this is an ionic compound. Calcium, as a group 2 metal, has a +2 charge: Ca^{2+}. Oxygen, as a group 16 nonmetal, has a –2 charge: O^{2-}. Since both ions have the same charge, they exist in a 1:1 ratio: CaO.

Row 2: For a neutral compound to exist, 4 F^- ions are needed for each Pb^{4+}: PbF_4. Lead can have multiple charges; so it must be named with a roman numeral. The fluorine atom becomes fluoride ion. Therefore, the name is lead(IV) fluoride.

Row 3: The prefixes are a tip-off that this is a molecular compound. Therefore, there are *no* ions. According to the prefixes, there is one sulfur and two chlorides.

Row 4: Copper(II) is the name of the Cu^{2+} ion. Hydroxide is OH^-. Two hydroxides will be needed to neutralize the charge on the copper: $Cu(OH)_2$.

Row 5: Xenon and fluorine are both nonmetals; so this is a molecular compound. There are no ions. *Tetra* is the appropriate prefix for 4: xenon tetrafluoride.

Formula	Cation	Anion	Name
CaO	Ca^{2+}	O^{2-}	calcium oxide
PbF_4	Pb^{4+}	F^-	lead(IV) fluoride
SCl_2	(none)	(none)	sulfur dichloride
$Cu(OH)_2$	Cu^{2+}	OH^-	copper(II) hydroxide
XeF_4	(none)	(none)	xenon tetrafluoride

THINK ABOUT IT Given either the name, the chemical formula, or the cation and anion, we can determine the other two.

EXAMPLE 10.
What are the names of the following acids?

a. HBr b. HNO_2 c. H_3PO_4 d. H_2S

COLLECT AND ORGANIZE We are given the molecular formulas of four acids and are asked to determine their names.

ANALYZE Binary acids are named in the following format: *hydro*[element root]*ic acid*. Oxyacids contain oxygen. The names of these acids are based on the anion the acid came from. If the anion has an *ate* ending, the *ate* is changed to *ic* and the word *acid* is added. If the anion has an *ite* ending, the *ite* is changed to *ous* and is followed by *acid.*

SOLVE

a. It is a binary acid. The root of bromine is *brom;* so the name is hydrobromic acid.

b. It is an oxyacid associated with the NO_2^-, nitrite, ion. So the name is nitrous acid.

c. It is an oxyacid associated with PO_4^{3-}, phosphate ion; so it is phosphoric acid. (Okay, maybe to follow the system rigorously, it should be phosphic acid, but phosphoric is what they call it.)

d. A binary acid. Remember that *binary,* in the context of naming, refers to two types of elements (or ions), not two atoms. So hydrosulfuric acid.

THINK ABOUT IT The naming convention of an acid depends on whether it's a binary acid or an oxyacid. Further distinctions are made for naming oxyacids depending on the nature of the anion.

EXAMPLE 11.
What is the formula of the following acids?

a. hydroiodic acid

b. carbonic acid

c. sulfurous acid

d. perchloric acid

COLLECT AND ORGANIZE From the names of the acids given, we are asked to determine their molecular formulas.

ANALYZE We use the prefixes and suffixes of the names of the acids to guide us as to how they are named.

SOLVE

a. The *hydro* tips us off that it is a binary acid. *Iod* is the root for iodine, which makes I^- ion. Therefore, it needs one hydrogen. The formula is HI.

b. The *ic* says that it came from an *ate* name. Changing the *ic* back to *ate,* it becomes carbonate, CO_3^{2-}. Since this ion has a –2 charge, it needs two hydrogens: H_2CO_3.

c. The *ous* implies it came from *ite.* So the base anion must be sulfurite—no, sulfite, which is SO_3^{2-}, which needs two hydrogens: H_2SO_3. When sulfur is used in naming, sometimes its root (sulf) is used and sometimes the whole name (sulfur) is used, depending on which sounds better.

d. Changing the *ic* to *ate,* you get perchlorate, which is ClO_4^-. Adding one hydrogen to neutralize the charge, you get $HClO_4$.

THINK ABOUT IT The names of acids contain all of the information needed to properly determine their molecular formulas.

BASIC NUCLEAR REACTIONS

EXAMPLE 1.
Fill in the blanks of the following nuclear equations:

a. $^{238}U \rightarrow \alpha +$ ____

b. $n + {}^{1}H \rightarrow$ ____

c. $^{17}O + \beta \rightarrow {}^{14}C +$ ____

COLLECT AND ORGANIZE We are given three nuclear equations and asked to complete them.

ANALYZE Balancing nuclear equations requires that the sum of the mass numbers be the same on each side of the equation. Charge is also conserved. The elemental symbol depends on the atomic number. It is normally easier to figure out the missing pieces in atomic number and mass number and then match the symbol to the atomic number.

SOLVE

a. The charges and mass numbers not given in the problem should be added. An α-particle has a charge of 2 and a mass number of 4. The uranium nucleus will have a charge equal to its atomic number of 92.

$$^{238}_{92}U \rightarrow {}^{4}_{2}\alpha + {}^{A}_{Z}\_\_\_\_$$

The mass number and charge of the substance in the blank can then be calculated. Filling in the atomic number of uranium and both the atomic and mass numbers for the α-particle, the atomic and mass numbers for the blank can be calculated.

$$238 = 4 + A, \text{ then } A = 234$$

$$92 = 2 + Z, \text{ then } Z = 90$$

Since 90 is the atomic number for thorium, the symbol *Th* is used.

$$^{238}_{92}U \rightarrow {}^{4}_{2}\alpha + {}^{234}_{90}Th$$

b. Filling in the appropriate values, the equation for mass number is

$$^{1}_{0}n + {}^{1}_{1}H \rightarrow {}^{A}_{Z}$$

$$1 + 1 = A, \text{ therefore } A = 2$$

The equation for atomic number is

$0 + 1 = Z$, therefore $Z = 1$, the atomic number of hydrogen.

$$^{1}_{0}n + {}^{1}_{1}H \rightarrow {}^{2}_{1}H$$

c. $^{17}_{8}O + {}^{0}_{-1}\beta \rightarrow {}^{14}_{6}C + {}^{A}_{Z}$____

For mass number, $17 + 0 = 14 + A$, so $A = 3$.
For atomic number, $8 - 1 = 6 + Z$, so $Z = 1$, the atomic number of hydrogen.

$$^{17}_{8}O + {}^{0}_{-1}\beta \rightarrow {}^{14}_{6}C + {}^{3}_{1}H$$

THINK ABOUT IT Often not all the information is included in the problem. For example, the atomic and mass numbers of an α particle are consistent; so it may not be included. Since the elemental symbol depends on the atomic number, the atomic number is not always included either. It will probably be easier to do the problem if you fill that information in.

d. This is another polyatomic with a two oxygen–element combination, this common combination has the higher number of oxygens (the highest number varies from element to element); arsenate ion.

e. This is a hydrogen variation of the common ion carbonate (*ate* means 3, the highest number of oxygens carbon combines with). So this ion can be called either hydrogen carbonate or bicarbonate.

THINK ABOUT IT You should quickly memorize the names of common polyatomic ions.

EXAMPLE 6.
What is the formula of each of the following ions?

a. sulfite ion

b. sulfide ion

c. dihydrogen phosphate

d. perchlorate ion

COLLECT AND ORGANIZE We are asked to determine the chemical formulas from the names of several anions.

ANALYZE It is best to memorize the names and formulas for common anions.

SOLVE

a. Sulfite has fewer oxygens than sulfate. The formula of sulfite is SO_3^{2-}.

b. *Sulfide* has the *ide* ending; so it doesn't combine with oxygen. It is *ide* added to the root for *sulfur;* so must be S^{2-}.

c. Recalling the other system, *di* means 2; so two hydrogens combine with a phosphate ion. The phosphate ion is PO_4^{3-} (*ate* is the "lots-of-oxygen" ending). So dihydrogen phosphate is $H_2PO_4^-$.

d. The *ate* of the ending implies many oxygens; however, chlorine has more than two oxygen combinations. The combination of the *per* and the *ate* implies that it has the most possible oxygens. Its formula is ClO_4^-.

THINK ABOUT IT You should quickly memorize the names of common polyatomic ions.

EXAMPLE 7.
What are the names of the following ionic compounds?

a. Na_2CO_3 b. $Co(NO_3)_2$ c. SnS_2

COLLECT AND ORGANIZE We are given the molecular formulas of several ionic compounds and are asked to determine their names.

ANALYZE The name of an ionic compound is the name of the cation followed by the name of the anion, just drop the term *ion* after each. It only requires recognizing each ion. Careful attention must be made to the charge of each ion. Because the compound must have a net charge of zero, if the charge of one ion is known, the other can be calculated.

SOLVE

a. The ions are Na^+ and CO_3^{2-}. Sodium is in group 1; so there is no roman numeral. The name is sodium carbonate.

b. Cobalt is a metal with several possible charges. However, nitrate (NO_3^-) is always –1. For the compound to be electrically neutral, the cobalt must have a +2 charge; so its name is cobalt(II) nitrate.

c. Tin is another metal with variable charges. However, sulfur always gains two electrons to make S^{2-}. Therefore, the charge on tin must be +4. The name is tin(IV) sulfide.

THINK ABOUT IT To name ionic compounds, simply start with the name for the element of the cation, followed by the name of the anion. If the charge on one ion is not known, use the charge on the other ion to figure it out.

EXAMPLE 8.
What is the formula of the following compounds?

a. potassium iodide

b. aluminum sulfate

c. lead(II) sulfite

d. ferrous hydroxide

e. ammonium phosphate

f. sulfur dioxide

COLLECT AND ORGANIZE From the names given for several compounds, we are asked to determine the molecular formula for each.

ANALYZE It is best to memorize the charges and names of common anions and cations. When more than one charged state is possible, a roman numeral following the ion designates the charge. The formula must be the simplest ratio of ions that results in a net charge of zero. Only use parentheses for polyatomic ions if a subscript, designating more than one of these ions, follows.

SOLVE

a. Potassium is K^+ and iodide is I^-. The formula is KI.

Self-Test

Key Vocabulary Completion Questions

__________ 1. SO_4^{2-}

__________ 2. Positively charged particle found in the nucleus

__________ 3. Substance on the right side of the periodic table

__________ 4. Same atomic number, different mass number

__________ 5. Uncharged particle found in an atom

__________ 6. Prefix meaning "six atoms"

__________ 7. Group 17

__________ 8. An atom that has lost an electron

__________ 9. Scientist credited with the nuclear atom

__________ 10. Contains protons and neutrons

__________ 11. Row of the periodic table

__________ 12. Elements in the column on the far right of the periodic table

__________ 13. Elements with atomic numbers 58–71

__________ 14. Name of the group 1 elements

__________ 15. Discovered by J. J. Thompson

__________ 16. Simplest whole-number ratio of elements in an ionic compound

__________ 17. Elements that always form +2 ions

__________ 18. Releases hydrogen ions in water

__________ 19. P_4O_{10}

__________ 20. Another name for hydrogen-2

__________ 21. Famous for his oil-drop experiment

__________ 22. Sum of protons and neutrons

__________ 23. High-energy electron

__________ 24. Nuclear reaction that creates a larger particle from smaller ones

__________ 25. $MgBr_2$

__________ 26. Connection between atoms sharing electrons

__________ 27. Elements in the center of the periodic table, including Fe and Ag

__________ 28. Substances that are shiny, malleable, and ductile

__________ 29. Substances composed of metals and nonmetals

__________ 30. A proton weighs about one of these

Multiple-Choice Questions

1. How many neutrons are in ^{11}B?
   a. 11
   b. 5
   c. 6
   d. 10

2. If ^{40}Ca captures a neutron, what is the isotope formed?
   a. ^{40}Sc
   b. ^{40}K
   c. ^{41}Sc
   d. ^{41}Ca

3. What is a product of the β decay of ^{98}Mo?
   a. ^{98}Nb
   b. ^{98}Tc
   c. ^{99}Mo
   d. ^{97}Mo

4. Which of the following is potassium an example of?
   a. metal
   b. nonmetal
   c. metalloid
   d. compound

5. How many neutrons are in ^{58}Ni?
   a. 58
   b. 28
   c. 30
   d. 58.7

6. How many electrons are in ^{127}I?
   a. 127
   b. 53
   c. 74
   d. 180

7. Gallium (Ga) has an average mass of 69.8 amu. In what proportion are the common isotopes of gallium?
   a. 50% ^{69}Ga and 50% ^{71}Ga
   b. 10% ^{69}Ga and 90% ^{71}Ga
   c. 40% ^{69}Ga and 60% ^{71}Ga
   d. 60% ^{69}Ga and 40% ^{71}Ga

8. If atom A has 10 protons, 11 neutrons, and 10 electrons, and atom B has 11 protons, 11 neutrons, and 10 electrons, which of the following statements is true?
   a. Atoms A and B are isotopes.
   b. Atoms A and B are different elements.
   c. Atoms A and B are ions.
   d. All of the above are true.

9. If an atom has 11 protons, 11 neutrons, and 10 electrons, it would be classifed as which of the following?
   a. anion
   b. cation
   c. isotope
   d. none of these

10. What is the atomic number of ^{59}Co?
    a. 59
    b. 27
    c. 32
    d. 58.93

11. Where are protons located?
    a. in the nucleus
    b. outside the nucleus
    c. throughout the whole atom
    d. in orbitals

12. Which of the following is a noble gas?
    a. sodium
    b. hydrogen
    c. neon
    d. nitrogen

13. What is a likely ion of calcium?
    a. Ca^+
    b. Ca^{2+}
    c. Ca^{2-}
    d. Ca^-

14. What is the name of $ZnCl_2$?
    a. zinc(II) chloride
    b. zinc chloride
    c. zinc dichloride
    d. dizinc chloride
    e. zinc(I) dichloride

15. What is the formula of sodium carbonate?
    a. Na_2C
    b. $NaCO_2$
    c. Na_2CO_3
    d. $NaCO_4$
    e. Na_3C_2

16. Which prefix is used for four atoms?
    a. four
    b. quad
    c. quar
    d. terti
    e. tetra

17. Which of the following is an alkali metal?
    a. Cs
    b. Cr
    c. Ca
    d. Cl
    e. Sn

18. What is the name of HI?
    a. iodic acid
    b. iodous acid
    c. periodic acid
    d. periodous acid
    e. hydroiodic acid

19. When sulfur makes an ion, what is the charge this ion and its name?
    a. 2+, sulfate
    b. 2–, sulfite
    c. 1+, sulfide
    d. 1–, sulfate
    e. 2–, sulfide

20. When Al^{3+} combines with ClO_4^-, what is the formula of the resulting compound and the name of the compound?
    a. $AlClO_4$, aluminum chlorate
    b. $Al(ClO_4)_3$, aluminum perchlorate
    c. $Al(ClO_4)_3$, aluminium trichlorate
    d. Al_3ClO_4, trialuminum chloride
    e. $AlClO_4$, aluminum chloride

21. Which of the following ions is in $Cr(ClO_4)_3$?
    a. Cl^-
    b. O_2^-
    c. Cr^{3+}
    d. ClO_4^{3-}
    e. $(ClO_4)_3^{3-}$

22. What is the name of K_3PO_4?
    a. potassium(III) phosphate
    b. tripotassium phosphite
    c. tripotassium phosphide
    d. potassium(III) phosphite
    e. potassium phosphate

23. What is an ion in NO_2?
    a. nitrate
    b. nitride
    c. nitrite
    d. nitrous
    e. There are no ions in this molecule.

24. Which of the following is NOT part of Dalton's atomic theory?
    a. Each element is composed of identical atoms.
    b. The ratio of atoms in an element is a simple whole number or simple fraction.
    c. The number of protons determines the identity of the atom.
    d. All of the above are part of the theory.
    e. None of the above is part of the theory.

25. What does an *ic* ending on the name of a transition metal mean?
    a. It has a +3 charge.
    b. It has a +1 charge.
    c. It has a +2 charge.
    d. It has the highest common charge for that element.
    e. It has the lowest common charge for that element.

Additional Practice Problems

1. What is the average atomic mass of an element with the following naturally occurring isotopes: 78.99% ^{24}Mg, 10.00% ^{25}Mg, and 11.01% ^{26}Mg?

2. How many protons, neutrons, and electrons are in the following isotopes?
   a. ^{65}Cu b. ^{22}Ne c. ^{131}Xe d. $^{27}Al^{3+}$

3. Name the following compounds:
   a. $CuSO_4$
   b. $ZnCl_2$
   c. ClF_5
   d. $Ca(NO_3)_2$
   e. I_2O_5

4. What is the formula of each of the following compounds?
   a. barium hydroxide
   b. oxygen difluoride
   c. lithium oxide
   d. chromium(III) carbonate
   e. dichlorine heptaoxide

5. Fill in the blanks for the following reactions:
   a. $^{60}Co \rightarrow \beta^- +$ ____
   b. $^{40}Ca + n \rightarrow$ ____

Self-Test Answer Key

Key Vocabulary Completion Answers

1. sulfate ion
2. proton
3. nonmetal
4. isotope
5. neutron
6. hexa
7. halogen
8. cation
9. Rutherford
10. nucleus
11. period
12. noble gases
13. lanthanoids
14. alkali metals
15. electron
16. formula unit
17. alkaline earth metals
18. acid
19. tetraphosphorus decaoxide
20. deuterium
21. Millikan
22. mass number
23. β particle
24. fusion
25. magnesium bromide
26. covalent bond
27. transition metals
28. metals
29. ionic compounds
30. atomic mass unit (amu)

Multiple-Choice Answers

For further information about the topics these questions address, you should consult the text. The appropriate section of the text is listed in parentheses after the explanation of the answer.

1. c. The mass number is 11, which is the sum of protons and neutrons. Since boron has $Z = 5$, or five protons, there are six neutrons. (2.4)

2. d. $^{40}_{20}Ca + n \rightarrow ^{41}_{20}Ca$. (2.9)

3. b. $^{98}_{42}Mo \rightarrow ^{0}_{-1}\beta + ^{98}_{43}Tc$. (2.9)

4. a. Potassium is an element on the left side of the periodic table "staircase"; therefore, it is a metal. (2.6)

5. c. Nickel has an atomic number of 28 (from the periodic table), which is the number of protons; 58 is the mass number, which is equal to proton + neutrons. (2.4)

6. b. Because the atom is electrically neutral, the numbers of protons and electrons must be the same. The atomic number, 53, is also the number of protons and electrons. (2.4)

7. d. 0.6(69) + (0.4)71 = 69.8 amu. (2.5)

8. b. The number of protons determines the identity of the element. (2.4)

9. b. Charged atoms for which the number of protons and electrons are not the same are ions. Since there are more positively charged protons than negatively charged electrons, this atom must have an overall positive charge. Thus it is a cation. (2.7)

10. b. The atomic number is also the number of protons, and characteristic of the element. (2.4)

11. a. Protons and neutrons are located in the nucleus. (2.3)

12. c. Noble gases are in the group in the rightmost column of the periodic table. (2.6)

13. b. As a group 2 metal, calcium loses 2 electrons for a charge of +2. (2.7)

14. b. It is an ionic compound; therefore, no prefixes are used. As a transition metal, normally there is a roman numeral. However, zinc is one of the exceptions. It does not use a roman numeral. (2.8)

15. c. Sodium and carbon constitute a metal–nonmetal combination and thus an ionic compound. As an ion from group 1, sodium ion is Na^+. Carbonate is the polyatomic ion: CO_3^{2-}. For charges to cancel, there must be two sodiums to one carbonate. (2.8)

16. e. While some of the other prefixes are used to mean 4, only tetra is used in chemical names. (2.8)

17. a. Alkali metals are group 1. Cs is the only choice in that column. (2.6)

18. e. Binary acids of halogens are named as hydro[halogen]ic acid. (2.8)

19. e. Since sulfur is in group 16, it typically gains 2 electrons. Since it gains electrons, it has a negative charge. As a monatomic anion, the suffix is changed to *ide.* (2.8)

20. b. As an ionic compound (metal–nonmetal combination), no prefixes are used. The cation is aluminum; the anion is perchlorate. There must be three perchlorates to balance the +3 charge on the aluminum. (2.8)

21. c. The ions that make up $Cr(ClO_4)_3$ are Cr^{3+} and ClO_4^-. The subscripted *3* means that there are three of the perchlorate ions for each chromium(III) ion. (2.8)

22. e. It was an ionic compound; so it doesn't need prefixes. Potassium is a group 1 metal; so no roman numeral is needed. The anion is phosphate. (2.8)

23. e. This is a molecular compound, not an ionic compound. Molecular compounds do not contain ions. (2.8)

24. c. Dalton's theory was developed before the discovery of protons. (2.2)

25. d. The old system of naming depends on knowing the common charges for each element. The *ic* ending is for the higher charge; *ous* is for the lower charge. Its complexity is the reason a new system is used. (2.8)

Additional Practice Problem Answers

1. Assuming the mass number is the mass of the atom in atomic mass units:

$$0.7899(24) + 0.1000(25) + 0.1101(26) = 24.3202 = 24 \text{ amu}$$

2. a. ^{65}Cu $p = e = 29, n = 36$
   b. ^{22}Ne $p = e = 10, n = 12$
   c. ^{131}Xe $p = e = 54, n = 77$
   d. $^{27}Al^{3+}$ $p = 13, n = 14, e = 10$

3. a. copper(II) sulfate or cupric sulfate
   b. zinc chloride
   c. chlorine pentafluoride
   d. calcium nitrate
   e. diiodine pentaoxide

4. a. $Ba(OH)_2$
   b. OF_2
   c. Li_2O
   d. $Cr_2(CO_3)_3$
   e. Cl_2O_7

5. a. $^{60}Co \rightarrow \beta^- + {}^{60}Ni$
   b. $^{40}Ca + n \rightarrow {}^{41}Ca$

CHAPTER 3 Chemical Reactions and Earth's Composition

OUTLINE

REVIEW

Chapter Overview

When substances mix with other substances, the atoms may rearrange to form new substances. This is called a **chemical reaction**. Chemists express what occurs in a chemical reaction using a **chemical equation**. Chemical equations list the formulas of the starting materials (**reactants**) on the left (in no particular order) and use an arrow pointing to the list of formulas of the substances created (**products**) on the right. The **law of conservation of mass** says that atoms can neither be created nor destroyed in a chemical reaction. Therefore, the chemical equation must also account for each atom on both sides of the arrow. When each atom is accounted for, the equation is balanced. Since changing the subscripts of the chemical formulas would change the type of compound, equations are balanced by describing how many of each type of compound are used or produced. To accomplish this, a number in front of the chemical formula, called the **stoichiometric coefficient** is used.

A balanced chemical equation describes the ratio of molecules needed in the reaction, which is why whole numbers are used for stoichiometric coefficients. However, since chemists work with macroscopic properties, they need to relate measurable quantities, such as mass, to the number of molecules, atoms, or ions. This factor, the SI unit for amount of substance, is a **mole**. Because atoms and molecules are so small, this amount must be very large to constitute a measurable quantity. The number of things in a mole is called **Avogadro's number** and has a value of 6.022×10^{23}. This number is used because, whereas the average mass of an atom is its atomic weight in atomic mass units, the mass of a mole of atoms is its atomic weight in grams. Since masses are additive, the mass of a molecule is the sum of the mass of each of its atoms in atomic mass units, and the mass of a mole of the molecule is the sum of the mass of each of its atoms in grams. The mass of a mole of any substance is called **molar mass**. Also, since the ratios are the same in a mole of molecules as in the molecule by itself, the molecular formula can be used to obtain mole ratios of the elements of a compound. For the same reason, stoichiometric coefficients from a chemical equation can also be used to relate moles of products and reactants. The process of using mole relationships from either formulas or chemical equations to relate quantities of different substances is called **stoichiometry**.

There are many types of chemical reactions. One of these is **hydrolysis**, which is the reaction of a substance with water. Nonmetal oxides, also called **acid anhydrides**, are one type of substance that undergoes hydrolysis reactions. The hydrolysis of nonmetal oxides produces an acid. **Hydrocarbons**, substances consisting of only hydrogen and carbon, can undergo **combustion**, or burning, another type of chemical reaction. In the complete combustion of hydrocarbons, the hydrocarbon reacts with oxygen gas (O_2) to produce carbon dioxide and water.

Stoichiometric relationships can be used to relate the percentage of an element in a compound to its formula. **Percent composition** is the grams of element per 100 g of compound. Because this value is a ratio, any quantity of compound has the same value. Therefore, when determining percent composition from the chemical formula, a convenient amount of substance to use is 1 mol. By using 1 mol, the grams of compound are equal to the molar mass of the compound. In addition, the grams of element are equal to the molar mass of the element, accounting for the number of moles of that element in the formula. Thus the molar ratio from the chemical formula is used to obtain the mass ratio of percent composition.

The process can also be done in reverse, using the percent composition to determine the formula. In this case, the convenient amount of substance to use is 100 g. If there is 100 g of compound, the grams of each element are that same value as the percentage of each element. Molar mass can then be used to determine the moles of each element from which the molar ratio of the chemical formula is determined. A chemical formula obtained in this manner is called an empirical formula. **Empirical** means that it was determined from experiment rather than from a theory. In this context, the **empirical formula** is the simplest ratio of elements in a compound. However, many molecules have atoms that are not in their simplest ratio. To determine the true molecular formula, more information, specifically the molar mass (molecular weight), is needed. Since the molar mass contains the mass of all elements in a compound, the ratio of the true molar mass to the molar mass obtained from the empirical formula provides the multiplication factor to convert the empirical formula to the molecular formula. One method of obtaining the true molar mass is to use an instrument called a **mass spectrometer**.

A method for obtaining chemical formulas of organic molecules is **combustion analysis**. With this method, the organic molecule is reacted completely with oxygen gas (combustion) and the products of carbon dioxide and water are collected. Stoichiometry is used to relate the mass of carbon dioxide obtained to the moles of carbon in the original sample. Similarly, water is used to determine the moles of hydrogen. If oxygen was present in the original sample, its mass can be determined by subtracting the mass of carbon (determined from the amount of CO_2 produced) and the mass of hydrogen (determined from the amount of H_2O produced) from the mass of the original sample. The mass of oxygen can then be stoichiometrically converted to moles, and the mole ratio of each element in the compound can be determined.

Chemical formulas and chemical equations are very valuable in relating amounts of different substances to each other. However, chemical equations express the amount of reactants needed for the reaction to occur; they do not imply that that amount of material is actually available. Since all reactants must be present for the reaction to occur, if one of the reactants gets used up, the reaction stops regardless of how much of the other reactants are available. The reactant that gets used up first is called the **limiting reactant**. Since the reaction stops when the limiting reactant is gone, the limiting reactant determines the amount of product made.

The amount of product made in a chemical reaction is called the **yield**. Yields are normally expressed as mass (in grams). The yield predicted from a stoichiometric calculation is called the **theoretical yield**, because it is based on stoichiometric theory. Unfortunately, real life is never quite as good as theory. In real chemistry the reactants are sometimes used up by other competing reactions. Sometimes the reactions are just too slow for it to be practical to wait for its completion. Also, if atoms can be rearranged to form products, cannot the atoms of the products be rearranged to reform the reactants? Some reactions are never complete because of this process. Consequently, the **actual yield**, the amount of product really produced by the reaction, is never more than the theoretical yield. Thus, the theoretical yield is sometimes also called the maximum yield. The relationship between the theoretical yield and the actual yield is expressed as percent yield where

$$\textbf{percent yield} = \frac{\text{actual yield}}{\text{theoretical yield}} \times 100$$

Worked Examples

Working with Moles

Example 1.
How many moles of copper are 5.44×10^{22} atoms of copper?

Collect and Organize We are given a number of copper atoms, 5.44×10^{22}, and are asked to convert to moles.

Analyze We will use Avogadro's number, which gives the number of particles that are contained in 1 mol, to convert the given number of copper atoms into moles of copper.

Solve

$$(5.44 \times 10^{22} \text{ atoms}) \times \frac{1 \text{ mol}}{6.022 \times 10^{23} \text{ atoms}} = 0.0903 \text{ mol Cu}$$

Three significant figures are the fewest from atoms of copper.

Think About It Avogadro's number allows us to express a very large number of atoms in a more convenient way.

EXAMPLE 2.
How many molecules of HCl are in 0.97 mol HCl?

COLLECT AND ORGANIZE We are asked to determine the number of HCl molecules in a given number of moles of HCl.

ANALYZE Avogadro's number provides the conversion between the number of particles and the number of moles.

SOLVE

$$0.97 \text{ mol} \times \frac{6.022 \times 10^{23} \text{ molecules}}{1 \text{ mol}} = 5.8 \times 10^{23} \text{ molecules}$$

THINK ABOUT IT Avogadro's number allows us to easily convert between moles and number of molecules.

EXAMPLE 3.
How many atoms of oxygen are in one molecule of sulfuric acid? How many moles of oxygen are in 1 mol sulfuric acid?

COLLECT AND ORGANIZE We are asked to determine how many oxygen atoms are in one molecule of sulfuric acid and how many moles of oxygen are in 1 mol of sulfuric acid.

ANALYZE We can use the molecular formula of sulfuric acid to answer both questions.

SOLVE Sulfuric acid has the formula of H_2SO_4. There are four atoms of oxygen in one molecule of H_2SO_4. There are also 4 mol of oxygen in 1 mol of sulfuric acid. These values are exact (infinite significant figures).

THINK ABOUT IT The ratio of oxygen to sulfuric acid is the same whether one considers it on a atom-to-atom basis or a mole-to-mole basis.

EXAMPLE 4.
How many atoms of hydrogen are in 0.5129 mol water?

COLLECT AND ORGANIZE We are given 0.5129 mol of water and are asked to determine the number of hydrogen atoms it contains.

ANALYZE We can use the molecular formula and Avogadro's number to make these determinations.

SOLVE Note that two changes are required in this question: a change from moles to atoms and a change from water molecules to hydrogen atoms.

$$0.5129 \text{ mol } H_2O \times \frac{2 \text{ mol H}}{1 \text{ mol } H_2O} \times \frac{6.022 \times 10^{23} \text{ atoms}}{1 \text{ mol H}} = 6.177 \times 10^{23} \text{ atoms H}$$

Or equally correct:

$$0.5129 \text{ mol } H_2O \times \frac{6.022 \times 10^{23} \text{ molecule}}{1 \text{ mol}} \times \frac{2 \text{ atoms H}}{1 \text{ molecule } H_2O} = 6.177 \times 10^{23} \text{ atoms H}$$

The starting number had four significant figures, Avogadro's number had four significant figures, and the atom-to-molecule ratio has infinite significant figures. Thus the answer has four significant figures.

THINK ABOUT IT Note that you change units and type of substance as separate steps. It is a good idea to keep track of both as you do these types of calculations.

EXAMPLE 5.
What is the molar mass of the following compounds?

a. $FeCl_3$ b. $Zn(NO_3)_2$ c. N_2O d. CH_4

COLLECT AND ORGANIZE We are given four compounds and are asked to determine their molar masses.

ANALYZE The molar mass of a compound is determined by multiplying the molar mass of each of the atoms making up the compound by the number of times that atom appears in the compound and adding them up.

SOLVE

a. There are one iron atom (molar mass = 55.845) and three chlorine atoms (molar mass = 35.453 each); so

$$\text{molar mass of } FeCl_3 = 55.845 + 3(35.453) = 162.204 \text{ g/mol}$$

b. There are one zinc, two nitrogens, and six oxygens; so

$$\text{molar mass} = 65.39 + 2(14.01) + 6(16.00) = 189.41 \text{ g/mol}$$

c. Two nitrogens and one oxygen:

$$\text{molar mass} = 2(14.01) + 16.00 = 44.02 \text{ g/mol}$$

d. One carbon and four hydrogens:

$$\text{molar mass} = 12.01 + 4(1.008) = 16.04 \text{ g/mol}$$

THINK ABOUT IT The molar mass of any compound can be determined using only its molecular formula and the periodic table.

EXAMPLE 6.
How many moles are in 0.355 g of $FeCl_3$?

COLLECT AND ORGANIZE We are given 0.355 g of $FeCl_3$ and are asked to convert to moles of $FeCl_3$.

ANALYZE We can use the molar mass of the compound to convert the mass of the compound into moles.

Solve The molar mass of $FeCl_3$ is 162.204 g/mol (see previous example). Therefore 1 mol $FeCl_3$ = 162.204 g.

$$0.355 \text{ g} \times \frac{1 \text{ mol}}{162.204 \text{ g}} = 2.18 \times 10^{-3} \text{ mol FeCl}_3$$

Think About It To determine the number of moles of a given amount of a compound, we must know the molar mass of the compound.

Example 7.
How many grams are 0.0608 mol of nickel(II) iodide?

Collect and Organize Given 0.0608 mol of nickel(II) iodide, we are asked to convert to grams.

Analyze Two steps are necessary to solve the problem. First we determine the molecular formula of nickel(II) iodide. Once we have the molecular formula, we can calculate its molar mass and use that value to make the conversion from moles to grams.

Solve The formula for nickel(II) iodide is NiI_2. Its molar mass is 58.69 + 2(126.90) = 312.49 g/mol. Starting with the given number of grams and using this molar mass:

$$0.0608 \text{ mol NiI}_2 \times \frac{312.49 \text{ g}}{1 \text{mol}} = 18.999392 \text{ g} = 19.0 \text{ g}$$

Think About It We can easily convert from moles to grams once we know the molar mass of the compound.

Chemical Equations

Example 1.
Balance the following chemical equations:

a. $K_2CO_3 + Al(ClO_4)_3 \rightarrow KClO_4 + Al_2(CO_3)_3$

b. $Zn + HCl \rightarrow ZnCl_2 + H_2$

c. $C_8H_{18} + O_2 \rightarrow CO_2 + H_2O$

d. $HF + Mg(OH)_2 \rightarrow MgF_2 + H_2O$

e. $NH_4Cl + Ba(OH)_2 \rightarrow NH_3 + H_2O + BaCl_2$

Collect and Organize We are give five chemical equations and are asked to balance them.

Analyze Although there are systematic methods for balancing chemical equations, they are normally more difficult than trial and error. However, a few tips sometimes prove useful.

- Start with the most complex compound and work to the simplest compound.
- If a fraction balances the reaction, use one. As a final step, multiply by the denominator of the fraction to get whole numbers.
- Treat polyatomic atoms as one group rather than as separate elements.
- Water (H_2O) tends to react as H and OH; so balance in those parts.
- Ammonium ions (NH_4) can react as H and NH_3; so watch for those parts.

Solve

a. Treat the polyatomic ions like one group. Note that the most complex one is $Al_2(CO_3)_3$. (It has the most subscripts.)

To balance the aluminums, put a 2 in front of aluminum perchlorate. To balance the carbonates, put a 3 in front of potassium carbonate. That means there are 6 potassiums; so a 6 in front of potassium perchlorate is needed. That results in the following balanced equation:

$$3\ K_2CO_3 + 2\ Al(ClO_4)_3 \rightarrow 6\ KClO_4 + Al_2(CO_3)_3$$

b. You need two chlorines on the left; so a 2 in front of HCl will do it.

$$Zn + 2\ HCl \rightarrow ZnCl_2 + H_2$$

c. Start with C_8H_{18} as the most complex and O_2 is the simplest.

To balance the carbons, put an 8 in front of the CO_2. To balance the hydrogens, put a 9 in front of the water. That makes 16 + 9 = 25 oxygens on the right. Since it is O_2 on the left, 25/2 would balance the number of oxygens. Do that as a temporary step:

$$C_8H_{18} + 25/2\ O_2 \rightarrow 8\ CO_2 + 9\ H_2O$$

Multiply each coefficient by 2:

$$2\ C_8H_{18} + 25\ O_2 \rightarrow 16\ CO_2 + 18\ H_2O$$

d. The groups in this equation are H, F, Mg, and OH. It is easier to see the relationships by noting that water acts like HOH.

$$HF + Mg(OH)_2 \rightarrow MgF_2 + H\text{–}OH$$

So the OH can be balanced with 2 waters, and the fluorine with 2 HF. That does it.

$$2\ HF + Mg(OH)_2 \rightarrow MgF_2 + 2\ H_2O$$

e. Treat ammonium ion as NH_3–H and water as H–OH.

$$NH_3\text{–}HCl + Ba(OH)_2 \rightarrow NH_3 + H\text{–}OH + BaCl_2$$

Balance the chlorine with 2 NH_4Cl. Balance the OH with 2 waters. Balance the ammonia with 2 NH_3. The final result is

$$2\ NH_4Cl + Ba(OH)_2 \rightarrow 2\ NH_3 + 2\ H_2O + BaCl_2$$

THINK ABOUT IT A balanced chemical equation describes both the ratio of molecules needed for a chemical reaction and the ratio of moles needed for a chemical reaction.

EXAMPLE 2.

How many grams of oxygen are in 0.1926 mol barium hydroxide?

COLLECT AND ORGANIZE We are given 0.1926 mol of barium hydroxide and are asked to determine the number of grams of oxygen it contains.

ANALYZE We need to determine the molecular formula of barium hydroxide. Once that is accomplished, we can use stoichiometry to relate the number of moles of oxygen to moles of barium hydroxide and finally use the molar mass of oxygen to convert to grams.

SOLVE The formula for barium hydroxide is $Ba(OH)_2$. The molar mass of barium hydroxide is 171.35 g/mol. The molar mass of oxygen (O) is 16.00 g/mol. The relationship between barium hydroxide and oxygen is either

$$1 \text{ molecule } Ba(OH)_2 = 2 \text{ atoms of oxygen}$$

or

$$1 \text{ mol } Ba(OH)_2 = 2 \text{ mol oxygen}$$

$$0.1926 \text{ mol } Ba(OH)_2 \times \frac{2 \text{ mol oxygen}}{1 \text{ mol } Ba(OH)_2} \times \frac{16.00 \text{ g O}}{1 \text{ mol O}} = 6.163 \text{ g}$$

The molar mass of barium hydroxide was not needed because barium hydroxide was already in units of moles and moles are the proper unit to change substance.

THINK ABOUT IT Stoichiometry is a very powerful tool in chemistry.

EXAMPLE 3.

How many moles of nitrogen are in 0.18 g of dinitrogen tetraoxide?

COLLECT AND ORGANIZE We are given 0.18 g of dinitrogen tetraoxide and are asked to determine the number of moles of nitrogen it contains.

ANALYZE We must determine the molecular formula and molar mass of dinitrogen tetraoxide. Once that is accomplished, we can use stoichiometry to relate moles of dinitrogen tetraoxide to moles of nitrogen.

SOLVE Dinitrogen tetraoxide is N_2O_4. This is determined from its name, which follows the systematic method for covalent compounds discussed in Chapter 2. Its molar mass is 2(14.01) + 4(16.00) = 92.02 g/mol.

$$0.18 \text{ g } N_2O_4 \times \frac{1 \text{ mol } N_2O_4}{92.02 \text{ g } N_2O_4} \times \frac{2 \text{ mol N}}{1 \text{ mol } N_2O_4} = 3.9 \times 10^{-3} \text{ mol}$$

THINK ABOUT IT Unlike the previous example, where we were given our compound in moles, here we are given the amount of dinitrogen tetraoxide in grams. It is, therefore, necessary first to convert to moles before applying stoichiometry.

EXAMPLE 4.

How many grams of $CaSO_4$ can be made with 0.742 g calcium?

COLLECT AND ORGANIZE We are asked to determine how many grams of $CaSO_4$ are possible if 0.742 g calcium is available?

ANALYZE After converting the given mass of calcium into moles of calcium and using stoichiometry to determine the number of moles of calcium contained in one mole of $CaSO_4$ we can determine the number of moles of $CaSO_4$ that would be possible with the given mass of calcium. Finally, the molar mass of $CaSO_4$ can be used to convert grams of $CaSO_4$ to moles of $CaSO_4$.

SOLVE The molar mass of calcium sulfate is 40.08 + 32.07 + 4(16.00) = 136.15 g/mol

$$0.742 \text{ g Ca} \times \frac{1 \text{ mol Ca}}{40.08 \text{ g}} \times \frac{1 \text{ mol } CaSO_4}{1 \text{ mol Ca}} \times \frac{136.15 \text{ g}}{1 \text{ mol } CaSO_4} = 2.52 \text{ g}$$

THINK ABOUT IT Stoichiometry provides the connection between moles of Ca and moles of $CaSO_4$ that is needed to determine the number of grams that are possible.

EXAMPLE 5.

How many grams of aluminum chloride can be made from 0.149 g of hydrochloric acid?

$$Al(OH)_3 + HCl \rightarrow H_2O + AlCl_3$$

COLLECT AND ORGANIZE We are asked to determine the maximum amount of product that can be formed from a given amount of starting material.

ANALYZE The following general procedure is used to solve stoichiometric problems.

1. Balance the reaction.
2. Convert starting unit to moles.
   a. Use molar mass to change grams to moles.
   b. Use Avogadro's number to change atoms or molecules to moles.

3. Convert moles of starting substance to moles of the substance you want.
   a. Use stoichiometric coefficients of balanced reaction.
   b. Use the chemical formula for an element-to-compound ratio.
4. Convert moles to unit you want.

   Use the methods in step 2.

SOLVE Step 1: Balance the reaction.

$$Al(OH)_3 + 3\ HCl \rightarrow 2\ H_2O + AlCl_3$$

Step 2: Convert the starting unit to moles.

The starting point is 0.149 g HCl. The method to convert grams to moles is by means of molar mass. Since the substance is HCl, the molar mass of HCl is required. The molar mass of HCl = 1.008 + 35.453 = 36.461 g/mol. So 36.461 g = 1 mol HCl.

$$0.149\text{ g HCl} \times \frac{1\text{ mol HCl}}{36.461\text{ g}}$$

Step 3: Convert the starting substance to the substance desired.

Starting with HCl, the question asks about $AlCl_3$. The chemical reaction says that 3 mol HCl makes 1 mol $AlCl_3$.

$$0.149\text{ g HCl} \times \frac{1\text{ mol HCl}}{36.461\text{ g}} \times \frac{1\text{ mol AlCl}_3}{3\text{ mol HCl}}$$

Step 4: Convert moles to units desired.

The question asks for grams; so molar mass is used. Since the substance is now $AlCl_3$, the molar mass of $AlCl_3$ is needed. The molar mass of $AlCl_3$ = 26.98 + 3(35.453) = 133.34 g/mol.

$$0.149\text{ g HCl} \times \frac{1\text{ mol HCl}}{36.461\text{ g}} \times \frac{1\text{ mol AlCl}_3}{3\text{ mol HCl}} \times \frac{133.34\text{ g}}{1\text{ mol AlCl}_3} = 0.182\text{ g AlCl}_3$$

The three significant figures in the answer were determined by the value of HCI, which has the fewest significant figures of any number used in the calculation. The molar masses have five significant figures, and the mole-to-mole ratio has infinite significant figures.

THINK ABOUT IT Examples 2–4 also followed these steps, using chemical formulas instead of chemical equations.

EXAMPLE 6.

How many milligrams of carbon tetrachloride can be made from 0.039 mol Cl_2?

$$Cl_2 + CO_2 \rightarrow CCl_4 + O_2$$

COLLECT AND ORGANIZE We are asked to determine the maximum amount of product that can be formed from a given amount of starting material.

ANALYZE We will use the procedure outlined in the previous example to solve the problem.

SOLVE Step 1: Balance the reaction.

$$2\ Cl_2 + CO_2 \rightarrow CCl_4 + O_2$$

Step 2: Convert the unit given to moles.

The starting point is 0.039 mol Cl_2. The unit is already moles; so move on to the next step.

Step 3: Convert the substance.

The balanced reaction says 2 mol Cl_2 are needed for 1 mol CCl_4

$$0.039\text{ mol Cl}_2 \times \frac{1\text{ mol CCl}_4}{2\text{ mol Cl}_2}$$

Step 4: Convert the units from moles.

The requested unit is milligrams, a variation on grams. To convert moles to grams, molar mass is used. The substance is now CCl_4.

molar mass = 12.01 + 4(35.453) = 153.82 g/mol

To convert grams to milligrams, use the SI relationships (Chapter 1) 1000 mg = 1 g.

$$0.039\text{ mol Cl}_2 \times \frac{1\text{ mol CCl}_4}{2\text{ mol Cl}_2} \times \frac{153.82\text{ g}}{1\text{ mol CCl}_4} \times \frac{1000\text{ mg}}{1\text{ g}} = 3.0 \times 10^3\text{ mg CCl}_4$$

THINK ABOUT IT Having a stoichiometric equation for a chemical reaction allows us to easily determine the amount of product that can be produced from a given amount of reactant.

EXAMPLE 7.

How many micrograms (μg) of CaO are made when 3.6×10^{16} molecules of O_2 are produced?

$$CaCO_3 \rightarrow CaO + O_2$$

COLLECT AND ORGANIZE We are asked to determine the amount of one product produced in the reaction based on information about a second product that is produced in the same reaction.

ANALYZE We can still use the procedure outlined in the previous two examples to solve the problem.

SOLVE Step 1: Balance the equation.

$$CaCO_3 \rightarrow CaO + O_2$$

Sometimes it is balanced as it stands. Check anyway.

Step 2: Convert the given units to moles.

The starting unit is molecules of O_2. Molecules are converted to moles with Avogadro's number, 1 mol = 6.022×10^{23} molecules.

$$3.6 \times 10^{16} \text{ molecules } O_2 \times \frac{1 \text{ mol } O_2}{6.022 \times 10^{23} \text{ molecules}}$$

Step 3: Change the substance.

The balanced equation says 1 mol CaO to 1 mol O_2

$$3.6 \times 10^{16} \text{ molecules } O_2 \times \frac{1 \text{ mol } O_2}{6.022 \times 10^{23} \text{ molecules}} \times \frac{1 \text{ mol CaO}}{1 \text{ mol } O_2}$$

Step 4: Convert to the desired units.

Desired units are micrograms, a variation of grams: 1 gram = 1,000,000 μg. Get to grams with molar mass. The substance is now CaO; so 40.08 + 16.00 = 56.08 g/mol.

$$3.6 \times 10^{16} \text{ molecules } O_2 \times \frac{1 \text{ mol } O_2}{6.022 \times 10^{23} \text{ molecules}} \times$$

$$\frac{1 \text{ mol CaO}}{1 \text{ mol } O_2} \times \frac{56.08 \text{ g}}{1 \text{ mol}} \times \frac{1{,}000{,}000 \text{ μg}}{1 \text{ g}} = 3.3525 \text{ g}$$

$$= 3.4 \text{ μg CaO}$$

THINK ABOUT IT Expressing reactants and products in moles allows us to easily perform calculations using stoichiometry.

CHEMICAL FORMULAS AND PERCENT COMPOSITION

EXAMPLE 1.
What is the percent composition of each element in MnS?

COLLECT AND ORGANIZE We are asked to determine the percent composition of each element in MnS.

ANALYZE The percent composition of each element in a compound can be found by dividing the molar mass of the element by the molar mass of the compound and multiplying by 100.

SOLVE The molar mass of MnS is 54.94 + 32.07 = 87.01 g/mol.

$$\% \text{ Mn} = (54.94/87.01) \times 100 = 63.14\%$$

$$\% \text{ S} = (32.07/87.01) \times 100 = 36.85\%$$

or

$$\% \text{ S} = 100\% - 63.14\% = 36.85\%$$

THINK ABOUT IT Percent composition implies mass percent. Because percentage is a ratio, any quantity of compound gives the same ratio and the same percent composition. Therefore, you can choose any convenient amount of compound to work with. When determining percent composition from a chemical formula, it is convenient to choose moles.

EXAMPLE 2.
What is the percent composition of each element in $Al(OH)_3$?

COLLECT AND ORGANIZE We are asked to determine the percent composition of each element in $Al(OH)_3$.

ANALYZE The percent composition of each element in a compound can be found by dividing the total molar mass of the element by the molar mass of the compound and then multiplying by 100. To determine the total molar mass of each element, you must take into account the number of moles of that element per mole of compound.

SOLVE The molar mass of $Al(OH)_3$ is 26.98 + 3(16.00) + 3(1.008) = 78.00 g/mol.

$$\% \text{ Al} = \frac{26.98}{78.00} \times 100 = 34.59\%$$

$$\% \text{ O} = \frac{3(16.00)}{78.00} \times 100 = 61.54\%$$

$$\% \text{ H} = \frac{3(1.008)}{78.00} \times 100 = 3.877\%$$

THINK ABOUT IT Even though there are 3 mol of H and only 1 mol of Al, the mass percent of Al is much greater than H due to its much greater molar mass.

EXAMPLE 3.
What is the empirical formula of a compound whose percent composition is 40.0% sulfur and 60.0% oxygen?

COLLECT AND ORGANIZE We are asked to determine the empirical formula of a compound given the percentages of each element comprising the compound.

ANALYZE For ease of calculation, we can assume that we have 100 g of compound. We can then directly convert the percentages of the constituents into grams. We then need to convert the grams of each element into moles using its molar mass. Finally, the smallest ratio among the moles of each of the elements is determined. The simplest molar ratio of elements in a compound is its empirical formula.

SOLVE If 100 g of compound is assumed, there are 40.0 g sulfur and 60.0 g oxygen. Using the molar mass of each to convert from grams to moles:

$$40.0 \text{ g S} \times \frac{1 \text{ mol}}{32.07 \text{ g}} = 1.247 \text{ mol}$$

$$60.0 \text{ g O} \times \frac{1 \text{ mol}}{16.00 \text{ g}} = 3.75 \text{ mol}$$

To get the smallest ratio, divide by the smallest mole value. In this case, that is 1.247.

$$\frac{1.247 \text{ mol S}}{1.247} = 1$$

$$\frac{3.75 \text{ mol O}}{1.247} = 3.007$$

Therefore, the ratio is one sulfur to three oxygens. Because the elements represent the number of atoms in the compound, the ratios must be whole numbers. It is appropriate to round the 3.007 value of oxygen to reflect this. The order of the elements in the formula can be obtained from their relative positions in the periodic table. Sulfur and oxygen are in the same group, and sulfur is lower. Therefore, sulfur is first in the formula: SO_3.

THINK ABOUT IT The empirical formula of a chemical compound is a simple expression of the relative number of each type of atom in it.

EXAMPLE 4.
What is the empirical formula of a molecule with percent composition of 15.8% Al, 28.1% S, and 56.1% O?

COLLECT AND ORGANIZE We are asked to determine the empirical formula of a compound given the percentage of each element comprising the compound.

ANALYZE For ease of calculation, we can assume that we have 100 g of compound. We can then directly convert the percentages of the constituents into grams. We then need to convert the grams of each element into moles using its molar mass. Finally, the smallest ratio among the moles of each of the elements is determined.

SOLVE In 100 g of compound, there are 15.8 g Al, 28.1 g S, and 56.1 g oxygen. To convert to moles:

$$15.8 \text{ g Al} \times \frac{1 \text{ mol}}{26.98 \text{ g}} = 0.5856 \text{ mol Al}$$

$$28.1 \text{ g S} \times \frac{1 \text{ mol}}{32.07 \text{ g}} = 0.8762 \text{ mol S}$$

$$56.1 \text{ O} \times \frac{1 \text{ mol}}{16.00 \text{ g}} = 3.506 \text{ mol O}$$

To get the simplest ratio, divide by the smallest number of moles.

$$\frac{0.5856 \text{ mol Al}}{0.5856} = 1 \text{ Al}$$

$$\frac{0.8762 \text{ mol S}}{0.5856} = 1.496 \text{ S}$$

$$\frac{3.506 \text{ mol O}}{0.5856} = 5.987 \text{ O}$$

The final formula must use whole numbers. However, you should notice that sulfur rounds better to 1.5 than to either 1 or 2. Since 1.5 is a simple fraction (1½), it is better to round to that than to either 1 or 2. So that gives the ratio of 1 Al : 1.5 S : 6 O. Since fractions are not acceptable in a formula, to get rid of the half, multiply each amount by 2. Thus the formula is $Al_2S_3O_{12}$.

Note the order of the elements. On the periodic table, Al is to the left of sulfur and oxygen; so it is first. Sulfur is lower on the periodic table than oxygen; so it is next.

THINK ABOUT IT Since empirical formulas are determined experimentally, they may not always produce nice whole numbers.

EXAMPLE 5.
What is the molecular formula of a compound that is 86% C and 14% H and that has a molar mass of 56 g/mol?

COLLECT AND ORGANIZE We are asked to determine the molecular formula of a compound given the percentages of each element comprising the compound and the compound's molar mass.

ANALYZE The first step in the determination of the molecular formula from the given information is to determine the empirical formula using the procedure of the previous example. Once this is accomplished the molar mass of the compound is divided by the molar mass obtained from the empirical formula to determine a multiplication factor, which is applied to the empirical formula to obtain the molecular formula.

SOLVE First determine the empirical formula in the usual way.

$$86 \text{ g C} \times \frac{1 \text{ mol}}{12.01 \text{ g}} = \frac{7.2 \text{ mol}}{7.2} = 1$$

$$14 \text{ g H} \times \frac{1 \text{ mol}}{1.008 \text{ g}} = \frac{13.88 \text{ mol}}{7.2} = 1.92$$

The empirical formula is CH_2. The apparent molar mass is 12.01 + 2(1.008) = 14.02 g. The multiplication factor is 56/14 = 4. So the actual molecular formula is C_4H_8.

THINK ABOUT IT The empirical formula determines only the simplest ratio of elements in a compound. It turns out that many molecules have atoms that are not in the simplest ratio. To determine the true molecular formula, more information is needed. That extra information is the molar mass. Since the molar mass contains the mass of all elements in a compound, the ratio of the molar mass of the true formula to the molar mass of the empirical formula provides the multiplication factor to determine the true molecular formula.

LIMITING REACTANTS

EXAMPLE 1A. METHOD 1

What is the limiting reactant when 0.549 g potassium metal is mixed with 0.669 g of Cl_2 to make KCl? How many grams of KCl will be produced?

$$K + Cl_2 \rightarrow KCl$$

COLLECT AND ORGANIZE We are asked to determine which of the reactants is the limiting reagent in the given reaction and the amount of KCl produced.

ANALYZE There are two ways to determine the limiting reactant. In method 1, you compare the two reactants (method 2 is shown in example 1b). Pick one reactant, and assume that this reactant will be all used up. Then use the procedure for solving stoichiometric problems to determine the amount of the other reactant needed. Compare your answer to the value given. If the answer is less than the amount given in the problem, the reactant you started with is the limiting reactant. If the answer is greater than the amount given in the problem, the reactant you ended up with is the limiting reactant. The limiting reactant is used to determine the amount of product produced.

SOLVE Step 1: Balance the equation.

$$2\ K + Cl_2 \rightarrow 2\ KCl$$

Test to see if K is the limiting reactant.
Steps 2–4: Starting with grams K, going to grams Cl_2, you need the molar masses of K (39.10 g/mol) and Cl_2 [2(35.453) = 70.906 g/mol].

$$0.549\text{ g K} \times \frac{1\text{ mol K}}{39.10\text{ g}} \times \frac{1\text{ mol Cl}_2}{2\text{ mol K}} \times \frac{70.906\text{ g}}{1\text{ mol}} = 0.498\text{ g Cl}_2$$

Since stoichiometry says 0.498 g Cl_2 is needed and 0.669 g Cl_2 is given in the problem (we need less than we have), then K is the limiting reactant.

Steps 2–4: To get the correct quantity of product, you must start with the limiting reactant for Steps 2–4:

$$0.549\text{ g K} \times \frac{1\text{ mol K}}{39.10\text{ g}} \times \frac{2\text{ mol KCl}}{2\text{ mol K}} \times \frac{74.55\text{ g}}{1\text{ mol KCl}} = 1.05\text{ g KCl}$$

THINK ABOUT IT The stoichiometric equation tells us the relationship between the moles of products and moles of reactants. By converting the masses of reactants into moles, we can determine whether one reactant will be used up before another. You can determine the limiting reactant by starting with either reactant. You get the same answer either way. Once one of the reactants runs out, the reaction stops and no more product is formed. The starting amount of limiting reactant determines the amount of product.

EXAMPLE 1B. METHOD 2

What is the limiting reactant when 0.549 g potassium metal is mixed with 0.669 g of Cl_2 to make KCl? How many grams of KCl will be produced?

COLLECT AND ORGANIZE We are asked to determine which of the reactants is the limiting reactant in the given reaction and the amount of KCl produced.

ANALYZE Using this alternate method, you calculate the amount of product with both reactants and choose the correct answer of the two, which is the lower answer. The limiting reactant will be the reactant that results in a smaller amount of product.

SOLVE Step 1: Balance the equation.

$$2\ K + Cl_2 \rightarrow 2\ KCl$$

Steps 2–4: Assuming K is the limiting reactant:

$$0.549\text{ g K} \times \frac{1\text{ mol K}}{39.10\text{ g}} \times \frac{2\text{ mol KCl}}{2\text{ mol K}} \times \frac{74.55\text{ g}}{1\text{ mol KCl}} = 1.05\text{ g KCl}$$

Steps 2–4: Assuming Cl_2 is the limiting reactant:

$$0.669\text{ g Cl}_2 \times \frac{1\text{ mol Cl}_2}{70.906\text{ g}} \times \frac{2\text{ mol KCl}}{1\text{ mol Cl}_2} \times \frac{74.55\text{ g}}{1\text{ mol KCl}} = 1.41\text{ g KCl}$$

Choose the right answer, which is the lower one: 1.05 g KCl. That makes K the limiting reactant.

THINK ABOUT IT This is the same question as in example 1A. However, it is solved in a different manner. Either method for determining limiting reagent provides the correct answer.

EXAMPLE 2.

How many grams of copper can you make from 0.094 mL of zinc (density = 7.13 g/mL) and 7.74 × 10^{22} molecules of copper(II) sulfate?

$$Zn + CuSO_4 \rightarrow ZnSO_4 + Cu$$

Collect and Organize We are asked to determine the amount of copper that can be produced from a given amount of two reactants.

Analyze We will use both methods shown in the previous example to determine which of the reactants is the limiting reagent. Once this is known, we can determine the maximum amount of product that can be produced. We must also be sure that, when comparing reactants, they have the same units.

Solve Using method 1, pick a reactant and compare it to the other one. The reaction is balanced. You can get to grams from milliliters by means of density.

$$0.094 \text{ mL Zn} \times \frac{7.13 \text{ g}}{1 \text{ mL}} \times \frac{1 \text{ mol Zn}}{65.39 \text{ g}} \times \frac{1 \text{ mol CuSO}_4}{1 \text{ mol Zn}} \times \frac{6.022 \times 10^{23} \text{ molecules}}{1 \text{ mol Cu}} = 6.4 \times 10^{21} \text{ molecules CuSO}_4$$

That is less than the amount given, 7.74×10^{22} molecules $CuSO_4$; so zinc is the limiting reactant.

To solve the problem, start with zinc and ignore the copper sulfate.

$$0.094 \text{ mL Zn} \times \frac{7.13 \text{ g}}{1 \text{ mL}} \times \frac{1 \text{ mol Zn}}{65.39 \text{ g}} \times \frac{1 \text{ mol Cu}}{1 \text{ mol Zn}} \times \frac{63.55 \text{ g}}{1 \text{ mol Cu}} = 0.65 \text{ g Cu}$$

Using method 2, solve for the amount of product twice, each time starting with a different reactant.

$$0.094 \text{ mL Zn} \times \frac{7.13 \text{ g}}{1 \text{ mL}} \times \frac{1 \text{ mol Zn}}{65.39 \text{ g}} \times \frac{1 \text{ mol Cu}}{1 \text{ mol Zn}} \times \frac{63.55 \text{ g}}{1 \text{ mol Cu}} = 0.65 \text{ g Cu}$$

$$7.74 \times 10^{22} \text{ molecules CuSO}_4 \times \frac{1 \text{ mol CuSO}_4}{6.022 \times 10^{23} \text{ molecules}} \times \frac{1 \text{ mol Cu}}{1 \text{ mol CuSO}_4} \times \frac{63.55 \text{ g}}{1 \text{ mol Cu}} = 8.17 \text{ g Cu}$$

The lower answer is correct so the correct answer is 0.65 g Cu.

Think About It Either method for determining limiting reagent provides the correct answer.

Actual, Theoretical, and Percent Yield

Example 1.
What is the percent yield of zinc chloride when 0.2583 g of ZnO with excess HCl produces 0.2999 g $ZnCl_2$?

$$ZnO + HCl \rightarrow H_2O + ZnCl_2$$

Collect and Organize We are asked to determine the percent yield of $ZnCl_2$ given that the actual yield is 0.2999 g $ZnCl_2$.

Analyze Not all reactions result in the maximum amount of product being produced. The percent yield gives a measure of the extent of the reaction and is given by

$$\% \text{ yield} = \frac{\text{actual}}{\text{theoretical}} \times 100$$

To use this relation, we must determine the theoretical maximum yield based on knowledge of the limiting reagent.

Solve The question asks for percent yield; the product discussed is $ZnCl_2$. Its actual yield = 0.2999 g. To find the theoretical yield, stoichiometry is used.

Step 1: Balance the reaction.

$$ZnO + 2\ HCl \rightarrow H_2O + ZnCl_2$$

Step 2: Start with the limiting reactant and convert to moles. The limiting reactant is ZnO. The problem itself tells you that HCl is in excess; therefore, ZnO is the limiting reactant by default.

Steps 2–4:

$$0.2583 \text{ g ZnO} \times \frac{1 \text{ mol ZnO}}{81.39 \text{ g}} \times \frac{1 \text{ mol ZnCl}_2}{1 \text{ mol ZnO}} \times \frac{136.30 \text{ g}}{1 \text{ mol}} = 0.4326 \text{ g ZnCl}_2 = \text{theoretical yield}$$

$$\% \text{ yield} = \frac{0.2999 \text{ g}}{0.4326 \text{ g}} \times 100 = 69.32\%$$

Think About It Percent yields greater than 100% are possible if the product contains impurities or is not completely free of solvent.

Example 2.
The reaction of 0.213 g $Mn(NO_3)_2$ and 0.407 g K_2CO_3 produces $MnCO_3$ with a reaction yield of 95.6%. How many grams of manganese(II) carbonate are made?

$$Mn(NO_3)_2 + K_2CO_3 \rightarrow MnCO_3 + KNO_3$$

Collect and Organize We are asked to determine how many grams of the product manganese(II) carbonate will be produced given the initial amounts of two starting reactants and the percent yield of the reaction.

Analyze This time percent yield is given and actual yield is requested. We must first determine which of the reactants is the limiting reagent. We can then calculate the theoretical yield of the reaction and, using the given percent yield, determine the grams of manganese(II) carbonate produced.

SOLVE Using method 2 from the section on limiting reactants, the amount of product is determined for each reactant.

Balance the equation.

$$Mn(NO_3)_2 + K_2CO_3 \rightarrow MnCO_3 + 2\ KNO_3$$

The amount of product based on manganese(II) nitrate is

$$0.213\text{ g } Mn(NO_3)_2 \times \frac{1\text{ mol } Mn(NO_3)_2}{178.96\text{ g}} \times \frac{1\text{ mol } MnCO_3}{1\text{ mol } Mn(NO_3)_2} \times \frac{114.95\text{ g}}{1\text{ mol } MnCO_3} = 0.137\text{ g } MnCO_3$$

The amount of product based on potassium carbonate is

$$0.407\text{ g} \times \frac{1\text{ mol } K_2CO_3}{138.21\text{ g}} \times \frac{1\text{ mol } MnCO_3}{1\text{ mol } K_2CO_3} \times \frac{114.95\text{ g}}{1\text{ mol } MnCO_3} = 0.339\text{ g}$$

Since manganese(II) nitrate makes a smaller amount of product, it must be the limiting reactant and 0.137 g $MnCO_3$ is the theoretical yield.

$$\%\text{ yield} = \frac{\text{actual}}{\text{theoretical}} \times 100$$

$$95.6\% = (\text{actual}/0.137\text{ g}) \times 100$$

$$0.956(0.137) = \text{actual yield}$$

$$0.131\text{ g} = \text{actual yield}$$

THINK ABOUT IT In a real lab setting, we would have to know how much product was produced for a percent yield to be reported.

EXAMPLE 3.
The following reaction produces iron metal at 74.2% yield. How much Fe_3O_4 is needed to produce 2.35 kg of Fe?

$$Fe_3O_4 + CO \rightarrow Fe + CO_2$$

COLLECT AND ORGANIZE We are given the percent yield and amount of product to be produced, and we are asked to determine the necessary amount of one of the reactants.

ANALYZE We can use the expression for percent yield to solve the problem.

SOLVE In this problem, the 2.35 kg of Fe is the actual yield. Percent yield is also given, and the question asks about a reactant. Since products and reactants are related theoretically by stoichiometry, the theoretical yield is used to determine the amount of reactant.

$$74.2\% = \frac{2.35\text{ kg}}{\text{theoretical yield}} \times 100$$

$$0.742 = \frac{2.35\text{ kg}}{\text{theoretical yield}}$$

$$0.742(\text{theoretical yield}) = 2.35\text{ kg}$$

$$\text{theoretical yield} = 3.17\text{ kg}$$

Since we are looking for a reactant, the limiting reactant issue is not relevant.

Balance the reaction.

$$Fe_3O_4 + 4\ CO \rightarrow 3\ Fe + 4\ CO_2$$

Starting with the theoretical yield, use stoichiometric relationships to solve for the amount of reactant.

$$3.17\text{ kg Fe} \times \frac{1000\text{ g}}{1\text{ kg}} \times \frac{1\text{ mol Fe}}{55.845\text{ g}} \times \frac{1\text{ mol } Fe_3O_4}{3\text{ mol Fe}} \times \frac{231.535\text{ g}}{1\text{ mol } Fe_3O_4}$$

$$= 4380.97\text{ g} = 4.38 \times 10^3\text{ g} \times (1\text{ kg}/1000\text{ g}) = 4.38\text{ kg}$$

THINK ABOUT IT Many reactions do not run to completion. Percent yields are useful if a certain amount of product is needed from such a reaction.

Self-Test

KEY VOCABULARY COMPLETION QUESTIONS

__________ 1. 6.022×10^{23}

__________ 2. Determined from experiments

__________ 3. SI unit for amount of substance

__________ 4. Grams of element per 100 g of compound

__________ 5. Mass, in grams, of one mole

__________ 6. Substance formed in a chemical equation

__________ 7. Molar relationships of reactants and products

__________ 8. Reaction with oxygen to produce carbon dioxide and water

__________ 9. Simplest whole-number ratio of elements in a compound

__________ 10. Reactant that determines the theoretical yield

__________ 11. Amount of product determined from a stoichiometric calculation

__________ 12. Amount of product determined experimentally

__________ 13. Mass is neither created nor destroyed in a chemical reaction

__________ 14. Reaction with water

__________ 15. C_3H_8, for example

__________ 16. Adding stoichiometric coefficients to satisfy the law of conservation of mass

__________ 17. Nonmetal oxide that produces hydrogen ion in water

__________ 18. Reacts with a hydrocarbon in a combustion reaction

__________ 19. Moles of hydrogen atoms in one mole of water

__________ 20. Starting material in a chemical reaction

MULTIPLE-CHOICE QUESTIONS

1. How many moles of fluorine are in 9.2×10^{23} molecules of carbon tetrafluoride?
   a. 1.5 mol
   b. 6.1 mol
   c. 55 mol
   d. 222 mol
   e. 3.0 mol

2. What is the molar mass of iron(III) fluoride?
   a. 112.85 g/mol
   b. 93.85 g/mol
   c. 186.55 g/mol
   d. 205.55 g/mol
   e. 74.85 g/mol

3. How many moles of nickel are in 1 mole of nickel(II) sulfate?
   a. 1
   b. 2
   c. 6.022×10^{23}
   d. 1.2×10^{24}
   e. 3.0×10^{23}

4. Which of the following is a product of the combustion of a hydrocarbon?
   a. hydrogen
   b. carbon
   c. carbonate
   d. carbon dioxide
   e. The specific hydrocarbon is needed to predict the answer.

5. Why is the actual yield always less than the theoretical yield?
   a. other reactions may be occurring
   b. the reaction takes longer to complete
   c. the reaction might never be complete
   d. all of the above
   e. actual yield is always more than theoretical yield

6. How many moles of $Br_2(\ell)$ are in 10.0 mL, which has a density of 3.12 g/mL?
   a. 0.0200 mol
   b. 0.0626 mol
   c. 0.195 mol
   d. 0.0975 mol
   e. 0.0100 mol

7. How many grams does 2.13 mol of Pt weigh?
   a. 0.0273 g
   b. 166 g
   c. 415 g
   d. 91.6 g
   e. 66.0 g

8. How many atoms are in 0.934 mol of copper?
   a. 5.62×10^{23} atoms
   b. 5.47×10^{23} atoms
   c. 5.62×10^{-23} atoms
   d. 8.85×10^{21} atoms
   e. 3.57×10^{25} atoms

9. What is the percent composition of oxygen in Li_2O?
   a. 33.3%
   b. 46.4%
   c. 53.6%
   d. 66.6%
   e. 23.2%

10. How many atoms of gold (Au) are in 1.12 mg of pure gold?
    a. 3.42×10^{18} atoms
    b. 3.42×10^{21} atoms
    c. 6.74×10^{23} atoms
    d. 6.74×10^{21} atoms
    e. 1.32×10^{26} atoms

11. What is the empirical formula of a compound that is 43% carbon and 57% oxygen?
    a. CO
    b. CO_2
    c. C_2O
    d. C_3O_3
    e. CO_3

12. If the empirical formula of a compound is CH_2O and its molecular weight is 180 g/mol, what is its molecular formula?
    a. $C_6H_6O_6$
    b. $C_{18}H_{36}O_{18}$
    c. $C_6H_8O_6$
    d. $C_6H_{12}O_6$
    e. $C_7H_9O_7$

13. The limiting reactant is
    a. the substance you run out of first
    b. the reactant that determines the amount of product
    c. the substance left over
    d. both a and b
    e. both b and c

14. Consider the unbalanced reaction:

$$Cr(ClO_4)_3 + K_2CO_3 \rightarrow KClO_4 + Cr_2(CO_3)_3$$

What is the stoichiometric coefficient on $KClO_4$ after it is balanced?
    a. 1 (understood)
    b. 2
    c. 3
    d. 4
    e. more than 4

15. Consider the unbalanced reaction:

$$Cr(ClO_4)_3 + K_2CO_3 \rightarrow KClO_4 + Cr_2(CO_3)_3$$

If 2 moles of chromium(III) perchlorate react completely, how many moles of chromium(III) carbonate are made?
    a. 1
    b. 2
    c. 3
    d. less than 1
    e. more than 3

16. Consider the unbalanced reaction:

$$Cr(ClO_4)_3 + K_2CO_3 \rightarrow KClO_4 + Cr_2(CO_3)_3$$

If 4 moles of chromium(III) perchlorate react, how many moles of potassium carbonate are also used up?
    a. 2
    b. 4
    c. 6
    d. 8
    e. none of these

17. Consider the unbalanced reaction:

$$Cr(ClO_4)_3 + K_2CO_3 \rightarrow KClO_4 + Cr_2(CO_3)_3$$

If 1 mole of both reactants is mixed, which is the limiting reactant?
    a. $Cr(ClO_4)_3$
    b. K_2CO_3
    c. both are completely used up
    d. neither are completely used up
    e. more information is needed to tell

18. Consider the unbalanced reaction:

$$Cr(ClO_4)_3 + K_2CO_3 \rightarrow KClO_4 + Cr_2(CO_3)_3$$

If 10.0 g of K_2CO_3 completely reacted, how many grams of chromium(III) carbonate are made?
    a. 1.62 g
    b. 6.85 g
    c. 14.6 g
    d. 20.5 g
    e. 61.6 g

19. Consider the unbalanced reaction:

$$Cr(ClO_4)_3 + K_2CO_3 \rightarrow K_2CO_3 + Cr_2(CO_3)_3$$

If 7.18×10^{10} molecules of $Cr(ClO_4)_3$ react, how many molecules of K_2CO_3 also react?
    a. 2.84×10^{10}
    b. 4.79×10^{10}
    c. 7.18×10^{10}
    d. 3.59×10^{10}
    e. 1.08×10^{11}

20. Consider the unbalanced reaction:

$$Cr(ClO_4)_3 + K_2CO_3 \rightarrow KClO_4 + Cr_2(CO_3)_3$$

If 5.00 g of $Cr(ClO_4)_3$ are mixed with 7.00 g of K_2CO_3, how much reactant is left over?
    a. 2.00 g
    b. 4.05 g
    c. 5.03 g
    d. 6.83 g
    e. 2.95 g

21. Consider the unbalanced reaction:

$$Cr(ClO_4)_3 + K_2CO_3 \rightarrow KClO_4 + Cr_2(CO_3)_3$$

If 7.00 g of K_2CO_3 completely reacts to make 3.00 g of $Cr_2(CO_3)$, what is the percent yield of the reaction?
    a. 6.95%
    b. 20.9%
    c. 31.3%
    d. 37.8%
    e. 62.5%

22. If a reaction has a 45% yield and a theoretical yield of 2.50 g, what is its actual yield?
    a. 1.13 g
    b. 1.38 g
    c. 3.63 g
    d. 0.18 g
    e. 5.56 g

23. Consider the unbalanced reaction:

$$KOH + H_2SO_4 \rightarrow H_2O + K_2SO_4$$

When the reaction is balanced, what is the stoichiometric coefficient of sulfuric acid?
a. 1 (understood)
b. 2
c. 3
d. 4
e. more than 4

24. Consider the unbalanced reaction:

$$KOH + H_2SO_4 \rightarrow H_2O + K_2SO_4$$

If 1.0 mol KOH is mixed with 1.0 mol H_2SO_4, what is the limiting reactant?
a. KOH
b. H_2SO_4
c. both are completely used up
d. neither are completely used up
e. more information is needed to tell

25. Consider the unbalanced reaction:

$$KOH + H_2SO_4 \rightarrow H_2O + K_2SO_4$$

If 1.0 mol KOH and 1.0 mol H_2SO_4 are mixed, how many moles of water are made?
a. 0.50 mol
b. 1.00 mol
c. 1.50 mol
d. 2.00 mol
e. 1.33 mol

26. Consider the unbalanced reaction:

$$KOH + H_2SO_4 \rightarrow H_2O + K_2SO_4$$

If 1.0 mol KOH and 1.0 mol H_2SO_4 are mixed, how many moles of reactant are left over?
a. 0.10 mol
b. 0.25 mol
c. 0.50 mol
d. 1.00 mol
e. 2.00 mol

27. Consider the unbalanced reaction:

$$KOH + H_2SO_4 \rightarrow H_2O + K_2SO_4$$

If there is an excess of KOH, how many moles of H_2SO_4 are needed to make 1.0 mol water?
a. 1.00 mol
b. 0.50 mol
c. 0.25 mol
d. 2.00 mol
e. 3.00 mol

28. What is the percentage of oxygen in manganese(II) nitrate?
a. 41.05%
b. 8.94%
c. 64.73%
d. 31.70%
e. 53.65%

Additional Practice Problems

1. What is the mass percent nitrogen in $Cu(NO_3)_2$?

2. A 1.00 g sample of a hydrocarbon was burned in excess oxygen to produce 3.14 g CO_2 and 1.29 g water. The molar mass of the compound was 56 g/mol. What is the molecular formula of the compound?

3. What is the empirical and molecular formula of a compound that is 40.2% carbon, 6.15% hydrogen, and 53.6% oxygen and has a molar mass of 358 g/mol?

4. Balance the following chemical reactions.
a. $Mg + HNO_3 \rightarrow H_2 + Mg(NO_3)_2$
b. $N_2 + H_2 \rightarrow NH_3$
c. manganese(II) iodide + beryllium sulfide → manganese(II) sulfide and beryllium iodide
d. silver nitrate and sodium chloride react to yield silver chloride and sodium nitrate
e. C_5H_{12} completely burns in oxygen.

5. How many grams of oxygen are in 2.13×10^{18} molec of perchloric acid?

6. How many grams of ammonia are made from 0.726 g of nitrogen and 0.828 g of hydrogen?

$$N_2 + H_2 \rightarrow NH_3$$

7. If the percent yield of nickel metal in the reaction between nickel(II) oxide and hydrogen gas is 33.5%, how many grams of nickel metal can be made from 0.4055 g of nickel(II) oxide with excess hydrogen?

$$NiO + H_2 \rightarrow H_2O + Ni$$

8. How many atoms of hydrogen are in 1.0 mg of ammonium phosphate?

9. When 8.90 g of C_9H_{20} is burned in excess oxygen, how many grams of carbon dioxide are produced?

10. If 0.969 g of copper(II) hydroxide are produced from the reaction of 1.486 g copper(II) chloride with 2.895 g lithium hydroxide, what is the percent yield of the reaction? (*Hint:* The other product is lithium chloride.)

Self-Test Answer Key

Key Vocabulary Completion Answers

1. Avogadro's number
2. empirical
3. mole
4. percent composition
5. molar mass
6. product
7. stoichiometric coefficients
8. combustion
9. empirical formula
10. limiting reactant
11. theoretical yield
12. actual yield
13. law of conservation of mass
14. hydrolysis
15. hydrocarbon
16. balancing
17. acid anhydride
18. oxygen gas
19. two
20. reactant

Multiple-Choice Answers

For further information about the topics these questions address, you should consult the text. The appropriate section of the text is listed in parentheses after the explanation of the answer.

1. b. There are 6.022×10^{23} molecules in a mole. So 9.2×10^{23} molec $CF_4 \times \left(\frac{1\text{ mol}}{6.022 \times 10^{23}\text{ molec}}\right) \times \left(\frac{4\text{ mol F}}{1\text{ mol }CF_4}\right) = 6.1$ mol. (3.2 and 2.8)

2. a. iron(III) fluoride = FeF_3. Molar mass is the sum of the mass of each element = 55.85 + 3(19.00) = 112.85 g/mol. (3.2 and 2.8)

3. a. The atom ratio in a compound is the same as the mole ratio. The formula is nickel(II) sulfate = $NiSO_4$, one nickel per formula unit. (3.2 and 2.8)

4. d. A combustion reaction reacts the hydrocarbon (which contains hydrogen and carbon) with oxygen. The products of such a reaction are always carbon dioxide and water. (3.4)

5. d. In real life, things never seem to work as well as in theory. (3.9)

6. c. $10.0\text{ mL} \times \left(\frac{3.12\text{ g}}{1\text{ mL}}\right) \times \left(\frac{1\text{ mol}}{159.8\text{ g}}\right) = 0.195\text{ mol}$ (3.2 and 1.4)

7. c. $2.13\text{ mol} \times \left(\frac{195.08\text{ g}}{1\text{ mol}}\right) = 415\text{ g}$ (3.2)

8. a. $0.934\text{ mol} \times \left(\frac{6.022 \times 10^{23}\text{ atom}}{1\text{ mol}}\right) = 5.62 \times 10^{23}\text{ atoms}$ (3.2)

9. c. Percent composition is g O/g $Li_2O \times 100 = \left(\frac{16.00}{29.88}\right) \times 100 = 53.6\%$ (3.6)

10. a. $1.12\text{ mg} \times \left(\frac{1\text{ g}}{1000\text{ mg}}\right) \times \left(\frac{1\text{ mol}}{196.97\text{ g}}\right) \times \left(\frac{6.022 \times 10^{23}\text{ atom}}{1\text{ mol}}\right) = 3.42 \times 10^{18}\text{ atoms}$ (3.2)

11. a. Assuming 100 g of compound, 43 g C and 57 g of O. Then $43\text{ g} \times \left(\frac{1\text{ mol}}{12\text{ g}}\right) = 3.58$ mol C and $57\text{ g} \times \left(\frac{1\text{ mol}}{16\text{ g}}\right) = 3.56$ mol of O, which gives a 1:1 mole ratio. An empirical formula is the simplest ratio. (3.6)

12. d. The empirical formula is the simplest ratio. The apparent molecular weight of CH_2O is about 30 g/mol. The actual molecular weight is 180. This is six times the apparent weight, so the formula must be six times the empirical formula. (3.7)

13. d. The limiting reactant limits how much product you get because you run out of it first. (3.9)

14. e. The balanced reaction is $2\ Cr(ClO_4)_3 + 3\ K_2CO_3 \rightarrow 6\ KClO_4 + Cr_2(CO_3)_3$ (3.3)

15. a. The stoichiometric coefficients are also mole ratios. The balanced equation ($2\ Cr(ClO_4)_3 + 3\ K_2CO_3 \rightarrow 6\ KClO_4 + Cr_2(CO_3)_3$) shows that 2 moles of $Cr(ClO_4)_3$ make 1 mole of $Cr_2(CO_3)_3$. (3.5)

16. c. From the balanced equation the mole ratio is $2\ Cr(ClO_4)_3$ to $3\ K_2CO_3$, so 4 moles of $Cr(ClO_4)_3$ use 6 moles of K_2CO_3. (3.5)

17. b. The stoichiometric coefficients in the balanced reaction show that 3 moles of potassium carbonate are needed for every 2 moles of chromium(III) perchlorate. Since we started with equal amounts of reactants, the potassium carbonate will be used up first. (3.9)

18. b. $10.0 \text{ g } K_2CO_3 \times \left(\frac{1 \text{ mol}}{138.21 \text{ g}}\right) \times \left(\frac{1 \text{ mol } Cr_2(CO_3)_3}{3 \text{ mol } K_2CO_3}\right) \times \left(\frac{284.03 \text{ g}}{1 \text{ mol}}\right) = 5.40 \text{ g}$. (3.5)

19. e. Stoichiometric coefficients represent molecule ratios as well as mole ratios, so 7.18×10^{10} molec $Cr(ClO_4)_3 \times \left(\frac{3 \text{ molec } K_2CO_3}{2 \text{ molec } Cr(ClO_4)_3}\right) = 1.08 \times 10^{11}$ molec. (3.5)

20. b. If $Cr(ClO_4)_3$ is the limiting reactant, then the amount of K_2CO_3 needed is: $5.00 \text{ g } Cr(ClO_4)_3 \times \left(\frac{1 \text{ mol}}{350.35 \text{ g}}\right) \times \left(\frac{3 \text{ mol } K_2CO_3}{2 \text{ mol } Cr(ClO_4)_3}\right) \times \left(\frac{138.21 \text{ g}}{1 \text{ mol}}\right) =$ 2.95 g of K_2CO_3 used up. So 7.00 g to start −2.95 g used up = 4.05 g left over. (3.9 and 3.5)

21. e. Theoretical yield is determined from $7.00 \text{ g } K_2CO_3 \times \left(\frac{1 \text{ mol}}{138.21 \text{ g}}\right) \times \left(\frac{1 \text{ mol } Cr_2(CO_3)_3}{3 \text{ mol } K_2CO_3}\right) \times \left(\frac{284.03 \text{ g}}{1 \text{ mol}}\right) = 4.80 \text{ g}$.
Percent yield = 3.00 g ÷ 4.80 g × 100 = 6.25%. (3.9 and 3.5)

22. a. Percent yield = actual/theory × 100, so actual = $\%\left(\frac{\text{theory}}{100}\right) = 45\left(\frac{2.5}{100}\right) = 1.13 \text{ g}$. (3.9)

23. a. The balanced reaction is $2\ KOH + H_2SO_4 \rightarrow 2\ H_2O + K_2SO_4$. (3.3)

24. a. According to the balanced reaction, 1 mole of sulfuric acid requires 2 moles of potassium hydroxide, so KOH will run out first. (3.9)

25. b. KOH is the limiting reactant, so $1.0 \text{ mol KOH} \times \left(\frac{2 \text{ mol } H_2O}{2 \text{ mol KOH}}\right) = 1.0$ mol water. (3.5 and 3.9)

26. c. $1.0 \text{ mol KOH} \times \left(\frac{1 \text{ mol } H_2SO_4}{2 \text{ mol KOH}}\right) = 0.50 \text{ mol } H_2SO_4$ used. 1.0 mol −0.50 mol = 0.50 mol. (3.9)

27. b. $1.0 \text{ mol water} \times \left(\frac{1 \text{ mol } H_2SO_4}{2 \text{ mol water}}\right) = 0.50$ mol. (3.5)

28. e. The formula of manganese(II) nitrate is $Mn(NO_3)_2$. Therefore its molar mass is 178.94 g. The mass of oxygen in this compound is 6(16) = 96 g. Percent oxygen $= \frac{96}{178.94} \times 100 = 53.65\%$. (3.6)

Additional Practice Problem Answers

1. One mole of $Cu(NO_3)_2$ weighs 187.57 g. In one mole of $Cu(NO_3)_2$ there are two moles of nitrogen, each weighing 14.01 g. So $\%\ N = \frac{2(14.01)}{187.57} \times 100 =$ 14.94%

2. $3.14 \text{ g } CO_2 \times \left(\frac{1 \text{ mol } CO_2}{44.01 \text{ g}}\right) \times \left(\frac{1 \text{ mol C}}{1 \text{ mol } CO_2}\right) =$ 0.0713 mol

$1.29 \text{ g } H_2O \times \left(\frac{1 \text{ mol } H_2O}{18.01 \text{ g}}\right) \times \left(\frac{2 \text{ mol H}}{1 \text{ mol } H_2O}\right) =$ 0.143 mol

Divide by the smaller number to get the simplest ratio:

$$\frac{0.0713}{0.0713} = 1 \text{ mol C}$$

$$\frac{0.143}{0.0713} = 2.01 \text{ mol H}$$

So the empirical formula = CH_2. Since $\frac{56}{14} = 4$, the molecular formula is 4 times the empirical formula, so the molecular formula is C_4H_8.

3. 40.2% C, 6.15% H, 53.6% O

In 100 g of compound,

$$40.2 \text{ g C} \times \left(\frac{1 \text{ mol}}{12.01 \text{ g}}\right) = 3.35 \text{ mol}$$

$$6.15 \text{ g H} \times \left(\frac{1 \text{ mol}}{1.008 \text{ g}}\right) = 6.10 \text{ mol}$$

$$53.6 \text{ g O} \times \left(\frac{1 \text{ mol}}{16.00 \text{ g}}\right) = 3.35 \text{ mol}$$

Dividing by the smallest value,

$$\left(\frac{3.35 \text{ mol C}}{3.35}\right) = 1$$

$$\left(\frac{6.10 \text{ mol H}}{3.35}\right) = 1.82$$

$$\left(\frac{3.35 \text{ mol O}}{3.35}\right) = 1$$

1.82 is about 2, and since the formula requires that whole numbers be used, the empirical formula is CH_2O.

The molar mass based on the empirical formula is 30 g.

The molar mass is 358 g/mol. The $\frac{\text{molar mass}}{\text{apparent molar mass}}$

$= \frac{358}{30} = 11.9$, about 12. Therefore, the molecular

formula is 12 $(CH_2O) = C_{12}H_{24}O_{12}$.

4. a. $Mg + 2\ HNO_3 \rightarrow H_2 + Mg(NO_3)_2$
   b. $N_2 + 3\ H_2 \rightarrow 2\ NH_3$
   c. $MnI_2 + BeS \rightarrow MnS + BeI_2$
   d. $AgNO_3 + NaCl \rightarrow AgCl + NaNO_3$
   e. $C_5H_{12} + 8\ O_2 \rightarrow 5\ CO_2 + 6\ H_2O$
5. Perchloric acid = $HClO_4$.

$$2.13 \times 10^{18} \text{ molecules } HClO_4 \times \left(\frac{4 \text{ atoms O}}{1 \text{ molec}}\right) \times$$

$$\left(\frac{1 \text{ mol}}{6.022 \times 10^{23} \text{ atoms}}\right) \times \left(\frac{16.00 \text{ g}}{1 \text{ mol}}\right) = 2.26 \times 10^{-4} \text{ g}$$

6. $N_2 + 3\ H_2 \rightarrow 2\ NH_3$

$$0.726 \text{ g } N_2 \times \left(\frac{1 \text{ mol}}{28.02 \text{ g}}\right) \times \left(\frac{2 \text{ mol } NH_3}{1 \text{ mol } N_2}\right) \times$$

$$\left(\frac{17.03 \text{ g}}{1 \text{ mol}}\right) = 0.882 \text{ g } NH_3$$

$$0.828 \text{ g } H_2 \times \left(\frac{1 \text{ mol}}{2.02 \text{ g}}\right) \times \left(\frac{2 \text{ mol } NH_3}{3 \text{ mol } H_2}\right) \times$$

$$\left(\frac{17.03 \text{ g}}{1 \text{ mol}}\right) = 4.65 \text{ g } NH_3$$

Since N_2 gives a smaller amount of product, it must be the limiting reactant and 0.882 g is how much NH_3 was made.

7. $NiO + H_2 \rightarrow H_2O + Ni$ (is balanced)

$$0.4055 \text{ g NiO} \times \left(\frac{1 \text{ mol}}{74.69 \text{ g}}\right) \times \left(\frac{1 \text{ mol Ni}}{1 \text{ mol NiO}}\right) \times$$

$$\left(\frac{58.69 \text{ g}}{1 \text{ mol}}\right) = 0.3186 \text{ g} = \text{theoretical yield}$$

$$\% \text{ yield} = 33.5\% = \left(\frac{\text{actual}}{\text{theory}}\right) \times 100 = \left(\frac{\text{actual}}{0.3186 \text{ g}}\right) \times$$

$$100 \text{ actual} = 0.107 \text{ g}$$

8. Ammonium phosphate = $(NH_4)_3PO_4$, so there are 12 H per molecule.

$$1.0 \text{ mg } (NH_4)_3PO_4 \times \left(\frac{1 \text{ g}}{10^3 \text{ mg}}\right) \times \left(\frac{1 \text{ mol}}{149.10 \text{ g}}\right) \times$$

$$\left(\frac{12 \text{ mol H}}{1 \text{ mol } (NH_4)_3PO_4}\right) \times \left(\frac{6.022 \times 10^{23} \text{ atoms}}{1 \text{ mol}}\right) =$$

$$4.8 \times 10^{19} \text{ atoms}$$

9. $C_9H_{20} + 14\ O_2 \rightarrow 9\ CO_2 + 10\ H_2O$

$$8.90 \text{ g } C_9H_{20} \times \left(\frac{1 \text{ mol}}{128.25 \text{ g}}\right) \times \left(\frac{9 \text{ mol } CO_2}{1 \text{ mol } C_9CH_{20}}\right) \times$$

$$\left(\frac{44.01 \text{ g}}{1 \text{ mol}}\right) = 27.5 \text{ g}$$

10. $CuCl_2 + 2\ LiOH \rightarrow Cu(OH)_2 + 2\ LiCl$

$$1.486 \text{ g } CuCl_2 \times \left(\frac{1 \text{ mol}}{134.45 \text{ g}}\right) \times \left(\frac{1 \text{ mol } Cu(OH)_2}{1 \text{ mol } CuCl_2}\right) \times$$

$$\left(\frac{97.57 \text{ g}}{1 \text{ mol}}\right) = 1.078 \text{ g}$$

$$2.895 \text{ g LiOH} \times \left(\frac{1 \text{ mol}}{23.95 \text{ g}}\right) \times \left(\frac{1 \text{ mol } Cu(OH)_2}{2 \text{ mol LiOH}}\right) \times$$

$$\left(\frac{97.57 \text{ g}}{1 \text{ mol}}\right) = 5.897 \text{ g}$$

So copper(II) chloride is the limiting reactant and the theoretical yield is 1.078 g.

$$\% \text{ yield} = \left(\frac{\text{actual}}{\text{theory}}\right) \times 100 = \left(\frac{0.969 \text{ g}}{1.078 \text{ g}}\right) \times 100 = 89.9\%$$

CHAPTER 4 Solution Chemistry and the Hydrosphere

OUTLINE

REVIEW

Chapter Overview

Solutions are homogeneous mixtures of two or more substances. Usually the solution is liquid, but this is not required. Gaseous and solid solutions do exist. The substance with the greatest number of moles is the **solvent**; other materials in the solutions are **solutes**. The solute is also normally the substance of interest, while the solvent is a convenient vehicle for it. The most common of all solvents is water. Solutions with water as the solvent are described as **aqueous**.

Since solutions are mixtures, the ratio of solute to solvent varies. The ratio of solute to solution is called **concentration**. Many units can be used to express concentration; these include **parts per million** (grams of solute per one million grams of solution) and **parts per billion** (grams of solute per one billion grams of solution). Both of these are used when the fraction of solute is extremely small. The most common unit for chemists, however, is molarity. **Molarity** is the moles of solute per liter of solution. Its popularity is due to its convenient use in stoichiometric problems.

Solutions of specific molarity can be made from pure forms of the solute and solvent. Alternatively, solutions with high fractions of solute (concentrated solutions) can have their concentration lowered by adding solvent. The process of decreasing the concentration of a solution by adding solvent is called **dilution**.

Substances that form ions when dissolved in solution are called **electrolytes**. **Strong electrolytes** are solutes that exist only as ions in solution. For aqueous solutions, there are two classes of strong electrolytes: strong acids and soluble salts. **Weak electrolytes** form some ions in solution, but exist primarily in their original form. The most common weak electrolytes in aqueous solutions are weak acids, weak bases, and moderately soluble salts. **Nonelectrolytes** do not form ions in solutions. Most molecular compounds other than acids and bases are nonelectrolytes.

Four important skills in working with chemical reactions, particularly those that occur in aqueous solutions, are to identify the type of reaction, predict the products of the reaction, write the reaction as a net ionic equation, and use the equation to stoichiometrically relate amounts of products and reactants.

An **acid–base reaction** is a reaction of an acid with a base; thus an acid *and* a base must be present as reactants. Using the **Brønsted–Lowry definitions** of acid and base, an **acid** is a proton (or H^+) donor and a **base** is a proton (H^+) acceptor. Thus an acid–base reaction is a proton transfer.

Because the H^+ exists in water, it is sometimes written as H_3O^+ and called the **hydronium ion**. When the acid donates its H^+ to a hydroxide (OH^-) base, water is created. In addition, the anion from the acid combines with the cation from the base, thereby creating a **salt**. This particular type of acid–base reaction is called a **neutralization**. The products of neutralization reactions can be predicted based on the stoichiometric relationships where one H^+ and one OH^- create one water, and cations and anions must create a salt where the charges cancel. An acid may be categorized as strong, when it completely ionizes in water, or weak, if it only partly ionizes. Similarly, a base may also be strong, if it completely ionizes in water or weak, if it only partly ionizes. Acids, such as HCl, that have one H^+ that reacts are called **monoprotic**. Acids, such as H_2SO_4, that have two H^+ ions to donate are called **diprotic**. A base may also be monoprotic or diprotic, depending on the number of H^+ ions it will accept.

When large quantities of ionic compounds, or salts, dissolve in water, they are called **soluble**. If only a small amount (very near zero) dissolves, the salt is called **insoluble**. Precipitation reactions produce an insoluble salt, called a **precipitate**, as a product. The general solubility of a salt can be determined from the solubility rules.

Oxidation–reduction (redox) reactions involve the transfer of electrons and are characterized by a change in oxidation number. **Oxidation numbers** are used to keep track of the number of electrons assigned to an element. They are not necessarily based in physical reality. Sometimes a change in oxidation number is obvious (e.g., when a substance appears in its elemental form on one side of the equation and as an ion on the other). At other times the oxidation number must be assigned based on the rules given in the chapter. Each element of a given compound has the same oxidation number. For ionic compounds, it is often convenient to consider each ion separately, even if it is not a strong electrolyte.

In a redox reaction, two elements will change oxidation number. For one of those elements the oxidation number increases (becomes more positive or less negative). This element has lost electrons and was **oxidized**. The oxidation number of the other element decreases (becomes less positive or more negative). This element has gained electrons and was **reduced**.

Balanced reactions must balance both mass (atoms) and charge. For most reactions, charge is automatically balanced when mass is balanced. However, this is not always true for redox reactions. A convenient way to balance redox reactions is to separate the element being oxidized from the element being reduced. The reactions are then balanced, using water, H^+ and OH^- if needed, and showing the electrons gained or lost. Each of these equations is called a **half-reaction**. The chemical equation wherein electrons are reactants is the reduction half-reaction. The chemical equation wherein electrons are products is the oxidation half-reaction. The reactant of the oxidation half-reaction is providing electrons and is therefore, called the **reducing agent**. Similarly, the reactant of the reduction half-reaction accepts the electrons and is called the **oxidizing agent**. Then the half-reactions are combined so that the electrons will cancel from the equation.

Net ionic equations include only the substances that are actually participating in a reaction. Since strong electrolytes exist as ions in aqueous solutions, these ions, rather than the compound itself, are used in the net ionic equations. Although weak electrolytes do make some ions in solution, most of the substance exists in its molecular form; thus the molecular form is used in the equation. Not every substance in the reaction solution must participate in the reaction. Some ions are present only to keep the net charge of the solution at zero. Ions that do not participate in the reaction are called **spectator ions**. Spectator ions are not included in a net ionic reaction. Equations that show all the species in the solution (including the spectator ions) are called **overall ionic equations**.

Stoichiometry relates quantities of reactants and products in a chemical reaction. In this chapter it is used for analysis of solutions using precipitation and titration. Ions in solution can be quantitatively precipitated. The solid product (precipitate) is weighed, and the formula of the precipitate is used to stoichiometrically relate the mass of the precipitate to the original amount of ions. This experiment requires that the ion being analyzed be the limiting reactant.

A **titration** is an experiment in which the volume of a reactant solution with known concentration, a **standard solution**, required to react exactly with a specific amount of another reactant is measured. The volume when both reactants have completely reacted, with no excess of either, is called the **equivalence point**. An **indicator** is used to signal this volume. The actual volume measured is called the **endpoint**. Since the goal of a titration is for the endpoint and the equivalence point to be the same value, the terms are often used interchangeably. Because it is the reactants that are stoichiometrically related and each reacts exactly, there is no limiting reactant in a titration.

Worked Examples

SOLUTION CONCENTRATION AND MOLARITY

EXAMPLE 1.
What is the molarity of a solution made by dissolving 0.479 g of Li_2CO_3 to make 250.0 mL of solution?

COLLECT AND ORGANIZE We are asked to determine the molarity of a solution made by dissolving a given amount of solute in a given amount of solvent.

ANALYZE Molarity is defined as moles of solute/liters of solution. We must convert the solute from grams to moles using its molar mass and convert the solvent from mL to L.

SOLVE

0.479 g of Li_2CO_3 = solute
250.0 mL = solution

Grams of Li_2CO_3 must be changed to moles; the gram–mole relationship is molar mass. The molar mass of lithium carbonate is

$2(6.941) + 12.011 + 3(15.999) = 73.890$ g/mol

$$0.479 \text{ g } Li_2CO_3 \times \left(\frac{1 \text{ mol}}{73.890 \text{ g}}\right) = 0.00648 \text{ mol}$$

Milliliters of solution must be changed to liters; the mL–L relationship is the metric prefix.

$$1000 \text{ mL} = 1 \text{ L}$$

$$250.0 \text{ mL} \times \left(\frac{1 \text{ L}}{1000 \text{ mL}}\right) = 0.2500 \text{ L}$$

Then the molarity formula can be used:

$$\text{Molarity} = \frac{0.00648 \text{ mol}}{0.2500 \text{ L}} = 0.0259\ M$$

The answer will have only three significant figures; it was limited by the mass of lithium carbonate.

THINK ABOUT IT Molarity allows for a convenient way to express the concentration of a given solute in solution.

EXAMPLE 2.
What is the molarity of a solution made by adding water to 0.33 mol of $CaBr_2$ until there is 500.0 mL of solution?

COLLECT AND ORGANIZE We are given an amount of solute that is diluted to a given volume and are asked to determine the resulting molarity of the solution.

ANALYZE We will use the expression for molarity given in the previous example.

SOLVE

$$\text{Molarity} = \frac{\text{moles of solute}}{\text{liters of solution}}$$

solute = 0.33 mol $CaBr_2$
solution = 500.0 mL aqueous

The solute is already in the correct units. (Avoid the temptation to change it!) The solution volume must be changed from mL to L. (This is typical, but not always the case.)

$$500.0 \text{ mL} \times \left(\frac{1 \text{ L}}{1000 \text{ mL}}\right) = 0.5000 \text{ L}$$

(Don't lose your significant figures.)

$$\text{Molarity} = \frac{0.33 \text{ mol}}{0.5000 \text{ L}} = 0.66\ M$$

THINK ABOUT IT We must be careful to note if we are given grams of solute or moles of solute.

EXAMPLE 3.
How many grams of $AlBr_3$ are required to make 500.0 mL of 0.20 *M* $AlBr_3(aq)$?

COLLECT AND ORGANIZE We are asked to determine the grams of solute required to make a given volume of solution with a given molarity.

ANALYZE To determine the amount of solute from a given molarity and solution volume, one of two equally correct methods can be used. Method 1: Plug the values into the formula and solve algebraically. Method 2: Use molarity as a dimensional analysis relationship.

SOLVE METHOD 1

Molarity = moles of solute/liters of solution
solute = $AlBr_3$
solution = 500.0 mL

$$500.0 \text{ mL} \times \left(\frac{1 \text{ L}}{1000 \text{ mL}}\right) = 0.5000 \text{ L}$$

$$0.20\ M = \frac{\text{mol } AlBr_3}{0.5000 \text{ L}}$$

$$\left(\frac{0.20 \text{ mol}}{\text{L}}\right)(0.5000 \text{ L}) = \text{mol } AlBr_3$$

0.10 = mol $AlBr_3$
molar mass of $AlBr_3 = 26.982 + 3(79.904) = 266.694$ g/mol

$$0.10 \text{ mol } AlBr_3 \times \left(\frac{266.694 \text{ g}}{1 \text{ mol}}\right) = 27 \text{ g } AlBr_3$$

METHOD 2
0.20 *M* can be interpreted as 0.20 mol $AlBr_3$ = 1 L solution.

$$500.0 \text{ mL solution} \times \left(\frac{1 \text{ L}}{1000 \text{ mL}}\right) \times \left(\frac{0.20 \text{ mol } AlBr_3}{1 \text{ L}}\right) \times \left(\frac{266.694 \text{ g}}{1 \text{ mol}}\right) = 27 \text{ g}$$

THINK ABOUT IT An advantage of method 2 is that it is very similar to the way molarity is used in stoichiometric problems.

EXAMPLE 4.
How many milligrams of silver nitrate are required to make 1.0 mL of 0.042 *M* solution?

COLLECT AND ORGANIZE We are asked to determine the milligrams of solute required to make a given volume of solution with a given molarity.

ANALYZE We will use the two methods described in the previous example for the determination of the amount of solute from a given molarity and solution volume.

SOLVE For either method you will need to know the formula for silver nitrate. Recall the naming rules from Chapter 4. As a metal and a nonmetal it would be an ionic compound. Silver ion always has the same charge: Ag^+. Nitrate is one of the polyatomic ions you should have memorized: NO_3^-. Since both ions have the same charge, the formula is $AgNO_3$. You will also need its molar mass: 107.87 + 14.007 + 3(15.999) = 169.87 g/mol.

METHOD 2
Using molarity as 0.042 mol $AgNO_3$ = 1 L solution,

$$1.0\text{ mL} \times \left(\frac{1\text{ L}}{1000\text{ mL}}\right) \times \left(\frac{0.042\text{ mol AgNO}_3}{1\text{ L}}\right) \times \left(\frac{169.87\text{ g}}{1\text{ mol AgNO}_3}\right) \times \left(\frac{1000\text{ mg}}{1\text{ g}}\right) = 7.1\text{ mg}$$

METHOD 1
Using the formula Molarity = mol/L,

$$1.0\text{ mL} \times \left(\frac{1\text{ L}}{1000\text{ mL}}\right) = 0.0010\text{ L}$$

$$0.042\ M = \frac{x\text{ mol}}{0.0010\text{ L}}$$

$$(0.042)(0.0010) = 4.2 \times 10^{-5}\text{ mol AgNO}_3$$

$$4.2 \times 10^{-5}\text{ mol AgNO}_3 \times \left(\frac{169.87\text{ g}}{1\text{ mol}}\right) = 7.1 \times 10^{-3}\text{ g}$$

$$7.1 \times 10^{-3}\text{ g} \times \left(\frac{1000\text{ mg}}{1\text{ g}}\right) = 7.1\text{ mg}$$

THINK ABOUT IT Its important to be able to identify molecular formulas and give the names of compounds.

DILUTIONS

EXAMPLE 1.
What is the molarity of a solution made by diluting 5.00 mL of 1.8 *M* H_2SO_4 to a volume of 250.0 mL?

COLLECT AND ORGANIZE We are asked to determine the molarity of a solution resulting from the dilution of 5.00 mL of 1.8 *M* H_2SO_4 to a volume of 250.0 mL.

ANALYZE We will use the relation that the initial molarity times the initial volume equals the final molarity times the final volume: $M_{\text{initial}}V_{\text{initial}} = M_{\text{final}}V_{\text{final}}$

SOLVE Rearrangement of the above expression for the final molarity gives:

$$M_{\text{final}} = \frac{(M_{\text{initial}}V_{\text{initial}})}{V_{\text{final}}}$$

$$M_{\text{final}} = \frac{(1.8\ M) \times (5.00\text{ mL})}{(250.0\text{ mL})} = 0.036\ M$$

Don't forget significant figures.

THINK ABOUT IT Since the units of volume cancel it is not necessary to convert from mL to L.

EXAMPLE 2.
What is the concentration when 10.00 mL of 1.36 *M* NaCl is diluted to make 1.00 L of solution?

COLLECT AND ORGANIZE We are asked to determine the molarity of a solution that is made when 10.00 mL of 1.36 *M* NaCl is diluted to 1.00 L.

ANALYZE We will use the relation $M_{\text{initial}}V_{\text{initial}} = M_{\text{final}}V_{\text{final}}$, but must ensure that the units are the same for the initial and final volumes.

SOLVE

$$1.00\text{ L} \times \left(\frac{1000\text{ mL}}{1\text{ L}}\right) = 1000\text{ mL}$$

$$M_{\text{final}} = \frac{(M_{\text{initial}}V_{\text{initial}})}{V_{\text{final}}}$$

$$M_{\text{final}} = \frac{(1.36\ M) \times (10.00\text{ mL})}{(1000\text{ mL})} = 0.0136\ M$$

THINK ABOUT IT We could have alternatively converted the initial volume to liters.

EXAMPLE 3.
How would you make 500.0 mL of a 0.10 *M* HNO_3 solution from 16 *M* HNO_3?

COLLECT AND ORGANIZE We are asked to make 500.0 mL of a 0.10 *M* HNO_3 solution by dilution of a stock solution of 16 *M* HNO_3.

ANALYZE This is still a dilution problem. The concentration is changing but not the identity of the solution. Solve for what you don't know, making sure to keep units the same and volumes and concentrations together as before.

SOLVE We need to determine the volume of the 16 *M* stock solution that is required for dilution to 500.0 mL to result in a solution that has a molarity of 0.10 *M*. Using the relation $M_{initial}V_{initial} = M_{final}V_{final}$ we can solve for $V_{initial}$, the initial volume that will be diluted.

$$V_{initial} = \frac{M_{final}V_{final}}{M_{initial}}$$

$$= \frac{(0.10\ M)(500.0\ mL)}{(16\ M)}$$

$$= 3.1\ mL$$

THINK ABOUT IT Notice that this does not really answer the question. The question asks for directions. In this example the directions should go something like this: "Add 3.1 mL of the 16 *M* acid to a 500.0-mL volumetric flask that already contains some water. Mix carefully. Slowly, with mixing, add solvent (water) until the level of liquid reaches the mark on the volumetric flask."

EXAMPLE 4.
What volume of 0.10 *M* KI(*aq*) is required to make 100.0 mL of 0.010 *M* KI(*aq*)?

COLLECT AND ORGANIZE We are asked to determine the initial volume of a 0.10 *M* KI solution that will be diluted to 100.0 mL to make a solution with a molarity of 0.010 *M*.

ANALYZE This is also a dilution problem that can be easily solved using $M_{initial}V_{initial} = M_{final}V_{final}$.

SOLVE Rearrangement of $M_{initial}V_{initial} = M_{final}V_{final}$ for the initial volume gives:

$$V_{initial} = \frac{M_{final}V_{final}}{M_{initial}}$$

Substituting the given variables gives

$$V_{initial} = \frac{(0.010\ M)(100.0\ mL)}{(0.1\ M)}$$

$$= 10\ mL$$

THINK ABOUT IT Any dilution problem is easily solved using $M_{initial}V_{initial} = M_{final}V_{final}$.

ELECTROLYTES AND NONELECTROLYTES

EXAMPLE 1.
Are the following compounds strong electrolytes? If so, what ions do they form in water?

a. $NaNO_3$ b. CO_2 c. HF d. MgO e. $FeCl_3$

COLLECT AND ORGANIZE We are given five compounds and are asked to determine which ones are strong electrolytes. For the compounds that are strong electrolytes, we are asked to identify the ions that they form in water.

ANALYZE Strong electrolytes come in two general categories: strong acids and soluble salts. Strong acids include: HCl, HBr, HI, HNO_3, $HClO_4$, $HClO_3$, and H_2SO_4. To determine whether the salt is soluble, you should learn the solubility rules given in the text.

SOLVE

a. $NaNO_3$ is a strong electrolyte. Ions: Na^+ and NO_3^-.
Logic: It is a salt with a metal, sodium, and a nonmetal: the polyatomic anion nitrate. Compounds containing either sodium or nitrate are soluble, so this certainly is. Sodium as a group 1, alkali metal, always has a +1 charge; nitrate is a common polyatomic anion with a formula that should already be memorized.

b. CO_2 is not a strong electrolyte; therefore, it does not form ions in water.
Logic: It is a molecular compound of two nonmetals and is not one of the strong acids.

c. HF is not a strong electrolyte.
Logic: Since hydrogen is the first element, it is an acid. However, it is not on the list of strong acids; therefore, it is not a strong electrolyte.

d. MgO is not a strong electrolyte.
Logic: It is a salt of magnesium metal and oxide. However, neither element appears on the list of solubility rules. By default, that makes it insoluble. To be a strong electrolyte it must both be a salt and be soluble.

e. $FeCl_3$ is a strong electrolyte. In water it will make Fe^{3+} and Cl^-.
Logic: It is a salt of iron and chloride. According to the solubility rules, chlorides (group 17, the halogens) are normally soluble. The exceptions do *not* include iron, so this compound is soluble. Iron is a transition metal, so its charge depends on how it is combined. Chloride, however, always has a charge of –1. Since there are three and the entire compound has a net charge of zero, iron must have a charge of +3.

THINK ABOUT IT Since there are only seven strong acids, the easiest way to identify strong acids is to learn the list.

EXAMPLE 2.
Which of the following are strong electrolytes? For each strong electrolyte, what are the ions it makes in aqueous solution?

a. HNO_3 b. $PbBr_2$ c. K_2CO_3 d. Na e. $Ba(OH)_2$

COLLECT AND ORGANIZE We are given five compounds and are asked to determine which ones are strong electrolytes. For the compounds that are strong electrolytes, we are asked to identify the ions that they form in water.

ANALYZE Strong electrolytes come in two general categories: strong acids and soluble salts. Strong acids include HCl, HBr, HI, HNO_3, $HClO_4$, $HClO_3$, and H_2SO_4. To determine whether the salt is soluble, you should learn the solubility rules given in the text.

SOLVE

a. HNO_3 is a strong electrolyte. Ions = H^+ and NO_3^-.

It is on the list of strong acids.

b. $PbBr_2$ is not a strong electrolyte.

Bromide is a halide, which is usually soluble, but Pb^{2+} is one of the exceptions.

c. K_2CO_3 is a strong electrolyte. Ions = two ions of K^+ and one ion of CO_3^{2-}.

Potassium is in group 1, is always soluble, and always has a +1 charge. Carbonate is a familiar polyatomic. If you forgot its charge, the two K^+'s were a hint!

d. Na is not a strong electrolyte.

It is not a strong acid. It is not a salt. It is sodium salts that are always soluble, this is sodium metal.

e. $Ba(OH)_2$ is a strong electrolyte. Ions = Ba^{2+} and two ions of OH^-.

It is one of the soluble hydroxides.

THINK ABOUT IT Substances that form ions when dissolved in solution are called electrolytes. The stronger the electrolyte, the more the substance ionizes when dissolved in solution.

EXAMPLE 3.
Write the balanced total and net ionic equations for the following reactions in aqueous solution.

a. $(NH_4)_3PO_4 + CaBr_2 \rightarrow Ca_3(PO_4)_2 + NH_4Br$

b. $NaC_2H_3O_2 + CrI_3 \rightarrow NaI + Cr(C_2H_3O_2)_3$

c. $H_2S + KOH \rightarrow H_2O + K_2S$

d. $Li_2CO_3 + HI \rightarrow LiI + H_2O + CO_2$

e. $HCl + Ca(OH)_2 \rightarrow H_2O + CaCl_2$

COLLECT AND ORGANIZE We are given molecular equations, asked to balance these equation, write the total ionic equation, and then write the net ionic equation.

ANALYZE To write the total ionic equation, each substance must be characterized as either a strong electrolyte or not. Strong electrolytes are either strong acids or soluble salts. If a substance fall in the category of a strong electrolyte, it should be written as its component ions in the total ionic equation. Otherwise it should be written as it appears in the molecular equation.

To write the net ionic equation, substances that appear in the same form on both sides of the equation (spectator ions) are removed. Recall that stoichiometric coefficients are how many, not part of the identity. Therefore, stoichiometric coefficients do not have to match to cancel some of the substance.

After the spectator ions have been removed, if the stoichiometric coefficients can be simplified, they should be.

SOLVE

a. First the molecular equation needs to be balanced

$$2\,(NH_4)_3PO_4 + 3\,CaBr_2 \rightarrow Ca_3(PO_4)_2 + 6\,NH_4Br$$

Next it must be determined which substances are strong electrolytes. In this reaction, $(NH_4)_3PO_4$, $CaBr_2$ and NH_4Br are soluble salts. Therefore, those substances are written as their component ions, being careful to keep track of the stoichiometry. For example, each ammonium phosphate formula unit contains three ammonium ions, according to its formula. According to the balanced molecular equation, there are two formula units of ammonium phosphate. Thus, six ammonium ions appear in the reaction. This creates the total ionic reaction below.

$$6\,NH_4^+ + 2\,PO_4^{3-} + 3\,Ca^{2+} + 6\,Br^- \rightarrow Ca_3(PO_4)_2 + 6\,NH_4^+ + 6\,Br^-$$

Next the spectator ions, NH_4^+ and Br^- are removed. Calcium is not a spectator ion, since it appears alone on the reactant side and as part of an insoluble salt on the product side.

$$2\,PO_4^{3-} + 3\,Ca^{2+} \rightarrow Ca_3(PO_4)_2$$

No further simplification is needed; thus this is the net ionic equation.

b. The balanced molecular equation is

$$3\,NaC_2H_3O_2 + CrI_3 \rightarrow 3\,NaI + Cr(C_2H_3O_2)_3$$

Based on the solubility rules, these compounds break into the following ions

$$3\,Na^+ + 3\,C_2H_3O_2^- + Cr^{3+} + 3\,I^- \rightarrow 3\,Na^+ + 3\,I^- + Cr^{3+} + 3\,C_2H_3O_2^-$$

All ions cancel; there is no reaction.

c. The balanced molecular equation is

$$H_2S + 2\ KOH \rightarrow 2\ H_2O + K_2S$$

H_2S is a weak acid, not a strong one. Therefore, it is NOT broken into ions. Water is a covalent compound, neither a strong acid nor a soluble salt; therefore, it is not shown as ions either. When soluble salts of K_2S and KOH are broken into their component ions, the total ionic reaction is

$$H_2S + 2\ K^+ + 2\ OH^- \rightarrow 2\ H_2O + 2\ K^+ + S^{2-}$$

Potassium is the only spectator ion; therefore, the net ionic reaction is

$$H_2S + 2\ OH^- \rightarrow 2\ H_2O + S^{2-}$$

d. The balanced molecular equation is

$$Li_2CO_3 + 2\ HI \rightarrow 2\ LiI + H_2O + CO_2$$

HI is a strong acid and therefore, it is written as its ions; lithium salts are soluble and are written as ions. Water and carbon dioxide are covalent compounds and should not be broken into ions. The total ionic reaction is

$$2\ Li^+ + CO_3^{2-} + 2\ H^+ + 2\ I^- \rightarrow 2\ Li^+ + 2\ I^- + H_2O + CO_2$$

Canceling the spectator ions, the net ionic reaction is

$$CO_3^{2-} + 2\ H^+ \rightarrow H_2O + CO_2$$

e. The balanced molecular equation is

$$2\ HCl + Ca(OH)_2 \rightarrow 2\ H_2O + CaCl_2$$

Breaking the strong acid of HCl, the strong base of $Ca(OH)_2$, and soluble salt of $CaCl_2$ into ions, the total ionic reaction is

$$2\ H^+ + 2\ Cl^- + Ca^{2+} + 2\ OH^- \rightarrow 2\ H_2O + Ca^{2+} + 2\ Cl^-$$

Canceling the spectator ions, you get

$$2\ H^+ + 2\ OH^- \rightarrow 2\ H_2O$$

Simplifying the stochiometric coefficients (without changing the ratios) gives you the net ionic equation of

$$H^+ + OH^- \rightarrow H_2O$$

THINK ABOUT IT Net ionic equations describe the primary form in which each substance exists in solutions. While weak acids do make ions, it is only a small amount of ions; the acid itself is the primary form. Net ionic reactions also describe which substances participate in a reaction. If none of the substances are reacting, that too can be seen!

ION EXCHANGE

EXAMPLE 1.
Predict the products for the reaction of HCl with $Ca(OH)_2$.

COLLECT AND ORGANIZE We are given two reactants and are asked to predict the products of their reaction.

ANALYZE In an ion-exchange reaction, the cation of one reactant combines with the anion of the other. The ratio of ions in the product will be such that a neutral compound is formed. The remaining cation and anion will combine in the same way. If one of the reactants is an acid, it is treated as if its cation is H^+.

SOLVE The anion from the acid is Cl^-. The cation associated with the hydroxide is Ca^{2+}. Therefore, when they combine it will be $CaCl_2$. The H^+ and OH^- combine to make water. After the products are determined, the reaction is balanced. Thus the reaction is

$$2\ HCl + Ca(OH)_2 \rightarrow 2\ H_2O + CaCl_2$$

THINK ABOUT IT In an ion-exchange reaction, the ions that make up each reactant do not change, only the ion each is paired with. When a cation is paired with an anion, the formula is the simplest ratio that produces a neutral compound.

ACID–BASE REACTIONS

EXAMPLE 1.
What is the acid and the base in the following acid–base reaction?

$$HCN + HPO_4^{2-} \rightarrow CN^- + H_2PO_4^-$$

COLLECT AND ORGANIZE We are asked to identify the acid and base in the given acid–base reaction.

ANALYZE According to the Brønsted–Lowry theory, an acid is defined as a proton (or H^+) donor. A base is defined as a proton (or H^+) acceptor.

SOLVE HCN is the acid. It loses that first hydrogen when it becomes CN^-. Notice that the charge also changes. HPO_4^{2-} is the base. It gains a hydrogen and the charge increases by 1.

THINK ABOUT IT Acids and bases are identified by considering the behavior of protons.

EXAMPLE 2.
Determine if the following are acid–base reactions. If so, identify the acid and the base.

a. $2\ HBr + K_2CO_3 \rightarrow H_2O + CO_2 + 2\ KBr$

b. $FeI_2 + 2\ NaOH \rightarrow Fe(OH)_2 + 2\ NaI$

c. $H_2S + Ni(OH)_2 \rightarrow 2\ H_2O + NiS$

d. $Sr(OH)_2 + C_6H_5COOH \rightarrow H_2O + Sr(C_6H_5COO)_2$

COLLECT AND ORGANIZE We are given four reactions and asked to identify which ones are acid–base reactions. For those reactions identified as acid–base, we are asked to identify the acid and the base.

ANALYZE Acid–base reactions require both an acid and a base as reactants. (Products are irrelevant!) Many other types of reactions include one or the other. For it to be an acid–base reaction it must have both.

SOLVE

a. It is acid–base. The HBr is the acid, hydrogen first. The K_2CO_3 is the base, a carbonate.

b. It is *not* acid–base. The NaOH is a base, but FeI_2 is not an acid.

c. This is acid–base. H_2S is the acid; two acidic hydrogens begin the formula. $Ni(OH)_2$ is the base, an hydroxide.

d. This is acid–base. $Sr(OH)_2$ is an hydroxide-type base. C_6H_5COOH is an organic acid. Notice that only the hydrogen at the end has moved in the product formula. This is the way organic acids react.

THINK ABOUT IT Don't worry about the products, consider only the reactants when trying to identify an acid–base reaction.

EXAMPLE 3.
Predict the products of the following reactions.

a. $H_2S + KOH \rightarrow$

b. $Li_2CO_3 + HI \rightarrow$

c. $CH_3COOH + NaHCO_3 \rightarrow$

COLLECT AND ORGANIZE We are given three acid–base reactions and asked to predict their products.

ANALYZE The products of acid–base reactions are easily determined by the nature of the base.

Hydroxide bases will produce water, when the hydroxide combines with the proton (H^+), and a salt, when the remaining cation combines with the anion.

Carbonate bases will produce carbon dioxide, water (from two H^+ ions donated by the acid and the CO_3^{2-} of the base) and a salt (from the cation of the carbonate and the anion of the acid).

SOLVE

a. It's an hydroxide base, so water will be one product. The other ions are K^+ and S^{2-}, so the reaction is

$$H_2S + 2\ KOH \rightarrow 2\ H_2O + K_2S$$

Don't forget to balance.

b. It's a carbonate base, so water and CO_2 are products. The other ions are Li^+ and I^-. Therefore,

$$Li_2CO_3 + 2\ HI \rightarrow 2\ LiI + H_2O + CO_2$$

c. It's a carbonate base, but it already has one hydrogen, so it needs only one more instead of two to make the water and CO_2. The acid is organic, so only the hydrogen on the end reacts. Remember that when you write ionic compounds, the cation comes first. Therefore,

$$CH_3COOH + NaHCO_3 \rightarrow H_2O + CO_2 + NaCH_3COO$$

THINK ABOUT IT When the acid loses its proton (or protons), an anion will remain. The proton will combine with the base.

PRECIPITATION REACTIONS

EXAMPLE 1.
Determine whether or not the following reactions are precipitation reactions. If they are, what is the precipitate?

a. $2\ HBr + K_2CO_3 \rightarrow H_2O + CO_2 + 2\ KBr$

b. $FeI_2 + 2\ NaOH \rightarrow Fe(OH)_2 + 2\ NaI$

c. $H_2S + Ni(OH)_2 \rightarrow 2\ H_2O + NiS$

d. $2\ KCl + I_2 \rightarrow 2\ KI + Cl_2$

e. $Na_2CO_3 + Ca(OH)_2 \rightarrow CaCO_3 + 2\ NaOH$

COLLECT AND ORGANIZE We are given five reactions and are asked to determine which ones are precipitation reactions and identify the precipitate.

ANALYZE We will use our knowledge of solubility rules to determine which, if any, of the products are insoluble. Reactions with insoluble products are precipitation reactions.

SOLVE

a. This is not a precipitation reaction. Water and carbon dioxide are molecular compounds; KBr is soluble.

b. This is a precipitation reaction. The precipitate is iron(II) hydroxide. Sodium salts are always soluble.

c. This is a precipitation reaction. The precipitate is NiS.

d. This is not a precipitation reaction. KI is soluble. Cl_2 is a molecular compound not an insoluble salt.

e. This is a precipitation reaction. Calcium carbonate is the precipitate.

THINK ABOUT IT Don't worry about the reactants, a precipitate is a product.

EXAMPLE 2.
Predict the products of the following reactions. Identify the precipitate, if any.

a. $H_2S + MnSO_4 \rightarrow$

b. $Hg_2(NO_3)_2 + KCl \rightarrow$

c. $NaC_2H_3O_2 + CrI_3 \rightarrow$

d. $Ba(OH)_2 + H_2SO_4 \rightarrow$

e. $(NH_4)_3PO_4 + CaBr_2 \rightarrow$

COLLECT AND ORGANIZE We are given five reactions and are asked to predict their products.

ANALYZE Predicting the products of precipitation reactions is just a matter of rearranging the cations and anions. The only tricky part is making sure the ions have the appropriate ratio. If one of the reactants is an acid, work with it as if it were H^+ and an anion (although this is not strictly true, it is how it reacts).

SOLVE

a. The "pieces" that make up the reactants are H^+, S^{2-}, Mn^{2+}, and SO_4^{2-}. So the reaction is

$$H_2S + MnSO_4 \rightarrow H_2SO_4 + MnS$$

Sulfuric acid is a strong acid and will make ions in solution; manganese(II) sulfide is an insoluble salt and the precipitate.

b. The "pieces" that make up the reactants are Hg_2^{2+}, NO_3^-, K^+, and Cl^-, so

$$Hg_2(NO_3)_2 + 2\ KCl \rightarrow Hg_2Cl_2 + 2\ KNO_3$$

Hg_2Cl_2 is the precipitate, since Hg_2^{2+} is an exception to the "chlorides are soluble" rule.

c. The "pieces" that make up the reactants are Na^+, $C_2H_3O_2^-$, Cr^{3+}, and I^-, so

$$3\ NaC_2H_3O_2 + CrI_3 \rightarrow 3\ NaI + Cr(C_2H_3O_2)_3$$

There are no precipitates in this reaction. Sodiums are always soluble, and so are acetates.

d. The "pieces" that make up the reactants are Ba^{2+}, OH^-, H^+, and SO_4^{2-}, so

$$Ba(OH)_2 + H_2SO_4 \rightarrow BaSO_4 + 3\ H_2O$$

Barium sulfate is the precipitate. It is an exception to the "sulfates are soluble" rule.

e. The "pieces" that make up the reactants are NH_4^+, PO_4^{3-}, Ca^{2+}, and Br^-, so

$$2\ (NH_4)_3PO_4 + 3\ CaBr_2 \rightarrow Ca_3(PO_4)_2 + 6\ NH_4Br$$

Calcium phosphate is the precipitate.

THINK ABOUT IT This method will get two products. Either, both, or neither might be a precipitate. You must refer to the solubility rules to be sure. Precipitates are insoluble salts. All these reactions are also examples of ion-exchange reactions.

OXIDATION–REDUCTION REACTIONS

EXAMPLE 1.
What is the oxidation number of each element in the following examples?

a. $Ca(ClO_4)_2$ b. $K_2Cr_2O_7$ c. PH_3 d. Br_2 e. LiH

COLLECT AND ORGANIZE We are asked to determine the oxidation numbers of each element in the given compounds.

ANALYZE Oxidation numbers are found using the following rules:

- The sum of all oxidation numbers in any compound or ion is equal to its charge. Therefore,
  - a. The oxidation number of a pure element is zero.
  - b. The oxidation number of a monatomic ion is equal to its charge.
- The oxidation number of hydrogen is
  - a. +1 if combined with a nonmetal
  - b. –1 if combined with a metal
  - c. 0 if combined only with itself
- The oxidation number of oxygen is –2 (unless that contradicts rule 1 or 2).
- Unless they are combined with oxygen or fluorine, the halogens are –1.

SOLVE

a. Since this is an ionic compound, it will be easier to consider each ion separately. The two ions that make up this compound are Ca^{2+} and ClO_4^-. That there are two ClO_4^- ions is not needed to calculate the oxidation numbers. As a monatomic ion, the oxidation number of calcium is equal to the charge = +2.

The oxygen in ClO_4^- is assigned an oxidation number of –2, since there is no contradictory rule. The oxidation number of the chlorine is determined using the rule that the sum of the oxidation numbers must equal the charge:

$$Cl + 4(-2) = -1$$

$$Cl = +7$$

b. This is also an ionic compound of K^+ and $Cr_2O_7^{2-}$. The potassium will have an oxidation number equal to its charge, +1. The oxygen will have an oxidation number of –2, since there is no contradictory rule. The chromiums, since they are in the same atom, must have the same charge and can be determined from the "sum of the oxidation numbers equals the charge" rule.

$$2Cr + 7(-2) = -2$$

$$Cr = +6$$

c. This is a molecular compound, so it cannot be divided into ions. Since hydrogen is combined with a nonmetal, it will have an oxidation number = +1. Using the summing rule,

$$P + 3(+1) = 0$$

$$P = -3$$

d. Bromine is an uncombined element. Therefore, its oxidation number is zero.

e. There are two ways to approach this. If you divide it into its component ions, Li^+ and H^-, and the oxidation number is equal to the charge. You could also note that hydrogen is combined with a metal, so its oxidation number = –1 and the summing rule makes lithium = +1.

THINK ABOUT IT Oxidation numbers are easily found by following the above guidelines.

EXAMPLE 2.
Characterize the following reactions as redox or nonredox.

a. $2\ HCl + Mg \rightarrow H_2 + MgCl_2$

b. $HCl + AgNO_3 \rightarrow HNO_3 + AgCl$

c. $KMnO_4 + 5\ FeCl_2 + 8\ HCl \rightarrow 5\ FeCl_3 + MnCl_2 + 4\ H_2O + KCl$

COLLECT AND ORGANIZE We are given three reactions and are asked to characterize them as redox or nonredox.

ANALYZE Redox reactions are characterized by changes in oxidation numbers. However, it is often possible to notice that oxidation numbers are changing without actually calculating them. Some clues that redox is occurring are the following:

- An element that is uncombined (with any different element) on one side of the reaction and combined on the other.
- The number of oxygens or hydrogens associated with element changes.
- The charge on the element changes.

Sometimes it is easier to spot that a reaction is *not* a redox reaction. Nonredox reactions (which include acid–base and precipitation reactions) are often characterized by reactants that each come as two pieces (like HCl with pieces of H^+ and Cl^- or K_2SO_4 with pieces of K^+ and SO_4^{2-}) and the pieces of the two reactants stay the same but are rearranged in the products.

SOLVE

a. Redox reaction. The magnesium was alone as a reactant and was combined with chlorine as a product. Hydrogen was combined as a reactant and was uncombined as a product. In this reaction, hydrogen and magnesium change oxidation numbers.

b. Nonredox reaction. The pieces here are H^+, Cl^-, Ag^+, and NO_3^-. None of the pieces change; they are just rearranged.

c. Redox. First, remember that if you can spot one change, it must be redox. If that is all the question asks, you can stop there. Don't get distracted by the extras. Notice the iron. It changes charge! In the reactants it has two Cl^- ions, so it must have a +2 charge, but in the products there are three Cl^- ions, so the charge is +3. Notice the manganese. It is combined with four oxygens in the anion but is combined with Cl^- in the products. The change in number of oxygens is a tip-off. So is the fact that it switched from the anion position to the cation position. Obviously, this is not just a simple rearrangement. In this reaction, the iron and the manganese change oxidation number.

THINK ABOUT IT Redox reactions require the presence of both oxidized and reduced substances.

EXAMPLE 3.

Balance the following redox reaction in acidic solution. Identify the oxidation half-reaction, the reduction half-reaction, the oxidizing agent, and the reducing agent.

$$U^{4+} + MnO_4^- \rightarrow UO_2^{2+} + Mn^{2+}$$

COLLECT AND ORGANIZE We are asked to balance the given redox reaction in acidic solution, identify the oxidation and reduction half-reactions, and identify the oxidizing and reducing agents.

ANALYZE Balancing redox reactions in acidic solutions is accomplished with the following steps:

a. Balance all elements except hydrogen and oxygen.

b. Add H_2O to balance water.

c. Add H^+ to balance hydrogen.

d. Add electrons (e^-) to balance charge.

e. Multiply each stoichiometric coefficient of either (or both) reaction to obtain the least common multiple of electrons.

f. Add the two reactions together, keeping all reactants on the reactant side and products on the product side.

g. Simplify by canceling compounds that are the same on each side of the reaction.

Hints: (i) If electrons don't cancel, you did it wrong (ii) Stoichiometric coefficients are number of molecules, not part of the compound formula; thus it is perfectly appropriate for $3\ H^+ + OH^- \rightarrow 2\ H^+ + H_2O$ to become $H^+ + OH^- \rightarrow H_2O$.

SOLVE Uranium and manganese are changing oxidation states so the two skeletal reactions are

$$U^{4+} \rightarrow UO_2^{2+}$$

$$MnO_4^- \rightarrow Mn^{2+}$$

a. Every element except oxygen is already balanced.

b. Add water to balance oxygen:

$$U^{4+} + 2\ H_2O \rightarrow UO_2^{2+}$$

$$MnO_4^- \rightarrow Mn^{2+} + 4\ H_2O$$

c. Add H^+ to balance hydrogen (which is now unbalanced due to the addition of water):

$$U^{4+} + 2\ H_2O \rightarrow UO_2^{2+} + 4\ H^+$$

$$MnO_4^- + 8\ H^+ \rightarrow Mn^{2+} + 4\ H_2O$$

d. Add electrons to balance charge. (Just add up the charges on ions, multiplied by their coefficients. Remember that balance is the same, not zero.) For the uranium half-reaction, charge on the reactants = +4; charge on the products = +6. For the manganese half, charge on the reactants = +7; charge on the products = +2.

$$U^{4+} + 2\ H_2O \rightarrow UO_2^{2+} + 4\ H^+ + 2\ e^-$$

$$MnO_4^- + 8\ H^+ + 5\ e^- \rightarrow Mn^{2+} + 4\ H_2O$$

These are the balanced half-reactions. The one for uranium is the oxidation reaction and U^{4+} is the reducing agent. The one for manganese is the reduction reaction and MnO_4^- is the oxidizing agent.

e. Multiply reactions to get the least common multiple of electrons. The least common multiple of electrons in this example is 10. So the uranium half will be multiplied by 5 and the manganese half by 2.

$$5\ U^{4+} + 10\ H_2O \rightarrow 5\ UO_2^{2+} + 20\ H^+ + 10\ e^-$$

$$2\ MnO_4^- + 16\ H^+ + 10\ e^- \rightarrow 2\ Mn^{2+} + 8\ H_2O$$

f. Add the two reactions together:

$$5\ U^{4+} + 10\ H_2O + 2\ MnO_4^- + 16\ H^+ + 10\ e^- \rightarrow 2\ Mn^{2+} + 8\ H_2O + 5\ UO_2^{2+} + 20\ H^+ + 10\ e^-$$

g. Simplify:

$$5\ U^{4+} + 2\ H_2O + 2\ MnO_4^- \rightarrow 2\ Mn^{2+} + 5\ UO_2^{2+} + 4\ H^+$$

THINK ABOUT IT It looks scary, but this is usually the easiest and most consistent way to balance redox reactions.

EXAMPLE 4.

Balance the following redox reaction in acidic solution. Identify the oxidation half-reaction, the reduction half-reaction, the oxidizing agent, and the reducing agent.

$$Cu + NO_3^- \rightarrow Cu^{2+} + NO_2$$

COLLECT AND ORGANIZE We are asked to balance the given redox reaction in acidic solution, identify the oxidation and reduction half-reactions, and identify the oxidizing and reducing agents.

ANALYZE We will follow the procedure outlined in the previous example to solve the problem.

SOLVE Copper and nitrogen are changing oxidation state, so the skeletal reactions are

$$Cu \rightarrow Cu^{2+}$$

$$NO_3^- \rightarrow NO_2$$

a. All elements except oxygen are balanced.

b. The oxygen is balanced for copper (at zero) so that is unchanged, but the nitrogen has a difference of one oxygen:

$$Cu \rightarrow Cu^{2+}$$

$$NO_3^- \rightarrow NO_2 + H_2O$$

c. Now we have to balance hydrogen on the nitrogen reaction; copper is still fine.

$$Cu \rightarrow Cu^{2+}$$

$$NO_3^- + 2\,H^+ \rightarrow NO_2 + H_2O$$

d. Now electrons: for copper, the reactant side charge = 0 and the product side charge = +2; for nitrogen the reactant side charge = +1 and the product side charge = 0.

$$Cu \rightarrow Cu^{2+} + 2\,e^-$$

$$NO_3^- + 2\,H^+ + e^- \rightarrow NO_2 + H_2O$$

These are the balanced half-reactions. The copper reaction represents oxidization, and copper is the reducing agent. The nitrogen half represents reduction, and nitrate ion is the oxidizing agent.

e. Determine the least common multiple of electrons (it's 2) and multiply the reactions so that each has that many electrons. The copper reaction is fine; the nitrogen must be multiplied by 2.

$$Cu \rightarrow Cu^{2+} + 2\,e^-$$

$$2\,NO_3^- + 4\,H^+ + 2\,e^- \rightarrow 2\,NO_2 + 2\,H_2O$$

f. Add the reactions:

$$Cu + 2\,NO_3^- + 4\,H^+ + 2\,e^- \rightarrow 2\,NO_2 + 2\,H_2O + Cu^{2+} + 2\,e^-$$

g. Simplify:

$$Cu + 2\,NO_3^- + 4\,H^+ \rightarrow 2\,NO_2 + 2\,H_2O + Cu^{2+}$$

Done!

THINK ABOUT IT Following the steps outlined in the solution provides an easy way to balance a given redox reaction.

EXAMPLE 5.

Balance the following redox reaction in basic solution. Identify the oxidation half-reaction, the reduction half-reaction, the oxidizing agent, and the reducing agent.

$$S_2O_6^{2-} + TeO_3^{2-} \rightarrow SO_4^{2-} + Te$$

COLLECT AND ORGANIZE We are asked to balance the given redox reaction in basic solution.

ANALYZE Balancing redox reactions in basic solutions is accomplished in the same way as for acids. However, an additional step is needed to remove the H^+ and make the solution basic. This step is done immediately after balancing the hydrogens, so we'll call it step c2.

c2. Add the same number of OH^- ions to BOTH sides, as there are H^+ ions. When H^+ and OH^- are on the same side of the equation, replace them with an equal number of waters (e.g., $2H^+$ and $2\,OH^-$ become $2\,H_2O$).

SOLVE The elements changing oxidation state are sulfur and tellurium, so the skeletal half-reactions are

$$S_2O_6^{2-} \rightarrow SO_4^{2-}$$

$$TeO_3^{2-} \rightarrow Te$$

a. Sulfurs aren't balanced! This is an easy step to forget because you don't need it all that often.

$$S_2O_6^{2-} \rightarrow 2\,SO_4^{2-}$$

$$TeO_3^{2-} \rightarrow Te$$

b. Balance oxygens with water. For sulfur there are six oxygens on the reactant side and eight on the product side; for tellurium there are three oxygens on the reactant side and none on the product side.

$$2\,H_2O + S_2O_6^{2-} + \rightarrow 2\,SO_4^{2-}$$

$$TeO_3^{2} \rightarrow Te + 3\,H_2O$$

c1. Balance the hydrogens by adding H^+.

$$2\,H_2O + S_2O_6^{2-} + \rightarrow 2\,SO_4^{2-} + 4\,H^+$$

$$6\,H^+ + TeO_3^{2-} \rightarrow Te + 3\,H_2O$$

c2. Remove H^+ by adding OH^-. Combine H^+ and OH^- on the same side of the equation to make water.

$$4\,OH^- + 2\,H_2O + S_2O_6^{2-} + \rightarrow 2\,SO_4^{2-} + 4\,H_2O$$

$$6\,H_2O + TeO_3^{2-} \rightarrow Te + 3\,H_2O + 6\,OH^-$$

d. Balance charge. For sulfur the reactant side charge = –6, the product side charge = –4. For tellurium, the reactant side charge = –2, the product side charge = –6. You can also simplify by canceling the extra waters in this step (as below) or wait until the final step.

$$S_2O_6^{2-} + 4\,OH^- \rightarrow 2\,SO_4^{2-} + 2\,H_2O + 2\,e^-$$

$$TeO_3^{2-} + 3\,H_2O + 4\,e^- \rightarrow Te + 6\,OH^-$$

With these balanced half-reactions, the sulfur half-reaction is the oxidation, with $S_2O_6^{2-}$ acting as the reducing agent. The tellurium reaction is the reduction and TeO_3^{2-} is the oxidizing agent.

e. Use the least common multiple of electrons, which in this case = 4.

$$2\,S_2O_6^{2-} + 8\,OH^- \rightarrow 4\,SO_4^{2-} + 4\,H_2O + 4\,e^-$$

$$TeO_3^{2-} + 3\,H_2O + 4\,e^- \rightarrow Te + 6\,OH^-$$

f. Add:

$$2\,S_2O_6^{2-} + 8\,OH^- + TeO_3^{2-} + 3\,H_2O + 4\,e^- \rightarrow Te + 6\,OH^- + 4\,SO_4^{2-} + 4\,H_2O + 4\,e^-$$

g. Simplify:

$$2\ S_2O_6^{2-} + 2\ OH^- + TeO_3^{2-} \rightarrow Te + 4\ SO_4^{2-} + H_2O$$

Done!

Think About It It is important to be sure of the conditions of the redox reaction (acidic or basic) before balancing.

Titrations

Example 1.
What is the molarity of a solution of hydrochloric acid if 0.1511 g of sodium carbonate in 100.0 mL of water required 27.31 mL of the acid solution to reach the equivalence point?

$$HCl + Na_2CO_3 \rightarrow NaCl + H_2O + CO_2$$

Collect and Organize We are asked to determine the molarity of an HCl solution given the volume used in the titration of a given amount of sodium carbonate.

Analyze Stoichiometry problems require you to relate amounts of different reactants and products. These problems normally involve the use of a chemical reaction. Specific to this chapter are titrations, where quantities of reactants are related and the reactants are in solution form. Because they are solutions (mixtures) you must use concentration (molarity) to relate volume of solution to moles of solute (pure substance). However, the basic steps for solving these types of problems are the same as in Chapter 4.

Steps for solving stoichiometry problems are

Step 1: Balance the chemical reaction

Step 2: Change unit given to moles

a. atoms or molecules to moles: use Avogadro's number

b. grams to moles: use molar mass

c. (new) volume of solution to moles: use molarity

Step 3: Change moles of given substance to moles of desired substance

a. stoichiometric coefficients of chemical reaction

Step 4: Change moles of desired substance to units desired (the reverse of 2a, b, and c).

Solve Before solving any stoichiometry problem, make sure the reaction is balanced (Step 1). In this example it is not. Balancing the reaction gives you

$$2\ HCl + Na_2CO_3 \rightarrow 2\ NaCl + H_2O + CO_2$$

Because this example asks you to find molarity (a formula or relationship) as your final answer, it is useful to write the formula and see what you have and what you need to find:

$$\text{molarity} = \frac{\text{mol solute}}{\text{L solution}}$$

Since the question asks for moles of hydrochloric acid solution, HCl is your solute. Looking closely at the question, you will also notice that the volume of HCl solution is given as 27.31 mL, since you need volume in liters for use in the formula

$$27.31\ \text{mL} \times \left(\frac{1\ \text{L}}{1000\ \text{mL}}\right) = 0.02731\ \text{L}$$

Now all you need is the moles of HCl. No further information is given directly about HCl; however, information (in grams) is given about sodium carbonate, and that can be related to HCl from the chemical reaction

$$0.1511\ \text{g}\ Na_2CO_3 \times \left(\frac{1\ \text{mol}\ Na_2CO_3}{105.99\ \text{g}}\right) \times \left(\frac{2\ \text{mol HCl}}{1\ \text{mol}\ Na_2CO_3}\right) = 2.851 \times 10^{-3}\ \text{mol HCl}$$

Note that step 4 was not needed, because the desired unit was moles! Now there is sufficient information to use the formula

$$\text{molarity} = \frac{\text{mol}}{\text{L}} = \frac{2.851 \times 10^{-3}\ \text{mol}}{0.2731\ \text{L}} = 0.1044\ M$$

Think About It Note that the volume of water was not used in these calculations. It is not a reactant, does not affect the moles of sodium carbonate, and was not part of the HCl solution the question asked about. It is not unusual to have extra information in problems of this type.

Example 2.
Magnesium hydroxide was titrated with standard 0.1478 M HCl(aq). It required 39.11 mL of the hydrochloric acid solution to reach the equivalence point. How many grams of magnesium hydroxide were titrated?

Collect and Organize We are asked to determine the amount of magnesium hydroxide in a sample that required 39.11 mL of 0.1478 M HCl(aq) to reach the equivalence point.

Analyze We must first determine the stoichiometric equation for the reaction described. We can then use dimensional analysis to relate the volume of titrant to the grams of magnesium hydroxide using the titrant concentration, the stoichiometry of the reaction, and the molecular weight of magnesium hydroxide.

SOLVE The reaction is not given, but titrations relate reactants. Therefore, the question tells you that $Mg(OH)_2$ and HCl are your reactants. For tips on predicting the products, see the section on acid–base reactions. The balanced chemical reaction will be

$$Mg(OH)_2 + 2\ HCl \rightarrow H_2O + MgCl_2$$

Using stoichiometric relationships including the reaction above:

$$39.11\ \text{mL HCl solution} \times \left(\frac{1\ \text{L}}{1000\ \text{mL}}\right) \times \left(\frac{0.1487\ \text{mol HCl}}{1\ \text{L}}\right) \times \left(\frac{1\ \text{mol}\ Mg(OH)_2}{2\ \text{mol HCl}}\right) \times \left(\frac{58.319\ \text{g}}{1\ \text{mol}\ Mg(OH)_2}\right) = 0.1696\ \text{g}\ Mg(OH)_2$$

THINK ABOUT IT In picking out what is "given," it helps to remember that molarity is a ratio rather than a specific value. Therefore, it is not your "given." Molarity is the ratio between volume of solution and moles. In this example: 0.1487 mol HCl = 1 L solution.

EXAMPLE 3.
15.27 mL of 0.113 *M* NaOH was used to neutralize 25.00 g of vinegar. What is the percentage of acetic acid (CH_3COOH) in the vinegar?

$$NaOH + CH_3COOH \rightarrow NaCH_3CO_2 + H_2O$$

COLLECT AND ORGANIZE We are asked to determine what percentage of a 25.00 g sample of vinegar is acetic acid from the volume and concentration of titrant.

ANALYZE Stoichiometry can be used to relate the moles of NaOH used in the titration with the moles of acetic acid in the vinegar sample. The number of moles can then be converted into grams using the molar mass of acetic acid and this mass compared to the given mass of vinegar to determine percentage of acetic acid.

SOLVE Note that the reaction is balanced. The question asks for a percent as the final answer. Recall that percent is a formula and the formula is % = part/whole × 100—where part and whole must have the same units. In this example the question asks for the percentage of acetic acid (part) in vinegar (whole). Note that the total amount of vinegar is given in the problem: grams of vinegar = 25.00 g. To complete the formula, grams of acetic acid are needed, but information is given about sodium hydroxide instead. Therefore, stoichiometry will be used.

$$15.27\ \text{mL NaOH solution} \times \left(\frac{1\ \text{L}}{1000\ \text{mL}}\right) \times \left(\frac{0.113\ \text{mol NaOH}}{1\ \text{L solution}}\right) \times \left(\frac{1\ \text{mol}\ CH_3COOH}{1\ \text{mol NaOH}}\right) \times \left(\frac{60.052\ \text{g}}{1\ \text{mol}\ CH_3COOH}\right) = 0.104\ \text{g}\ CH_3COOH$$

Then use the formula:

$$\text{part} = \text{grams of acetic acid} = 0.104\ \text{g}\ CH_3COOH$$

$$\% = \left(\frac{0.104\ \text{g}}{25.00\ \text{g}}\right) \times 100 = 0.414\%$$

THINK ABOUT IT Acetic acid, the active ingredient of vinegar, is present in only a small amount.

Self-Test

KEY VOCABULARY COMPLETION QUESTIONS

__________ 1. Water is the solvent

__________ 2. Reactant that collects electrons in a redox reaction

__________ 3. Proton donor

__________ 4. KOH, for example

__________ 5. mg/kg

__________ 6. Chemical equation showing only substances that actually react

__________ 7. Reactant that undergoes oxidation

__________ 8. Substance in greatest quantity in a homogeneous mixture

__________ 9. Substance that forms ions in an aqueous solution

__________ 10. Experiment relating moles of reactants by measuring the volume of a solution

__________ 11. Volume at which reactants are stoichiometrically equivalent

__________ 12. Moles of solute per liters of solution

__________ 13. Adding solvent to decrease concentration

__________ 14. Way of assigning electrons to each element of a compound

__________ 15. Insoluble reaction product

__________ 16. Substance that is present but does not react

__________ 17. One mole of this substance reacts with two moles of OH^-

__________ 18. Ionic compound

__________ 19. Contains the maximum amount of solute that will dissolve

__________ 20. Can act as either an acid or a base

__________ 21. An acid–base reaction producing water

__________ 22. Electron-transfer reaction

__________ 23. Substance that does not make ions in water

__________ 24. Theory describing a base as a "proton acceptor"

__________ 25. Solution of known concentration

Multiple-Choice Questions

1. What type of reaction is:
$2\ K + 2\ H_2O \rightarrow 2\ KOH + H_2$?
   a. acid–base
   b. precipitation
   c. redox
   d. both acid–base and precipitation
   e. none of these

2. In the reaction: $2\ K + 2\ H_2O \rightarrow 2\ KOH + H_2$, what type of reactant is K?
   a. oxidizing agent
   b. reducing agent
   c. acid
   d. base
   e. precipitate

3. In the reaction: $2\ K + 2\ H_2O \rightarrow 2\ KOH + H_2$, which of the following is a spectator ion in the reaction?
   a. K^+
   b. OH^-
   c. H^+
   d. all are
   e. none is

4. In the reaction: $2\ K + 2\ H_2O \rightarrow 2\ KOH + H_2$, the oxidation number of hydrogen in H_2 is
   a. +1
   b. 0
   c. –1
   d. –2
   e. none of these

5. In the reaction: $2\ K + 2\ H_2O \rightarrow 2\ KOH + H_2$, potassium is
   a. oxidized
   b. reduced
   c. neutralized
   d. precipitated
   e. none of these

6. When KCl reacts with $Pb(NO_3)_2$, the resulting precipitate will be
   a. KNO_3
   b. PbCl
   c. $PbCl_2$
   d. $K(NO_3)_2$
   e. there will be no precipitate

7. How many milliliters of 0.10 *M* HCl are required to react with 25 mL of 0.10 *M* NaOH?
   a. 25 mL
   b. 50 mL
   c. 12.5 mL
   d. 36.5 mL
   e. 40 mL

8. How many milliliters of 0.10 *M* KOH are required to react with 2.0 mmol H_2SO_4?
   a. 20 mL
   b. 10 mL
   c. 200 mL
   d. 40 mL
   e. 1.0 mL

9. How many grams of KOH are needed to make 500 mL of a 0.20 *M* solution?
   a. 5.6 g
   b. 22.5 g
   c. 2.25 g
   d. 0.0018 g
   e. 140 g

10. A solution of 0.22 *M* CaI_2 has a density of 1.213 g/mL. How many grams of water are in 1 L of this solution?
    a. 1000 g
    b. 1213 g
    c. 220 g
    d. 65 g
    e. 1148 g

11. How much 1.0 *M* $Mg(NO_3)_2$ is needed to make 500 mL of 0.50 *M* $Mg(NO_3)_2$?
    a. 1000 mL
    b. 250 mL
    c. 74 mL
    d. 297 mL
    e. 148 mL

12. Which of the following is a strong electrolyte?
    a. Li_2CO_3
    b. $CrPO_4$
    c. $BaSO_4$
    d. all are
    e. none is

13. If $(NH_4)_3PO_4$ is added to water, which of the following ions will form?
a. $(NH_4)_3^{3+}$
b. NH_4^+
c. 3 N^{3-}
d. H^+
e. no ions form

14. What is the precipitate that forms when an aqueous solution of $FeCl_3$ reacts with an aqueous solution of $(NH_4)_2CO_3$?
a. $FeCO_3$
b. NH_4Cl
c. $Fe_3(CO_3)_2$
d. $Fe_2(CO_3)_3$
e. no precipitate forms

15. How many grams of $CoCl_2$ are in 1.0 L of 0.10 *M* $CoCl_2(aq)$?
a. 129.84 g
b. 13 g
c. 0.1 g
d. 0.077 g
e. 9.9 g

16. Which of the following is a strong electrolyte?
a. $Mg(OH)_2$
b. HNO_2
c. SO_3
d. all of the above
e. none of the above

17. Which of the following aqueous solutions has the highest conductivity?
a. 0.10 *M* $NaNO_3$
b. 0.10 *M* K_2S
c. 0.10 *M* HF
d. 0.10 *M* NH_4Cl
e. all have the same conductivity

18. When 10.0 mL of 0.5 *M* HCl is diluted to make 1.0 L of solution, the concentration of the resulting solution is
a. 0.005 *M*
b. 5 *M*
c. 20 *M*
d. 0.05 *M*
e. 0.02 *M*

19. What is the oxidation number of carbon in CO_3^{2-}?
a. +2
b. −2
c. −4
d. +4
e. +6

20. When zinc metal (Zn) reacts with hydrochloric acid to become zinc ion (Zn^{2+}), the zinc has been
a. oxidized
b. reduced
c. precipitated
d. neutralized
e. none of the above

21. What type of reaction is $Ni(OH)_2 + 2\ HNO_3 \rightarrow Ni(NO_3)_2 + 2\ H_2O$?
a. redox
b. acid–base
c. precipitation
d. both acid–base and precipitation
e. none of the above

22. In the reaction: $Ni(OH)_2 + 2\ HNO_3 \rightarrow Ni(NO_3)_2 + 2\ H_2O$, which of the following is a spectator ion?
a. Ni^{2+}
b. OH^-
c. NO_3^-
d. both Ni^{2+} and NO_3^-
e. all the ions are spectator ions

23. In the reaction: $Ni(OH)_2 + 2\ HNO_3 \rightarrow Ni(NO_3)_2 + 2\ H_2O$, nickel hydroxide is classified as
a. a reducing agent
b. an oxidizing agent
c. an acid
d. a base
e. both a base and an oxidizing agent

24. In the reaction: $Ni(OH)_2 + 2\ HNO_3 \rightarrow Ni(NO_3)_2 + 2\ H_2O$, what volume of 0.10 *M* nitric acid is required to react with 0.200 g of nickel(II) hydroxide?
a. 43 mL
b. 21.5 mL
c. 2.7×10^3 mL
d. 86 mL
e. 0.43 mL

25. For the reaction: $Ni(OH)_2 + 2\ HNO_3 \rightarrow Ni(NO_3)_2 + 2\ H_2O$, which of the following ions appear in the total ionic equation?
a. Ni^+
b. Ni^{2+}
c. OH^-
d. both Ni^+ and OH^-
e. none of the above

Additional Practice Problems

1. For each of the following reactions
   a. predict the products
   b. identify as acid–base or precipitation
   c. write the balanced net ionic equation

   $HNO_3 + Ca(OH)_2 \rightarrow$

   $(NH_4)_2SO_4 + BaCl_2 \rightarrow$

   $K_2CO_3 + FeBr_3 \rightarrow$

   $H_2S + NiSO_4 \rightarrow$

   $HCl + CuCO_3 \rightarrow$

2. For each reaction, write the balanced net ionic equation and identify the type of reaction.
   a. $Na + H_2O \rightarrow NaOH + H_2$
   b. $AgNO_3 + CuCl_2 \rightarrow AgCl + Cu(NO_3)_2$
   c. $CH_3COOH + NaOH \rightarrow NaCH_3CO_2 + H_2O$
   d. $HNO_3 + CoCl_2 \rightarrow HCl + Co(NO_3)_2$
   e. $KCl + I_2 \rightarrow KI + Cl_2$

3. When 0.431 g of a chloride-containing sample was titrated with 0.015 *M* aqueous silver nitrate, 43.71 mL of titrant was required to reach the equivalence point. What is the percentage of Cl in the sample?

4. What is the oxidation number for the underlined element in each of the following compounds?

   a. $\underline{C}H_4$ b. $\underline{Fe}CO_3$ c. $H\underline{N}O_3$ d. $\underline{P}_4$ e. $\underline{Cl}O_4^-$

5. Consider the reaction of aluminum metal with hydrochloric acid to make aluminum chloride and a hydrogen gas. Write and balance each net-ionic half-reaction. Identify the half-reaction as reduction or oxidation and then add the two half-reactions to make one complete redox reaction.

6. When 35.0 mL of 0.010 *M* H_2S is mixed with 25.00 mL of 0.030 *M* $FeCl_3$, how many mg of precipitate are made? What is the formula and name of the precipitate?

7. Balance the following redox reactions:
   a. in acidic solution

   $Fe^{2+} + Cr_2O_7^{2} \rightarrow Fe^{3+} + Cr^{3+}$

   b. in basic solution

   $MnO_4^- + Zn \rightarrow Zn^{2+} + MnO_2$

8. What is the molarity of chloride ion in the following solutions?
   a. 0.25 *M* $MgCl_2$
   b. 2.98 g NaCl diluted with water to a total volume of 250.0 mL
   c. 15.00 mL of 1.00 *M* $FeCl_3$ diluted to a total volume of 500.0 mL

9. How do you make 1.00 L of a 0.200 *M* $NH_4F(aq)$ from solid ammonium fluoride?

10. How many grams of iron(III) carbonate can be made from 50.00 mL of 0.180 *M* $FeCl_3$ with excess sodium carbonate?

Self-Test Answer Key

Key Vocabulary Completion Answers

1. aqueous
2. oxidizing agent
3. acid
4. strong base (although strong electrolyte is also appropriate)
5. ppm (parts per million)
6. net-ionic equation
7. reducing agent
8. solvent
9. strong electrolyte
10. titration
11. equivalence point
12. molarity
13. dilution
14. oxidation numbers
15. precipitate
16. spectator ion
17. diprotic acid
18. salt
19. saturated solution
20. amphiprotic
21. neutralization
22. oxidation–reduction
23. nonelectrolytes
24. Brønsted–Lowry
25. standard solution

Multiple-Choice Answers

For further information about the topics these questions address, you should consult the text. The appropriate section of the text is listed in parentheses after the explanation of the answer.

1. c. Note that potassium does not have a charge as a reactant but is part of an ionic compound, with a charge of +1, as a product. The change of oxidation number is characteristic of redox reactions. Hydrogen is the other element changing oxidation number. (4.8)

2. b. In the redox reaction, potassium loses electrons, providing them to the hydrogen. That makes it the reducing agent. Note also that it undergoes oxidation. (4.8)

3. e. The net ionic and total ionic reaction is

   $$2\ K + 2\ H_2O \rightarrow 2\ K^+ + 2\ OH^- + H_2\ (4.4)$$

4. b. An uncombined element has an oxidation number of zero. (4.8)

5. a. K becomes K^+, losing electrons. Therefore, it is oxidized. (4.8)

6. c. Potassium and nitrate ions are always soluble, so they will not precipitate. Since nitrate has a charge of –1, the lead must have a charge of +2, so that the compound is neutral. Thus when Pb^{2+} combines with Cl^-, the formula is $PbCl_2$, which is one of the insoluble halides. (4.6)

7. a. The reaction between HCl and NaOH is HCl + NaOH→ H_2O + NaCl, so

   $$25\text{ mL NaOH} \times \left(\frac{1\text{ L}}{1000\text{ mL}}\right) \times \left(\frac{0.10\text{ mol NaOH}}{1\text{L}}\right) \times \left(\frac{1\text{ mol HCl}}{1\text{ mol NaOH}}\right) \times \left(\frac{1\text{ L}}{0.10\text{ mol HCl}}\right) \times \left(\frac{1000\text{ mL}}{1\text{ L}}\right) = 25\text{ mL. (4.5 and 4.9)}$$

8. d. The balanced reaction is: 2 KOH + H_2SO_4 → 2 H_2O + K_2SO_4. Therefore,

   $$2.0\text{ mmol } H_2SO_4 \times \left(\frac{2\text{ mmol KOH}}{1\text{ mmol } H_2SO_4}\right) \times \left(\frac{1\text{ mL}}{0.10\text{ mmol}}\right) = 40\text{ mL. (4.5 and 4.9)}$$

9. a. M = mol/L so $M \times$ L = mol and 0.20 $M \times$ 0.5 L =

   $$0.1\text{ mol} \times \left(\frac{56.18\text{ g}}{1\text{ mol}}\right) = 5.6\text{ g (4.2)}$$

10. e. Assuming 1 L (1000 mL) of solution,

    $1000\text{ mL} \times \left(\frac{1.213\text{ g}}{1\text{ mL}}\right) = 1213$ g solution. In 1 L there are 0.22 mol $CaI_2 \times \left(\frac{293.90\text{ g}}{1\text{ mol}}\right) = 65$ g of CaI_2

    1213 g solution – 65 g solute = 1148 g solvent. (4.2)

11. b. It is a dilution problem, so use $M_{\text{initial}}V_{\text{initial}} = M_{\text{final}}V_{\text{final}}$. $(1.0\ M)(V_{\text{initial}}) = (0.5)(500\text{ mL})$; then $V_{\text{initial}} = 250$ mL. (4.3)

12. a. All salts of group 1 ions are soluble. Ba^{2+} is an exception to the "sulfates are soluble" rule, and neither chromium nor phosphate appears on the solubility rules. (4.4)

13. b. Ammonium salts are always soluble, so ions will form. NH_4^+ is a common polyatomic ion, which does not further dissociate in water. Three ions needed to balance the charge on the phosphate. The 3 tells how many ammonium ions per phosphate ion there are and is not part of the formula of the ion. (4.4)

14. d. Since ammonium is always soluble and chloride usually is (neither ammonium nor iron is an exception), a precipitate must form between the iron and the carbonate, neither of which is soluble on its own. The charge on the iron must be +3, since chloride is always –1. Carbonate always has a charge of –2. For the compound to be neutral, the formula must be $Fe_2(CO_3)_3$. (4.6)

15. b. The molar mass of $CoCl_2$ is 129.84 g/mol. 0.10 mol × 129.84 g/mol = 12.984 g, which rounded to the appropriate significant figure is 13 g. (4.2)

16. e. Strong electrolytes are soluble salts and strong acids. $Mg(OH)_2$ is a salt but is not soluble; HNO_2 is an acid but not a strong one; and SO_3 is a molecular compound (not a salt), which is also not a strong acid. (4.4)

17. b. The solution with the highest conductivity will have the most ions. Although all solutions have the same concentration, K_2S produces three ions for each mole that is dissolved, $NaNO_3$ produces only two, and HF, as a weak acid, does not completely ionize. (4.4)

18. a. Since this is a dilution the formula, $M_{initial}V_{initial} = M_{final}V_{final}$ is used. However, volume must be in the same units, so 1.0 L = 1000 mL. Using numbers, (0.5 *M*)(10.0 mL) = *M*(1000 mL) and solving for *M* gives $(0.5\ M)\frac{(10.0\ mL)}{(1000\ mL)} = 0.005\ M$. (4.3)

19. d. The sum of the oxidation numbers must add up to the charge on the ion, which in this example is –2. Oxygen normally has an oxidation number of –2, and this is not an exception. Consequently: C + 3(–2) = –2 and C = +4. (4.8)

20. a. When zinc metal becomes zinc ion, it loses two electrons. Loss of electrons is oxidation. (4.8)

21. b. An acid–base reaction requires an acid and a base as reactants. In the example, HNO_3 is the acid and $Ni(OH)_2$ is the base. Since the product $Ni(NO_3)_2$ is a soluble salt, this reaction does not form a precipitate. In addition, no element changes oxidation number, so it is not a redox reaction. (4.5)

22. c. The total ionic equation for the reaction is $Ni(OH)_2 + 2\ H^+ + 2\ NO_3^- \rightarrow Ni^{2+} + 2\ NO_3^- + 2\ H_2O$. Nitrate is the only ion that is the same on both sides. (4.4 and 4.5)

23. d. Since this is not a redox reaction, no oxidizing or reducing agent is involved. Metal hydroxides are classic bases. (4.5)

24. a. $0.200\ g\ Ni(OH)_2 \times \left(\frac{1\ mol\ Ni(OH)_2}{92.85\ g}\right) \times$

$$\left(\frac{2\ mol\ HNO_3}{1\ mol\ Ni(OH)_2}\right) \times \left(\frac{1\ L}{0.10\ mol\ HNO_3}\right) \times \left(\frac{1000\ mL}{1\ L}\right)$$

= 43 mL (4.9)

25. b. Since hydroxide has a –1 charge, the nickel must have a charge of +2. Hydroxide ion does not appear in the total ion equation, since it is bound up as a solid in the nickel(II) hydroxide and as a molecular compound in water. However, since nickel(II) nitrate is a soluble salt, it will dissolve into its component ions, Ni^{2+} and NO_3^-. (4.4)

Additional Practice Problem Answers

1. a. $2\ HNO_3 + Ca(OH)_2 \rightarrow 2\ H_2O + Ca(NO_3)_2$
   b. acid–base reaction
   c. $2\ H^+ + 2\ NO_3^- + Ca^{2+} + 2\ OH^- \rightarrow 2\ H_2O + Ca^{2+} + 2\ NO_3^-$ (total ionic)
   $2\ H^+ + 2\ OH^- \rightarrow 2\ H_2O$ (net ionic, but not simplest ratio)
   $H^+ + OH^- \rightarrow H_2O$ (net ionic)

   a. $(NH_4)_2SO_4 + BaCl_2 \rightarrow 2\ NH_4Cl + BaSO_4$
   b. precipitation
   c. $2\ NH_4^+ + SO_4^{2-} + Ba^{2+} + 2\ Cl^- \rightarrow 2\ NH_4^+ + 2\ Cl^- + BaSO_4$ (total ionic)
   $SO_4^{2-} + Ba^{2+} \rightarrow BaSO_4$ (net ionic)

   a. $3\ K_2CO_3 + 2\ FeBr_3 \rightarrow 6\ KBr + Fe_2(CO_3)_3$
   b. precipitation
   c. $6\ K^+ + 3\ CO_3^{2-} + 2\ Fe^{3+} + 6\ Br^- \rightarrow 6\ K^+ + 6\ Br^- + Fe_2(CO_3)_3$
   $3\ CO_3^{2-} + 2\ Fe^{3+} \rightarrow Fe_2(CO_3)_3$

   a. $H_2S + NiSO_4 \rightarrow H_2SO_4 + NiS$
   b. precipitation
   c. $H_2S + Ni^{2+} + SO_4^{2-} \rightarrow 2\ H^+ + SO_4^{2-} + NiS$
   $H_2S + Ni^{2+} \rightarrow 2\ H^+ + NiS$

   a. $2\ HCl + CuCO_3 \rightarrow H_2O + CO_2 + CuCl_2$
   b. acid–base
   c. $2\ H^+ + 2\ Cl^- + CuCO_3 \rightarrow H_2O + CO_2 + Cu^{2+} + 2\ Cl^-$
   $2\ H^+ + CuCO_3 \rightarrow H_2O + CO_2 + Cu^{2+}$

2. a. $2\ Na + 2\ H_2O \rightarrow 2\ NaOH + H_2$
   $2\ Na + 2\ H_2O \rightarrow 2\ Na^+ + 2\ OH^- + H_2$ (total and net ionic equation)
   This is a redox reaction. Sodium is oxidized; hydrogen is reduced.

   b. $2\ AgNO_3 + CuCl_2 \rightarrow 2\ AgCl + Cu(NO_3)_2$
   $2\ Ag^+ + 2\ NO_3^- + Cu^{2+} + 2\ Cl^- \rightarrow 2\ AgCl + Cu^{2+} + 2\ NO_3^-$ (total ionic)
   $Ag^+ + Cl^- \rightarrow AgCl$ (net ionic)
   This is a precipitation reaction, with AgCl as the precipitate.

   c. $CH_3COOH + NaOH \rightarrow NaCH_3CO_2 + H_2O$
   $CH_3COOH + Na^+ + OH^- \rightarrow Na^+ + CH_3CO_2^- + H_2O$ (total ionic)
   $CH_3COOH + OH^- \rightarrow CH_3CO_2^- + H_2O$ (net ionic)
   This is an acid–base reaction. CH_3COOH is the acid; NaOH is the base.

   d. $2\ HNO_3 + CoCl_2 \rightarrow 2\ HCl + Co(NO_3)_2$
   $2\ H^+ + 2\ NO_3^- + Co^{2+} + 2\ Cl^- \rightarrow 2\ H^+ + 2\ Cl^- + Co^{2+} + 2\ NO_3^-$ (total ionic)
   All the ions cancel; there is *no* reaction.

   e. $2\ KCl + I_2 \rightarrow 2\ KI + Cl_2$
   $2\ K^+ + 2\ Cl^- + I_2 \rightarrow 2\ K^+ + 2\ I^- + Cl_2$
   $2\ Cl^- + I_2 \rightarrow 2\ I^- + Cl_2$
   This is a redox reaction. Chlorine is oxidized; iodine is reduced.

3. The reaction between chloride and silver nitrate is

$$Cl^- + AgNO_3 \rightarrow AgCl + NO_3^-$$

$$43.71 \text{ mL } AgNO_3(aq) \times \left(\frac{1 \text{ L}}{1000 \text{ mL}}\right) \times \left(\frac{0.015 \text{ mol } AgNO_3}{1 \text{ L}}\right) \times \left(\frac{1 \text{ mol Cl}}{1 \text{ mol } AgNO_3}\right) \times \left(\frac{35.453 \text{ g}}{1 \text{ mol Cl}}\right) = 0.023 \text{ g Cl}$$

$$\text{percentage} = \frac{\text{g Cl}}{\text{g sample}} \times 100 = \frac{0.023 \text{ g Cl}}{0.431 \text{ g sample}} \times 100 = 5.4\%$$

4. a. $\underline{C}H_4$; H = +1, so C + 4(+1) = 0 and C = –4.
   b. $\underline{Fe}CO_3$; split into ions, Fe^{2+} and CO_3^{2-} so Fe = +2.
   c. $H\underline{N}O_3$; H = +1, O = –2 and so +1 + N + 3(–2) = 0 so N = +5.
   d. $\underline{P}_4$; uncombined element oxidation number is zero.
   e. $\underline{Cl}O_4^-$; O = –2 so Cl + 4(–2) = –1 so Cl = +7

5. $Al + HCl \rightarrow AlCl_3 + H_2$
   Start with the net ions to make life easier!

   $Al + H^+ + Cl^- \rightarrow Al^{3+} + Cl^- + H_2$

   $Al \rightarrow Al^{3+} + 3\ e^-$ is the oxidation half-reaction.

   $2\ H^+ + 2\ e^- \rightarrow H_2$ is the reduction half-reaction.

   Six electrons is the least common multiple, so multiplying and combining gives:

$$2\ Al + 6\ H^+ \rightarrow 2\ Al^{3+} + 3\ H_2$$

6. The reaction is

$$3\ H_2S + 2\ FeCl_3 \rightarrow Fe_2S_3 + 6\ HCl$$

   The precipitate is Fe_2S_3 and is called iron(III) sulfide. Since a specific quantity of each reactant is given, the limiting reactant must be determined.

$$35.00 \text{ mL } H_2S \times \left(\frac{1 \text{ L}}{1000 \text{ mL}}\right) \times \left(\frac{0.010 \text{ mol } H_2S}{1 \text{ L}}\right) \times \left(\frac{2 \text{ mol } FeCl_3}{3 \text{ mol } H_2S}\right) \times \left(\frac{1 \text{ L}}{0.030 \text{ mol}}\right) \times \left(\frac{1000 \text{ mL}}{1 \text{ L}}\right) = 7.77 \text{ mL of } FeCl_3$$

   This many milliliters of $FeCl_3$ is needed. Since 25.00 mL are available, there is an excess of iron(III) chloride; H_2S is the limiting reactant.

$$35.00 \text{ mL } H_2S \times \left(\frac{1 \text{ L}}{1000 \text{ mL}}\right) \times \left(\frac{0.010 \text{ mol}}{1 \text{ L}}\right) \times \left(\frac{1 \text{ mol } Fe_2S_3}{3 \text{ mol } H_2S}\right) \times \left(\frac{207.87 \text{ g}}{1 \text{ mol}}\right) \times \left(\frac{1000 \text{ mg}}{1 \text{ g}}\right) = 24 \text{ mg}$$

7. a. In acidic solution:

$$Fe^{2+} + Cr_2O_7^{2-} \rightarrow Fe^{3+} + Cr^{3+}$$

   Half-reactions:
   Oxidation: $Fe^{2+} \rightarrow Fe^{3+} + e^-$
   Reduction: $Cr_2O_7^{2-} + 14\ H^+ + 6\ e^- \rightarrow 2\ Cr^{3+} + 7\ H_2O$
   Combined: $Cr_2O_7^{2-} + 14\ H^+ + 6\ Fe \rightarrow 2\ Cr^{3+} + 7\ H_2O + 6\ Fe^{3+}$

   b. In basic solution:

$$MnO_4^- + Zn \rightarrow Zn^{2+} + MnO_2$$

   Half-reactions:
   Oxidation: $Zn \rightarrow Zn^{2+} + 2\ e^-$
   Reduction: $MnO_4^- + 2\ H_2O + 3\ e^- \rightarrow MnO_2' + 4\ OH^-$
   Combined: $2\ MnO_4^- + 4\ H_2O + 3\ Zn \rightarrow 2\ MnO_2 + 8\ OH^- + 3\ Zn^{2+}$

8. What is the molarity of chloride ion in the following solutions?

   a. 0.25 *M* $MgCl_2$
   $MgCl_2 \rightarrow Mg^{2+} + 2\ Cl^-$

$$\frac{0.25 \text{ mol } MgCl_2}{\text{L}} \times \left(\frac{2 \text{ mol } Cl^-}{1 \text{ mol } MgCl_2}\right) = 0.50 \text{ mol/L} = 0.50\ M$$

   b. 2.98 g NaCl diluted with water to a total volume to 250.0 mL

$$2.98 \text{ g NaCl} \times \left(\frac{1 \text{ mol NaCl}}{58.44 \text{ g}}\right) \times \left(\frac{1 \text{ mol } Cl^-}{1 \text{ mol NaCl}}\right) = 0.0510 \text{ mol } Cl^-$$

$$250.0 \text{ mL} \times \left(\frac{1 \text{ L}}{1000 \text{ mL}}\right) = 0.2500 \text{ L}$$

$$[Cl^-] = \frac{0.510 \text{ mol}}{0.2500 \text{ L}} = 0.204\ M$$

   c. 15.00 mL of 1.00 *M* $FeCl_3$ diluted to a total volume of 500.0 mL

$$M_{\text{initial}}V_{\text{initial}} = M_{\text{final}}V_{\text{final}}$$

$$(1.00\ M)(15.00 \text{ mL}) = M(500.0 \text{ mL})$$

$$M = 0.0300\ M\ FeCl_3 \times \left(\frac{3 \text{ mol } Cl^-}{1 \text{ mol } FeCl_3}\right) = 0.0900\ M\ Cl^-$$

9. How do you make 1.00 L of a 0.200 *M* $NH_4F(aq)$ from solid ammonium fluoride?

0.200 mol/L × (1.00 L) = 0.200 mol NH_4F needed.

$$0.200 \text{ mol} \times \left(\frac{37.05 \text{ g}}{1 \text{ mol}}\right) = 7.41 \text{ g}$$

Weigh 7.41 g of ammonium fluoride into a 1.00 L volumetric flask. Dilute to the mark with water.

10. How many grams of iron(III) carbonate can be made from 50.00 mL of 0.180 *M* $FeCl_3$ with excess sodium carbonate?

The reaction will be

$$2\, FeCl_3 + 3\, Na_2CO_3 \rightarrow Fe_2(CO_3)_3 + 3\, NaCl$$

$$50.00 \text{ mL} \times \left(\frac{1 \text{ L}}{1000 \text{ mL}}\right) \times \left(\frac{0.180 \text{ mol } FeCl_3}{1 \text{ L}}\right) \times$$

$$\left(\frac{1 \text{ mol } Fe_2(CO_3)_3}{2 \text{ mol } FeCl_3}\right) \times \left(\frac{291.73 \text{ g}}{1 \text{ mol}}\right) = 1.31 \text{ g}$$

CHAPTER 5 Thermochemistry

OUTLINE

REVIEW

Chapter Overview

Although not a form of matter itself, energy changes accompany both chemical and physical changes of matter. The study of energy and its transformations is called **thermodynamics**. The study of the relationship between energy and chemical reactions is called **thermochemistry**. **Energy**, the ability to do work (w) or transfer heat (q), can exist as either **kinetic energy**, the energy of motion, or **potential energy**, the energy of position.

Heat transfer occurs as heat energy moves from a region of high temperature to a region of low temperature. When all regions are the same temperature, the system is in **thermal equilibrium**. At the atomic level, the kinetic energy of the atoms is proportional to their absolute temperature (temperature in Kelvin). This thermal energy is associated with the random motion of molecules and the number of molecules.

Work occurs when a force changes the position of an object. For example, $P\Delta V$ work is a change in volume at a constant pressure. At the atomic level, molecules have potential energy due to the columbic interaction of the charge of atoms or molecules with the charges of other particles.

Energy is a **state function**; its change depends only on its initial and final values not the path taken. It is also governed by the **first law of thermodynamics**, which states that in a chemical reaction, energy can be neither created nor destroyed. Thus changes in energy can be measured by carefully measuring how a chemical reaction changes the energy of its environment. This is generally done by comparing the energy of the **system**, the limited part of the universe being studied, with its **surroundings**. Experiments can be designed so that the system is **isolated**, exchanging neither matter nor energy with its surroundings; or **open**, exchanging both matter and energy with its surroundings; or, most commonly, **closed**, exchanging energy with its surroundings but not matter.

When heat energy (q) flows from the system to the surroundings, the change is called **exothermic**. With the increase of heat in the surroundings, their temperature increases. Since measuring devices are always defined as part of the surroundings rather than the system, an increase in temperature is observed in an exothermic change and q of the system has a negative value.

When heat energy (q) flows from the surroundings into the system, the change is called **endothermic**. With the removal of heat from the surroundings, a decrease in temperature is observed and q of the system has a positive value.

Internal energy (E) is the sum of the kinetic and potential energy of system components. It is substantially easier to measure changes in the energy of a system rather than its actual internal energy. Because of the first law of thermodynamics, energy gained or lost by the system must be transferred to or from the surroundings. Thus changes in temperature or position of the surroundings can be used to measure the change in energy of the system.

One common unit of energy is the **calorie (cal)**, which is the amount of energy needed to increase the temperature of one gram of water one degree Celsius. For food, energy is reported as **Calories**, with the capital "C" making this unit equivalent to kilocalories or 1000 calories. However, the SI unit of energy is the **joule (J)**, and is the energy unit commonly used in this chapter. One calorie is equal to 4.184 J. $P\Delta V$ work often has the unit of atm•L. One atm•L is equal to 101.32 J.

Many studies in thermochemistry are designed to measure **enthalpy (*H*)**, the heat flow at constant pressure. Changes in enthalpy (ΔH) are easily determined by measuring the temperature change of the surroundings when a chemical or physical change occurs. Negative values of ΔH indicate an exothermic change in the system; positive ΔH values indicate an endothermic change of the system.

The magnitude of the temperature changes with heat energy depends on the system being studied. The relationship between heat (q) and temperature change (ΔT) is called the **heat capacity (C_p)**, which is the quantity of heat needed to raise the temperature of an object 1°C. Thus, $q = C_p\Delta T$. However, the heat capacity depends on both the identity and the size of the object. Therefore, **molar heat capacity (*c*)**, the quantity of heat required to raise the temperature of one mole of substance 1°C and the **specific heat (c_s)**, the quantity of heat required to raise the temperature of one gram of substance 1°C, are often more convenient relationships. Therefore, heat may also be calculated as

$$q = nc\Delta T \qquad \text{or} \qquad q = mc_s\Delta T$$

During a change in physical state, temperature is constant. However, heat flow is still involved in the transformation. In this case, the energy is being used to change the position of the molecules rather than the temperature. The **molar heat of fusion (ΔH_{fus})** is the heat required to melt one mole of a solid. Since the molecules of a solid are in a fixed position, energy is required so that these molecules can move from those positions, achieving a liquid state. Thus melting is an endothermic process ($+\Delta H_{fus}$). However, the first law of thermodynamics indicates that to do exactly the opposite, and freeze the substance, would require that the same amount of energy be removed ($-\Delta H_{fus}$). Thus the value of ΔH_{fus} is the same for melting or freezing; just the direction, as indicated by the sign, would change. Similarly, the **molar heat of vaporization (ΔH_{vap})** is the amount of energy required to change one mole of substance from a liquid to a gas.

An experimental approach to measuring heat is called **calorimetry**. In calorimetry, the process being measured is conducted in a calorimeter of known heat capacity called the **calorimeter constant (C_{calor})**. The temperature change of the calorimeter and surroundings is measured and the heat generated or absorbed by the process determined. Using a **bomb calorimeter**, which is designed to withstand high pressure so that no $P\Delta V$ work can occur, allows determination of the **heat of reaction (ΔH_{rxn})** also known as **enthalpy of reaction**. Calorimeter constants can be determined using systems with known heats of reaction.

Enthalpy is a state function, so its value is independent of path. A consequence of this fact is **Hess's law**, which states that the heat of a reaction which is the sum of two or more reactions is equal to the sum of the changes of the enthalpies of the constituent reactions. One reaction particularly useful in Hess's law calculations is a formation reaction. A **formation reaction** is the production of one mole of substance, at standard state, from its constituent elements at standard state. **Standard state** is the most stable (lowest energy) physical form of the substance at 25°C and 10^5 Pa (about 1 atm). The heat change involved in this reaction is called the **standard enthalpy of formation ($\Delta H°_f$)**. Since the elements balance in a reaction, any reaction can be written as a variation of formation reactions. Therefore,

$$\Delta H°_{rxn} = \Sigma\, n\Delta H°_{f,\,products} - \Sigma\, n\Delta H°_{f,\,reactants}$$

where n represents the number of moles of each product and reactant in the balanced reaction.

There are many uses of energy produced from a chemical reaction. The **fuel value** is the energy per gram of a substance when burned completely. Similarly, the **food value** is the energy of the material when completely burned by an organism for sustenance.

Worked Examples

Systems, Surroundings, and the Flow of Energy

Example 1.
What is the work (in joules) required to expand a gas from 100.0 mL to 1.00 L at 750 torr?

Collect and Organize We are asked to determine the work needed to expand a gas from an initial volume of 100.0 mL to a final volume of 1.00 L at a pressure of 750 torr.

Analyze The work involved in the expansion or compression of a gas is given by:

$$w = P\Delta V$$

$$w = P(V_{final} - V_{initial})$$

SOLVE To use the expression for work (w), volume must be in liters and pressure in atmospheres, so we must convert units first.

$$750 \text{ torr} \times \left(\frac{1 \text{ atm}}{760 \text{ torr}}\right) = 0.987 \text{ atm}$$

$$100.0 \text{ mL} \times \left(\frac{1 \text{ L}}{1000 \text{ mL}}\right) = 0.1000 \text{ L}$$

$$\Delta V = 1.00 \text{ L} - 0.1000 \text{ L} = 0.90 \text{ L}$$

$$w = P\Delta V$$

$$= (0.987 \text{ atm})(0.90 \text{ L}) = 0.89 \text{ L}\bullet\text{atm} \times \left(\frac{101.3 \text{ J}}{1 \text{ L}\bullet\text{atm}}\right) = 90 \text{ J}$$

THINK ABOUT IT Expansion of a gas results in a final volume that is larger than the initial volume. In this case the work involved is (+), and work is done by the gas. If the gas is compressed, the final volume will be smaller than the initial volume, work will be (–) and work will be done on the gas.

ENTHALPY (H) AND ENTHALPY CHANGES (ΔH)

EXAMPLE 1.

How much energy is involved if 0.263 g K reacts in the following manner? (Is the energy produced or used up?)

$$4 \text{ K} + O_2 \rightarrow 2 \text{ } K_2O \qquad \Delta H = -722 \text{ kJ/mol}$$

COLLECT AND ORGANIZE We are given a thermochemical equation (a stoichiometric equation and its heat of reaction) and are asked to determine how much energy is produced or used up from a given mass of reactant.

ANALYZE We must convert the given mass of K to moles of K using its molar mass. Then we can relate the heat associated with the reaction of the given quantity of K to the given thermochemical equation. The ΔH value relates the energy to the quantity of each substance in the reaction. When the stoichiometric coefficients refer to moles, the entire equation can be referred to as a "mole of reaction." Thus there are 4 moles of K per mole of reaction and 722 kJ per one mole of reaction.

SOLVE The stoichiometric coefficient of potassium metal is 4, so there are 4 moles of potassium for each mole of reaction. ΔH gives the relationship between energy and moles of reaction. Dimensional analysis can be used to determine energy.

$$0.263 \text{ g K} \times \left(\frac{1 \text{ mol K}}{39.10 \text{ g}}\right) \times \left(\frac{1 \text{ mol rxn}}{4 \text{ mol K}}\right) \times \left(\frac{722 \text{ kJ}}{1 \text{ mol rxn}}\right) = 1.21 \text{ kJ}$$

The negative sign on the given value of ΔH indicates that energy is produced rather than used up.

THINK ABOUT IT Notice that the negative sign is dropped in the dimensional analysis. The sign indicates direction; it does not indicate negative energy.

EXAMPLE 2.

How much energy is involved in the production of 4.00 g NO from its elements? Is energy used up or produced in this process?

$$N_2 + O_2 \rightarrow 2 \text{ NO} \qquad \Delta H = +180.4 \text{ kJ/mol}$$

COLLECT AND ORGANIZE From the given thermochemical equation we are asked to determine the amount of energy that is produced or used up in the production of a given quantity of product.

ANALYZE Since we are given the quantity of product in moles, we can use the thermochemical equation directly to determine the amount of heat involved. The value of ΔH tells us that there are 180.4 kJ per mole of reaction. The equation tells us that there are 2 moles of NO per mole of reaction.

SOLVE

$$4.00 \text{ g} \times \left(\frac{1 \text{ mol NO}}{30.01 \text{ g}}\right) \times \left(\frac{1 \text{ mol rxn}}{2 \text{ mol NO}}\right) \times \left(\frac{180.4 \text{ kJ}}{1 \text{ mol rxn}}\right) = 12.0 \text{ kJ}$$

The positive sign indicates energy is used up.

THINK ABOUT IT Thermochemical equations can be used to determine heats of reaction whether we are given the amount of a reactant used or the amount of a product formed.

EXAMPLE 3.

How much energy is required to melt 50.00 g water at its melting point?

COLLECT AND ORGANIZE We are asked to determine the amount of energy that is required when a given quantity of water undergoes a phase change from a solid to a liquid.

ANALYZE The energy involved in changing from solid to liquid is called the *heat of fusion* (ΔH_{fus}), which is given in units of kJ/mol. Converting the given mass of water to moles of water and using the tabulated value of ΔH_{fus} for water (6.01 kJ/mol) allows us to solve the problem.

SOLVE

$$50.00 \text{ g water} \times \left(\frac{1 \text{ mol}}{18.01 \text{ g}}\right) \times \left(\frac{6.01 \text{ kJ}}{1 \text{ mol}}\right) = 16.7 \text{ kJ}$$

Since we are melting ice this amount of heat must be added.

THINK ABOUT IT A change in physical state involves a change in energy. Because the energy is used in changing physical state, it is not used to change the temperature, which remains constant during the process.

CALORIMETRY

EXAMPLE 1.
How much heat is lost when 36.0 g platinum (c = 25.94 J/mol•°C) at 35.0°C is cooled to 27.4°C?

COLLECT AND ORGANIZE We are asked to determine the amount of heat that a given mass of platinum at a given temperature will lose when it is cooled to some other temperature given the heat capacity of platinum.

ANALYZE The change in energy is determined from the change in temperature, which are related by: $q = nc\Delta T$ in which ΔT is the temperature difference between the initial and final states of the substance, n is the number of moles of substance, and c is the molar heat capacity of the substance.

SOLVE

$$q = nc\Delta T$$

The heat lost is q.

$$n = \text{moles Pt} = 36.0 \text{ g} \times \left(\frac{1 \text{ mol}}{195.1 \text{ g}}\right) = 0.184 \text{ mol}$$

c = heat capacity of platinum, given in problem = 25.94 J/mol•°C

$\Delta T = 35.0 - 27.4 = 7.6°\text{C}$

$q = (0.184 \text{ mol})(25.94 \text{ J/mol•°C})(7.6°\text{C}) = 36 \text{ J}$

The use of two significant figures is due to the ΔT value.

THINK ABOUT IT Since the heat capacity depends upon the number of moles, a sample of greater mass of Pt will evolve more heat than a sample of lesser mass when cooling to the same temperature.

EXAMPLE 2.
If 72.5 g of gold (c = 25.41 J/mol•°C) at 64.2°C is added to 100.0 g of water (c = 75.3 J/mol•°C) at 23.0°C, what is the final temperature of the water, assuming no energy is lost to the container?

COLLECT AND ORGANIZE We are given the mass, initial temperature, and heat capacity of a sample of gold that is added to water of a given mass and temperature. From the known heat capacity of water we are asked to deduce the final temperature of the water after the gold is added.

ANALYZE Using $q = nc\Delta T$, we will first derive an expression for the amount of heat lost by the gold as a function of its final temperature. A similar expression can be derived for the water. Using the first law of thermodynamics, we will then be able to equate the two heat values and solve for the final temperature.

SOLVE All the heat energy is lost by the gold and gained by the water. The final temperature of the gold and water will be the same (T) when all the heat is transferred. The temperature of the water will increase, so its $\Delta T = T -$ 23.0°C. The temperature of the gold will decrease, so its ΔT = 64.2°C – T. By writing the ΔT so that the value is positive, the signs will all work out correctly.

$$n \text{ of gold} = 72.5 \text{ g} \times \left(\frac{1 \text{ mol}}{197.0 \text{ g}}\right) = 0.368 \text{ mol Au}$$

$$n \text{ of water} = 100.0 \text{ g} \times \left(\frac{1 \text{ mol}}{18.01 \text{ g}}\right) = 5.55 \text{ mol water}$$

q lost by gold = $nc\Delta T$ = (0.368 mol)(25.94 J/mol•°C)(64.2 – T)

q gained by water = $nc\Delta T$ = (5.55 mol)(75.3 J/mol•°C)(T – 23.0)

According to the first law of thermodynamics, q lost by gold = q gained by water.

$$(0.368)(25.41)(64.2 - T) = (5.55)(75.3)(T - 23.0)$$

$$9.35(64.2 - T) = 417.915(T - 23.0)$$

$$600.27 - 9.35T = 417.915T - 9612.045$$

$$10212.315 = 427.265T$$

$$23.9°\text{C} = T$$

THINK ABOUT IT Since the experiment is constructed so that no work is done, we can relate the heat gained by the water to the heat lost by the gold using the first law of thermodynamics.

EXAMPLE 3.
What is the heat capacity of the calorimeter if 1.395 g benzoic acid changes the temperature of the calorimeter from 21.9°C to 39.9°C?

COLLECT AND ORGANIZE We are asked to determine the heat capacity of a calorimeter given that the combustion of 1.395 g of benzoic acid results in a temperature change from 21.9°C to 39.9°C.

ANALYZE We can determine the heat that evolves from the given quantity of benzoic acid using the amount of heat it releases during combustion (26.38 kJ/g), which is a known quantity. The same amount of heat will flow to the calorimeter. We can re-express $q = c_P\Delta T$ and solve for the heat capacity of the calorimeter, c_P, after substitution of the appropriate variables.

SOLVE Heat evolved by the benzoic acid = heat absorbed by the calorimeter

Heat evolved by benzoic acid = 1.395 g × (26.38 kJ/g) = 36.80 kJ = q

Change in temperature = 39.9 – 21.9 = 18.0°C

$q = c_P\Delta T$

36.80 kJ = c_P(18.0°C)

2.04 kJ/°C = c_P = heat capacity of calorimeter

THINK ABOUT IT The heat capacity of a calorimeter must be determined experimentally, since even instruments of the same type will not have the same heat capacity. This requires that a known reaction be used. The traditional reaction is the combustion of benzoic acid ($C_7H_6O_2$). The combustion of benzoic acid will produce 26.38 kJ for each gram of benzoic acid.

EXAMPLE 4.
Using the preceding calorimeter, how much heat changes the temperature from 23.4°C to 20.3°C?

COLLECT AND ORGANIZE We are asked to use the calorimeter in the previous example to determine the amount of heat needed to change the temperature from 23.4°C to 20.3°C.

ANALYZE From the heat capacity of the calorimeter determined in the previous example we know how the temperature of the calorimeter will be changed as a function of heat flow.

SOLVE The question asks for heat energy, $q = c_P\Delta T$. The heat capacity was calculated in the previous example = 2.04 kJ/mol. Since ΔT = 3.1°C,

$$q = (2.04\ \text{J/°C})(3.1\text{°C}) = 6.3\ \text{kJ}$$

THINK ABOUT IT Since the temperature of the calorimeter cools during the process, heat is flowing out of the calorimeter.

ENTHALPIES OF FORMATION AND ENTHALPIES OF REACTION

EXAMPLE 1.
If 1.902 g of hydrogen gas reacting with excess oxygen raises the temperature of 150.0 g of water from 21.2°C to 57.5°C, what is the ΔH_{rxn} (in kJ/mol)? Assume no energy is lost to the surroundings.

$$2\ H_2 + O_2 \rightarrow 2\ H_2O$$

COLLECT AND ORGANIZE We are asked to calculate the heat of reaction when the heat transferred from the reaction of a given mass of hydrogen with excess oxygen gas results in a given mass of water in the calorimeter to undergo a given increase in temperature.

ANALYZE The ΔH_{rxn} can be determined from q, moles of reaction, and the direction of the temperature change. The value of q can be calculated from either $q = nc\Delta T$ or $q = mc_s\Delta T$, depending on the data given. The moles of reaction can be determined from the amount of limiting reactant. The direction of the temperature change indicates the sign on ΔH.

SOLVE $\Delta H_{rxn} = -q/mol$. Because the temperature increased, the ΔH_{rxn} value must be negative.

$$\text{mol } O_2 = 0.1902\ \text{g}\ H_2 \times \left(\frac{1\ \text{mol}\ H_2}{2.02\ \text{g}}\right) \times \left(\frac{1\ \text{mol rxn}}{2\ \text{mol}\ H_2}\right) = 0.0471\ \text{mol rxn}$$

q lost by reaction = q gained by water = $nc_{water}\Delta T$

$$n_{water} = 150.0\ \text{g} \times \left(\frac{1\ \text{mol}}{18.01\ \text{g}}\right) = 8.329\ \text{mol water}$$

c_{water} = 75.3 J/mol•K (value from text)

ΔT = 57.5 – 21.2 = 36.3 (remember to use the absolute value)

$$q = (8.329)(75.3)(36.3) = 2.28 \times 10^4\ \text{J} \times \left(\frac{1\ \text{kJ}}{1000\ \text{J}}\right) = 22.8\ \text{kJ}$$

$$\Delta H = -q/\text{mol rxn} = \frac{-22.8\ \text{kJ}}{0.0471\ \text{mol}} = -484\ \text{kJ/mol}$$

THINK ABOUT IT The sign is determined from the direction of the temperature change. If the temperature increases, the reaction must have given off energy; therefore, the sign on ΔH is negative. If the temperature decreases, the sign on ΔH is positive.

EXAMPLE 2.

What is the ΔH_{rxn} if the dissolution of 1.645 g of ammonium chloride in 200.0 mL water decreases the temperature from 23.4°C to 18.3°C? (Assuming no energy is lost to the surroundings.)

COLLECT AND ORGANIZE We are asked to determine ΔH_{rxn} when 1.645 g of ammonium chloride is dissolved in 200.0 mL water and the temperature of the water decreases from 23.4°C to 18.3°C.

ANALYZE We will use the heat capacity of water and its temperature change to calculate the resulting heat flow. Then we can use $\Delta H = -q/mol$ to determine the enthalpy of the dissolution.

SOLVE $\Delta H = +q/mol$. Because the temperature of the surroundings decreased, the reaction must have taken energy from the surroundings and ΔH is positive. The dissolution reaction is

$$NH_4Cl \rightarrow NH_4^+ + Cl^-$$

$$\text{mole of reaction} = 1.645 \text{ g } NH_4Cl \times \left(\frac{1 \text{ mol}}{53.495 \text{ g}}\right) \times \left(\frac{1 \text{ mole of reaction}}{1 \text{ mole of } NH_4Cl}\right) = 0.03075 \text{ mole of reaction}$$

$$\text{moles of water} = 200.0 \text{ g} \times \left(\frac{1 \text{ mol}}{18.01 \text{ g}}\right) = 11.10 \text{ mol}$$

$c = 75.3$ J/mol•°C for water

$\Delta T = 23.4 - 18.3 = 5.1$°C

$$q_{NH_4Cl} = q_{water} = n_{water} c\Delta T = (11.10)(75.3)(5.1) = 4262.73 \text{ J}$$

(only two digits are significant)

$$\Delta H = \frac{+(4262.73)}{(0.03075)} = \frac{+138625 \text{ J}}{\text{mol}} \times \left(\frac{1 \text{ kJ}}{1000 \text{ J}}\right) = +138.6 \text{ kJ/mol}$$

$$\Delta H = +1.4 \times 10^5 \text{ J/mol} = +1.4 \times 10^2 \text{ kJ/mol}$$

THINK ABOUT IT The dissolution of ammonium chloride in water is an endothermic reaction. The beaker will feel cold during the process.

EXAMPLE 3.

What is the ΔH_{rxn} if the reaction of 0.556 g Na raises the temperature of a calorimeter (C_P = 3.145 kJ/°C) from 19.23°C to 20.65°C?

$$2 \text{ Na} + 2 \text{ H}_2\text{O} \rightarrow 2 \text{ NaOH} + \text{H}_2$$

COLLECT AND ORGANIZE We are asked to calculate ΔH_{rxn} if the reaction of 0.556 g Na raises the temperature of a calorimeter (C_P = 3.145 kJ/°C) from 19.23°C to 20.65°C.

ANALYZE We will use the given information to solve for q using $q = C_P \Delta T$. Once we have q we can use $\Delta H = -q$ to determine the ΔH_{rxn}.

SOLVE

$\Delta H = -q/mol$

$$\text{moles of reaction} = 0.556 \text{ g} \times \left(\frac{1 \text{ mol}}{22.99 \text{ g}}\right) \times \left(\frac{1 \text{ mol rxn}}{2 \text{ mol Na}}\right) = 0.0121 \text{ mol}$$

$C_P = 3.145$ kJ/°C given in the problem

$\Delta T = 20.65 - 19.23 = 1.42$°C

$q = C_P\Delta T = (3.145 \text{ kJ/°C})(1.42) = 4.46$ kJ

$$\Delta H_{rxn} = \frac{-(4.46 \text{ kJ})}{(0.0121 \text{ mol})} = -369 \text{ kJ/mol}$$

THINK ABOUT IT Since the temperature increases, the reaction is exothermic and ΔH is negative.

EXAMPLE 4.

Write formation reactions for the following compounds.

a. $H_3PO_4(\ell)$ b. $C_2H_6O(\ell)$ c. $Zn(NO_3)_2(s)$ d. $XeF_4(g)$

COLLECT AND ORGANIZE We are asked to write formation reactions for the given molecules.

ANALYZE A formation reaction forms one mole of product from its elements in standard state. This requires knowing the standard state of the elements.

SOLVE

a. The standard state of hydrogen is $H_2(g)$. The standard state of oxygen is $O_2(g)$. The standard state of phosphorus is $P_4(s)$. Therefore, the formation reaction is

$$H_2(g) + P_4(s) + O_2(g) \rightarrow H_3PO_4(\ell)$$

The reaction must be balanced so that there is one mole of product. Therefore, the balanced reaction is

$$\frac{3}{2}H_2(g) + \frac{1}{4}P_4(s) + 2\,O_2(g) \rightarrow H_3PO_4(\ell)$$

b. The standard state of carbon is C(*graphite*). The standard state of hydrogen is $H_2(g)$. The standard state of oxygen is $O_2(g)$. Therefore, the formation reaction is

$$C(graphite) + H_2(g) + O_2(g) \rightarrow C_2H_6O(\ell)$$

Balanced, the reaction is

$$2\ C(graphite) + 3\ H_2(g) + \frac{1}{2}O_2(g) \rightarrow C_2H_6O(\ell)$$

c. The standard state of zinc is Zn(*s*). The standard state of nitrogen is $N_2(g)$. The standard state of oxygen is $O_2(g)$. Therefore, the reaction, balanced, is

$$Zn(s) + N_2(g) + 3\ O_2(g) \rightarrow Zn(NO_3)_2(s)$$

d. The standard state of xenon is Xe(*g*). The standard state of fluorine is $F_2(g)$. Therefore, the balanced formation reaction is

$$Xe(g) + 2\ F_2(g) \rightarrow XeF_4(g)$$

THINK ABOUT IT The other important part of formation reactions is that only one mole of product is formed. Because it is *one mole,* the stoichiometric coefficients can be and often are fractions.

EXAMPLE 5.
Using heats of formation, what is the ΔH_{rxn} for the following? (All reactants and products are solids.)

$$Al + Fe_2O_3 \rightarrow Al_2O_3 + Fe$$

COLLECT AND ORGANIZE We are asked to use heats of formation to determine ΔH°_{rxn} for the given reaction.

ANALYZE If the ΔH is determined solely from enthalpies of formation, the following formula can be used:

$$\Delta H^\circ_{rxn} = \Sigma\ n\Delta H^\circ_{f,\ products} - \Sigma\ m\Delta H^\circ_{f,\ reactants}$$

where n is the stoichiometric coefficient of each product and m is the stoichiometric coefficient of each reactant. Values for ΔH°_f are listed in the appendix.

SOLVE Since Al and Fe are elements in their standard states, their ΔH°_f is zero. The ΔH°_{rxn} for ferric oxide is –824.2 kJ/mol and that for aluminum oxide is –1676 kJ/mol. Therefore,

$$\Delta H^\circ_{rxn} = [-1676 + 0] - [-824.2 + 0] = -851.8\ kJ/mol$$

THINK ABOUT IT It is necessary to know the states of all reactants and products so that the correct value of ΔH°_f can be determined from the tables.

EXAMPLE 7.
What is the ΔH_{rxn} of the following reaction based on heats of formation?

$$H_2SO_4(aq) + 2\ NaOH(aq) \rightarrow 2\ Na^+(aq) + SO_4^{-2}(aq) + 2\ H_2O(\ell)$$

COLLECT AND ORGANIZE We are asked to use heats of formation to determine ΔH_{rxn} for the given reaction.

ANALYZE We will use $\Delta H^\circ_{rxn} = \Sigma\ n\Delta H^\circ_{f,\ products} - \Sigma\ m\Delta H^\circ_{f,\ reactants}$ with tabulated values of the heats of formation.

SOLVE From the appendix:

Compound	ΔH°_f (kJ/mol)
H_2SO_4	–909.3
NaOH	–470.1
Na^+	–240.1
SO_4^{2-}	–909.3
H_2O	–285.8

$$\Delta H_{rxn} = [2(-240.1) + (-909.3) + 2(-285.8)] - [-909.3 + 2(-470.1)]$$

$$= [-1961.1] - [-1849.5]$$

$$= -111.6\ kJ/mol$$

THINK ABOUT IT Be sure—particularly with hydrochloric acid, sodium hydroxide, and water—that you use the ΔH_f appropriate to the physical state. There is often more than one listing for these compounds.

FUEL VALUES

EXAMPLE 1.
The ΔH for the combustion of CH_2O is –518 kJ/mol. What is the fuel value of CH_2O?

$$CH_2O + O_2 \rightarrow CO_2 + H_2O$$

COLLECT AND ORGANIZE We are asked to determine the fuel value of CH_2O given its enthalpy and stoichiometry of combustion.

ANALYZE The fuel value is the amount of energy produced per gram of fuel. Since a combustion reaction is normally used to produce the energy from a fuel, this reaction is assumed for fuel values.

SOLVE The fuel value should have units of kJ/g. The negative sign is not included. In ΔH it represents direction, but since a fuel must be exothermic, that information is not needed.

$$\frac{518 \text{ kJ}}{\text{mol}} \times \left(\frac{1 \text{ mol}}{30.03 \text{ g}}\right) = 17.3 \text{ kJ/g}$$

THINK ABOUT IT Since energy of the reaction comes from bonds being formed and broken, the ratio of bonds to molar mass is one estimate of fuel value. Consequently, molecules with lower molar mass tend to be molecules with a large fraction of hydrogens and fewer of the higher molar mass atoms like oxygen.

HESS'S LAW

EXAMPLE 1.

What is the ΔH_{rxn} for $C_2H_4 + H_2O \rightarrow C_2H_6O$ based on the following reactions:

$H_2 + C_2H_4 \rightarrow C_2H_6$	$\Delta H = -137$ kJ/mol
$2\ C_2H_6 + O_2 \rightarrow 2\ C_2H_6O$	$\Delta H = -300$ kJ/mol
$2\ H_2 + O_2 \rightarrow 2\ H_2O$	$\Delta H = -572$ kJ/mol

COLLECT AND ORGANIZE We are asked to determine ΔH_{rxn} for $C_2H_4 + H_2O \rightarrow C_2H_6O$ based on a number of given reactions.

ANALYZE To determine the enthalpy of a reaction from other reactions, the other reactions must be manipulated so that the sum of the reactions is the reaction of interest. Reactions may be reversed or the stoichiometric coefficients may be multiplied by a factor to achieve this.

SOLVE For clarity, let's label the reactions

1. $H_2 + C_2H_4 \rightarrow C_2H_6$	$\Delta H = -137$ kJ/mol
2. $2\ C_2H_6 + O_2 \rightarrow 2\ C_2H_6O$	$\Delta H = -300$ kJ/mol
3. $2\ H_2 + O_2 \rightarrow 2\ H_2O$	$\Delta H = -572$ kJ/mol
4. $C_2H_4 + H_2O \rightarrow C_2H_6O$	

Reaction 1 contains a reactant (C_2H_4) from the desired equation (#4), which is also in the reactant position. Since the stoichiometric coefficient of C_2H_4 for reaction 1 is the same as for reaction 4, it can remain unchanged. Reaction 2 contains the product (C_2H_6O) of reaction 4. The product is in the appropriate position, but the stoichiometric coefficient is incorrect. To make that correct, the entire equation needs to be multiplied by $\frac{1}{2}$. If the equation is multiplied by $\frac{1}{2}$, the ΔH must also be multiplied by $\frac{1}{2}$:

2b. $C_2H_6 + \frac{1}{2} O_2 \rightarrow C_2H_6O$ $\quad \Delta H = -150$ kJ/mol

Reaction 3 contains the other reactant (H_2O), but it is in the product position, so the reaction should be reversed. If the reaction is reversed, the sign is changed:

3b. $2\ H_2O \rightarrow 2\ H_2 + O_2$ $\quad \Delta H = +572$ kJ/mol

In addition, the stoichiometric coefficient must be changed, so the equation should also be multiplied by $\frac{1}{2}$:

3c. $H_2O \rightarrow H_2 + \frac{1}{2} O_2$ $\quad \Delta H = +286$ kJ/mol

Next, reactions 1, 2b, and 3c are added together:

$$H_2O + C_2H_6 + \frac{1}{2} O_2 + H_2 + C_2H_4 \rightarrow C_2H_6 + C_2H_6O + H_2 + \frac{1}{2} O_2$$

Substances that are the same on each side can be canceled, giving the desired equation:

$$C_2H_4 + H_2O \rightarrow C_2H_6O$$

As the equations are added, so are the ΔH values, so

$$\Delta H = -137 - 150 + 286 = -1 \text{ kJ/mol}$$

THINK ABOUT IT Since Hess's law determines ΔH_{rxn} by the appropriate addition and subtraction of other reactions, it can be used to determine ΔH_{rxn} for reactions that may be impossible or much too dangerous to carry out.

EXAMPLE 2.

What is the ΔH_{rxn} for $Pb^{2+} + 2\ Br^- \rightarrow PbBr_2$ based on the following reactions:

$Pb + Br_2 \rightarrow PbBr_2$	$\Delta H = -279$ kJ/mol
$Pb \rightarrow Pb^+ + e^-$	$\Delta H = +716$ kJ/mol
$Pb^+ \rightarrow Pb^{2+} + e^-$	$\Delta H = +844$ kJ/mol
$Br + e^- \rightarrow Br^-$	$\Delta H = -324$ kJ/mol
$Br_2 \rightarrow 2\ Br$	$\Delta H = +111$ kJ/mol

COLLECT AND ORGANIZE We are asked to determine ΔH_{rxn} for $Pb^{2+} + 2\ Br^- \rightarrow PbBr_2$ based on the given reactions.

ANALYZE We will use the procedure outlined in the previous example to solve the problem.

SOLVE Labeling the reactions for convenience,

1. $Pb + Br_2 \rightarrow PbBr_2$ $\Delta H = -279$ kJ/mol
2. $Pb \rightarrow Pb^+ + e^-$ $\Delta H = +716$ kJ/mol
3. $Pb^+ \rightarrow Pb^{2+} + e^-$ $\Delta H = +844$ kJ/mol
4. $Br + e^- \rightarrow Br^-$ $\Delta H = -324$ kJ/mol
5. $Br_2 \rightarrow 2\ Br$ $\Delta H = +111$ kJ/mol

Reaction 1 contains the product with the appropriate stoichiometric coefficient, so it remains unchanged. Reaction 3 contains a reactant (Pb^{2+}), but in the product position, so it should be reversed:

3b. $Pb^{2+} + e^- \rightarrow Pb^+$ $\Delta H = -844$ kJ/mol

Reaction 4 contains a reactant (Br^-), but in the product position, so it should be reversed:

4b. $Br^- \rightarrow Br + e^-$ $\Delta H = +324$ kJ/mol

The stoichiometric coefficient should also be changed by multiplying reaction 4b by 2:

4c. $2\ Br^- \rightarrow 2\ Br + 2e^-$ $\Delta H = +648$ kJ/mol

Adding reactions 1, 3b, and 4c, you get

$$2\ Br^- + Pb^{2+} + e^- + Pb + Br_2 \rightarrow PbBr_2 + Pb^+ + 2\ Br + 2e^-$$

$$\Delta H = -279 - 844 + 648 = -475 \text{ kJ/mol}$$

Only one electron cancels from this reaction:

6. $2\ Br^- + Pb^{2+} + Pb + Br_2 \rightarrow PbBr_2 + Pb^+ + 2\ Br + e^-$

$$\Delta H = -475 \text{ kJ/mol}$$

To remove the Pb, the reverse of reaction 2 is needed:

2b. $Pb^+ + e^- \rightarrow Pb$ $\Delta H = -716$ kJ/mol

To remove Br_2, the reverse of equation 5 is needed:

5b. $2\ Br \rightarrow Br_2$ $\Delta H = -111$ kJ/mol

Adding these to equation 6:

$$Pb^+ + e^- + 2\ Br + 2\ Br^- + Pb^{2+} + Pb + Br_2 \rightarrow PbBr_2 + Pb^+ + 2\ Br + e^- + Br_2 + Pb$$

The extra substances cancel to get the correct equation:

$$Pb^{2+} + 2\ Br^- \rightarrow PbBr_2$$

$$\Delta H = -475 - 716 - 111 = -1302 \text{ kJ/mol}$$

THINK ABOUT IT If any substances remain that are not in the desired reaction, more reactions may be needed to cancel those compounds.

Self-Test

KEY VOCABULARY COMPLETION QUESTIONS

__________ 1. Heat energy required to raise 1 g of a substance 1°C

__________ 2. The energy involved in forming 1 mole of substance from its elements at standard state

__________ 3. Study of energy transfer in a chemical reaction

__________ 4. Energy of movement

__________ 5. Heat required to melt one mole of substance

__________ 6. The energy of a reaction can be determined from the sum of the energies of reactions for the constituent step in the process

__________ 7. Energy required to move an object over a distance

__________ 8. Heat transfer from the system to the surroundings

__________ 9. Heat flow at a constant pressure in a chemical reaction

__________ 10. Capacity to do work or transfer heat

__________ 11. Energy of position

__________ 12. Most stable form at 25°C and 10^5 Pa

__________ 13. Energy per gram obtained from combustion

__________ 14. ΔH_f° of an element in its standard state

__________ 15. Experiment to measure heat energy

__________ 16. Heat required to raise the temperature of an object 1°C

__________ 17. $+\Delta H$

__________ 18. Energy is neither created nor destroyed.

__________ 19. Value that is independent of path

__________ 20. System where energy is exchanged, but matter is not

__________ 21. System and surroundings are at the same temperature

__________ 22. SI unit of energy

__________ 23. Device to measure heat exchange at a constant pressure

__________ 24. Type of energy associated with the random motion of molecules

__________ 25. Components under study

Multiple-Choice Questions

1. Which requires more work on a 1 lb object?
   a. to move it 1 ft to the right
   b. to move it up 1 ft
   c. to move it down 1 ft
   d. all require the same amount of work

2. Which of the following is a state function?
   a. ETA (estimated time of arrival)
   b. mileage
   c. altitude
   d. all are state functions

3. Which change will lead to the highest increase in kinetic energy?
   a. doubling the mass
   b. doubling the velocity
   c. doubling the altitude
   d. all these would have the same effect

4. If 5.0 kJ is gained by the surroundings, then
   a. the system lost 5.0 kJ
   b. the system gained 5.0 kJ
   c. the system gained 10 kJ
   d. the system gained 12 kJ

5 Boiling water is
   a. exothermic and increasing in temperature
   b. endothermic and increasing in temperature
   c. exothermic and constant temperature
   d. endothermic and constant temperature

6. How much work is done when a balloon is inflated to 1.00 L at 760 torr?
   a. 101 J
   b. 1.00 J
   c. 760 J
   d. 7.60 J

7. The energy of melting is called
   a. the heat of fission
   b. the heat of liquefication
   c. the heat of vaporization
   d. the heat of fusion

8. How much energy is required to melt 10.0 g of water at 0°C? (ΔH°_{fus} = 6.01 kJ/mol; ΔH°_{vap} = 40.76 kJ/mol; c = 75.3 kJ/mol•°C)
   a. none
   b. 6.01 J
   c. 60.1 kJ
   d. 3.33 kJ

9. How much energy is required to heat 5.00 g water from 4°C to 37°C? (ΔH°_{fus} = 6.01 kJ/mol; ΔH°_{vap} = 40.76 kJ/mol; c = 75.3 J/mol•°C)
   a. 12425 J
   b. 691 J
   c. 75.3 J
   d. 376.5 J

10. How much energy is needed to turn 10.0 g of water at 25°C to steam? (ΔH°_{fus} = 6.01 kJ/mol; ΔH°_{vap} = 40.76 kJ/mol; c = 75.3 J/mol•°C)
    a. 3.13 kJ
    b. 22.62 kJ
    c. 25.75 kJ
    d. 19.5 kJ

11. What is the heat capacity of a calorimeter if the combustion of 5.00 g of benzoic acid causes a temperature change of 14.2°C? ($\Delta H_{combustion}$ = –26.38 kJ/g)
    a. 1070 J/°C
    b. 297 J/°C
    c. 1.85 kJ/°C
    d. 9.29 kJ/°C

12. The specific heat of water is
    a. 75.3 J/mol•°C
    b. 23.38 J/mol•°C
    c. 4.18 J/g•°C
    d. 601 kJ/mol

13. Which will have the greater temperature change if the same amount of energy is applied to 1 mole of each of the following?
    a. iron (c = 25.19 J/mol•°C)
    b. gold (c = 25.41 J/mol•°C)
    c. magnesium (c = 24.79 J/mol•°C)
    d. platinum (c = 25.95 J/mol•°C)

14. Which will have the smallest temperature change for a given addition of energy?
    a. 5.00 g water (c = 75.3 J/mol•°C)
    b. 10.00 g water (c = 75.3 J/mol•°C)
    c. 5.00 g gold (c = 25.4 J/mol•°C)
    d. 10.00 g gold (c = 25.4 J/mol•°C)

15. The heat capacity of aluminum is 25.4 J/mol•°C. How much energy is released when 5.00 g Al is cooled from 50°C to 30°C?
    a. 2440 J
    b. 90 J
    c. 488 J
    d. 3.7 J

16. Experimental determination of energy is called
    a. work
    b. enthalpy
    c. calorimetry
    d. Hess's law

17. What is the heat capacity of a bomb calorimeter if the combustion of 0.3550 g of benzoic acid ($\Delta H° =$ –26.38 kJ/g) raises the temperature of the calorimeter from 25.13°C to 33.30°C?
   a. 9.40 J/°C
   b. 1.15 kJ/°C
   c. 1.149 J/°C
   d. 23.0°C

18. The heat capacity of a calorimeter is 2.22 kJ/°C. What is the temperature change if 997 J of energy is absorbed by a reaction in that calorimeter?
   a. 2.21°C higher
   b. 2.21°C lower
   c. 0.449°C higher
   d. 0.449°C lower

19. Which of the following is a formation reaction?
   a. $K^+(g) + Br^-(g) \rightarrow KBr(s)$
   b. $2\ K(s) + Br_2(\ell) \rightarrow 2\ KBr(s)$
   c. $K(g) + Br(g) \rightarrow KBr(s)$
   d. none of the above

20. Which element is not in its standard state?
   a. $H_2(g)$
   b. $Na(s)$
   c. $Ne(g)$
   d. $Cl(g)$

21. The $\Delta H_f°$ of $Ar(g)$ is
   a. positive
   b. negative
   c. zero
   d. must be determined experimentally

22. If a reaction has a $\Delta H = +369$ kJ/mol, the ΔH of the reverse reaction is
   a. +369 kJ/mol
   b. –369 kJ/mol
   c. +185 kJ/mol
   d. –185 kJ/mol

23. The reaction $K + \frac{1}{2} Cl_2 + 2\ O_2 \rightarrow KClO_4$ has a $\Delta H =$ –432 kJ/mol. What is the ΔH of $2\ KClO_4 \rightarrow 2\ K + Cl_2 + 4\ O_2$?
   a. –432 kJ/mol
   b. +432 kJ/mol
   c. –216 kJ/mol
   d. +864 kJ/mol

24. A product of a combustion reaction is
   a. gaseous water
   b. hydrocarbon
   c. saturated hydrocarbon
   d. all of the above

25. The first law of thermodynamics is
   a. heat lost equals heat gained
   b. energy is neither created nor destroyed
   c. conservation of energy
   d. all of the above

Additional Practice Problems

1. How much work (in joules) is done when 22.4L of helium at STP expands to 50.0 L at constant pressure? (101.32 J = 1 L • atm)

2. How much energy is required to freeze 100.0 mL of water at 25°C? ($\Delta H°_{fus} = 6.01$ kJ/mol; $\Delta H°_{vap} =$ 40.76 kJ/mol; $c = 75.3$ J/mol•°C)

3. If the formation reaction of sodium chloride has a $\Delta H_f° = -411.2$ kJ/mol, write the formation reaction. How much energy is involved when of 4.99 g of chlorine react to make salt? Is energy evolved or used up?

4. When 1.638 g of benzoic acid (ΔH of combustion = –26.38 kJ/g) was combusted in a bomb calorimeter, the temperature rose from 17.8°C to 43.1°C. When 1.142 g of hexene (C_6H_{12}) underwent combustion, the temperature rose from 20.8°C to 49.3°C. What is the ΔH_{rxn} for the combustion of hexene?

5. When 10.00 g of water at 41.1°C is mixed with 50.00 g of water at 20.3°C in an insulated container (no heat lost to the container), what will the final temperature of the water be?

6. Based on enthalpies of formation, what is the ΔH_{rxn} for the following reactions? (*Note:* First balance the reactions and assume all reactants and products are gases.)
   a. $C_4H_{10} + O_2 \rightarrow CO_2 + H_2O$
   b. $CO + O_2 \rightarrow CO_2$
   c. $C_2H_6 + O_2 \rightarrow CO + H_2$

7. Calculate the energy for the reaction

$$Fe_2O_3 + 3\ CO \rightarrow 2\ Fe + 3\ CO_2$$

from ΔH_f for $Fe_2O_3 = -824.2$ kJ/mol and for $CH_4 = -74.8$ kJ/mol and for $H_2O = -241.8$ kJ/mol

$CH_4 + 2\ O_2 \rightarrow CO_2 + 2\ H_2O$ $\quad \Delta H = -782.4$ kJ/mol

$C + H_2O \rightarrow CO + H_2$ $\quad \Delta H = +131$ kJ/mol

Self-Test Answer Key

Key Vocabulary Completion Answers

1. specific heat
2. enthalpy of formation (or heat of formation)
3. thermochemistry
4. kinetic energy
5. molar heat of fusion
6. Hess's law
7. work
8. exothermic
9. enthalpy
10. energy
11. potential
12. standard state
13. fuel value
14. zero
15. calorimetry
16. heat capacity
17. endothermic
18. first law of thermodynamics
19. state function
20. closed
21. thermal equilibrium
22. joule
23. bomb calorimeter
24. heat
25. system

Multiple-Choice Answers

For further information about the topics these questions address, you should consult the text. The appropriate section of the text is listed in parentheses after the explanation of the answer.

1. b. Work is force times distance. Since distance is the same, the variable is force. By moving upward, you work against the force of gravity. Consequently, more force and thus more work is required. (5.1)
2. c. A state function is independent of path. Mileage and time of arrival depend on the route; however, altitude is distance from the ground (sea level), regardless of path. (5.3)
3. b. Kinetic energy is $\frac{1}{2}mu^2$, where m = mass and u = velocity. Because of the squared term the velocity effect is much greater. (5.1)
4. a. Because of the first law of thermodynamics, energy cannot be created or destroyed. Therefore, any energy gained by the surroundings must have been lost from somewhere else (the system). (5.2)
5. d. Boiling is endothermic because it requires the input of energy to boil water. However, the temperature of boiling water remains constant because the energy is used to turn the liquid water to a gas instead of to increase temperature. (5.3)
6. a. Work at constant pressure = $P\Delta V$. Since 760 torr is 1 atm and the change in volume is from 0 to 1.0 L, the work done is 1.0 L•atm. Using the conversion factor 101.32 J = 1 L•atm, inflating the balloon requires 101 J of work. (5.1)
7. d. The heat of fusion is the energy required to melt, or the energy released by freezing. The name probably comes from the freezing part. (5.4)
8. d. The heat of fusion is 601 kJ/mol for water. $10.0\text{ g H}_2\text{O} \times \left(\frac{1\text{ mol}}{18.01\text{ g}}\right) = 0.555\text{ mol}$, so to melt 10.0 g water, (0.555 mol)(601 kJ/mol) = 3.33 kJ. (5.3)
9. b. The heat required, $q = nc\Delta T$. n = number of moles = $5.00\text{ g} \times \left(\frac{1\text{ mol}}{18.01\text{ g}}\right) = 0.278\text{ mol}$.
   c = heat capacity of water = 75.3 J/mol,
   $\Delta T = 37 - 4 = 33°\text{C}$. Therefore,
   $q = (0.278\text{ mol})(75.3\text{ J/mol}\cdot°\text{C})(33°\text{C}) = 691\text{ J}$.
   (5.5)

10. c. To turn the water to steam, first energy must be used to raise the temperature to the boiling point of water, 100°C. Next, energy must be used to turn the water to steam. The energy required to raise the temperature from 25°C to 100°C is $q = nc\Delta T$.

n = moles of water = $10.0 \text{ g} \times \left(\frac{1 \text{ mol}}{18.01 \text{ g}}\right) =$

0.555 mol; c = 75.3 J/mol•°C; ΔT = 100 – 25 = 75°C, so (0.555)(75.3)(75) = 3134 J. The energy required to turn the water to steam (gas) is the heat of vaporization (40.76 kJ/mol). Thus for 0.555 mol it is (0.555 mol)(40.76 kJ/mol) = 22.62 kJ. To add to the energy used to raise the temperature, the units must be the same. 22.62 kJ = 22620 J. The total energy required is 22,620 J + 3134 J = 25,754 J = 25.75 kJ. (5.4)

11. d. The combustion of 1.00 g benzoic acid generates 26.38 kJ of heat. Therefore, 5.00 g × (26.38 kJ/g) = 131.9 kJ raises the temperature of the calorimeter

14.2°C. The heat capacity = $\frac{131.9 \text{ kJ}}{14.2 \text{ mol} \bullet °\text{C}}$ =

9.29 kJ/°C. (5.5)

12. c. The heat capacity of water is 75.3 J/mol•°C, specific heat is J/g•°C, so the conversion is

$75.3 \text{ J/mol} \bullet °\text{C} \times \left(\frac{1 \text{ mol}}{18.01 \text{ g}}\right) = 4.18 \text{ J/g} \bullet °\text{C}$. (5.5)

13. c. $q = nc\Delta T$. Since the number of moles and the amount of heat is the same, the change in temperature is inversely proportional to the heat capacity. Consequently, the substance with the lowest heat capacity will have the highest temperature change. (5.5)

14. b. $\Delta T = \frac{q}{nc}$. So a smaller temperature change will occur with a higher number of moles and higher heat capacity. Since gold has a higher molar mass than water, there are fewer moles of gold than water. Gold also has a lower heat capacity. Both factors will make gold increase in temperature more than water. The higher mass (therefore, higher number of moles) will have a smaller temperature change. (5.5)

15. b. $q = nc\Delta T$, so $5.00 \text{ g Al} \times \left(\frac{1 \text{ mol}}{26.98 \text{ g}}\right) = 0.185 \text{ mol}$; c = heat capacity = 24.4 J/mol•°C; ΔT = 50 – 30 = 20.

Therefore, q = (0.185)(24.4)(20) = 90 J. (5.5)

16. c. Calorie is an older unit of energy, so calorimetry means "measurement of energy." Hess's law is a theoretical relationship between different types of reaction, work is a specific type of energy (not a measurement) and enthalpy is the general name for heat flow (ΔH). (5.5)

17. b. Heat capacity = $q/\Delta T$. $q = 0.3550 \text{ g} \times \left(\frac{26.38 \text{ kJ}}{1 \text{ g}}\right) =$ 9.365 kJ. ΔT = 8.17°C. So heat capacity = $\frac{9.365 \text{ kJ}}{8.17 \text{ °C}} = 1.15 \text{ kJ/°C}$. (5.5)

18. d. $997 \text{ J} \times \left(\frac{1 \text{ kJ}}{1000 \text{ J}}\right) \times \left(\frac{1°\text{C}}{2.22 \text{ kJ}}\right) = 0.449°\text{C}$. Because energy is absorbed, the temperature will decrease. (5.5)

19. d. A formation reaction is the formation of 1 mole of substance from its elements at standard state.

Therefore, the correct answer is $K(s) + \frac{1}{2}Br_2(\ell) \rightarrow KBr(s)$. (5.6)

20. d. All halogens are diatomic in their standard state. (5.6)

21. c Argon gas is an element in its standard state; therefore, its enthalpy of formation is zero. (5.6)

22. b. Reversing the reaction reverses the sign on ΔH. (5.8)

23. d. The equation was reversed and multiplied by a factor of 2. Therefore, the sign on ΔH changes and its value is multiplied by two. (5.8)

24. a. A combustion reaction is the reaction of an organic substance (like a hydrocarbon) with oxygen to make the gases carbon dioxide and water. (5.7)

25. d. The first law of thermodynamics is also the law of conservation of energy, which can be stated as "energy is neither created nor destroyed." Thus energy must go somewhere (heat lost = heat gained). (5.2 and 5.5)

ADDITIONAL PRACTICE PROBLEM ANSWERS

1. $\Delta V = 50 - 22.4 = 27.6$ L

   $P = 1$ atm (standard pressure)

   work = $P\Delta V$ at constant pressure = (1.00 atm)(27.6 L)

   $= 27.6 \text{ L} \bullet \text{atm} \times \left(\frac{101.32 \text{ J}}{1 \text{ L} \bullet \text{atm}}\right) = 2.80 \times 103 \text{ J} = 2.80 \text{ kJ}$

   Note: The three significant figures are from the standard molar volume and the final volume of helium.

2. $100.0 \text{ mL} \times \left(\frac{1.00 \text{ g}}{1 \text{ mL}}\right) \times \left(\frac{1 \text{ mol}}{18.01 \text{ g}}\right) = 5.552 \text{ mol}$

   $\Delta H_{fus} = 601$ kJ/mol

   $c_{water} = 75.3$ J/mol•°C

   First, the energy removed to cool the water from 25°C to 0°C:

   $q = nc\Delta T = (5.552 \text{ mol})(75.3 \text{ J/mol}\bullet°\text{C})(25°\text{C}) =$

   $10452.5 \text{ J} = 1.0 \times 10^4$ J. (Only two digits are significant, limited by temperature.)

   $10452.5 \text{ J} \times \left(\frac{1 \text{ kJ}}{1000 \text{ J}}\right) = 10 \text{ kJ}$

   Then the energy removed to change the liquid to a solid is

   $q = (\Delta H_{fus})n = (6.01 \text{ kJ/mol})(5.552 \text{ mol}) = 43.8 \text{ kJ}$

   The energy removed for freezing is the sum. When adding, the values must have the same units:

   $43.8 \text{ kJ} + 10 \text{ kJ} = 44 \text{ kJ}$

   This is the amount of energy removed, not added!

3. The formation reaction is

   $\text{Na}(s) + \frac{1}{2}\text{Cl}_2(g) \rightarrow \text{NaCl}(s)$

   $4.99 \text{ g Cl}_2 \times \left(\frac{1 \text{ mol Cl}_2}{70.90 \text{ g}}\right) \times \left(\frac{1 \text{ mol reaction}}{0.5 \text{ mol Cl}_2}\right) \times$

   $\left(\frac{411.62 \text{ kJ}}{1 \text{ mol}}\right) = 57.9$ kJ evolved. (The negative sign on the ΔH denotes that energy is evolved and not consumed, but energy itself is not truly negative.)

4. The heat capacity of the calorimeter is determined from the reaction of benzoic acid (BA) is

   C_P = heat capacity = $q/\Delta T$

   $1.638 \text{ g BA} \times \left(\frac{26.38 \text{ kJ}}{1 \text{ g}}\right) = 43.21 \text{ kJ}$

   $\Delta T = 25.3°\text{C}$

   $C_P = \frac{43.21 \text{ kJ}}{25.3°\text{C}} = 1.71 \text{ kJ/°C}$

   The ΔH_{rxn} will be negative since heat is evolved (the temperature goes up):

   $\Delta H_{rxn} = -q/mol$

   heat evolved = $q = c\Delta T = (1.71 \text{ kJ/°C})(25.3°\text{C}) =$ 48.7 kJ

   moles of $C_6H_{12} = 1.142 \text{ g} \times \left(\frac{1 \text{ mol}}{84.16 \text{ g}}\right) = 0.0136 \text{ mol}$

   $\Delta H_{rxn} = \frac{-48.7 \text{ kJ}}{0.0136 \text{ mol}} = -3580 \text{ kJ/mol} =$

   -3.58×10^3 kJ/mol

5. According to the first law of thermodynamics, heat lost = heat gained. So all the energy lost by the hot water is absorbed by the cold water.

   heat lost = heat gained

   $q_{lost} = q_{gained}$

   $10.0 \text{ g} \times \frac{1 \text{ mol}}{18.01 \text{ g}} = 0.555 \text{ mol}$

   $50.0 \text{ g} \times \frac{1 \text{ mol}}{18.01 \text{ g}} = 2.78 \text{ mol}$

   $(0.555 \text{ mol})\,c\,(41.1 - T) = (2.78)\,c\,(T - 20.3)$

   Since it is water for both, c is the same and cancels:

   $(0.555)(41.1 - T) = 2.78(T - 20.3)$

   $22.81 - 0.555T = 2.78T - 56.434$

   $79.24 = 3.335T$

   $23.8 = T$

   The final temperature is 23.8°C.

6. $\Delta H_{rxn} = \Sigma\, n\Delta H^\circ_{f,product} - \Sigma\, n\Delta H^\circ_{f,reactant}$

   a. $2\ C_4H_{10} + 13\ O_2 \rightarrow 8\ CO_2 + 10\ H_2O$

   $\Delta H = [(8)(-393.5) + (10)(-241.8)] - [(2)(-125.7) + (13)(0)] = [-3148 - 2418] - [-251.4] = -5314.6$ kJ/mol

   b. $2\ CO + O_2 \rightarrow 2\ CO_2$

   $\Delta H = [2(-393.5)] - [(2)(-110.5) + 0] = [-787.0] - [-221.0] = -566.0$ kJ/mol

   c. $C_2H_6 + O_2 \rightarrow 2\ CO_2 + 3\ H_2$

   $\Delta H = [2(-110.5) + 3(0)] - [-84.7 + 0] = -221.0 + 84.7 = -136.3$ kJ/mol

7. To make the reaction

   $$Fe_2O_3 + 3\ CO \rightarrow 2\ Fe + 3\ CO_2$$

   from the following reactions with their given values of ΔH

$2\ Fe + 3/2\ O_2 \rightarrow Fe_2O_3$	$\Delta H = -824.2$ kJ/mol
$C + 2\ H_2 \rightarrow CH_4$	$\Delta H = -74.8$ kJ/mol
$H_2 + 1/2\ O_2 \rightarrow H_2O$	$\Delta H = -241.8$ kJ/mol
$CH_4 + 2\ O_2 \rightarrow CO_2 + 2\ H_2O$	$\Delta H = -782.4$ kJ/mol
$C + H_2O \rightarrow CO + H_2$	$\Delta H = +131$ kJ/mol

   Change the reactions by

   reverse $Fe_2O_3 \rightarrow 2\ Fe + 3/2\ O_2$ $\Delta H = +824.2$ kJ/mol

   reverse and 3× $3\ CO + 3\ H_2 \rightarrow 3\ C + 3\ H_2O$
   $\Delta H = -393$ kJ/mol

   3× $3\ CH_4 + 12/2\ O_2 \rightarrow 3\ CO_2 + 6\ H_2O$
   $\Delta H = -2347.2$ kJ/mol

   3× $3\ C + 6\ H_2 \rightarrow CH_4$ $\Delta H = -224.4$ kJ/mol

   reverse and 9× $9\ H_2O \rightarrow 9\ H_2 + 9/2\ O_2$
   $\Delta H = -2176.2$ kJ/mol

   add together $Fe_2O_3 + 3\ CO \rightarrow 2\ Fe + 3\ CO_2$
   $\Delta H = +35.8$ kJ/mol

CHAPTER 6 Properties of Gases: The Air We Breathe

OUTLINE

REVIEW

Chapter Overview

Gases, substances that have variable volume and shape, have several other properties that make them distinctive from liquids and solids. The volume of a gas is very dependent on conditions of temperature and pressure. Gases are always completely **miscible** (will mix in any proportions) in each other. Gases have a much lower density than liquids and solids. These properties are explained with the atomic view of gases as molecules that are far apart from each other and in constant motion.

One particular property of gases is the force they exert on their containers called pressure. **Pressure** is the force per unit area. With gases, this force comes from gas molecules hitting the side of the container. The most common gas, air, is actually a mixture of gases. Its pressure is measured with a **barometer**. Barometers compare the force being applied by air to the force applied by a column of liquid, normally mercury. The higher the column, the more force applied by both the liquid and the air. If the column of liquid is used to measure a gas other than air, it is called a **manometer**.

Because of this method of measuring gas pressure, one unit of pressure is **mmHg**, referring directly to the height of the mercury column. Another name for this unit is **torr**. The average atmospheric pressure at sea level is 760 torr. This leads to another unit of pressure, **atmosphere (atm)** where 1 atm is exactly equal to 760 torr. **Pascals (Pa)** are the SI unit of pressure and based on the definition of pressure as force per unit area so are equivalent to the unit kg/m•s^2. 1 atm = 101.325 kPa. A related unit is a bar. **Bars or millibars** are the unit of pressure typically used in meteorology. 10 mbar = 1 kPa; 1 atm = 1013.25 mbar.

Changing the one condition under which a gas exists can dramatically affect the others. The conditions generally related to each other are pressure (P), volume (V), number of molecules (n), and temperature (T), and the relationships apply regardless of the identity of the gas. If the temperature and amount of gas are kept constant, and pressure is increased, volume will decrease. This is **Boyle's law**, that pressure and volume are inversely proportional. **Charles' law** states that volume is proportional to **absolute temperature** (Kelvin units) when amount of gas and pressure is kept constant. **Amonton's law** states that pressure is proportional to absolute temperature when volume and amount of gas is kept constant. **Avogadro's law** states that amount of gas is proportional to volume under conditions of constant pressure and temperature.

These laws can be combined into the **general gas equation**, also called the **combined gas law**. Because all the variables are included in this gas law, it can be rearranged into the **ideal gas law**: $PV = nRT$, where R represents the gas constant. This a universal constant and its value depends only on its units. By rearranging the ideal gas law, these experimental parameters can be rearranged to solve for density and molar mass. The relationship to moles allows this equation to be used in stoichiometric relationships.

A useful reference point in studying the properties of gases is called **standard temperature and pressure (STP)**. Standard temperature is 0°C (273.15 K) and standard pressure is officially 10^5 Pa, but 1 atm is generally used instead. At STP, the volume of one mole of gas is 22.4 L. This is referred to as the **standard molar volume**.

Since the identity of the gas is irrelevant to the gas laws, the laws work as well for mixture of gases as for a single gas. Thus the total pressure is proportional to the total number of moles. Since all molecules of gas in a mixture must have the same temperature and volume, the gases are differentiated by their pressure. The pressure due to each gas in a mixture is called its partial pressure. The sum of the partial pressures is equal to the total pressure **(Dalton's law)**. A variation on Dalton's law can be used to relate partial pressures to mole fraction. The **mole fraction (X_A)** is the ratio of moles of gas A to the total number of moles in the mixture. The partial pressure of gas A is the product of the mole fraction and the total pressure, $P_A = X_A \cdot P_{total}$.

A gas formed by molecules escaping from a liquid is called a **vapor**. In a closed container containing a pure liquid, the partial pressure of the vapor above its liquid **(vapor pressure)** depends only on the type of vapor and temperature. Many gases are collected over water, thus the gas collected also contains water vapor. However, Dalton's law can be used to differentiate the vapor pressure of water from the pressure of the gas. Since the vapor pressure of water depends only on temperature, its value can be obtained from the appropriate table and subtracted from the total pressure giving the partial pressure of the collected gas. The ideal gas law can then be used to relate the partial pressure of the gas to moles of that gas.

All the gas laws simply state the observed relationships between pressure, volume, temperature, and moles. The theory that explains these relationships is the **kinetic molecular theory**. The kinetic molecular theory describes gases as molecules in independent constant motion. The molecules are neither attracted to nor repelled from each other and do not take up a significant amount of the volume. In addition, the molecules have **elastic collisions**, meaning that energy is not lost in collisions with either the container or another gas molecule. Finally, the temperature is proportional to the average kinetic energy of the molecules.

The relationship between temperature and kinetic energy has several implications. Kinetic energy depends on both the mass and velocity of the molecules. However, not all molecules have the same kinetic energy; therefore, not all molecules will have the same velocity. The distribution of energy (or velocity) is not symmetric, so there are several ways to express the velocity of molecules. These include the most probable speed (u), the velocity at which most molecules are traveling; the average speed (u_{avg}), which is the actual arithmetic average; and the **root-mean-square speed (u_{rms})**, which is the velocity of a molecule having the average kinetic energy. Since there is no such thing as a negative kinetic energy or a negative velocity, the Kelvin temperature scale is *always* used in gas law calculations. The velocity can be determined by the rate of **effusion**, the rate at which gas escapes through a pinhole or **diffusion**, the rate of spreading of a gas. At a constant temperature, the rate of effusion or diffusion is proportional to the square root of the molar mass. The relationship between molar mass and rate of effusion (or u_{rms}) is called **Graham's law of effusion**.

Gases as described by the kinetic molecular theory perfectly follow the relationships stated in all the gas laws and are called **ideal gases**. However, molecules of real gases are often attracted to each other. This attraction is greater at low temperatures (when molecular speeds are slow) and high pressures (when there are frequent collisions between molecules). The molecules do take up some space as well. That space would be significant at low overall volumes. These effects are accounted for with **van der Waals equation**, which introduces correction terms based on the identity and quantity of gas present.

Worked Examples

Atmospheric Pressure

Example 1.
Convert 501 mmHg into torr, atmospheres, pascals, and millibars.

Collect and Organize We are asked to convert a pressure given in units of mmHg into several other units typically used to express pressure.

Analyze
In order to perform the requested unit conversion we use the following relations:

$$1 \text{ torr} = 1 \text{ mmHg}$$

$$1 \text{ atm} = 760 \text{ mmHg}$$

$$1 \text{ atm} = 1.01325 \times 10^5 \text{ Pa}$$

$$1 \text{ bar} = 1.0 \times 10^5 \text{ Pa}$$

Solve 501 mmHg = 501 torr,

$$501 \text{ mmHg} \times \left(\frac{1 \text{ atm}}{760 \text{ mmHg}}\right) = 0.659 \text{ atm}$$

Significant figures matter. There are three in the initial value, and the conversion is exact so it has infinite significant figures, resulting in three significant figures in the final answer. For the other units, the conversions are

$$501\text{ mmHg} \times \left(\frac{1\text{ atm}}{760\text{ mmHg}}\right) \times \left(\frac{1.01325 \times 10^5\text{ Pa}}{1\text{ atm}}\right) = 6.68 \times 10^4\text{ Pa}$$

$$501\text{ mmHg} \times \left(\frac{1\text{ atm}}{760\text{ mmHg}}\right) \times \left(\frac{1.01325 \times 10^5\text{ Pa}}{1\text{ atm}}\right) \times \left(\frac{1\text{ bar}}{1.0 \times 10^5\text{ Pa}}\right) \times \left(\frac{1000\text{ mbar}}{1\text{ bar}}\right) = 668\text{ mbar}$$

THINK ABOUT IT There are a number of units that can be used to express pressure. Care must be taken to use consistent units when undertaking calculations.

EXAMPLE 2.
Convert 0.511 atm into torr, Pa, and millibar.

COLLECT AND ORGANIZE We are asked to convert a pressure given in units of atm into several other units typically used to express pressure.

ANALYZE In order to perform the requested unit conversion we use the following relations:

$$1\text{ atm} = 760\text{ torr}$$

$$1\text{ atm} = 1.01325 \times 10^5\text{ Pa}$$

$$1\text{ bar} = 1.0 \times 10^5\text{ Pa}$$

SOLVE

$$0.511\text{ atm} \times \left(\frac{760\text{ torr}}{1\text{ atm}}\right) = 388\text{ torr}$$

$$0.511\text{ atm} \times \left(\frac{1.01325 \times 10^5\text{ Pa}}{1\text{ atm}}\right) = 5.18 \times 10^4\text{ Pa}$$

$$0.511\text{ atm} \times \left(\frac{1.01325 \times 10^5\text{ Pa}}{1\text{ atm}}\right) \times \left(\frac{1\text{ kPa}}{1000\text{ Pa}}\right) \times \left(\frac{10\text{ mbar}}{1\text{ kPa}}\right) = 518\text{ mbar}$$

THINK ABOUT IT The choice pressure unit typically depends on the magnitude of the pressure. For moderate pressures it is much "cleaner" to use atm rather than Pa, which contains an exponent. The reverse would be true of very low pressures where it might be more convenient to use Pa rather than the large negative exponent that would be necessary if atm were used.

THE GAS LAWS

EXAMPLE 1.
If the pressure on 100.0 mL of gas is changed from 1.0 atm to 680 torr, what will the volume be?

COLLECT AND ORGANIZE Given an initial volume of gas (V_{initial} = 100.0 mL) at an initial pressure (P_{initial} = 1.0 atm), we are asked to determine the new volume, V_{final}, resulting from changing the pressure to P_{final} = 680 torr. One assumes for this problem that the temperature and composition (number of moles) remain constant. If anything other than pressure and volume were changing, it would be mentioned in the problem. It is not practical to list everything that doesn't happen!

ANALYZE The combined gas law $\frac{P_{\text{initial}}V_{\text{initial}}}{T_{\text{initial}}} = \frac{P_{\text{final}}V_{\text{final}}}{T_{\text{final}}}$, can be used to analyze the change in the given variables. Because T is assumed to remain constant it cancels from each side of the relation giving $P_{\text{initial}}V_{\text{initial}} = P_{\text{final}}V_{\text{final}}$. We must also be sure that consistent units are used for each variable. Therefore, it is also necessary to first express the initial and final pressures in the same pressure unit.

SOLVE We must first convert the given pressures into the same units.

$$P_{\text{final}} = 680\text{ torr} \times \left(\frac{1.0\text{ atm}}{760\text{ torr}}\right) = 0.895\text{ atm}$$

Rearranging $P_{\text{initial}}V_{\text{initial}} = P_{\text{final}}V_{\text{final}}$ to solve for V_{final} and inserting the given values of P and V we obtain

$$V_{\text{final}} = \frac{P_{\text{initial}}V_{\text{initial}}}{P_{\text{final}}} = \frac{(1.0\text{ atm})(100.0\text{ mL})}{(0.895\text{ atm})} = 111.7\text{ mL}$$

THINK ABOUT IT The equation obtaining by cancelling T from the combined gas law is a variation of Boyle's law. Since Boyle's law states that pressure and volume are inversely proportional you can quickly check your answer. Since the pressure decreased, the volume should increase. Since the final volume is greater than the initial volume, there is some assurance that the math was done correctly.

EXAMPLE 2.
At what temperature will 0.860 L of gas at 9°C expand to 1.00 L at constant pressure?

COLLECT AND ORGANIZE Given an initial volume of gas (V_{initial} = 0.860 L) at an initial temperature (T_{initial} = 9°C), we are asked to determine the temperature required, T_{final}, to cause an increase in the volume to V_{final} = 1.00 L while keeping pressure constant.

ANALYZE The combined gas law, $\frac{P_{initial}V_{initial}}{T_{initial}} = \frac{P_{final}V_{final}}{T_{final}}$, can be used to analyze the change in the given variables. Since P is constant during the process, it can be eliminated from both sides of the equation yielding $\frac{V_{initial}}{T_{initial}} = \frac{V_{final}}{T_{final}}$. We must also express temperatures in the Kelvin unit, so it will be necessary to convert the initial temperature from °C to K using the relation: K = °C + 273.15.

SOLVE We must first convert the initial temperature into the proper units.

$$T(\text{K}) = 9°\text{C} + 273 = 282 \text{ K}$$

Rearranging $\frac{V_{initial}}{T_{initial}} = \frac{V_{final}}{T_{final}}$ to solve for T_{final} and inserting the given values of T and V we obtain

$$T_{final} = \frac{T_{initial}V_{final}}{V_{initial}} = \frac{(282 \text{ K}) \times (1.00 \text{ L})}{(0.860 \text{ L})} = 328 \text{ K}$$

The answer has three significant figures because each number used in the equation had three significant figures. Although the temperature value has only one significant figure in Celsius, it is converted to Kelvin by adding. The significant figure rule for addition counts decimal places, not significant figures. Therefore, the temperature value in Kelvin really does have three significant figures. It is the value with three significant figures that is multiplied and divided in the gas law.

THINK ABOUT IT Check the answer by recalling that volume and temperature are directly proportional. Since volume increased, temperature should also increase.

Whenever temperature is used in a gas law, it must be in units of Kelvin.

EXAMPLE 3.
What is the final pressure when the conditions on 20.8 mL of neon gas at STP are changed to 0.800 L and 23°C?

COLLECT AND ORGANIZE Given 20.8 mL of gas initially at STP (standard temperature and pressure, $T_{initial}$ = 273.15 K, $P_{initial}$ = 1 atm) we are asked to determine the final pressure P_{final} when the volume and temperature change to V_{final} = 0.800 L and T_{final} = 23°C.

ANALYZE After the appropriate conversion of units, the combined gas law, $\frac{P_{initial}V_{initial}}{T_{initial}} = \frac{P_{final}V_{final}}{T_{final}}$, can be used to analyze the change in the given variables. Temperature must be Kelvin. The units of pressure and volume must be the same for initial and final conditions.

SOLVE Rearranging the combined gas law

$\frac{P_{initial}V_{initial}}{T_{initial}} = \frac{P_{final}V_{final}}{T_{final}}$ to solve for P_{final} yields:

$$P_{final} = \frac{P_{initial}V_{initial}T_{final}}{T_{initial}V_{final}}$$

Converting units leads to:

$$V_{initial} = 20.8 \text{ mL} \times \left(\frac{1 \text{ L}}{1000 \text{ mL}}\right) = 0.0208 \text{ L}$$

$$T_{final} = 23°\text{C} + 273.15 = 296 \text{ K}$$

Using the given values in the expression for P_{final} obtained by rearranging the combined gas law,

$$P_{final} = \frac{(1 \text{ atm})(0.0208 \text{ L})(296 \text{ K})}{0.800 \text{ L}}$$

$$P_{final} = 0.282 \text{ atm}$$

The volume and temperature measurements all have three significant figures, so the final answer has three significant figures. STP is a defined value and is, therefore, exact and has infinite significant figures.

THINK ABOUT IT The volume increases, so the prediction is that pressure will decrease. Temperature increased. Since temperature is proportional to pressure, the pressure should increase. Since one change predicts one direction and the other change predicts a different direction, the actual change depends on which effect is more important. In this case, you just have to do the math and see. The direction cannot be predicted without it.

THE IDEAL GAS LAW

EXAMPLE 1.
What is the pressure of 0.363 g of N_2 in 750.0 mL at 24°C?

COLLECT AND ORGANIZE We are asked to determine the pressure exerted by 0.363 g of N_2 contained in a volume of 750.0 mL at a temperature of 24°C. Assuming that the gas behaves ideally we can use the ideal gas law, $PV = nRT$, to evaluate the pressure. The most commonly used value of the gas constant R is 0.08206 L•atm/mol•K. When this number is used, it is necessary to express the amount of gas in moles, the volume in liters, and the temperature in Kelvin.

ANALYZE After the appropriate conversion of units, the ideal gas law, $PV = nRT$, can be used to determine the pressure of the system.

SOLVE Making the necessary conversions results in:

$$V = 750.0 \text{ mL} \times \left(\frac{1 \text{ L}}{1000 \text{ mL}}\right) = 0.7500 \text{ L}$$

$$n = 0.363 \text{ g } N_2 \times \left(\frac{1 \text{ mol}}{28.03 \text{ g}}\right) = 0.012955 \text{ mol}$$

(Only three digits are significant.)

$$T = 24°\text{C} + 273.15 = 297 \text{ K}$$

(Significant figures depend on decimal places.)

Rearranging the ideal gas law to solve for P and inserting the values above gives:

$$P = \frac{nRT}{V}$$

$$= (0.012955 \text{ mol})(0.08206 \text{ L}\cdot\text{atm/mol}\cdot\text{K})\frac{(297 \text{ K})}{(0.7500 \text{ L})} = 0.421 \text{ atm}$$

The temperature and the moles have three significant figures; R and V have four. Therefore, the answer has three significant figures.

THINK ABOUT IT The ideal gas law is used when there is one set of conditions. The combined gas law is used when one or more of the conditions are changing. When working with the ideal gas law, we must always ensure that the units match the units of the gas constant.

EXAMPLE 2.
What is the molar mass of 0.7314 g of gas in 100.0 mL at STP?

COLLECT AND ORGANIZE We are asked to determine the molar mass of a 0.7314 g sample of gas contained in a volume of 100.0 mL at STP (standard temperature and pressure, $T = 273.15$ K, $P = 1$ atm).

ANALYZE We are given P, V, and T, so the ideal gas law can be used to solve for n, the number of moles, after the appropriate unit conversion of the volume. The molar mass of a substance is defined as the number of grams per mole. We can determine the molar mass of the substance using the given mass of gas, 0.7314 g, and the number of moles determined using the ideal gas law.

SOLVE

$$V = 100.0 \text{ mL} \times \left(\frac{1 \text{ L}}{1000 \text{ mL}}\right) = 0.1000 \text{ L}$$

$$PV = nRT$$

$$n = \frac{PV}{RT} = \frac{(1 \text{ atm})(0.1000 \text{ L})}{(0.08206 \text{ L}\cdot\text{atm/mol}\cdot\text{K})(273.15 \text{ K})}$$

$$n = 4.461 \times 10^{-3} \text{ mol}$$

$$\text{molar mass} = \frac{0.7314 \text{ g}}{4.461 \times 10^{-3} \text{ mol}} = 163.9 \text{ g/mol}$$

The final answer is expressed with four significant figures since n, R, and V have four. T and P have infinite significant figures since STP is a defined value.

THINK ABOUT IT You can use the formula for molar mass with the ideal gas law to derive a single equation to calculate this value.

EXAMPLE 3.
When calcium carbonate is strongly heated, it will produce calcium oxide and carbon dioxide gas. If 0.5707 g calcium carbonate completely reacts, what volume of carbon dioxide will be produced at 99.40°C and 731.8 torr?

COLLECT AND ORGANIZE We are asked to determine the volume of CO_2 that will be produced at 99.40°C and 731.8 torr if 0.5707 g calcium carbonate completely reacts according to the reaction:

$$CaCO_3 \rightarrow CaO + CO_2$$

ANALYZE The conditions refer to the gas, which is carbon dioxide. The gas laws assume that all variables refer to the same substance. Only one set of conditions is given, so the ideal gas law is used. We must convert the given temperature and pressure to the appropriate units for the ideal gas equation and determine the number of moles of CO_2 produced from the given amount of $CaCO_3$ using stoichiometry.

SOLVE Converting pressure and temperature to the appropriate units results in:

$$P = 731.8 \text{ torr} \times \left(\frac{1 \text{ atm}}{760 \text{ torr}}\right) = 0.9629 \text{ atm}$$

$$T = 99.40°\text{C} + 273.15 = 372.55 \text{ K}$$

Moles of CO_2 are not given directly but can be determined from the stoichiometric relationships:

$$0.5707 \text{ g } CaCO_3 \times \left(\frac{1 \text{ mol } CaCO_3}{100.09 \text{ g}}\right) \times \left(\frac{1 \text{ mol } CO_2}{1 \text{ mol } CaCO_3}\right) = 0.005702 \text{ mol}$$

Using these values in the ideal gas law $PV = nRT$,

$$(0.9629 \text{ atm})V = (0.005702 \text{ mol})(0.08206 \text{ L}\cdot\text{atm/mol}\cdot\text{K})(372.55 \text{ K})$$

$$V = 0.1810 \text{ L}$$

Four significant figures is correct, since P, n, and R have four. Temperature has five significant figures.

THINK ABOUT IT This problem involves a reaction. It gives information about one substance and asks about a different one. This means that stoichiometric relationships will have to be used.

Gas Density

Example 1.
What is the density of F_2 gas at 513 torr and 33°C?

Collect and Organize We are asked to determine the density of a sample of F_2 gas given that the sample has a pressure of 513 torr and a temperature of 33°C.

Analyze The density (d) of a material is defined as the mass (m) of sample per volume ($d = m/V$). The fact that we are given the identity (F_2), temperature, and pressure of a gas and are asked to determine the density, a quantity which contains the variables of amount and volume, suggests that the ideal gas law might be relevant.

Solve The molar mass ($\mathcal{M}$) of a substance, $\mathcal{M} = m/n$, can be solved for n, the number of moles, and used in the ideal gas equation.

$$PV = (m/\mathcal{M})RT$$

Solving for m/V gives:

$$P\mathcal{M}/RT = m/V = \text{density}$$

From the periodic table we find for the molar mass of F_2:

$$\mathcal{M} = 19.00(2) = 38.00 \text{ g/mol}$$

Performing the necessary unit conversions gives:

$$P = 513 \text{ torr} \times \left(\frac{1 \text{ atm}}{760 \text{ torr}}\right) = 0.675 \text{ atm}$$

$$T = 33°\text{C} + 273.15 = 306 \text{ K}$$

Finally, inserting these values into the expression derived for the density, $d = m/V$

$$= \frac{P(\mathcal{M})}{RT} = \frac{(0.675\text{atm})(38.00 \text{ g/mol})}{(0.08206 \text{ L}\cdot\text{atm/mol}\cdot\text{K})(306 \text{ K})}$$

$$= 1.02 \text{ g/L}$$

Both pressure and temperature limit the number of significant figures to three.

Think About It The ideal gas law provides an experimental way to determine the molar mass or the density of a sample of pure unknown gas.

Example 2.
What is the molar mass of a gas with a density of 1.39 g/L at STP?

Collect and Organize We are asked to determine the molar mass of a gas given its density under the conditions of standard temperature and pressure (P = 1 atm, T = 273.15 K).

Analyze There are two ways to solve this problem. Method 1 is to use the expression for density derived in the previous example, $d = P\mathcal{M}/RT$, to get $\mathcal{M} = dRT/P$. Method 2 would be to recall that the volume of 1 mole of gas at STP = 22.4 L.

Solve

Method 1 $\quad \mathcal{M} = dRT/P$

$$= \frac{(1.39 \text{ g/L})(0.08206 \text{ L}\cdot\text{atm/mol}\cdot\text{K})(273.15 \text{ K})}{(1 \text{ atm})}$$

$$= 31.2 \text{ g/mol}$$

Method 2 By assuming 1 mole of gas at STP,
1 mol of gas = its molar mass in grams
1 mol gas = 22.4 L

$$\text{Therefore, density} = \frac{\text{molar mass}}{22.4 \text{ L}}$$

$$1.39 \text{ g/L} = \mathcal{M}/22.4 \text{ L}$$

$$31.1 \text{ g/mol} = \mathcal{M}$$

Think About It The second method will work only at STP, the first at any temperature and pressure.

Dalton's Law and Mixtures of Gases

Example 1.
What is the total pressure of a mixture of oxygen, nitrogen, and helium if their partial pressures are P_{O_2} = 0.057 atm, P_{N_2} = 0.535 atm, and P_{He} = 592 torr?

Collect and Organize We are asked to determine the total pressure of a mixture of oxygen, nitrogen, and helium, given the partial pressures of each of the constituent gases.

Analyze The total pressure of a mixture of gases is given as

$$P_T = P_A + P_B + P_C + \cdots$$

where P_A, P_B, P_C represent the partial pressures of gases A, B, C comprising the gas mixture. This is Dalton's law.

Solve We must ensure that the pressure units for each of the constituent gases are the same. In this example, the pressure of nitrogen and oxygen are in atmospheres and the pressure of helium is in torr. Since it is easier to change one pressure than two

$$P_{He} = 592 \text{ torr} \times \left(\frac{1 \text{ atm}}{760 \text{ torr}}\right) = 0.779 \text{ atm}$$

$$P_T = 0.535 \text{ atm} + 0.057 \text{ atm} + 0.779 \text{ atm} = 1.371 \text{ atm}$$

Since each of the values has three decimal places, the answer should have three decimal places.

THINK ABOUT IT Dalton's law is used for mixtures of gases. Different components of a mixture are measured with either the moles of that gas or the pressure of the gas. Because gases fill all the available space, each component occupies the entire volume. The various components of the mixture will all exist at the same temperature.

EXAMPLE 2.
For a mixture of 0.258 g CO_2, 0.935 g Ne, and 0.682 g O_2, what is the mole fraction of each component?

COLLECT AND ORGANIZE We are given a mixture containing three gases. The masses of each of the three components are given and we are asked to calculate the mole fraction of each component.

ANALYZE The mole fraction of component i in a mixture is $X_i = n_i/n_{\text{total}}$.

SOLVE Since the mole fraction is the ratio of moles of each substance, each substance must be converted to moles.

$$0.258 \text{ g } CO_2 \times \left(\frac{1 \text{ mol}}{44.01 \text{ g}}\right) = 0.00586 \text{ mol}$$

$$0.935 \text{ g Ne} \times \left(\frac{1 \text{ mol}}{20.18 \text{ g}}\right) = 0.0463 \text{ mol}$$

$$0.682 \text{ g } O_2 \times \left(\frac{1 \text{ mol}}{32.00 \text{ g}}\right) = 0.0213 \text{ mol}$$

Total number of moles = 0.00586 + 0.0463 + 0.0213 = 0.0735 mol (To determine significant figures for addition count decimal places.)

$$X_{CO_2} = \frac{n_{CO_2}}{n_{\text{total}}} = \frac{0.00586}{0.0735} = 0.0797$$

(Mole fraction does not have units. Moles cancel.)

$$X_{\text{Ne}} = \frac{n_{\text{Ne}}}{n_{\text{total}}} = \frac{0.046}{0.0735} = 0.630$$

$$X_{O_2} = \frac{n_{O_2}}{n_{\text{total}}} = \frac{0.0213}{0.0735} = 0.290$$

THINK ABOUT IT Mole fractions are a different type of concentration unit.

EXAMPLE 3.
If a mixture of 5.08 g N_2 and 6.29 g F_2 has a total pressure of 881 torr, what is the partial pressure of each gas?

COLLECT AND ORGANIZE For a mixture of two gases we are given both the total pressure of the mixture and the mass of each component comprising the mixture. We are asked to calculate the partial pressure of each component.

ANALYZE The partial pressure of a component making up a mixture is given by $P_A = X_A P_T$, where X_A is the mole fraction of component A.

SOLVE In this example P_T = 881 torr. The mole fraction of each gas is also needed.

$$5.08 \text{ g } N_2 \times \left(\frac{1 \text{ mol}}{28.02 \text{ g}}\right) = 0.181 \text{ mol } N_2$$

$$6.29 \text{ g } F_2 \times \left(\frac{1 \text{ mol}}{38.00 \text{ g}}\right) = 0.166 \text{ mol } F_2$$

total number of moles = 0.181 + 0.166 = 0.347 mol

$$\text{partial pressure of nitrogen} = X_{N_2} P_T = \left(\frac{0.181}{0.347}\right)(881 \text{ torr}) = 460 \text{ torr}$$

Since the total pressure is the sum of the partial pressures,

$$P_T = P_{N_2} + P_{F_2}$$

$$881 \text{ torr} = 460 \text{ torr} + P_{F_2}$$

$$421 \text{ torr} = P_{F_2}$$

Alternatively, we could have solved the problem using,

$$P_{F_2} = X_{F_2} P_T = \left(\frac{0.166}{0.347}\right)(881) = 421 \text{ torr}$$

THINK ABOUT IT
The equation $P_A = X_A P_T$ is a variation of Dalton's law.

EXAMPLE 4.
How many moles of oxygen were produced when 54.9 mL of oxygen gas were collected over water at 746 torr and 30°C?

COLLECT AND ORGANIZE
We are asked to determine the number of moles of oxygen that are contained in 54.9 mL of a mixture of oxygen and water vapor collected at a given temperature and pressure.

ANALYZE
The partial pressures of a component making up a mixture is given by $P_A = X_A P_T$, where X_A is the mole fraction of component A. Since the question gives only one set of conditions, the ideal gas law ($PV = nRT$) is used. Since the question asks about moles of oxygen, the partial pressure of oxygen must be used for the variable P. Since the oxygen was collected "over water," the gas has both oxygen and water vapor. Dalton's law of partial pressures can be used to separate the two. By consulting the table of water vapor pressures found in the text, the vapor pressure of water at 30°C is 31.8 mmHg (31.8 torr).

SOLVE
According to the problem, the total pressure is 746 torr. Dalton's law says

$$P_T = P_{O_2} + P_{H_2O}$$

$$746 \text{ torr} = P_{O_2} + 31.8 \text{ torr}$$

$$714 \text{ torr} = P_{O_2}$$

Using $R = 0.08206$ L•atm/mol•K, the other variables are

$$V = 54.9 \text{ mL} \times \left(\frac{1 \text{ L}}{1000 \text{ mL}}\right) = 0.0549 \text{ L}$$

$$T = 30°\text{C} + 273.15 = 303 \text{ K}$$

For n to represent moles of oxygen, the pressure must be the partial pressure of oxygen

$$P = 714 \text{ torr} \times \left(\frac{1 \text{ atm}}{760 \text{ torr}}\right) = 0.940 \text{ atm}$$

Using these values in the ideal gas law results in

$$PV = nRT$$

$$(0.949 \text{ atm})(0.0549 \text{ L}) = n(0.08206 \text{ L•atm/mol•K})(303 \text{ K})$$

$$0.00208 \text{ mol} = n = \text{moles of oxygen}$$

Think About It Dalton's law of partial pressures allows the partial pressure of water to be separated from the partial pressure of the gas collected. The gas laws can be used to relate pressure of one gas to moles of the same gas.

Example 5.
Lithium metal was added to water and the resulting hydrogen gas collected. If 86.0 mL of hydrogen gas was collected over the water at 25°C and 1.04 atm, how many grams of lithium reacted?

$$\text{Li} + \text{H}_2\text{O} \rightarrow \text{LiOH} + \text{H}_2$$

Collect and Organize We are asked to determine the amount of lithium consumed in the reaction described by the given stoichiometric equation if a sample of 86.0 mL of hydrogen gas and water vapor is collected at the given temperature and pressure.

Analyze Hydrogen gas was collected and the question was asked about lithium. Different substances are related by balanced chemical reactions. Since the gas was "collected over water," the pressure given must be the pressure of both hydrogen and water vapor.

Solve First, balance the equation.

$$2 \text{ Li} + 2 \text{ H}_2\text{O} \rightarrow 2 \text{ LiOH} + \text{H}_2$$

The information about hydrogen is volume, temperature, and pressure, so the ideal gas law can be used to get information about moles. To use the ideal gas law to obtain moles of hydrogen, the pressure must be partial pressure of hydrogen.

$$P_\text{T} = P_{\text{H}_2} + P_{\text{H}_2\text{O}}$$

The vapor pressure of water at 25°C is 23.8 mmHg, according to the table of water vapor pressures. The total pressure is 1.04 atm, according to the problem.

$$23.8 \text{ mmHg} \times \left(\frac{1 \text{ atm}}{760 \text{ mmHg}}\right) = 0.0313 \text{ atm} = P_{\text{H}_2\text{O}}$$

$$P_{\text{H}_2} = P_\text{T} - P_{\text{H}_2\text{O}} = 1.04 - 0.0313 = 1.01 \text{ atm}$$

$$R = 0.08206 \text{ L•atm/mol•K}$$

$$T = 25°\text{C} + 273.15 = 298 \text{ K}$$

$$V = 86.0 \text{ mL} \times \left(\frac{1 \text{ L}}{1000 \text{ mL}}\right) = 0.0860 \text{ L}$$

$$P_{\text{H}_2}V = n_{\text{H}_2}RT$$

$$(1.01 \text{ atm})(0.0860 \text{ L}) = n_{\text{H}_2}(0.08206 \text{ L•atm/mol•K})(298 \text{ K})$$

$$0.00355 \text{ mol} = n_{\text{H}_2}$$

Using the stoichiometric equation:

$$2 \text{ Li} + 2 \text{ H}_2\text{O} \rightarrow 2 \text{ LiOH} + \text{H}_2$$

$$0.00355 \text{ mol H}_2 \times \left(\frac{2 \text{ mol Li}}{1 \text{ mol H}_2}\right) \times \left(\frac{6.941 \text{ g}}{1 \text{ mol Li}}\right) = 0.0493 \text{ g Li}$$

Think About It Because many gases are collected over water, the gas collected is actually a mixture of water vapor and the gas collected. Therefore, the vapor pressure of water is of particular interest and values are tabulated.

The Kinetic Molecular Theory of Gases

Example 1.
If hydrogen (H_2) effuses at a rate of 1.5 m/s, at what rate does carbon dioxide effuse?

Collect and Organize We are given a rate of effusion of H_2 gas and are asked to determine the rate of effusion of CO_2 under the same conditions.

Analyze The relative rates of effusion of two gases x and y is calculated using:

$$\frac{r_x}{r_y} = \left(\frac{\mathcal{M}_y}{\mathcal{M}_x}\right)^{1/2}$$

SOLVE The molar mass of hydrogen is 2.02 g/mol. The molar mass of CO_2 is 44.01 g/mol. If you assign the lighter (or faster) molecule as *x*, you won't have to work with fractions.

$$\frac{1.5}{r_{CO_2}} = \left(\frac{44.01}{2.02}\right)^{1/2}$$

$$\frac{1.5}{r_{CO_2}} = 4.67$$

$$1.5 = 4.67 r_{CO_2}$$

$$0.32 \text{ m/s} = r_{CO_2}$$

THINK ABOUT IT The rate of effusion is related to molar mass and the average kinetic energy of the molecules. The equation compares two gases that are at the same temperature.

EXAMPLE 2.
What is the molar mass of a molecule if it effuses at one-eighth the rate of helium?

COLLECT AND ORGANIZE We are given the rate of effusion of unknown gas relative to helium and are asked to determine its molar mass.

ANALYZE We can solve the rate of effusion equation given in the previous example for the molar mass of the unknown gas.

SOLVE The molar mass of helium is 4.00 g/mol. Since the other molecule is slower, assign it as *y*. The problem says rate of $y = (1/8)r_{He}$, so

$$\frac{r_{He}}{\left(\frac{1}{8}\right)r_{He}} = 8 = \left(\frac{\mathcal{M}_y}{4.00}\right)^{1/2}$$

$$64 = \frac{\mathcal{M}_y}{4.00}$$

$$256 \text{ g/mol} = \mathcal{M}_y$$

THINK ABOUT IT The slower the rate of effusion the higher the molar mass of the gas.

REAL GASES

EXAMPLE 1.
What is the pressure of 3.45 g of nitrogen in a 10.00-mL container if it acts as a real gas at –15°C.

COLLECT AND ORGANIZE We are asked to calculate the pressure of a given mass of nitrogen gas contained in a given volume at a given temperature assuming that the gas deviates from ideal behavior. The *a* and *b* coefficients for nitrogen are $a = 1.39$ and $b = 0.0391$.

ANALYZE We will use the van der Waals equation which describes the nonideal behavior of real gases after finding the appropriate values of the a and b coefficients.

SOLVE

$$R = 0.082058 \text{ L} \cdot \text{atm/mol} \cdot \text{K}$$

P = unknown (what we're looking for)

$$n = 3.45 \text{ g} \times \left(\frac{1 \text{ mol}}{28.02 \text{ g}}\right) = 0.123 \text{ mol}$$

$$V = 10.00 \text{ mL} \times \left(\frac{1 \text{ L}}{1000 \text{ mL}}\right) = 0.01000 \text{ L}$$

$$T = -15°\text{C} + 273.15 = 258 \text{ K}$$

$$a = 1.39$$

$$b = 0.0391$$

$$\left[\frac{P + (0.123)^2(1.39)}{(0.01)^2}\right][0.01 - (0.123)(0.0392)] = (0.123)(0.082058)(258)$$

$$(P + 210.3)(0.00518) = 2.60$$

$$P + 210.3 = 502.7$$

$$P = 292 \text{ atm}$$

THINK ABOUT IT The van der Waals equation accounts for the nonideal behavior of real gases. The value *a* takes into account the attraction of molecules to each other and the value *b* reflects the amount of space taken up by the molecule.

Self-Test

KEY VOCABULARY COMPLETION QUESTIONS

_________ 1. Gas escaping through a hole

_________ 2. $PV = k$

_________ 3. Unit exactly equal to mmHg

_________ 4. Gas that takes up significant space

_________ 5. Unit of volume if the value of $R = 0.08206$

_________ 6. Gas spreading

_________ 7. 0°C and 1 atm

_________ 8. Pressure caused by one gas in a mixture of gases

__________ 9. Caused by the escape of molecules from the surface of a liquid

__________ 10. Proportional to the root-mean-square speed of gas molecules

__________ 11. SI unit for pressure

__________ 12. For a gas at a specific temperature, it is proportional to the square root of the gas's molar mass

__________ 13. 22.4 L

__________ 14. Force per unit area

__________ 15. Temperature scale without negative values

__________ 16. Gas that perfectly follows $PV = nRT$

__________ 17. Moles of gas divided by the total moles of gas in a mixture

__________ 18. Avogadro's law states this is proportional to volume

__________ 19. Proportional to temperature at constant pressure and moles

__________ 20. Unit of pressure when $R = 0.08206$

Multiple-Choice Questions

1. What is the density of xenon gas at STP?
   a. 22.4 L/mol
   b. 5.36 g/mL
   c. 5.86 g/L
   d. 0.171 g/mL

2. If you increase the temperature of a gas at constant pressure, the volume will
   a. increase
   b. decrease
   c. stay the same
   d. be impossible to predict without more information

3. If you increase the pressure of a gas at constant temperature, the volume will
   a. increase
   b. decrease
   c. stay the same
   d. be impossible to predict without more information

4. If you increase both pressure and temperature of a gas, the volume will
   a. increase
   b. decrease
   c. stay the same
   d. be impossible to predict without more information

5. If you increase the temperature and add more gas to a container at constant volume, the pressure will
   a. increase
   b. decrease
   c. stay the same
   d. be impossible to predict without more information

6. If the temperature increased from 15°C to 30°C at a constant volume, pressure will change from 1.00 atm to
   a. 2.00 atm
   b. 1.05 atm
   c. 0.950 atm
   d. 0.500 atm

7. How many moles of CO_2 are in 75.00 mL at 25.00°C and 771 torr?
   a. 0.00311 mol
   b. 2.36 mol
   c. 28.2 mol
   d. 0.0371 mol

8. A mixture of 0.14 g CO and 2.78 g CO_2 has a pressure of 1 atm. What is the partial pressure of CO?
   a. 0.93 atm
   b. 0.0050 atm
   c. 0.068 atm
   d. 0.073 atm

9. If helium effuses at a rate of 1.0 m/s, how fast will xenon effuse?
   a. 5.7 m/s
   b. 0.17 m/s
   c. 32.5 m/s
   d. 0.031 m/s

10. According to the kinetic molecular theory, gas molecules
    a. do not interact with each other
    b. take up a significant amount of space
    c. have sticky collisions
    d. all of the above

11. Which gas will behave most ideally?
    a. one at high pressure and high temperature
    b. one at low pressure and low temperature
    c. one at low pressure and high temperature
    d. one at high pressure and low temperature

12. A gas escaping through a small hole is called
    a. effusion
    b. diffusion
    c. solubility
    d. ideality

13. The weatherman says that the hurricane has a pressure of 920 millibar. What is the pressure in atmospheres?
    a. 0.920 atm
    b. 0.908 atm
    c. 9.1 atm
    d. 0.092 atm

14. If you add 2.00 g CO_2 to 5.00 g CO_2 at 500 torr, what is the new pressure?
    a. 199 torr
    b. 697 torr
    c. 700 torr
    d. 1255 torr

15. When the pressure is changed from 800 torr to 1.5 atm, the volume will change from 9.0 L to
    a. 4800 L
    b. 133 L
    c. 12.8 L
    d. 6.3 L

16. What is the mole fraction of neon in a mixture of 0.709 mol neon, 0.180 mol argon, and 0.963 mol xenon?
    a. 0.383
    b. 0.617
    c. 2.61
    d. 0.709

17. At higher temperatures, the speed of molecules
    a. increases
    b. decreases
    c. stays the same
    d. it depends on the pressure

18. The density of gas at STP is 1.25 g/L. What is its molar mass?
    a. 28.0 g/mol
    b. 30.6 g/mol
    c. 17.9 g/mol
    d. 125 g/mol

19. What halogen occupies 1.9 L at STP with a mass of 3.21 g?
    a. Ar
    b. Cl_2
    c. O_2
    d. F_2

20. The vapor pressure of water depends on
    a. the volume of liquid water
    b. the volume of space above the liquid
    c. temperature
    d. all of the above

21. If the pressure of O_2 in a 1.0-L container at 20°C is 0.37 atm, what will be the pressure on the container if 0.010 mol CO_2 is added to it?
    a. 0.24 atm
    b. 0.61 atm
    c. 0.37 atm
    d. 0.38 atm

22. If 50.00 mL of oxygen is collected over water at 20°C and 1.0 atm, how many moles of oxygen were collected?
    a. 0.0021 mol
    b. 0.0020 mol
    c 0.23 mol
    d. 1.54 mol

23. If neon has a volume of 50.00 mL at STP, what volume will it occupy at 20°C and 500 torr?
    a. 0.107 L
    b. 237 mL
    c. 0.313 mL
    d. 81.5 mL

24. Standard molar volume is the volume
    a. at 0°C
    b. at 760 torr
    c. 22.4 L
    d. all of the above

25. Which statement is not true of an ideal gas?
    a. It conforms to the ideal gas law.
    b. It contains molecules in constant motion.
    c. Molecules are attracted to each other.
    d. all of the above

26. Which statement is not true of a real gas?
    a. It conforms to the ideal gas law.
    b. It contains molecules in constant motion.
    c. Molecules are attracted to each other.
    d. all of the above

27. The mean free path of a gas molecule does *not* depend on
    a. the volume of the container
    b. the temperature
    c. moles of gas in the container
    d. any of the above

28. The density of a gas depends on
    a. temperature
    b. moles
    c. volume
    d. all of the above

29. The velocity of a gas does *not* depend on
    a. temperature
    b. pressure
    c. molar mass
    d. it depends on all of these

30. van der Waals law accounts for
    a. the size of the gas molecule
    b. dependence of kinetic energy on temperature
    c. differences in the paths of molecules
    d. variations in speed of molecules

Additional Practice Problems

1. What is the volume of a gas at 20°C and 701 torr if its volume is 50.0 mL at 25°C and 1.00 atm?
2. What is the volume of 2.8 g of helium at 37°C and 107 kPa?
3. How many grams of $KClO_3$ reacted if 27.3 mL of oxygen was collected over water at 30°C and 738 torr? ($KClO_3 \rightarrow KCl + O_2$)
4. What pressure does 7.38 g O_2 produce at 10°C in a 500.0-mL container when it acts as an ideal gas? When it acts as a real gas?
5. If a gas occupies 370.1 mL at STP, what is its pressure in a 500.0-mL container at 25°C?
6. At a given temperature, a noble gas effuses at half the rate of O_2. What is the gas?
7. What is the partial pressure of oxygen and neon in a mixture of 5.14 g O_2, 31.20 g N_2, and 0.94 g Ne at 1.00 atm?
8. What is the molar mass of a gas if it 0.58 g occupies 157 mL at STP?
9. When 0.048 g Mg react with 25.00 mL of 0.24 *M* HCl at 94.13% yield, what volume of hydrogen gas will be collected over water at 20°C and 764.0 torr?

$$Mg + HCl \rightarrow MgCl_2 + H_2$$

10. What is the volume of krypton gas at 1.00 atm if 50.0 mL has a pressure of 0.50 atm?

Self-Test Answer Key

Key Vocabulary Completion Answers

1. effusion
2. Boyle's law
3. torr
4. real gas
5. liter
6. diffusion
7. STP (standard temperature and pressure)
8. partial pressure
9. vapor
10. temperature (or average kinetic energy)
11. pascal
12. speed
13. standard molar volume
14. pressure
15. Kelvin
16. ideal
17. mole fraction
18. moles
19. volume
20. atmospheres

Multiple-Choice Answers

For further information about the topics these questions address, you should consult the text. The appropriate section of the text is listed in parentheses after the explanation of the answer.

1. c. The volume of a mole any gas at STP is 22.4 L. The mass of 1 mole is its molar mass. The molar mass of xenon is 131.3 g/mol. Therefore, the density of xenon is 131.3 g/22.4 L = 5.86 g/L. (6.5)
2. a. Volume is directly proportional to temperature (Charles's law). So if temperature increases, volume increases. (6.3)
3. b. Volume is inversely proportional to pressure (Boyle's law). So if pressure increases, volume decreases. (6.3)
4. d. Volume will increase with the increase in temperature and decrease with an increase in pressure. Numbers are needed to determine which change will have a greater effect. (6.3)
5. a. Temperature and moles are proportional to pressure, so if both increase, so will pressure. (6.3)
6. b. Pressure and temperature are proportional, so $\frac{P_{initial}}{T_{initial}} = \frac{P_{final}}{T_{final}}$. Temperature *must* be in Kelvin, so 1 atm/288 K = *P*/303 K. *P* = 1.05 atm. (6.3)

7. a. One set of conditions suggests that the ideal gas law should be used. $PV = nRT$. If $R = 0.08206$, the units must be liter, mole, atmosphere, and Kelvin or 0.075 L = V, 1.01 atm = P, 298.15 K = T. Therefore $n = 0.00311$ mol. (6.4)

8. d. $0.14 \text{ g CO} \times \left(\frac{1 \text{ mol}}{28.01 \text{ g}}\right) = 0.0050 \text{ mol}$ and

$$2.78 \text{ g } CO_2 \times \left(\frac{1 \text{ mol}}{44.01 \text{ g}}\right) = 0.0632 \text{ mol}$$

$$X_{CO} = \frac{0.0050}{(0.0050 + 0.0632)} = 0.073;$$

$$P_{CO} = X_{CO}\, P_T = 0.073(1) = 0.073 \text{ atm. } (6.6)$$

9. b. $\frac{r_{He}}{r_{Xe}} = \left(\frac{\mathcal{M}_{Xe}}{\mathcal{M}_{He}}\right)^{1/2} = \left(\frac{131.3}{4.0}\right)^{1/2} = 5.7$

$$\frac{1.0}{r_{Xe}} = 5.7, \text{ so } r_{Xe} = 0.17 \text{ m/s. } (6.7)$$

10. a. The kinetic–molecular theory describes gas molecules as being in constant, independent movement with elastic collisions and holds that the molecules take up almost no space. (6.7)

11. c. The molecules of an ideal gas described by the kinetic–molecular theory behave independently. This is most likely for fast-moving (high-temperature molecules) undergoing few collisions (low pressure). (6.7)

12. a. The definition of effusion is "gas escaping through a small hole." (6.7)

13. b. $920 \text{ mbar} \times \left(\frac{1 \text{ Pa}}{10 \text{ mbar}}\right) \times \left(\frac{1 \text{ atm}}{101.325 \text{ Pa}}\right) =$ 0.908 atm. (6.2)

14. b. $5.00 \text{ g } CO_2 \times \left(\frac{1 \text{ mol}}{44.01 \text{ g}}\right) = 0.114$ mol. 2.00 g CO_2 $\times \left(\frac{1 \text{ mol}}{44.01 \text{ g}}\right) = 0.0454$ mol. Since the 2.00 g are added to the 5.00 g already there, the final number of moles is 0.159. Pressure is proportional to number of moles, so $\frac{P_{initial}}{n_{initial}} = \frac{P_{final}}{n_{final}}$.

So $\frac{500}{0.114} = \frac{P_{final}}{0.159}$ and $P_{final} = 697$ torr. (6.3)

15. d. $P_{initial}V_{initial} = P_{final}V_{final}$. Pressure units must be the same, so $1.5 \text{ atm} \times \left(\frac{760 \text{ torr}}{1 \text{ atm}}\right) = 1140$ torr;

$(800)(9.0) = (1140)V$; $V = 6.3$ L. (6.3)

16. a. total moles = 0.709 mol + 0.180 mol + 0.963 mol = 1.852 mol. $X_{Ne} = \frac{0.709}{1.852} = 0.383$. (6.6)

17. a. Average speed $= \left(\frac{3RT}{\text{molar mass}}\right)^{1/2}$ (6.7)

18. a. Molar mass $= \frac{m}{n} = \frac{(\text{density})RT}{P}$. At STP,

$$\mathcal{M} = \frac{1.25(0.08206)(273.15)}{1} = 28.0. \ (6.5)$$

19. d. Halogens are group 17 and exist as diatomic molecules. $n = \frac{m}{\mathcal{M}}$. So $PV = \left(\frac{m}{\mathcal{M}}\right)RT$ and molar mass $= \frac{mRT}{PV}$. At STP and conditions stated, molar mass $= \frac{(3.21)(0.08206)(273.15)}{(1)(1.9)} = 37.9$ g/mol. If the element is diatomic, each element must have a mass of 18.95 g/mol, quite close to the 19.00 g/mol of fluorine. (6.4)

20. c. As observed from the table of vapor pressures, the only factor listed for vapor pressure is temperature. The water vapor escapes until at a certain pressure as much is recaptured as escapes. More can escape and is less likely to be captured at higher temperatures. (6.6)

21. b. $P_{CO_2} = \frac{nRT}{V} = \frac{(0.010)(0.08206)(293)}{1.0} = 0.24$ atm.

The total pressure is the sum of the partial pressures, so $P_T = 0.37 + 0.24 = 0.61$ atm. (6.6)

22. b. The vapor pressure of water at 20°C is 17.5 torr; the total pressure is 1 atm or 760 torr. So the partial pressure of oxygen is 760 – 17.5 = 742.5 torr $\times \left(\frac{1 \text{ atm}}{760 \text{ torr}}\right) = 0.977$ atm. To get moles, use the ideal gas law where pressure is the partial pressure of oxygen so (0.977 atm)(0.050 L) = n(0.08206)(293). Solving, $n = 2.0 \times 10^{-3}$ mol. (6.6)

23. d. Since conditions are changed, the combined gas law is used. $\frac{P_{initial}V_{initial}}{T_{initial}} = \frac{P_{final}V_{final}}{T_{final}}$,

$\frac{(760)(50.00)}{273.15} = \frac{(500)V}{293}$, so $V = 81.5$ mL. (6.3)

24. d. Standard molar volume is the volume of 1 mole of a gas at standard temperature and pressure, 0°C, and 1 atm (760 torr); it has a value of 22.4 L. (6.3)

25. c. An ideal gas is defined as conforming to the ideal gas law. It does so because its molecules act independently (are not attracted or repulsed) and are in constant motion with elastic collisions. (6.8)

26. a. A real gas does not conform to the ideal gas law. That gas law must be adjusted for the amount of space the molecules take up and the attraction of molecules for each other. (6.8)

27. b. The mean free path is the distance a gas molecule travels before colliding with another molecule or the container. With fewer moles of gas, it is less likely to collide. The temperature is the rate of speed, not how far it travels. (The time between collisions might be shorter but not the distance.) (6.7)

28. a. Density is the ratio of mass to volume. If the ideal gas law is written to include density it is P($\mathcal{M}$) = (density)RT. Thus density is affected by pressure and temperature and the identity of the gas, which determines $\mathcal{M}$. (6.5)

29. b. Kinetic energy $= \frac{1}{2}mu^2$, where m = mass and u = speed, and the speed $= \left(\frac{3RT}{\mathcal{M}}\right)^{1/2}$. (6.7)

30. a. van der Waals law for real gases has correction factors for both the size of the gas particle and attractive forces between molecules. (6.8)

Additional Practice Problem Answers

1. $V_1 = ?,\ T_1 = 20°\text{C} = 293\ \text{K},\ P_1 = 701\ \text{torr}$

   $V_2 = 50.0\ \text{mL},\ T_2 = 25°\text{C} = 298\ \text{K},\ P_2 = 1.00\ \text{atm} = 760\ \text{torr}$

   $$\frac{P_{\text{initial}}V_{\text{initial}}}{T_{\text{initial}}} = \frac{P_{\text{final}}V_{\text{final}}}{T_{\text{final}}},$$

   $$\frac{(701)(V)}{293} = \frac{(760)(50.0)}{298}$$

   $V = 53.3\ \text{mL}$

2. $n = 2.8\ \text{g} \times \left(\frac{1\ \text{mol}}{4.003\ \text{g}}\right) = 0.70\ \text{mol}$

   $T = 37°\text{C} + 273.15 = 310\ \text{K}$

   $P = 107\ \text{kPa} \times \left(\frac{1\ \text{atm}}{101.325\ \text{kPa}}\right) = 1.06\ \text{atm}$

   $R = 0.08206\ \text{L}\cdot\text{atm/mol}\cdot\text{K}$

   $PV = nRT$

   $(1.06\ \text{atm})V = (0.70\ \text{mol})(0.08206\ \text{L}\cdot\text{atm/mol}\cdot\text{K})(310\ \text{K})$

   $V = 17\ \text{L}$

3. Balance equation

   $2\ KClO_3 \rightarrow 2\ KCl + 3\ O_2$

   P_{H_2O} @ $30°\text{C} = 31.8\ \text{torr}$

   $P_{O_2} = P_{\text{atm}} - P_{H_2O} = 738\ \text{torr} - 31.8\ \text{torr} = 706\ \text{torr}$

   $PV = nRT$

   $P = 706\ \text{torr} \times \left(\frac{1\ \text{atm}}{760\ \text{torr}}\right) = 0.929\ \text{atm}$

   $V = 27.3\ \text{mL} \times \left(\frac{1\ \text{L}}{1000\ \text{mL}}\right) = 0.0273\ \text{L}$

   $R = 0.08206\ \text{L}\cdot\text{atm/mol}\cdot\text{K}$

   $T = 30°\text{C} + 273.15 = 303\ \text{K}$

   $(0.929)(0.0273) = n(0.08206)(303)$

   $0.00102\ \text{mol} = n$

   $0.00102\ \text{mol}\ O_2 \times \left(\frac{2\ \text{mol}\ KClO_3}{3\ \text{mol}\ O_2}\right) = \left(\frac{122.55\ \text{g}}{1\ \text{mol}}\right) = 0.08354\ \text{g}$

4. Ideal gas: $PV = nRT$

   $n = 7.83\ \text{g}\ O_2 \times \left(\frac{1\ \text{mol}}{32.00\ \text{g}}\right) = 0.231\ \text{mol}$

   $T = 10°\text{C} + 273.15 = 283\ \text{K}$

   $R = 0.08206\ \text{L}\cdot\text{atm/mol}\cdot\text{K}$

   $V = 500.0\ \text{mL} \times \left(\frac{1\ \text{L}}{1000\ \text{mL}}\right) = 0.5000\ \text{L}$

   $P(0.5000) = (0.231)(0.08206)(283)$

   $P = 10.7\ \text{atm}$

   real gas: $\left(\frac{P - n^2a}{V^2}\right)(V - nb) = nRT$

   for oxygen, $a = 1.36$ and $b = 0.0318$

   $$\left[\frac{P - (0.231)^2(1.36)}{(0.5)^2}\right][0.5 - (0.231)(0.0318)] = (0.231)(0.08206)(283)$$

   $(P - 0.2903)(0.4926) = 5.3645$

   $(P - 0.2903) = 10.89$

   $P = 11.2\ \text{atm}$

5. $V_1 = 370.1\ \text{mL},\ P_1 = 1\ \text{atm},\ T_1 = 273.15\ \text{K}$

   $V_2 = 500.0\ \text{mL},\ P_2 = ?,\ T_2 = 25°\text{C} = 298\ \text{K}$

   $$\frac{P_{\text{initial}}V_{\text{initial}}}{T_{\text{initial}}} = \frac{P_{\text{final}}V_{\text{final}}}{T_{\text{final}}}$$

   $$\frac{(1)(370.1)}{273.15} = \frac{P(500.0)}{298}$$

   $0.808\ \text{atm} = P$

6. Calling the unknown gas x and oxygen y, then $r_x = \left(\frac{1}{2}\right)r_y$. The molar mass of O_2 is 32.00 g/mol. Graham's law states

$$\frac{r_x}{r_y} = \left(\frac{\mathcal{M}_y}{\mathcal{M}_x}\right)^{1/2}$$

Substituting in values given,

$$\frac{\frac{1}{2}r_x}{r_y} = \left(\frac{32.00\text{ g/mol}}{\mathcal{M}_x}\right)^{1/2}$$

$$\frac{1}{2} = \left(\frac{32\text{ g/mol}}{\mathcal{M}_x}\right)^{1/2}$$

$$\frac{1}{4} = \frac{32\text{ g/mol}}{\mathcal{M}_x}$$

$$\frac{1}{4}\mathcal{M}_x = 32$$

$$\mathcal{M}_x = 128\text{ g/mol}$$

Since the problem says this is a noble gas, looking at the molar masses of the rightmost column of the periodic table, the value that comes closest is that for xenon. Since this technique has a fairly high experimental error, it is reasonable to conclude that the unknown gas is xenon.

7. Dalton's law states that the sum of the partial pressures is equal to the total pressure. Another way to state the law is that the partial pressure of a gas is equal to the mole fraction of that gas times the total pressure. So

$P_{O_2} = X_{O_2}P_T$ and $P_{Ne} = X_{Ne}P_T$

$P_T = 1.00\text{ atm}$

$$5.14\text{ g } O_2 \times \left(\frac{1\text{ mol}}{32.00\text{ g}}\right) = 0.161\text{ mol}$$

$$31.20\text{ g } N_2 \times \left(\frac{1\text{ mol}}{28.02\text{ g}}\right) = 1.113\text{ mol}$$

$$0.94\text{ g Ne} \times \left(\frac{1\text{ mol}}{20.18\text{ g}}\right) = 0.047\text{ mol}$$

total mass = 1.321 mol

$$P_{O_2} = \left(\frac{0.161}{1.321}\right)(1.00\text{ atm}) = 0.122\text{ atm}$$

$$P_{Ne} = \left(\frac{0.047}{1.321}\right)(1.00\text{ atm}) = 0.036\text{ atm}$$

8. Molar mass = g/mol. The gas weighs 0.58 g.

$PV = nRT$

$P = 1$ atm (from STP)

$$V = 157\text{ mL} \times \left(\frac{1\text{ L}}{1000\text{ mL}}\right) = 0.157\text{ L}$$

$R = 0.08206\text{ L}\cdot\text{atm/mol}\cdot\text{K}$

$T = 273.15$ (from STP)

$(1)(0.157) = n(0.08206)(273.15)$

$7.00 \times 10^{-3}\text{ mol} = n$

$$\text{molar mass} = \frac{0.58\text{ g}}{7.00 \times 10^{-3}\text{ mol}} = 83\text{ g/mol}$$

9. Balance the equation!

$$Mg + 2\ HCl \rightarrow MgCl_2 + H_2$$

Determine the limiting reactant.

$$0.048\text{ g Mg} \times \left(\frac{1\text{ mol Mg}}{24.305\text{ g}}\right) \times \left(\frac{1\text{ mol } H_2}{1\text{ mol Mg}}\right) \times \left(\frac{2.02\text{ g } H_2}{1\text{ mol}}\right) = 4.0 \times 10^{-3}\text{ g}$$

$$25.00\text{ mL HCl} \times \left(\frac{1\text{ L}}{1000\text{ mL}}\right) \times \left(\frac{0.24\text{ mol HCl}}{1\text{ L}}\right) \times \left(\frac{1\text{ mol } H_2}{2\text{ mol HCl}}\right) \times \left(\frac{2.02\text{ g}}{1\text{ mol}}\right) = 6.1 \times 10^{-3}\text{ g}$$

So Mg is the limiting reactant, and the theoretical yield is 4.0×10^{-3} g. To determine the actual yield:

$$\%\text{ yield} = \left(\frac{\text{actual}}{\text{theoretical}}\right) \times 100$$

$$94.13\% = \left(\frac{\text{actual}}{0.0040}\right) \times 100$$

$$0.0038\text{ g} = \text{actual yield}$$

To determine volume, we need $PV = nRT$.

$$n = 0.0038\text{ g } H_2 \times \left(\frac{1\text{ mol}}{2.02\text{ g}}\right) = 0.0019\text{ mol}$$

$$P = 764\text{ torr} \times \left(\frac{1\text{ atm}}{760\text{ torr}}\right) = 1.005\text{ atm}$$

$R = 0.08206\text{ L}\cdot\text{atm/mol}\cdot\text{K}$

$T = 20 + 273 = 293\text{ K}$

$(1.005)V = (0.0019)(0.08206)(293)$

$V = 0.045$ L or 45 mL

10. Only pressure and volume are changing so

$$P_{\text{initial}}V_{\text{initial}} = P_{\text{final}}V_{\text{final}}$$

$$(1.00\text{ atm})(50.0\text{ mL}) = (0.50\text{ atm})V$$

$$25\text{ mL} = V$$

CHAPTER 7 Electrons in Atoms and Periodic Properties

OUTLINE

REVIEW

Chapter Overview

In Chapter 2, the theories describing the arrangement of protons, neutrons, and electrons within an atom were developed from various experiments. Rutherford's nuclear model of the atom proposed that the protons and neutrons are concentrated in a very tiny volume of the atom, called the nucleus. Electrons were somewhere outside the nucleus and taking up most of the space of an atom. However, it took experiments involving the interaction of the electrons with light to determine the properties and behavior of electrons in an atom.

A more general term for light is **electromagnetic radiation**, since light is a form of energy that consists of oscillating electric and magnetic fields. Electromagnetic radiation has particle–wave duality, properties of both particles and waves. When radiation is described as a wave, it is generally described in terms of **amplitude**, the height of the wave, and **wavelength**, the distance between waves. Sometimes it is convenient to use the term **frequency**, the rate at which waves vibrate, instead of wavelength. Frequency and wavelength give the same information about a wave, but do so in different ways. Frequency (ν) and wavelength (λ) are related to each other and the speed of light (c) by $\lambda\nu = c$. The **electromagnetic spectrum** divides light into groups by wavelength and gives each group a name. The energies that can be seen with human eyes are in the visible region, slightly shorter wavelengths are in the ultraviolet region, and wavelengths slightly longer than visible wavelengths are in the infrared region. The electromagnetic spectrum also includes gamma, x-ray, microwave, and radiowave regions.

When two waves encounter each other they interact in various ways. This interaction of electromagnetic radiation is called interference. **Constructive interference** is when the waves interact in such as way as to produce higher amplitude waves. **Destructive interference** produces waves with lower amplitude. Light waves also interact with matter in several ways. When light passes from one medium to another, its direction may change. This is called **refraction**. When the path is bent by passing around an edge or through an opening, it is called **diffraction**.

Light can also be considered a particle, and that particle is called a **photon**. A photon is characterized by its energy (E). The energy of a photon is an equivalent idea to the wavelength or frequency of light. The two values are related by Planck's constant (h) and $E = h\nu = hc/\lambda$. Therefore, high frequencies and short wavelengths represent higher-energy light.

A graph of the intensity of light versus its energy or wavelength is called a **spectrum**. When atoms are excited, they produce light at wavelengths specific to the identity of the atom. These are called **emission line spectra**. When atoms are exposed to a source of light containing all energies, the same wavelengths (from that atom's emission spectrum) are removed, producing an **absorption line spectrum**. These line spectra show that the interaction between atoms and light is **quantized**, that it only occurs at specific energies. These and other experiments led to the development of **quantum theory**, that energy is transferred only in tiny units. The smallest possible unit of energy that can be transferred is called a **quanta**.

Quantum theory was also useful in explaining the photoelectric effect. The **photoelectric effect** is a process by which metals emit electrons when exposed to light. The light must have a minimum frequency for this process to occur. If the light has this frequency (or higher), then the number of electrons emitted is proportional to the intensity of the light. Einstein explained this by treating light as a particle (photon). The photon transfers all its energy to an atom in the metal and, if the energy is sufficient, an electron can escape. This minimum value of energy is called the threshold frequency (ν_0) or work function (Φ). Any energy beyond this is translated into the kinetic energy of the emitted electron. The intensity, which correlates to the number of photons, determines how many electrons are emitted.

Since hydrogen, with only one electron, is the simplest of all atoms, most theory was developed around its line spectrum. The lines in its spectrum were found to follow a pattern described by the Balmer formula, which was later revised by Johannes Rydberg. Niels Bohr explained this pattern by proposing that electrons orbit the nucleus as a planet orbits the sum. The lines then represent the change in energy of the electrons as they move from one orbit to another, called an **electronic transition**. Because only specific orbits are allowed, the quantum nature of the spectra was also explained. The orbit closest to the nucleus is lowest in energy and called the **ground state**. Orbits farther out were higher in energy and called **excited states**. It requires energy (the energy of a photon) to move an electron from the ground state to an excited state. Since all the energy of the photon is transferred to the electron, the energy of the photon must exactly match the energy difference between the allowed orbits. Those energies correspond to the lines in the spectrum.

Although the Bohr model of the atom makes a nice picture (in fact, it is the one normally used to represent an atom), it is no longer considered a good description of the behavior of electrons. There are two major reasons: (1) it only works for one-electron atoms and (2) the movement of electrons in an atom is not so clearly defined.

De Broglie proposed that **particle–wave duality** could explain the behavior of electrons (as it does the behavior of light). Thus, any particle can be treated as a wave and the relationship is:

$$\lambda = h/mv$$

where λ = wavelength, h = Planck's constant, m = mass of particle and ν = velocity of particle. Large particles have such small wavelengths that their wave properties are not apparent. However, electrons are small enough that their wavelengths are similar to that of light, thus the observed interactions.

De Broglie's equation inspired Erwin Schrödinger to describe the energy of a hydrogen electron as a wave. This equation not only matched the line spectrum of hydrogen, but also explained its behavior.

The Schrödinger equation uses a wave function (ψ) to describe the electron. Unfortunately, this does not have a physical meaning. However, the square of the wave function (ψ^2) represents the **probability density**, which is the probability of finding an electron at any point in space. A denser region represents a higher probability of finding an electron. A region where there is zero probability of finding an electron is called a **node**. The region of space where the probability of finding an electron is high is called an **orbital**.

One of the reasons that quantum mechanics describes electrons in terms of probability is stated in the Heisenberg uncertainty principle. This principle recognizes that making an observation can change the system being observed. Specifically, the **Heisenberg uncertainty principle** states that the position and momentum of an electron cannot be determined simultaneously.

Based on Schrödinger's equation, there are three values that describe an orbital. Later a fourth number, to more specifically describe the electron in the orbital, was added. These four values, called **quantum numbers**, completely describe an electron. For each quantum number there is a set of possible values, and each quantum number describes some aspect of the electron or orbital.

The **principal quantum number (*n*)** can have values of the counting numbers (1, 2, 3, etc.). It describes the maximum distance from the nucleus that an electron is likely to be found. Larger values are farther from the nucleus. The region of space described by n is called the **shell**. The principal quantum number is also the same n as found in the Rydberg equation and thus also describes the energy of the electron. Larger values of n represent higher energies.

The **angular momentum quantum number (ℓ)** has values that depend on the value of the principal quantum number. The possible values of ℓ are integers from zero to n-1. The angular momentum quantum number describes the shape of the orbital and, when combined with n, is called the **subshell**. As ℓ increases, the shapes get more complicated. Usually ℓ is described with a letter instead of a number. If $\ell = 0$, it is also called s and has a spherical shape. If $\ell = 1$, it is called p and the shape is two spheres on opposite sides of the nucleus. If $\ell = 2$, it is called *d* and the shape usually consists of four spheres arranged in a clover shape with the nucleus as the center. When $\ell = 3$, it is called an *f* subshell; $\ell = 4$ is a *g* subshell and the designations continue alphabetically. The higher values of ℓ are not generally involved in chemistry at this level, and so their shapes are not discussed. The numeric designation for the principal quantum number and alphabetic designation of the angular momentum quantum numbers are often combined. Thus an electron with $n = 1$ and $\ell = 0$ is also called a *1s* electron. Increasing values of ℓ also increase energy, but the differences are smaller than energy differences associated with *n*.

The **magnetic quantum number (m_ℓ)** has integer values from $-\ell$ to $+\ell$ (including zero) and in combination with *n* and ℓ it is called the **orbital**. It describes the orientation of the orbital in space. Because orientation depends on perspective, it is generally more important to determine the number of possible orientations rather than actual positions. For an *s* subshell ($\ell = 0$), the only possible value of m_ℓ is zero. Therefore, there is only one orientation of an *s* orbital, centered on the nucleus. If $\ell = 1$ (*p* subshell), m_ℓ may have values of –1, 0, and +1, so there are three possible orientations. There are five possible orientations of orbitals in a *d* subshell and seven orbitals in an *f* subshell. There is no energy difference with orientation, so the various orbitals within a subshell are called **degenerate**.

The **spin quantum number (m_s)** may have a value of +1/2 or –1/2. It describes the spin of the electron within the orbital. Another common way to represent oppositely spinning electrons is with up and down arrows.

In multielectron atoms, the electrons prefer the lowest energy level available. However, as stated by the **Pauli exclusion principle**, no two electrons in an atom can have the same four quantum numbers. The list of electrons in each orbital is called the **electron configuration**. For example, the electron configuration for the nickel atom is $1s^22s^22p^63s^23p^63d^84s^2$. The first number represents the shell (*n*), the letter represents the subshell (ℓ), and the superscripted number represents the number of electrons in that subshell. Since the *s* subshell has only one orbital and only two electrons fit in that orbital, two is the maximum superscript for that subshell. However, the *d* subshell has five orbitals so a maximum of ten electrons can fit in that subshell. Because the superscripted number is 8, that subshell is not full. Electrons in the outermost shell are called **valence electrons**.

When the electrons fill all the lowest energy levels available, the atom is in its **ground state**. When electrons fill degenerate orbitals, the lowest energy arrangement is for the electrons to fill separate orbitals with the same direction spin (**Hund's rule**). Although the energy difference for ℓ quantum numbers is smaller than that for *n* quantum numbers, several small steps turn out to be higher in energy than one large step. Consequently, the 4*s* subshell is lower in energy than the 3*d* subshell.

Properties of atoms are closely related to their electron configuration. Elements in the same group have the same arrangement of valence electrons; different regions of the periodic table can be defined by the subshell in the process of being filled. Ions generally form so that the atom will have the same electron configuration as a noble gas. Thus group 1 atoms, such as sodium with an electron configuration of $1s^22s^22p^63s^1$, will lose one electron to have the same configuration as neon, $1s^22s^22p^6$ and group 17 atoms such as fluorine, $1s^22s^22p^5$, will gain one electron. Atoms and ions with the same electron configuration are called **isoelectronic**. When transition metals become ions, the electrons come from the orbital farthest from the nucleus, thus *ns* electrons are lost before *(n*-1)*d* electrons.

Because most of the volume of an atom is occupied by the electrons, the size of an atom is determined by its electron configuration. As *n* increases, the distance of the electron from the nucleus increases, thus atoms that have electrons with higher values of *n* are larger. Because *n* increases for each row of the periodic table, size increases for elements lower on the periodic table. However, from left to right across the periodic table, size decreases. In this case, *n* remains unchanged, but the number of protons is increasing. Since the electron feels a stronger attraction to the nucleus, the probability is that it is closer to the nucleus and the atom is smaller. The **effective nuclear charge (Z_{eff})**, the charge actually felt by an electron, is not just a function of number of protons, but is also affected by **orbital penetration**, when an electron in an outer shell has some probability of being as close to the nucleus as an electron in a lower-lying shell, and **shielding**, the extent to which other electrons shield it from the charge of the nucleus.

The most important feature in the size of ions is a comparison to the size of the atom from which it was made. Cations, having lost electrons, are smaller than their parent atom. Anions, with extra electrons but no more protons to attract them, are larger than their parent atom. For isoelectronic ions, the higher the atomic number (number of protons), the smaller the ion.

Ionization energy is the energy required to remove an electron from an atom in the gaseous state. The energy required for removal of a second electron is called the **second ionization energy**. Since atoms farther down the periodic table have electrons that are farther from the nucleus and with more shielding electrons, they are easier to remove and have lower ionization energies. Across the periodic table (from left to right), ionization energy increases, since the electrons experience a greater effective nuclear charge because of the increased number of protons and smaller size. There are a couple of anomalies in this trend. These occur when the removal of an electron recreates a full or half-full orbital. Because such orbitals are more stable than would normally be predicted, the ionization energy is lower than the trend predicts. For multi-electron atoms, the energy needed to remove a second electron is always greater than that needed to remove the first, because a second, negatively charged electron is being removed from an ion that already has a positive charge. Similarly, the third ionization energy is always greater than the second. Superimposed on this trend is a much more dramatic increase in ionization energy when the outer-shell electrons have been removed and the next electron must come from an inner shell. The core electrons are held much more tightly by the atom, causing the increase in ionization energy.

Worked Examples

Waves of Light

Example 1.
What is the frequency (in Hz) of light with a wavelength of 650 nm?

Collect and Organize We are given the wavelength of a light wave (650 nm) and are asked to determine its frequency in Hz.

Analyze The frequency and wavelength of a wave of light are related to the speed of light by $\lambda\nu = c$. In this equation the λ stands for the wavelength, ν stands for the frequency, and c stands for speed. The speed of light is constant and has a value of 2.998×10^8 m/s.

Solve

$$\lambda = 650 \text{ nm} \times \left(\frac{1 \text{ m}}{10^9 \text{ nm}}\right) = 6.50 \times 10^{-7} \text{ m}$$

$$\lambda\nu = c$$

$$\nu = \frac{c}{\nu} = \frac{(2.998 \times 10^8 \text{ m/s})}{(6.50 \times 10^{-7} \text{ m})} = 4.61 \times 10^{14} \text{ s}^{-1}$$

$$= 4.61 \times 10^{14} \text{ Hz}$$

Think About It The product of the frequency and wavelength of light is constant and equal to its speed.

Example 2.
What is the wavelength of a 7.47-MHz wave?

Collect and Organize We are given the frequency of a wave and are asked to determine its wavelength.

Analyze We can use the fact that the product of frequency and wavelength is constant to determine the wavelength from the frequency given. Care must be taken, however, to ensure that the units used for each variable are consistent.

Solve

$$\nu = 7.47 \text{ MHz} \times \left(\frac{10^6 \text{ Hz}}{1 \text{ MHz}}\right) \times \left(\frac{1 \text{ s}^{-1}}{1 \text{ Hz}}\right) = 7.47 \times 10^6 \text{ s}^{-1}$$

$$\lambda\nu = c$$

$$\lambda = \frac{c}{\nu} = \frac{(2.998 \times 10^8 \text{ m/s})}{(7.47 \times 10^6 \text{ s}^{-1})} = 40.1 \text{ m}$$

Think About It The wavelength and frequency of light are inversely related. Light with very high frequency has a very short wavelength while light with a very low frequency has a very long wavelength.

Example 3.
What is the energy of a wave with a frequency of 969 MHz?

Collect and Organize We are given the frequency of a wave and are asked to calculate its energy.

Analyze The energy of a wave is directly proportional to its frequency. The constant of proportionality is equal to Planck's constant.

Solve

$$E = h\nu$$

$$\nu = 969 \text{ MHz} \times \left(\frac{10^6 \text{ Hz}}{1 \text{ MHz}}\right) = 9.69 \times 10^8 \text{ Hz}$$

$$= 9.69 \times 10^8 \text{ s}^{-1}$$

so

$$E = (6.626 \times 10^{-34} \text{ J}\cdot\text{s})(9.69 \times 10^8 \text{ s}^{-1})$$

$$= 6.42 \times 10^{-25} \text{ J}$$

Both Planck's constant and frequency have limited significant figures. Be sure your answer has the same number of significant figures as the value that has the fewest.

Think About It Rainbows are the result of the separation of sunlight into different frequencies after passing through water droplets in the atmosphere. Each color in a rainbow corresponds to a different frequency (and therefore, energy) of the separated sunlight.

Example 4.
What is the energy of a wave with a wavelength of 335.1 cm?

Collect and Organize We are asked to determine the energy of a wave given its wavelength.

Analyze Determination of the energy from the wavelength is accomplished by combining the following relations between the energy and frequency of a wave and the wavelength and frequency of a wave: $E = h\nu$ and $\lambda = c/\nu$.

SOLVE

$$E = \frac{hc}{\lambda}$$

$$\lambda = 335.1 \text{ cm} \times \left(\frac{1 \text{ m}}{100 \text{ cm}}\right) = 3.351 \text{ m}$$

$$E = \frac{(6.626 \times 10^{-34} \text{ J}\bullet\text{s})(2.998 \times 10^{8} \text{ m/s})}{(3.351 \text{ m})}$$

$$= 5.958 \times 10^{-26} \text{ J}$$

THINK ABOUT IT The energy of an electromagnetic wave is proportional to its frequency and inversely proportional to its wavelength. The constant of proportionality is Planck's constant h.

EXAMPLE 5.
What is the wavelength and frequency of a wave with an energy of 2.99×10^{-19} J?

COLLECT AND ORGANIZE We are given the energy of a wave (2.99×10^{-19} J) and asked to determine the wave's frequency and wavelength.

ANALYZE The frequency of a wave of given energy is easily determined using the relation $E = h\nu$. The wavelength of the wave is calculated using $\lambda = c/\nu$.

SOLVE

$$E = h\nu$$

$$\frac{E}{h} = \nu$$

$$\nu = \frac{(2.99 \times 10^{-19} \text{ J})}{(6.626 \times 10^{-34} \text{ J}\bullet\text{s})} = 4.51 \times 10^{14} \text{ s}^{-1}$$

$$\lambda = \frac{c}{\nu} = \frac{(2.998 \times 10^{8} \text{ m/s})}{(4.51 \times 10^{14} \text{ s}^{-1})} = 6.65 \times 10^{7} \text{ m}$$

THINK ABOUT IT Energy (E), wavelength (λ), and frequency (ν) are interrelated. Knowledge of one allows for determination of the other two.

PARTICLES OF LIGHT

EXAMPLE 1.
What is the work function for a metal that exhibits the photoelectric effect at 500 nm?

COLLECT AND ORGANIZE When a metal is exposed to light with a wavelength of 500 nm, an electron is ejected. We are asked to determine the work function for these conditions.

ANALYZE The photoelectric effect is described by the equation $E = hc/\lambda$. E represents the energy required to remove an electron from a metal. In this case it is the work function of the metal, Φ.

SOLVE

$$E = \frac{hc}{\lambda}$$

$$\lambda = 500 \text{ nm} \times \left(\frac{1 \text{ m}}{10^{9} \text{ nm}}\right) = 5.00 \times 10^{-7} \text{ m}$$

$$E = (6.626 \times 10^{-34} \text{ J}\bullet\text{s})(2.998 \times 10^{8} \text{ m/s}) \div 5.00 \times 10^{-7} \text{ m}$$
$$= 3.97 \times 10^{-19} \text{ J}$$

THINK ABOUT IT The energy calculated above is a threshold energy, so more energy is acceptable. Remember that higher energy photons have shorter wavelengths.

EXAMPLE 2.
At what λ will a metal with a work function of $\Phi = 8.20 \times 10^{-19}$ J exhibit the photoelectric effect?

COLLECT AND ORGANIZE We are asked to determine the wavelength of light that ejects an electron from the surface of a metal that has a work function of $\Phi = 8.20 \times 10^{-19}$ J.

ANALYZE In order to determine the wavelength of light that will eject an electron from the surface of a metal, we need to first determine the energy required to remove the electron.

SOLVE

$$E = \Phi = 8.20 \times 10^{-19} = (6.626 \times 10^{-34} \text{ J}\bullet\text{s})(2.998 \times 10^{8} \text{ m/s})/\lambda$$

$$(8.20 \times 10^{-19})\lambda = 1.9865 \times 10^{-25}$$

$$\lambda = 2.42 \times 10^{-7} \text{ m} \times \left(\frac{10^{9} \text{ nm}}{1 \text{ m}}\right) = 242 \text{ nm}$$

THINK ABOUT IT Since the energy calculated above represents a threshold energy, the photoelectric effect will be observed when the wavelength of the incident light is shorter than 242 nm.

THE HYDROGEN SPECTRUM

EXAMPLE 1.
What is the wavelength of the transition from $n = 2$ to $n = 4$?

COLLECT AND ORGANIZE We are given the initial and final states of an electronic transition and are asked to determine the wavelength of light that will cause a transition between the two.

ANALYZE The wavelength of an electronic transition from an initial state n_1 to a final state n_2 is given by:

$$\frac{1}{\lambda} = 1.097 \times 10^{-2}\ \text{nm}^{-1}\left(\frac{1}{n_1^2} - \frac{1}{n_2^2}\right)$$

The n refers to the principal quantum number. Note that λ in this equation is in nanometers rather than meters.

SOLVE

$$\frac{1}{\lambda} = 0.01097\left(\frac{1}{2^2} - \frac{1}{4^2}\right)$$

$$\frac{1}{\lambda} = 0.01097(0.25 - 0.0625)$$

$$\frac{1}{\lambda} = 0.01097(0.1875)$$

$$\frac{1}{\lambda} = 2.056 \times 10^{-3}$$

$$\lambda = \frac{1}{(2.056 \times 10^{-3})} = 486.2\ \text{nm}$$

THINK ABOUT IT As long as we know the principal quantum numbers of the initial and final states for electronic transitions in hydrogen, we can calculate the wavelength of light required to cause the transition using the equation given above.

EXAMPLE 2.
What wavelength of light is required to remove an electron from the $n = 3$ energy level.

COLLECT AND ORGANIZE We are asked to determine the wavelength of light that is required to remove an electron from the $n = 3$ energy level.

ANALYZE We will use the equation for the wavelength of an electronic transition from an initial state n_1 to a final state n_2 given in the previous example noting that when the electron is removed the final state $n_2 = \infty$. Note that $\infty^2 = \infty$ and $1/\infty = 0$.

SOLVE

$$\frac{1}{\lambda} = 0.01097\left(\frac{1}{3^2} - \frac{1}{\infty}\right)$$

$$= 0.01097\left(\frac{1}{9}\right) = 1.219 \times 10^{-3}$$

$$\lambda = \frac{1}{(1.219 \times 10^{-3})} = 820.4\ \text{nm}$$

THINK ABOUT IT The wavelength of light (λ = 820.4 nm) calculated to remove an electron from the n = 3 energy level of an hydrogen atom lies on the red end of the visible spectrum.

ELECTRONS AS WAVES

EXAMPLE 1.
What is the wavelength of an object weighing 20.0 lb and moving at 4.0 mph?

COLLECT AND ORGANIZE We are given the mass and velocity of an object and are asked to calculate its wavelength.

ANALYZE The de Broglie wavelength (λ) gives the relationship between an object's momentum mass (m) × velocity (ν) and its wavelength. It is given by the expression $\lambda = h/m\nu$.

SOLVE In order to use the relation for the de Broglie wavelength

$$\lambda = \frac{h}{m\nu}$$

we must first convert the given mass and speed into the appropriate units (kg and m/s).

$$\text{mass} = 20.0\ \text{lb} \times \left(\frac{454\ \text{g}}{1\ \text{lb}}\right) \times \left(\frac{1\ \text{kg}}{1000\ \text{g}}\right) = 9.08\ \text{kg}$$

$$\text{velocity} = 4.0\ \frac{\text{mi}}{\text{hr}} \times \left(\frac{1.61\ \text{km}}{1\ \text{mi}}\right) \times \left(\frac{1000\ \text{m}}{1\ \text{km}}\right) \times \left(\frac{1\ \text{hr}}{3600\ \text{s}}\right) = 1.8\ \text{m/s}$$

$$\lambda = \frac{(6.626 \times 10^{-34})}{(9.08\ \text{kg})(1.8\ \text{m/s})} = 4.1 \times 10^{-35}\ \text{m}$$

THINK ABOUT IT Heavier objects have wavelengths that are incredibly small and are not perceptible. For very light objects, such as electrons, much larger wavelengths occur.

EXAMPLE 2.
At what velocity must a neutron (mass = 1.67×10^{-24} g) be moving to have a wavelength of 1.73 cm?

COLLECT AND ORGANIZE We are given the wavelength and mass of a moving particle and are asked to determine its velocity.

ANALYZE We can rearrange the equation for the de Broglie wavelength λ of a particle to solve for velocity.

$$\lambda = \frac{h}{m\nu}$$

$$\nu = \frac{h}{m\lambda}$$

Mass must be expressed in kilograms and wavelength must be expressed in meters to match the units of the Plank constant h.

SOLVE

$$\lambda = \frac{h}{mv}$$

$$\text{mass} = 1.67 \times 10^{-24}\ \text{g} \times \left(\frac{1\ \text{kg}}{1000\ \text{g}}\right) = 1.67 \times 10^{-27}\ \text{kg}$$

$$\lambda = 1.73\ \text{cm} \times \left(\frac{1\ \text{m}}{100\ \text{cm}}\right) = 0.0173\ \text{m}$$

$$(0.0173) = \frac{(6.626 \times 10^{-34})}{(1.67 \times 10^{-27})\nu}$$

$$(0.0173)(1.67 \times 10^{-27})\nu = 6.626 \times 10^{-34}$$

$$\nu = \frac{(6.626 \times 10^{-34})}{(0.0173)(1.67 \times 10^{-27})}$$

$$\nu = 2.29 \times 10^{-5}\text{m/s}$$

THINK ABOUT IT We can calculate the velocities of objects if we know the mass and wavelength of the object.

QUANTUM NUMBERS

EXAMPLE 1.
Rank the following electrons with quantum numbers (n, ℓ, m_ℓ, m_s) from lowest energy to highest energy.

$A = \left(2, 1, 1, +\frac{1}{2}\right)$ $B = \left(1, 0, 0, -\frac{1}{2}\right)$

$C = \left(4, 1, -1, +\frac{1}{2}\right)$ $D = \left(4, 2, -1, +\frac{1}{2}\right)$

$E = \left(3, 2, -1, +\frac{1}{2}\right)$ $F = \left(4, 1, 0, +\frac{1}{2}\right)$

$G = \left(2, 1, -1, +\frac{1}{2}\right)$ $H = \left(3, 1, 0, +\frac{1}{2}\right)$

COLLECT AND ORGANIZE We are given the quantum numbers for 8 electrons (A–H) and are asked to rank them in order of increasing energy.

ANALYZE The first consideration is the principal quantum number n. Electrons with smaller values of n will be lower in energy. For electrons having the same value of n, one needs to consider the value of the angular momentum quantum number ℓ. The electron with the smaller value of ℓ will be lower in energy.

SOLVE Electron B has the lowest n quantum number and therefore, has the lowest energy. Electrons A and G will have higher energies than B, since they have a higher principal quantum number. However, since they both have the same value of n and ℓ, they both have the same energy. Electron H will be lower in energy than electron E, since it has a lower ℓ quantum number. Finally, based on the values of the ℓ quantum number, electron F is equal to C and both will have a lower energy than D.

$$B < A = G < H < E < C = F < D$$

THINK ABOUT IT Electrons can have different energies based on the orbitals they occupy. We can easily determine the relative energies of a set of electrons if we are provided the quantum numbers of each.

EXAMPLE 2.
Refer to the electrons in Example 1 to answer the following questions:

a. Which electron is spinning in a direction different from that of the others?

b. Which electron is in a spherically shaped orbital?

c. Which electron is in a *p* orbital?

d. Which electron is in a *d* orbital?

e. Which electron is farthest from the nucleus?

f. Which two electrons are in the same orbital?

g. Which two electrons differ only by the orientation of their orbitals?

h. Which two electrons cannot exist in the same atom?

i. Which electrons are degenerate?

j. Which electrons are in an *f* orbital?

COLLECT AND ORGANIZE We are asked to analyze a number of properties of electrons that depend upon their quantum numbers.

ANALYZE Using the values of n, ℓ, m_ℓ, and m_s we can determine which of the electrons A–H given in the previous example meet the conditions listed in a–j above.

SOLVE

a. The m_s quantum number refers to spin. Only electron B has a value of m_s that is different from that of the others.

b. The ℓ quantum number refers to shape. An *s* orbital ($\ell = 0$) is spherically shaped; therefore, any electrons with $\ell = 0$ will have a spherically shaped orbital. This is only electron B.

c. When $\ell = 1$, it is also called a *p* orbital. $\ell = 1$ for electrons A, C, F, G, and H.

d. A *d* orbital is $\ell = 2$. That is electrons D and E.

e. The distance from the nucleus is determined by the n quantum number. The larger the number, the farther from the nucleus. The highest n among our choices is 4, and three electrons have that value. These three electrons—C, D, and F—are equally distant from the nucleus.

f. For electrons to be in the same orbital, the first three quantum numbers must be the same. The first two give the type of orbital; the third gives its orientation. Two electrons spinning in opposite directions can exist in the same orbital. In this example there are no two electrons in the same orbital.

g. The orientation of the orbital is the third quantum number, m_ℓ. Electrons A and G differ only by this quantum number. Electrons C and F also differ in orientation. The comparison is not valid if the first two quantum numbers are not also the same.

h. The Pauli exclusion principle says that each electron in an atom must have a unique set of quantum numbers. Since each of these electrons has a unique set of quantum numbers, they all may exist in the same atom.

i. Atoms with the same energy (same n and ℓ values) are degenerate. In this example, A and G are degenerate. Electrons C and F are also degenerate.

j. An f orbital has $\ell = 3$. There are no electrons in this example with an f-type orbital.

Think About It The four quantum numbers (n, ℓ, m_ℓ, and m_s) completely describe the electron.

Example 3.
What is wrong with the quantum numbers (n, ℓ, m_ℓ, and m_s) of the following electrons?

a. $\left(2, 2, 0, +\frac{1}{2}\right)$

b. $\left(3, 1, -1, -\frac{1}{2}\right)$

c. $(3, 1, -2, 1)$

d. $\left(4, 0, 1, +\frac{1}{2}\right)$

e. $\left(\frac{1}{2}, 1, 1, 1\right)$

Collect and Organize We are given the quantum numbers for five electrons (a–e) and are asked to determine if any of the quantum numbers are invalid and why.

Analyze Certain restrictions exist for acceptable values of each of the four quantum numbers (n, ℓ, m_ℓ, m_s). Possible values for each of the quantum numbers are

$$n = 1, 2, 3, 4, \ldots\ldots$$

$$\ell = 0, 1, 2, 3, \ldots.. (n-1)$$

$$m_\ell = -\ell \ldots -2, -1, 0, +1, +2, \ldots\ldots + \ell$$

$$m_s = +\frac{1}{2} \text{ or } -\frac{1}{2}$$

Solve

a. The possible values for ℓ only extend to $n - 1$. Since $n = 2$ in this electron, ℓ can have a value of 1 or 0; 2 is not acceptable.

b. There is nothing wrong with this set of quantum numbers.

c. m_ℓ may only have values from $-\ell$ to $+\ell$. In this case, $-1, 0, +1$; -2 is not a valid choice. In addition, the only choices for m_s are $+1/2$ and $-1/2$, not 1.

d. Since $\ell = 0$, the only valid choice for m_ℓ is 0.

e. n must be a counting number. Fractions are not permitted (nor is zero). With that wrong, it is impossible for the other values to have any valid choices, with the possible exception of m_s (with choices of $+1/2$ and $-1/2$), but even that is wrong.

Think About It There are restrictions on the values that the different quantum numbers can take.

Example 4.
Rank the following from highest energy to lowest energy.

$3s$, $5p$, $4d$, $1s$, $5d$, $3p$

Collect and Organize We are given six atomic orbitals and are asked to rank them in order of decreasing energy.

Analyze Recall that s is $\ell = 0$, p is $\ell = 1$, and d is $\ell = 2$. We can therefore, use the rules of energy ordering which state that the lowest energy orbital will have the smallest value of the principal quantum number n and that in cases of where orbitals have the same value of n, the orbital with the smaller value of ℓ will be lower in energy.

Solve
(high energy) $5d > 5p > 4d > 3p > 3s > 1s$ (low energy)

Think About It Only n and ℓ effect the energies of orbitals.

Periodic Table and Filling in the Orbitals

Example 1.
What is the electron configuration of Ru?

Collect and Organize We are asked to write the electron configuration of Ru.

Analyze We will use the position of the element in the periodic table to build up its electron configuration.

Solve
$1s^2$ (the first two electrons in the positions of H and He)
$1s^22s^2$ (Li and Be positions)
$1s^22s^22p^6$ (B through Ne)
$1s^22s^22p^63s^2$ (moved to the next period, positions of Na and Mg)
$1s^22s^22p^63s^23p^6$ (Al through Ar positions)
$1s^22s^22p^63s^23p^64s^2$ (K and Ca are in the fourth period)
$1s^22s^22p^63s^23p^64s^23d^{10}$ (For the first time the *d* group is included. Remember the n = period – 1.)
$1s^22s^22p^63s^23p^64s^23d^{10}4p^6$ (The *p* group has n = period.)
$1s^22s^22p^63s^23p^64s^23d^{10}4p^65s^2$ (next period, *s* group)
$1s^22s^22p^63s^23p^64s^23d^{10}4p^65s^24d^6$ (The *d* group has the lower *n*, and the last electron of ruthenium is in the *d* group; there are six in that group to get to Ru.)

Think About It Care must be taken when dealing with the *d* and *f* electrons. The *d* electrons have a principal quantum number one less than the period. The *f* electrons have a principal quantum number two less than the period.

Example 2.
What is the abbreviated electron configuration of Sn?

Collect and Organize We are asked to determine the abbreviated electron configuration of Sn.

Analyze We can use the position of the element in the periodic table to build up its electron configuration. However, in the interest of a more compact notation we can use the symbol of the noble gas in the row previous to the element of interest to represent all preceding electrons.

Solve Sn is in period 5; the noble gas in period 4 is Kr. The rest of the electron configuration starts with the *s* section of period 5, then the *d* section (where n = period – 1 = 4), then two electrons in the *p* section, so $[Kr]5s^24d^{10}5p^2$.

Think About It Abbreviated electron configurations convey the same information but in a much more compact notation. This becomes especially useful as one considers heavier and heavier elements.

Example 3.
What is the abbreviated electron configuration of platinum (Pt)?

Collect and Organize We are asked to determine the abbreviated electron configuration of platinum.

Analyze We will use the position of the element in the periodic table to build up its electron configuration. Care must be taken, however, to include elements 58–71, the lanthanides, which are traditionally set at the bottom of the table as a space-saving device.

Solve

$$[Xe]6s^24f^{14}5d^8$$

Think About It Remember that the *d* electrons have principal quantum numbers that are one lower than the period and that the *f* electrons have principal quantum numbers that are two lower than the period.

Example 4.
What are the electron configurations of Mo and Au?

Collect and Organize We are asked to determine the electron configurations of Mo and Au.

Analyze We will use the position of the element in the periodic table to determine its electron configuration. The elements under consideration demonstrate an exception that occurs for the energy ordering of atomic orbitals. Because the *ns* and $(n-1)$ *d* energy levels are very close, occasionally an electron will move from the *s* to the *d* orbital. This occurs when it will create a full or half full *d* orbital (d^5 and d^{10}).

Solve Using the periodic table, the electron configuration for Mo and Au would be given as:

$$Mo = [Kr]5s^24d^4 \qquad Au = [Xe]6s^24f^{14}5d^9$$

However, since the promotion of an *s* electron into the *d* orbital results in a half-filled *d* orbital for Mo and a full *d* orbital for Au, the actual configurations would be:

$$Mo = [Kr]5s^14d^5 \qquad Au = [Xe]6s^1f^{14}5d^{10}$$

Think About It Remember that situations in which an *s* electron can be promoted to a *d* orbital such that the *d* orbital becomes half filled or full represent a special case.

Electron Configuration of Ions

Example 1.
What is the electron configuration of Na^+?

Collect and Organize We are asked to determine the electron configuration of the sodium ion.

Analyze We must modify the electron configuration obtained using the periodic table to take into account the loss of an electron that occurs when Na forms Na^+.

Solve The atom Na has the electron configuration of $[Ne]3s^1$. When it loses one electron, the electron configuration is [Ne].

Think About It Ions result from the addition or removal of electrons. Positively charged ions (cations) result when electrons are removed from the neutral element. Negatively charged ions (anions) result when electrons are added to the neutral element. We must take this into account when determining the electron configuration of ions.

Example 2.
What are the electron configurations of Fe^{2+} and Fe^{3+}?

Collect and Organize We are asked to determine the electron configurations of Fe^{2+} and Fe^{3+}.

Analyze We will use the element's position in the periodic table to build up its electron configuration. It will then be necessary to account for the loss of electrons when Fe ionizes to form Fe^{2+} and Fe^{3+}.

Solve The electron configuration of the atom Fe is $[Ar]4s^23d^6$.

The ion Fe^{2+} loses two electrons from the highest n value of 4, so

$$Fe^{2+} = [Ar]3d^6$$

The ion Fe^{3+} loses three electrons, the two from the $4s$ orbital and one more, so

$$Fe^{3+} = [Ar]3d^5$$

Think About It After writing the electron configuration for the neutral atom, we determine the electron configuration for cations by removing the required number of electrons starting with orbitals farther from the nucleus.

Example 3.
What are the electron configurations of Pb^{2+} and Pb^{4+}?

Collect and Organize We are asked to determine the electron configurations of Pb^{2+} and Pb^{4+}.

Analyze We will use the element's position in the periodic table to build up its electron configuration. It will then be necessary to account for the loss of electrons when Pb ionizes to form Pb^{2+} and Pb^{4+}.

Solve The electron configuration of the atom Pb is

$$[Xe]6s^24f^{14}5d^{10}6p^2$$

For Pb^{2+} the first two electrons lost are from the $6p$ orbital.

$$Pb^{2+} = [Xe]6s^24f^{14}5d^{10}$$

For Pb^{4+} the s electrons are also lost.

$$Pb^{4+} = [Xe]4f^{14}5d^{10}$$

Think About It Electrons in the orbital farthest from the nucleus are removed first. If there are two orbitals the same distance from the nucleus, then the higher energy electrons are removed first.

Example 4.
What are the electron configurations of I^- and N^{3-}?

Collect and Organize We are asked to determine the electron configurations of I^- and N^{3-}.

Analyze In order to form negatively charged ions, electrons must be added to the neutral element. We will use the position of the element in the periodic table to build up its electron configuration and add electrons accordingly to account for the ion's negative charge.

Solve The electron configuration of the I atom is $[Xe]6s^24f^{14}5d^{10}6p^5$.

To make I^-, one electron is added.

$$I^- = [Xe]6s^24f^{14}5d^{10}6p^6$$

The electron configuration of the N atom is $[He]2s^22p^3$.

To make N^{3-} means that three electrons have been added to the atom.

$$N^{3-} = [He]2s^22p^6$$

Think About It Electron configurations of anions are determined by placing the extra electrons in the lowest energy orbitals first.

EXAMPLE 5.
What ions will be produced from the following atoms?

a. Ca b. Li c. O d. Bi e. Al

COLLECT AND ORGANIZE We are given five elements and are asked to determine what ions they will form.

ANALYZE The main group elements will lose or gain electrons so that the resulting electron configuration will be the same as the nearest noble gas. Metals will always lose electrons. Nonmetals normally gain electrons.

SOLVE

a. Ca is a metal. Losing two electrons will give it the same number of electrons as argon: Ca^{2+}.

b. Li is a metal. Losing one electron will give it the same number of electrons as He: Li^{+}. (When writing charges of ions do not include a 1 in the formula!)

c. O is a nonmetal, so it will gain electrons. Two will give it the same number of electrons as Ne and O^{2-}.

d. Bi is a metal, so it will lose electrons. Its electron configuration is $6s^24f^{14}5d^{10}6p^3$. Recall that these atoms lose their *p*, and then their *s* electrons. Bi might have a charge of Bi^{3+} or Bi^{5+}. With so many valence electrons, it will not end up with a noble-gas electron configuration, but loses electrons in such a way as to empty the subshells.

e. Al is a metal, so it will lose electrons. Losing three will give it the same number of electrons as Ne and Al^{3+}.

THINK ABOUT IT The noble-gas configuration is a particularly stable configuration. As a result, elements are very happy to lose or gain electrons if it means they can have an electron configuration that is "noble gas like."

EXAMPLE 6.
How many unpaired electrons are in the following atoms or ions?

a. Co b. Cl c. Sr d. Mn^{2+} e. Si

COLLECT AND ORGANIZE We are given four elements and one ion and are asked to determine the number of unpaired electrons in each.

ANALYZE In order to determine the number of unpaired electrons we need only consider the highest energy orbital obtained from the element or ion's electron configuration. Using Hund's rule, electrons are distributed into empty orientations of the orbitals before any pairing up occurs.

SOLVE

a. Its electron configuration is $[Ar]4s^23d^7$. 3*d* is the highest-energy orbital. There are five orientations of *d* orbitals and seven electrons to distribute in these orbitals.

↓↑ ↓↑ ↑ ↑ ↑ There are three unpaired electrons.

b. The electron configuration of Cl is $[Ne]3s^23p^5$. 3*p* is the highest-energy orbital. There are three orientations of *p* orbitals and five electrons to distribute.

↑↓ ↑↓ ↑ There is one unpaired electron.

c. Sr has the electron configuration of $[Kr]5s^2$. 5*s* is the highest-energy orbital. It has one orientation, two electrons ↑↓, and zero unpaired electrons.

d. Mn^{2+} has the electron configuration of $[Ar]3d^5$. (The *s* electrons are lost in forming the ion.) There are five orientations of the *d* orbital, with one electron in each; there are five unpaired electrons.

e. Si has the electron configuration of $[Ne]3s^23p^2$. There are three orientations of *p* orbitals. Therefore, the two *p* electrons can each have their own orbital, and there are two unpaired electrons.

THINK ABOUT IT Significant energy is required to bring two particles with the same charge in close proximity. As a result, the electrons in an orbital will spread themselves out among the available orientations before pairing up.

IONIZATION ENERGIES

EXAMPLE 1.
Rank the following atoms from highest to lowest ionization energy.

F, S, He, Se, O

COLLECT AND ORGANIZE We are asked to rank the given elements in order of decreasing ionization energy.

ANALYZE Ionization energy is the energy required to remove an electron from an atom in the gaseous state. In general, the ionization energy increases from left to right and from bottom to top in the periodic table.

SOLVE He has the highest ionization energy; it is a noble gas and at the top of the periodic table. F is next at the right and in period 2. O is next to F. S is closer to the bottom but the same distance to the right as oxygen. Se is below sulfur. The order is He > F > O > S > Se.

THINK ABOUT IT Relative ionization energy is a particular property that can be determined using the element's position in the periodic table.

Self-Test

Key Vocabulary Completion Questions

_________ 1. Wave property proportional to energy

_________ 2. Atomic model where electrons orbit the nucleus

_________ 3. Quantum number reflecting the orientation of the orbitals

_________ 4. Region of zero electron density

_________ 5. Particle of light

_________ 6. Quantum number related to the most likely distance of an electron from the nucleus

_________ 7. It increases with increasing values of n or increasing values of ℓ

_________ 8. Positive charge the outermost electron experiences

_________ 9. Orbitals with the same energy

_________ 10. Unit of frequency

_________ 11. Lowest energy state of an atom

_________ 12. Speed of light divided by frequency

_________ 13. Person who proposed that no two electrons in an atom can have the same four quantum numbers

_________ 14. Quantum number with a value of either +1/2 or –1/2

_________ 15. Metals emitting electrons when exposed to light of a particular frequency

_________ 16. Having the same electron configuration

_________ 17. Equation that describes an electron as a wave (from theory)

_________ 18. Energy required to remove an electron from an atom in the gaseous state

_________ 19. The position and momentum of an electron cannot be determined simultaneously.

_________ 20. Having discrete rather than continuous values

_________ 21. State after an electron has absorbed energy

_________ 22. Space in which it is likely to find an electron

_________ 23. Mathematical equation describing the pattern of lines in hydrogen's atomic emission spectrum

_________ 24. System to determine electron distribution in a multielectron atom

_________ 25. Change in direction of light because of passing from one medium to a different one

Multiple-Choice Questions

1. What is the frequency of light with a wavelength of 299 nm?
   a. 10 MHz
   b. 1 MHz
   c. 1×10^{15} Hz
   d. 1×10^{12} Hz

2. Which is the highest energy?
   a. 200 nm
   b. 1000 nm
   c. 700 Hz
   d. 1500 Hz

3. Short wavelengths are also
   a. high energy and high frequency
   b. low energy and low frequency
   c. high energy and low frequency
   d. low energy and high

4. The quantum number that reflects distance from the nucleus is
   a. n
   b. ℓ
   c. m_ℓ
   d. m_s

5. The photoelectric effect is
   a. light of exactly the right wavelength moving an electron to a higher energy level
   b. light interacting with a metal and producing a current
   c. the pattern of dark spots in a spectrum characteristic of an element
   d. the pattern of bright lines in a spectrum characteristic of an element

6. All the following are characteristics of a quantum electron *except*
   a. a position that is described in terms of probabilities
   b. specific choices rather than a continuum of choices
   c. uncertainty of position or momentum
   d. orbits that are simple circles

7. What wavelength of light is required to remove an electron from Cu with $\Phi = 7.45 \times 10^{-19}$ J?
   a. greater than 267 nm
   b. less than 267 nm
   c. greater than 375 nm
   d. less than 375 nm

8. Higher-energy photons have
   a. shorter wavelengths
   b. lower frequencies
   c. higher values of n
   d. higher amplitudes

9. An electron has a mass of 9.1×10^{-28} g. What velocity does it need to be visible (have a wavelength of 700 nm)?
   a. 1.0 m/s
   b. 1.0×10^{-6} m/s
   c. 1.0×10^{-9} m/s
   d. 1.0×10^{3} m/s

10. What wavelength does a neutron (mass = 1.67×10^{-24} g) moving at 10% of the speed of light have?
    a. 1.32×10^{-15} m
    b. 1.32×10^{-17} m
    c. 1.32×10^{-14} m
    d. 1.32×10^{-18} m

11. Which of the following is a noble gas?
    a. sodium
    b. hydrogen
    c. neon
    d. nitrogen

12. Actinides are characterized by
    a. a charge of –1
    b. an incomplete 4*f* subshell
    c. none of the above; they do not naturally exist
    d. all of the above

13. Which species is isoelectronic with Ne?
    a. Ar
    b. Fe
    c. Fe^{3+}
    d. O^{2-}

14. What is a likely ion of magnesium?
    a. Mg^{+}
    b. Mg^{2+}
    c. Mg^{2-}
    d. Mg^{-}

15. Why can't an electron have a quantum number of $n = 3$, $\ell = 0$, $m_\ell = -1$, and $m_s = +1/2$?
    a. ℓ cannot be equal to zero
    b. m_ℓ cannot be negative
    c. m_ℓ cannot have a value of –1 if $\ell = 0$.
    d. m_s cannot be a fraction

16. Assuming electrons with quantum numbers (n, ℓ, m_ℓ, m_s), which electron is farthest from the nucleus?
    a. $\left(2, 0, 0, +\frac{1}{2}\right)$
    b. $\left(3, 1, 1, -\frac{1}{2}\right)$
    c. $\left(1, 0, 0, +\frac{1}{2}\right)$
    d. $\left(2, 1, 1, +\frac{1}{2}\right)$

17. Assuming electrons with quantum numbers (n, ℓ, m_ℓ, m_s), which electron has the highest energy?
    a. $\left(3, 1, 1, +\frac{1}{2}\right)$
    b. $\left(2, 1, -1, +\frac{1}{2}\right)$
    c. $\left(3, 0, 0, -\frac{1}{2}\right)$
    d. $\left(2, 0, 0, -\frac{1}{2}\right)$

18. Assuming electrons with quantum numbers (n, ℓ, m_ℓ, m_s), which has a *d* electron?
    a. $\left(3, 1, 1, +\frac{1}{2}\right)$
    b. $\left(2, 1, -1, +\frac{1}{2}\right)$
    c. $\left(3, 2, 0, -\frac{1}{2}\right)$
    d. $\left(2, 0, 0, -\frac{1}{2}\right)$

19. Assuming electrons with quantum numbers (n, ℓ, m_ℓ, m_s), which has a spherically shaped probability density?
    a. $\left(3, 1, 1, +\frac{1}{2}\right)$
    b. $\left(2, 1, -1, +\frac{1}{2}\right)$
    c. $\left(3, 2, 0, -\frac{1}{2}\right)$
    d. $\left(2, 0, 0, -\frac{1}{2}\right)$

20. Assuming electrons with quantum numbers (n, ℓ, m_ℓ, m_s), which is degenerate with (2, 1, 1, +1/2)?
    a. $\left(3, 1, 1, +\frac{1}{2}\right)$
    b. $\left(2, 1, -1, +\frac{1}{2}\right)$
    c. $\left(3, 2, 0, -\frac{1}{2}\right)$
    d. $\left(2, 0, 0, -\frac{1}{2}\right)$

21. Assuming electrons with quantum numbers (n, ℓ, m_ℓ, m_s), which is lowest in energy?
   a. $\left(3, 1, 1, +\frac{1}{2}\right)$
   b. $\left(2, 1, -1, +\frac{1}{2}\right)$
   c. $\left(3, 2, 0, -\frac{1}{2}\right)$
   d. $\left(2, 0, 0, -\frac{1}{2}\right)$

22. Assuming electrons with quantum numbers (n, ℓ, m_ℓ, m_s), how does an electron with quantum numbers (2, 1, 1, +1/2) differ from an electron with quantum numbers (2, 1, –1, +1/2)?
   a. spin
   b. size
   c. energy
   d. orientation

23. How many unpaired electrons are in a bromine atom?
   a. 0
   b. 1
   c. 2
   d. more than 2

24. What is the highest-energy electron in Au?
   a. 6*s*
   b. 5*d*
   c. 4*f*
   d. 6*p*

25. Which of the following was Heisenberg uncertain of?
   a. position
   b. shape of probability density
   c. spin
   d. wavelength

26. Which of the following has the highest ionization energy?
   a. Mn
   b. F
   c. Ne
   d. Cs

27. A transition metal is characterized by
   a. malleability
   b. incomplete subshell of *d* electrons
   c. charge of +2
   d. photoelectricity

28. $1s^22s^22p^63s^2$ is the electron configuration for
   a. Mg
   b. Ca
   c. Ti
   d. Si

29. When manganese becomes Mn^{2+} ion, it
   a. loses two 4*s* electrons
   b. gains two 4*d* electrons
   c. loses two 3*d* electrons
   d. gains two 3*p* electrons

30. Which electron energy transition is the largest?
   a. $1s \rightarrow 2s$
   b. $2s \rightarrow 3s$
   c. $2s \rightarrow 2p$
   d. $3p \rightarrow 3d$

31. Which electron configuration follows Hund's rule?
   a. ↑ ↑ ↑
   b. ↑ ↓ ↓
   c. ↑↑ ↑ __
   d. ↑↓ ↑ __

32. Which electron configuration represents an excited state?
   a. $1s^22s^22p^63s^23p^64s^1$
   b. $1s^22s^22p^63s^23p^63d^{10}$
   c. $1s^22s^22p^63s^23p^5$
   d. $1s^22s^22p^63s^23p^64s^23d^2$

33. Which of the following provides experimental evidence for the quantum nature of electrons?
   a. the photoelectric effect
   b. the spectrum of sodium
   c. the constant speed of light
   d. the transmission of light through a vacuum

34. In the Balmer series, electrons are excited from the
   a. $n = 1$ state
   b. $n = 2$ state
   c. $n = 3$ state
   d. $n = 4$ state

35. How many nodes are in an *s* orbital?
   a. zero
   b. one
   c. two
   d. three

Additional Practice Problems

1. What is the work function of a metal that emits electrons at a wavelength of 250 nm?

2. What is the wavelength of the following objects?
   a. an α particle with mass = 6.6×10^{-24} g and a velocity of 10% of the speed of light.
   b. a truck with mass = 2000 lb at 60 mph
   c. a runner with mass = 70 kg running at 3.5 m/s

3. What is the energy of the following electronic transitions?
   a. $n = 1 \rightarrow n = 3$
   b. $n = 2 \rightarrow n = 3$
   c. $n = 3 \rightarrow n = 4$
   d. $n = 3 \rightarrow n = \infty$

4. What is the electron configuration of the following?
   a. Mg b. Pd c. W^{4+} d. O^{2-} e. Fe^{3+}

5. How many unpaired electrons are in the following?
   a. Be b. Pt c. Ni^{2+} d. F^- e. Zn^{2+}

6. Rank the following elements from highest to lowest ionization energy.

   Sn, Ge, P, N, O

7. In an atom, how many electrons can have the described quantum numbers?
   a. $n = 1$
   b. $n = 2$
   c. $n = 2$ and $\ell = 0$
   d. $n = 3$ and $\ell = 1$
   e. $n = 3$, $\ell = 1$, and $m_l = 1$
   f. $n = 3$, $\ell = 3$, and $m_l = 3$

8. What is the ion formed by the following atoms?
   a. Na b. Cl c. P d. Ar e. Sr

9. If
   a. $n = 1$, then $\ell = ?$
   b. $n = 4$, $\ell = 3$, then $m_l = ?$
   c. $n = 3$, $\ell = 0$, then $m_l = ?$
   d. $n = 2$, then $\ell = ?$
   e. $n = 2$, $\ell = 1$, $m_l = 0$, then $m_s =$

10. Which quantum number relates to
    a. energy
    b. shape of the probability density
    c. orientation of the probability density
    d. size of the probability density
    e. spin of the electron

Self-Test Answer Key

Key Vocabulary Completion Answers

1. frequency
2. Bohr model
3. magnetic (m_ℓ)
4. node
5. photon
6. principal (n)
7. energy
8. effective nuclear charge (Z_{eff})
9. degenerate
10. Hertz (Hz or s^{-1})
11. ground
12. wavelength
13. Pauli
14. spin (m_s)
15. photoelectric effect
16. isoelectronic
17. Schrödinger
18. ionization energy
19. Heisenberg uncertainty principle
20. quantum
21. excited
22. probability density (or orbital)
23. Balmer series
24. aufbau principle
25. refraction

Multiple-Choice Answers

For further information about the topics these questions address, you should consult the text. The appropriate section of the text is listed in parentheses after the explanation of the answer.

1. c. $\lambda\nu = c$. 300 nm $\times \left(\frac{1\text{ m}}{10^9\text{ nm}}\right) = 2.99 \times 10^{-7}$ m.

   $c = 2.998 \times 10^8$ m/s. So, $\nu = \frac{c}{\lambda}$

   $= \frac{2.998 \times 10^8\text{ m/s}}{2.99 \times 10^{-7}\text{ m}} = 1.00 \times 10^{15}\text{ s}^{-1}$

   $= 1.00 \times 10^{15}$ Hz. (7.1)

2. a. Short wavelengths and high frequencies are higher energy, so the correct answer must be 1500 Hz or 200 nm. To convert Hz (ν) to wavelength (λ), the equation is $\lambda\nu = c$ or $\nu = \frac{c}{\lambda} = \frac{2.998 \times 10^8\text{ m/s}}{1500\text{ s}^{-1}}$

   $= 2.0 \times 10^5$ m, much longer than 200 nm. (7.1)

3. a. Frequency and energy are proportional to each other. Wavelength is inversely proportional to frequency. (7.1 and 7.3)

4. a. The principal quantum number (n) reflects energy and the distance from the nucleus. (7.8)

5. b. (a) is excitation, (c) is an absorption spectrum, (d) is an emission spectrum. (7.2)

6. d. The quantum model describes complex shapes in which an electron will probably exist. (7.8)

7. b. The energy to remove an electron is $\Phi = h\nu_0 =$

7.45×10^{-19}. Also $\nu_0 = \frac{c}{\lambda}$ so $\Phi = \frac{hc}{\lambda}$

$= \frac{(6.626 \times 10^{-34})(2.998 \times 10^{8})}{\lambda}$ so λ

$= \frac{(6.626 \times 10^{-34})(2.998 \times 10^{8})}{(7.45 \times 10^{-19})}$

$= 2.666 \times 10^{-7}\ \text{m} \times \left(\frac{10^{9}\ \text{nm}}{1\ \text{m}}\right) = 267\ \text{nm}$. This is

the minimum energy for the photoelectric effect; therefore, it is the maximum wavelength. (7.3)

8. a. Energy is inversely proportional to wavelength. (7.3)

9. d. The relationship between wavelength, mass, and

velocity is the de Broglie equation $\lambda = \frac{h}{m\nu}$. The

required units are λ in meters, m in kilograms, ν in meters per second, and $h = 6.626 \times 10^{-34}$ J•s. To rearrange the equation and solve for velocity,

$\nu = \frac{h}{m\lambda} = \frac{(6.626 \times 10^{-34}\ \text{J}\bullet\text{s})}{(9.1 \times 10^{-31}\ \text{kg})(7.00 \times 10^{-7}\ \text{m})}$

$= 1040$ m/s. (7.6)

10. c. According to the de Broglie equation, $\lambda = \frac{h}{m\nu}$, with

mass in kilograms and ν in meters per second, giving λ in meters. Thus,

$\lambda = \frac{(6.626 \times 10^{-34}\ \text{J}\bullet\text{s})}{(1.67 \times 10^{-27}\ \text{kg})(2.998 \times 10^{7}\ \text{m/s})}$

$= 1.32 \times 10^{-14}$ m. (7.6)

11. c. Noble gases are in the group in the rightmost column of the periodic table. (2.6 and 7.11)

12. b. The actinides are elements 58–71, which are naturally occurring metals. (The lanthanides tend to be synthetic.) As metals, these elements form cations. (7.11)

13. d. Isoelectronic means that the element would have the same electron configuration as neon. With two extra electrons, O^{2-} will have the same configuration as Ne. (7.11)

14. b. The loss of two electrons will give magnesium the electron configuration of Ne. (7.11)

15. c. m_ℓ values may be $-\ell, \ldots -1; 0; 1, \ldots +\ell$ (7.8)

16. b. The principal quantum number n reflects distance from the nucleus. The larger the value, the greater the distance. (7.8)

17. a. The first two quantum numbers reflect energy. Both the n and ℓ quantum numbers refer to energy. The n represents larger energy changes than the ℓ. However, larger changes of ℓ may be larger than small changes in n. As it turns out, the higher the value $n + \ell$, the higher the energy. If $n + \ell$ is the same, a higher n is a higher energy. (7.7 and 7.10)

18. c. The ℓ electrons can also be represented with letters; $\ell = 2$ is the same as a d electron. (7.8)

19. d. The ℓ quantum number represents the shape of the probability density. $\ell = 0$ or s electrons are in spherically shaped orbitals. (7.8)

20. b. Degenerate means that the energy is the same. Since both n and ℓ refer to energy, those two values must be the same for an electron to be degenerate. (7.8 and 7.10)

21. d. The first two quantum numbers reflect energy. Both the n and ℓ quantum numbers refer to energy. The n represents larger energy changes than the ℓ. However, larger changes of ℓ may be larger than small changes in n. As it turns out, the higher the value $n + \ell$, the higher the energy. If $n + \ell$ is the same, a higher n is a higher energy. Thus, electrons with the lowest sum value of $n + \ell$ have the lowest energy. (7.7 and 7.10)

22. d. The magnetic quantum number, m_ℓ, reflects orientation of the orbital. (7.8)

23. b. The electron configuration of bromine is $[\text{Ar}]4s^2 3d^{10} 4p^5$. Only the last group, if any, will have unpaired electrons. There are three p orbitals, for the five electrons, which will organize as $\underline{\uparrow\downarrow}\ \underline{\uparrow\downarrow}\ \underline{\uparrow}$.

24. b. The last electron is also the highest-energy electron; that was a $5d$ electron. Although $6p$ is higher energy, Au does not have $6p$ electrons. (7.10)

25. a. Heisenberg uncertainty principle says that position and momentum cannot be determined simultaneously. (7.6)

26. C. Ionization energy increases from left to right across the period and is higher for lower periods. (7.13)

27. b. Malleability is a property of all metals, not just transition metals. Although many transition metals form ions with a +2 charge, they also form ions with other charges; group II metals also form +2 ions. The *d* electrons are what characterize transition metals. (7.10)

28. a. The last electron determines the identity—in this case, a 3*s* orbital, so the element is in the third period. Since there are two electrons at this level, the element is two spaces from the left. (7.10)

29. a. You lose electrons to get a positive charge. The electrons lost are from the orbital with the highest *n*, in this case, the 4*s* electrons. (7.11)

30. a. Electron transitions get progressively smaller as the values of *n* and ℓ increase. Differences in *n* are larger than differences in ℓ. (7.10)

31. a. Hund's rule says that the lowest-energy arrangement of electrons is for the electrons to occupy separate orbitals and have parallel spins. (7.10)

32. b. An excited state promotes an electron to a higher energy level. In choice b the next highest energy level after 3*p* is 4*s*. Since it skipped to the higher level of 3*d*, that electron configuration is an excited state. (7.10)

33. b. The sodium spectrum is evidence of the quantum nature of light. The line spectra of the sodium spectrum is due to quantum transitions of electrons. (7.2)

34. b. The Balmer series, which has lines in the visible portion of the spectrum, excites electrons from $n = 2$, which is why the frequency is proportional to $c\left(\frac{1}{2^2} - \frac{1}{n^2}\right)$. (7.4)

35. a. The *s* orbital is spherical. Therefore, there are no regions where an electron cannot exist. (7.8)

ADDITIONAL PRACTICE PROBLEM ANSWERS

1. $250 \text{ nm} = \lambda$

$$E = \Phi = \frac{hc}{\lambda}$$

$$h = 6.626 \times 10^{-34} \text{ J}\bullet\text{s}$$

$$c = 2.998 \times 10^{8} \text{ m/s}$$

$$\lambda = 250 \text{ nm} \times \left(\frac{1 \text{ m}}{10^9 \text{ nm}}\right) = 2.50 \times 10^{-7} \text{ m}$$

$$\Phi = \frac{(6.626 \times 10^{-34} \text{ J}\bullet\text{s})(2.998 \times 10^{8} \text{ m/s})}{(2.50 \times 10^{-7} \text{ m})}$$

$$= 7.95 \times 10^{-19} \text{ J}$$

2. These problems require the de Broglie equation of $\lambda = h/mv$. The units on the equation are $\lambda = m$, $h = \text{J}\bullet\text{s}$, $m = \text{kg}$, $v = \text{m/s}$ and h is a constant of 6.0626×10^{-34} J•s.

a. $\text{mass} = 6.6 \times 10^{-24} \text{ g} \times \left(\frac{1 \text{ kg}}{1000 \text{ g}}\right) = 6.6 \times 10^{-27} \text{ kg}$

$$\text{velocity} = 10\% \text{ of } c = 0.1(2.998 \times 10^8 \text{ m/s}) = 2.998 \times 10^7 \text{ m/s}$$

$$\lambda = \frac{(6.626 \times 10^{-34})}{(6.6 \times 10^{-27})(2.998 \times 10^7)} = 3.3 \times 10^{-14} \text{ m}$$

b. $\text{mass} = 2000 \text{ lb} \times \left(\frac{454 \text{ g}}{1 \text{ lb}}\right) \times \left(\frac{1 \text{ kg}}{1000 \text{ g}}\right) = 908 \text{ kg}$

$$\text{velocity} = 60 \text{ mi/hr} \times \left(\frac{1.61 \text{ km}}{1 \text{ mi}}\right) \times \left(\frac{1000 \text{ m}}{1 \text{ km}}\right) \times \left(\frac{1 \text{ hr}}{3600 \text{ s}}\right) = 27 \text{ m/s}$$

$$\lambda = \frac{(6.626 \times 10^{-34})}{(908)(27)} = 2.7 \times 10^{-38} \text{ m}$$

c. mass = 70 kg velocity = 3.5 m/s

$$\lambda = \frac{(6.626 \times 10^{-34})}{(70)(3.5)} = 2.7 \times 10^{-36} \text{ m}$$

3. $\nu = 3.29 \times 10^{15} \text{ s}^{-1}\left(\frac{1}{n_1^2} - \frac{1}{n_2^2}\right)$

$$E = h\nu$$

so

$$E = (6.626 \times 10^{-34} \text{ J}\bullet\text{s})(3.29 \times 10^{15} \text{ s}^{-1})\left(\frac{1}{n_1^2} - \frac{1}{n_2^2}\right)$$

$$E = (2.18 \times 10^{-18} \text{ J})\left(\frac{1}{n_1^2} - \frac{1}{n_2^2}\right)$$

a. $E = (2.18 \times 10^{-18} \text{ J})\left(\frac{1}{1^2} - \frac{1}{3^2}\right) =$

$$(2.18 \times 10^{-18})(1 - 0.11111) = 1.94 \times 10^{-18} \text{ J}$$

b. $E = (2.18 \times 10^{-18})\left(\frac{1}{2^2} - \frac{1}{3^2}\right) =$

$$(2.18 \times 10^{-18})(0.25 - 0.1111) = 3.03 \times 10^{-19} \text{ J}$$

c. $E = (2.18 \times 10^{-18})\left(\frac{1}{3^2} - \frac{1}{4^2}\right) =$

$$(2.18 \times 10^{-18})(0.1111 - 0.0625) = 1.06 \times 10^{-19} \text{ J}$$

d. $E = (2.18 \times 10^{-18})\left(\frac{1}{3^2} - \frac{1}{\infty^2}\right) =$

$$(2.18 \times 10^{-18})(0.1111 - 0) = 2.42 \times 10^{-19} \text{ J}$$

4. a. Mg. $[Ne]3s^2$
   b. Pd. $[Kr]5s^24d^8$
   c. W^{4+}. $[Xe]4f^{14}5d^2$
   d. O^{2-}. $[He]2s^22p^6$
   e. Fe^{3+}. $[Ar]3d^5$

5. a. Be. Electron configuration = $[He]2s^2$ $2s$ = ↓↑ zero unpaired electrons
   b. Pt. Electron configuration = $[Xe]6s^24f^{14}5d^8$
   $5d$ = ↑↓ ↑↓ ↑↓ ↑ ↑, two unpaired electrons
   c. Ni^{2+}. Electron configuration = $[Ar]3d^8$
   $3d$ = ↑↓ ↑↓ ↑↓ ↑ ↑, two unpaired electrons
   d. F^-. Electron configuration = $[He]2s^22p^6$
   $2p$ = ↑↓ ↑↓ ↑↓, zero unpaired electrons
   e. Zn^{2+}. Electron configuration = $[Ar]3d^{10}$
   $3d$ = ↑↓ ↑↓ ↑↓ ↑↓ ↑↓, zero unpaired electrons

6. The atom with the highest ionization energy is N. Periodic trends generally predict that oxygen would be higher, but because removing an electron from N creates a half-full orbital, it actually has a lower ionization energy.
   The order is N, O, P, Ge, Sn.

7. a. If $n = 1$, then $\ell = 0$ and $m_\ell = 0$ and $m_s = \pm 1/2$; there are two electrons.
   b. If $n = 2$, then $\ell = 0$ and 1, and $m_\ell = 0$ and $-1, 0, +1$. With two electrons in each m_ℓ, there are eight electrons.
   c. If $n = 2$ and $\ell = 0$, then $m_\ell = 0$ and there are two electrons.
   d. *If* $n = 3$, $\ell = 1$, then $m_\ell = -1, 0, +1$, with two electrons each; that is six electrons.
   e. If $n = 3$, $\ell = 1$, and $m_\ell = 1$, there are two electrons for each m_ℓ, so only two electrons.
   f. $n = 3$, $\ell = 3$, and $m_\ell = 3$; there are no electrons. ℓ must be less than n; it's not. This orbital does not exist.

8. a. Na forms Na^+
   b. Cl forms Cl^-
   c. P forms P^{3-}
   d. Ar does not form an ion
   e. Sr forms Sr^{2+}

9. a. If $n = 1$, then $\ell = 0$
   b. If $n = 4$ and $\ell = 3$, then $m_\ell = -3, -2, -1, 0, 1, 2,$ or 3
   c. If $n = 3$ and $\ell = 0$, then $m_\ell = 0$
   d. If $n = 2$, then $\ell = 0$ or 1
   e. If $n = 2$, $\ell = 1$, and $m_\ell = 0$, then $m_s = +1/2$ or $-1/2$ (which it always does)

10. The quantum number that relates to
   a. energy is both principal (n) and angular momentum (ℓ)
   b. shape of probability density is angular momentum (ℓ)
   c. orientation of probability density is magnetic number (m_ℓ)
   d. size of the probability density is principal quantum number (n)
   e. spin of the electron is the spin quantum number (m_s)

CHAPTER 8 Chemical Bonding and Atmospheric Molecules

OUTLINE

REVIEW

Chapter Overview

Atoms are attached to each other by **bonds**. When atoms share valence electrons, this is called a **covalent bond**, the type of bond typically found between nonmetals. When valence electrons are transferred from one atom to another, the resulting ions are held together by electrostatic attraction; this is called an **ionic bond** and is typically found between metals and nonmetals. When atoms pool their valence electrons, the positive nuclei are held together by a sea of unconnected electrons; this is called a **metallic bond** and is typically found in metal–metal bonding. Because the sharing of electrons is very specific, formulas for covalent (or molecular) compounds are generally the actual molecular formulas that include all atoms in a molecule. However, ionic and metallic bonds have a general attraction, so there is no specific beginning and ending of the molecule; thus, the empirical (simplest ratio of atoms) formula is used.

Lewis structures are two-dimensional representations of how electrons are distributed in an atom, ion, or a molecule. Lewis structures are only used for the representative elements. Each valence electron is represented with a dot; covalent bonds are represented with lines, where each line represents two electrons. Ionic bonds are not shown with Lewis structures, although the individual ions that make up the bond can be. Similarly, metallic bonds are not represented through Lewis structures.

Electron distribution in Lewis structures is based on the octet rule. Electrons try to arrange themselves like a noble gas. If you consider just the *s* and *p* electrons, which are the valence electrons for a representative element, a noble gas would have eight valence electrons, hence the **octet rule**. Thus atoms tend to arrange themselves either through the formation of ions or of covalent bonds, so that each atom has an octet of valence electrons. In the covalent bond, both atoms get to count the shared electrons as part of their octet. Hydrogen, however, will never have more than a duet of electrons (two), since that will give it the electron configuration of the noble gas helium.

The electrons involved in the bonding of the representative elements are the outermost, or **valence**, electrons. The number of valence electrons contributed by an atom to the Lewis structure can be determined by counting across a period, skipping the transition metals if necessary. Since ions are made by adding (anions) or subtracting (cations) electrons, charge must also be considered when counting valence electrons.

For atoms and monoatomic ions, the valence electrons are represented as dots next to the atomic symbol. The possible positions are left, right, over, and under the atomic symbol. One dot (electron) is placed in each position, and then a second can be added to each position until all the available valence electrons are used up. The positions are all equivalent, so the order does not matter. However, the number of paired and unpaired electrons must be appropriate.

For molecules and polyatomic ions, the number of valence electrons is the sum of the valence electrons of each atom and the charge (adding electrons for a negative charge and subtracting them for a positive charge). Two shared electrons are represented by a line connecting the appropriate atoms. If only two electrons are shared between two atoms, it is called a **single bond**. If four electrons are shared, two lines are used and it is called a **double bond**. If six electrons are shared, three lines are used and it is called a **triple bond**. If electrons are not shared, it is called a **lone pair** and is represented by a pair of dots.

While electrons are shared, they are only shared equally if the two atoms are the same. Electrons that are shared equally are **nonpolar bonds**. Shared electrons that are more attracted to one atom than the other are called **polar bonds** or the bond is said to have a **dipole**. The atom that has the greatest share of electrons (and approximately how great a share) can be determined from the atom's electronegativity.

The **electronegativity** of an atom is its ability to attract electrons to itself in a covalent bond. The periodic trend of electronegativity is for electronegativity to increase from left to right across a period (with the exception of noble gases) and decrease from top to bottom across a group. Thus fluorine is the most electronegative atom, followed by oxygen, nitrogen, and then chlorine. Hydrogen is a special, but important case. If the periodic table were organized strictly by electronegativity, hydrogen would fit between carbon and boron.

If the electronegativity is large (difference > 2.0), the less electronegative atom (usually a metal) will give electrons to the other, more electronegative atom (usually a nonmetal). This creates ions that are then held together by electrostatic attraction and form an ionic bond. However, when the electronegativity of atoms is similar and high (as with nonmetals), neither atom is willing to completely give up electrons, so the electrons are shared in a covalent bond. The unequal sharing can be represented with an arrow next to the bond that points to the more electronegative atom. Alternately, lower case deltas (δ^+ or δ^-) can be used to represent the partial charge created by the unequal sharing.

For some substances it is possible to write a Lewis structure in more than one way. To determine which organization of atoms and electrons is correct, formal charges can be used. **Formal charge** compares the number of valence electrons contributed by the atom to the number of valence electrons assigned to the atom in the Lewis structure. Assigned electrons are all the electrons in the lone pairs associated with the atom and half the electrons in the bonds associated with the atom. The sum of the formal charges will be zero for a molecule or the charge for an ion. The appropriate Lewis structure is the one in which all formal charges are closest to zero and negative formal charges are on the more electronegative atom.

For some structures, it is possible for a double bond to have more than one exactly equivalent position (formal charges the same in each position). These equivalent structures are called **resonance structures**. The true arrangement of atoms and electrons is not any one of these structures, but the average of all of them. (For example, if there are two possible resonance structures, at each location where the double bond could exist, there is a 1.5 bond.)

Not all atoms strictly follow the octet rule. Atoms that contribute 3 or less valence electrons may have **deficient octets**. The most common example of this is boron, which contributes 3 electrons to the Lewis structure. Boron forms compounds where it is surrounded by only 6 (instead of 8) electrons. However, it also forms compounds where it is surrounded by 8 electrons, if they are available. Aluminum will also make compounds with deficient octets (6 e^-), as will beryllium (4 e^-). Other atoms will form compounds that require them to expand their octet and be surrounded by as many as 12 electrons. Atoms in the second period will never expand their octet. Only the central atom will expand its octet, and that usually occurs when it is bonded to a highly electronegative atom. Fulfilling the octet rule and being surrounded by eight electrons is a possibility for all atoms except hydrogen (which *always* has a maximum of 2 electrons).

A few molecules have an odd number of valence electrons. In this case it is impossible for all atoms to fulfill the octet rule. One of the atoms will be surrounded by only seven electrons. Use formal charges to determine which atom will have the seven electrons. Compounds with an unpaired electron are called **free radicals** and are very reactive.

Bond length is the distance between nuclei of bonded atoms. It depends on the type of atom as well as the type of bond. The type of bond can be described by its **bond order**, where a single bond has a bond order of one, a double bond has a bond order of two, and a triple bond has a bond order of three. As bond order increases, bond length decreases.

Bond enthalpy is the energy required to break a bond. It also depends on the type of atoms and the bond order. As bond order increases, bond enthalpy increases. Because of conservation of energy, the energy required to break a bond is the same quantity of energy released when a bond is formed. Because chemical reactions are essentially the forming and breaking of bonds, bond enthalpy can be used to estimate the enthalpy of reaction (ΔH_{rxn}). The enthalpy of reaction can be calculated from the sum of the bond enthalpy of each bond that is broken (bonds in the reactants) minus the sum of the bond enthalpy of each bond formed (bonds in products).

Worked Examples

Ionic and Covalent Bonds

Example 1.
When bonds form between the following atoms, will the bond be ionic, polar covalent, or nonpolar covalent?

a. Ca and F

b. H and F

c. F and F

d. O and F

Collect and Organize We are given pairs of atoms to combine into compounds and are asked to characterize the resulting bond as either ionic or covalent. Furthermore, for cases of covalent bonding, we are asked to determine whether the resulting bond is polar or nonpolar. The difference in electronegativity, $\Delta\chi$, between the atoms of the compound will allow us to distinguish between the three choices. Values of the electronegativity for the given elements are Ca (1.0), F (4.0), H (2.1), and O (3.5). (These values are from the table in your text.)

Analyze Bonds are considered ionic when $\Delta\chi > 2.0$. Nonpolar covalent bonds have $\Delta\chi = 0$. Polar covalent bonds result when $\Delta\chi$ lies between these two extremes.

Solve

a. $\Delta\chi = 4.0 - 1.0 = 3.0$, therefore, it is an ionic bond.

b. $\Delta\chi = 4.0 - 2.1 = 1.9$, therefore, it is a polar covalent bond.

c. $\Delta\chi = 4.0 - 4.0 = 0$, therefore, it is a nonpolar covalent bond.

d. $\Delta\chi = 4.0 - 3.5 = 0.5$, therefore, it is a polar covalent bond.

Think About It We can use the magnitude of the difference in electronegativities to predict the type of bonding that will occur between pairs of atoms. A reasonable way to estimate whether a bond is ionic or covalent is to consider whether the nonmetal is combining with a metal or another nonmetal. Metals tend to have low electronegativities and a metal–nonmetal bond is usually an ionic bond.

Lewis Theory

Example 1.
How many valence electrons are in each of the following?

a. N b. H_2S c. CO_3^{2-} d. NH_4^+

Collect and Organize We are given an element and three compounds and are asked to determine the number of valence electrons in each.

Analyze The number of valence electrons for an element is equal to its total of *s* and *p* valence electrons in the periodic table. In order to determine the number of valance electrons for compounds one simply needs to sum the number of valence electrons for each element in the compound.

Solve

a. Nitrogen has an electron configuration of $[He]2s^22p^3$. It has five valence electrons.

b. Hydrogen has one valence electron, and sulfur has six. The total for the molecule is $2(1) + 6 = 8$.

c. Carbon has four valence electrons; oxygen has six; then two for the charge. So, $4 + 3(6) + 2 = 24$.

d. Nitrogen has five valence electrons; hydrogen has one, minus one for the charge. So, $5 + 4(1) - 1 = 8$.

Think About It The power of the periodic table has been demonstrated once again.

Example 2.
Draw the Lewis structures for the following.

a. N b. Mg c. Mg^{2+} d. F^- e. Ne

Collect and Organize We are given five elements (two of which are ions) and are asked to draw the Lewis structure for each. Lewis structures for elements consist of the symbol for the element surrounded by one dot for each valence electron.

Analyze The number of dots to place around neutral elements can be determined by counting across the period, skipping the transition metals in the middle. For ions, you must add one dot for each negative charge and remove one for each positive charge.

Solve

a. N has five valence electrons, so the Lewis structure is

·N̈· (with a pair of dots below)

b. Mg has two valence electrons.

Ṁg·

c. Mg^{2+} has no valence electrons, so its Lewis structure is

Mg^{2+}

d. F^- has eight valence electrons, so its Lewis structure is

:F̤̈:⁻

e. Ne has eight valence electrons, so its Lewis structure is

:N̤̈e:

THINK ABOUT IT Lewis structures display the valence electrons of the atom. These outermost electrons are the ones available for reactions.

EXAMPLE 3.

What are the Lewis structures for the following molecules?

a. PCl_3 b. H_2CO_3 c. CH_3NH_2

COLLECT AND ORGANIZE We are given three molecules and are asked to determine their Lewis structures.

ANALYZE We can use electronic bookkeeping to predict the Lewis structure. First we must determine how may electrons we have; then how many electrons are needed. Most elements need 8; hydrogen only needs two. The difference is the number of electrons that must be shared. The number of bonds can be determined by dividing the difference by two.

SOLVE

a. 1. Count the number of valence electrons (have): P = 5, Cl = 7, so 5 + 3(7) = 26.
2. Determine the number of electrons needed (need): eight for each atom = 4(8) = 32.
3. Determine the number of bonds from the difference: 32 – 26 = 6, so there are three bonds.
4. Determine the skeletal structure. (The skeletal structure shows the way the atoms are connected to each other.) Phosphorus is the central atom. It is surrounded by chlorines
5. The chlorines are connected to the phosphorus using all three bonds.
6. The rest of the electrons are distributed around the atoms so that each has an octet.
Hint 1: Distribute in pairs working from the outside in.
Hint 2: Two electrons on each side (up, down, left, right) is an octet.

```
  :C̈l:
   |
  :P—C̤̈l:
   |
  :C̤l:
```

b. 1. Count the number of valence electrons: H = 1, C = 4, O = 6, so 2(1) + 4 + 3(6) = 24.
2. Number of electrons needed: H needs 2; the others need 8. So, 2(2) + 4(8) = 36.
3. Number of bonds needed: 36 – 24 = 12, so six bonds.
4. Skeletal structure: This is an oxyacid, so carbon is the central atom. It is surrounded by oxygen, and the hydrogen is connected to one of the oxygens. Each oxygen is the same, so it doesn't matter which one.
5. Since the skeletal structure only uses five of the six bonds, another bond must be somewhere.

```
H—O—C—O—H
    |
    O
```

The hydrogens already have their maximum number of bonds. The Lewis structure of the oxygen atom has two unpaired electrons, which implies that oxygen likes to bond twice. The oxygens bonded to the hydrogens already have two bonds, so a second (double) bond is most likely on the other oxygen. The rest of the electrons can then be distributed.

```
H—Ö̤—C—Ö̤—H
     ||
    :O̤:
```

c. 1. Count valence electrons: C = 4, H = 1, N = 5, so 4 + 3(1) + 5 + 2(1) = 14.
2. Count electrons needed: two for hydrogen, eight for others, so 2(8) + 5(2) = 26.
3. Determine number of bonds: 26 – 14 = 12, or six bonds.
4. Skeletal structure. Hydrogen cannot be a connecting atom, so the carbon must be connected to the nitrogen. The way the formula is written it implies three hydrogens on the carbon and two on the nitrogen. (Formulas are often written so that the Lewis structure is implied by the way the atoms are listed.)
5. Distribute the rest of the electrons. When all the atoms are connected, there are no more bonds, and only two more electrons. The only atom without an octet is nitrogen.

```
   H  H
   |  |
 H—C—N:
   |  |
   H  H
```

THINK ABOUT IT Following the procedure outlined above makes the determination of Lewis structures for molecules an easy task.

EXAMPLE 4.

What are the Lewis structures for the following ions?

a. SO_3^{2-} b. NH_4^+ c. CN^-

COLLECT AND ORGANIZE We are give three ions and are asked to determine their Lewis structures.

ANALYZE We can use the electronic bookkeeping to determine the Lewis structures of polyatomic ions, but with the exception that the number of valence electrons must be modified by the consideration of the charge of the ion.

SOLVE

a. 1. The number of valence electrons is 6 + 3(6) + 2 = 26.
   2. The number of electrons needed is 4(8) = 32.
   3. The number of bonds is 32 – 26 = 6, so there are three bonds.
   4. Three bonds are needed to connect the oxygens to the sulfur in the skeletal structure, so the rest of the electrons must be lone pairs.

:Ö–S̈–Ö:
    |
   :Ö:

b. 1. The number of valence electrons is 5 + 4(1) – 1 = 8.
   2. The number of needed electrons is 8 + 4(2) = 16.
   3. The number of bonds is 16 – 8 = 8, so there are four bonds.
   4. To connect the hydrogens to the nitrogen, four bonds are needed. That also uses up all the electrons.

```
    H
    |
H—N—H
    |
    H
```

c. 1. The number of valence electrons is 4 + 5 + 1 = 10.
   2. The number of needed electrons is 2(8) = 16.
   3. The number of bonds is 16 – 10 = 6, so three are three bonds.
   4. With only two atoms all the bonds must be between these two atoms. One lone pair for each atom will complete each atom's octet.

:N≡C:

THINK ABOUT IT It is necessary to account for the additional electrons found in anions and the loss of electrons upon cation formation when determining the Lewis structures on ions.

UNEQUAL SHARING, ELECTRONEGATIVITY, AND PERIODIC PROPERTIES

EXAMPLE 1.
Rank the following atoms from highest to lowest ionization energy.

Ca F Ne Ge

COLLECT AND ORGANIZE We are given four elements and are asked to list them according to decreasing ionization energy. Ionization energy is the energy required to remove an electron from a gaseous atom. The locations of atoms within the periodic table can be used to determine their relative ionization energies.

ANALYZE Ionization energy increases from left to right across a row and from the bottom to top of a column of the periodic table.

SOLVE The element whose position is the highest and farthest right in the periodic table is neon. Therefore, it has the highest ionization energy. The one whose position is closest to the bottom and farthest left is calcium. Therefore, it has the lowest ionization energy. Fluorine is very near neon; it should be next highest. Germanium is between fluorine and calcium and so is its ionization energy. The order from highest to lowest is Ne > F > Ge > Ca.

THINK ABOUT IT The increase in ionization energy from left to right across a row is the result of a larger force of attraction between the nucleus and an electron as the number of protons in the nucleus of the atom becomes larger.

The increase in ionization energy from bottom to top of a column results from the fact that the principal quantum number of the orbital holding the outermost electron becomes larger as we go down a column of the periodic table. Although the number of protons in the nucleus also becomes larger, the electrons in smaller shells and subshells tend to screen the outermost electron from some of the force of attraction of the nucleus. Furthermore, the electron being removed when the first ionization energy is measured spends less of its time near the nucleus of the atom, and it therefore, takes less energy to remove this electron from the atom.

EXAMPLE 2.
Rank the following from highest electronegativity to lowest electronegativity.

N P Ge As Cl

COLLECT AND ORGANIZE We are given five elements and are asked to list them according to decreasing electronegativity. Electronegativity is the tendency of an atom to draw electrons to itself when in a covalent bond. We can use the position of the elements within the periodic table to determine their relative electronegativities.

ANALYZE In general, electronegativity increases on passing from left to right along a period, and decreases on descending a group. The most electronegative atom is fluorine, followed by oxygen then nitrogen. The high electronegativity of these elements often influences the properties of the molecule, so it is helpful to know these top three.

SOLVE Nitrogen is on the list of the top three, so it is the highest. Chlorine is in the rightmost column (not counting noble gases) and still fairly high up, so it is next. Organizing the rest of the elements based on how close they are to fluorine, the ranking from highest electronegativity to lowest electronegativity is N > Cl > P > As > Ge.

THINK ABOUT IT There are some exceptions to this general rule. The most important exception is hydrogen which is slightly less electronegative than carbon.

RESONANCE

EXAMPLE 1.
Draw all the resonance structures of the following.

a. SO_2 b. NO_3^-

COLLECT AND ORGANIZE We are asked to draw resonance structures for SO_2 and NO_3^-.

ANALYZE Lewis structures where a double bond can have two or more exactly equivalent positions are resonance structures. Each resonance structure shows a different equivalent position of the double bond. The true structure is an average of all the possible resonance structures.

SOLVE

a.
```
 ..  ..  ..        ..  ..  ..
 O = S — O:      :O — S = O
 ..      ..        ..      ..
```
Each sulfur-oxygen bond is actually 1 1/2 bond rather than either a double or a single bond.

b.
```
  ..      ..     ..      ..      ..      ..
:O — N = O       O = N — O:    :O — N — O:
  ..  |   ..     ..  |   ..      ..  ||  ..
     :O:            :O:             :O:
      ..             ..
```
Each nitrogen-oxygen bond is actually a 1 1/3 bond rather than either a single or double bond.

THINK ABOUT IT Resonance structures stabilize molecules by delocalizing electrons in different parts of the molecule.

FORMAL CHARGES

EXAMPLE 1.
What is the formal charge of each atom in the following structures of CH_4O? Which structure is correct?

```
 H  H              H                 H
 |  |              |                 |
:C — O — H     H — C — O — H     H — C — O — H
 |    ..           |   ..            ..  |
 H                 H                     H
```

COLLECT AND ORGANIZE We are asked to determine the formal charge of each atom in the three structures of CH_4O.

ANALYZE Formal charge is calculated by subtracting the number of electrons assigned to an atom from the number of valence electrons contributed by the atom. Both electrons of a lone pair and half the electrons of a bond are "assigned." Formal charge = valence electrons – assigned electrons.

Formal charges are a method for determining the most appropriate Lewis structure. There is a formal charge for each atom in a structure. The sum of the formal charges will be the same as the charge on the compound. The best Lewis structure will have the lowest formal charges. Any structure containing an atom having a formal charge of more than +1 or less than –1 is unlikely to be correct.

SOLVE In all three structures, hydrogen has one bond. Therefore it is assigned one electron. It also has one valence electron, so its formal charge is zero in all the structures.

Carbon has four valence electrons. In the first structure, carbon has one lone pair (2) and three bonds (3), so five assigned electrons. Valance electrons – assigned electrons = –1 = formal charge. In the second structure the carbon has no lone pairs and four bonds, so four assigned electrons. Formal charge = 4 – 4 = 0. In the third structure carbon has two lone pairs and two bonds, so six assigned electrons. Formal charge = 4 – 6 = –2.

Oxygen has six valence electrons. In the first structure the oxygen has one lone pair and three bonds, so five assigned electrons. Formal charge = 6 – 5 = +1. In the second structure the oxygen has two lone pairs and two bonds, so six assigned electrons. Formal charge = 6 – 6 = 0. In the third structure the oxygen has no lone pairs and four bonds, thus four assigned electrons. Formal charge = 6 – 4 = +2.

The correct structure is the middle one. The formal charge of each atom is zero.

THINK ABOUT IT Formal charge is a way of keeping track of electrons in a molecule by assuming that electrons in a chemical bond are shared equally between atoms, regardless of relative electronegativity.

EXCEPTIONS TO THE OCTET RULE

EXAMPLE 1.
Will the following have an expanded octet?

a. ClO_3^- b. SCl_2

COLLECT AND ORGANIZE We are asked if either ClO_3^- or SCl_2 will have an expanded octet.

ANALYZE Only atoms with *d* orbitals can expand their octet. This requires that the atom have a principal quantum number of 3 or more. After drawing a structure in the normal way, if the formal charges on the molecule are decreased by creating a double bond, the double bond will form.

SOLVE

a. Using electronic bookkeeping, the molecule needs 32 electrons and has 7 + 3(6) + 1 = 26, so there are three bonds. This leads to a Lewis structure of

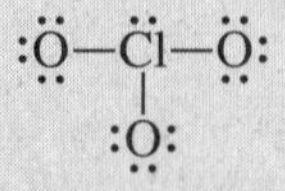

```
:Ö—Cl—Ö:
    |
   :O:
```

However, the formal charge on chlorine is +2 and that on each oxygen is –1. If the chlorine double-bonds to two of the oxygens, the formal charge on chlorine and those two oxygen atoms is zero. Therefore, the more likely structure is

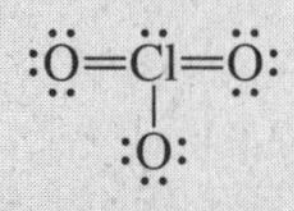

```
:Ö=Cl=Ö:
    |
   :O:
```

b. Using electronic bookkeeping, this molecule needs 24 electrons and has 20 electrons. This predicts two bonds. So the structure is

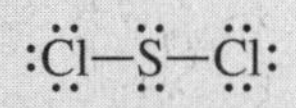

```
:Cl—S—Cl:
```

The formal charge on each atom is zero. There is no point in changing the structure.

THINK ABOUT IT Although some atoms can expand their octet, they do not always do so. Only the central atom will expand its octet.

EXAMPLE 2.

What are the Lewis structures for the following compounds?

a. XeF_4 b. PCl_5 c. I_3^-

COLLECT AND ORGANIZE We are given three compounds and are asked to determine their Lewis structures.

ANALYZE We use the procedure outlined in the Lewis structure examples.

SOLVE

a. The number of valence electrons "needed" is 8(5) = 40 and they have 8 + 4(7) = 36, so the difference is four, so there are two bonds. With a formula of XeF_4, having two bonds is obviously impossible. This is a clue that this molecule probably expands its octet. The best way to approach this is to connect everything to the central atom, fill the octets of the terminal (outside) atoms, and add any leftover electrons to the central atom.

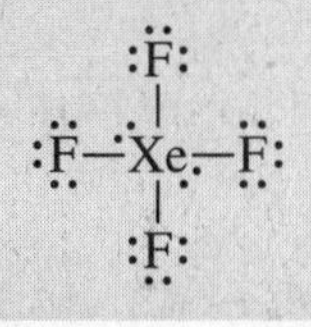

```
      :F:
       |
:F—Xe—F:
       |
      :F:
```

b. Unless they are the central atom, halogens only bond once. Consequently, it seems likely that all five chlorines bond to the phosphorus. The molecule needs 6(8) = 48 electrons and has 5 + 5(7) = 40 electrons; the difference is 8 electrons, or four bonds. That isn't going to work. Since phosphorus is in the third period of the periodic table, it can expand its octet. So the Lewis structure is

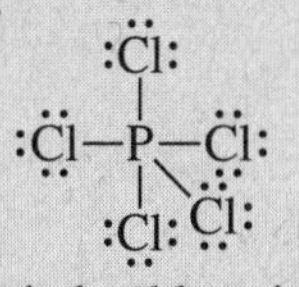

```
      :Cl:
        |
:Cl—P—Cl:
        | \
     :Cl: Cl:
```

c. If you use electronic bookkeeping, this molecule needs 24 electrons and has only 22, which predicts one bond. This won't work, so use the alternate method of filling the octets of the outside atoms and putting the extra electrons on the central atom.

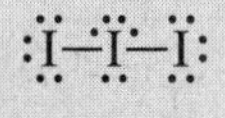

```
:I—I—I:
```

THINK ABOUT IT Electronic bookkeeping is helpful, but doesn't always work, especially with expanded octets.

EXAMPLE 3.

What is the Lewis structure for $AlCl_3$?

COLLECT AND ORGANIZE We are asked to determine the Lewis structure for $AlCl_3$

ANALYZE Boron, aluminum, and beryllium may have deficient octets. Aluminum is stable when surrounded by six electrons.

SOLVE This compound has 3 + 3(7) = 24 electrons. The Lewis structure of $AlCl_3$ is

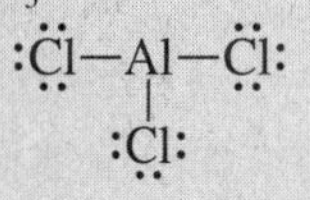

```
:Cl—Al—Cl:
      |
     :Cl:
```

THINK ABOUT IT Electronic bookkeeping does not work for deficient octets.

EXAMPLE 4.

What is the Lewis structure of ClO_2?

COLLECT AND ORGANIZE We are asked to determine the Lewis structure of ClO_2.

ANALYZE Some molecules do not have an even number of electrons. For this situation it is impossible for all atoms to have an octet and even for all electrons to be paired. Electronegativity is used to determine which atom will have seven instead of eight electrons.

Solve The total number of electrons is 7 + 2(6) = 19 electrons. The atom that is the least electronegative in this case is chlorine. Therefore, it is the atoms assigned seven electrons instead of eight.

:Ö–C̤̈l–Ö:

Think About It Since electronegative atoms attract electrons, it will be the least electronegative atom that is shorted an electron.

Bond Energies

Example 1.
What is the ΔH_{rxn} of the following reactions, based on bond energies?

a. $2\ HCl + F_2 \rightarrow Cl_2 + 2\ HF$

b. $C_2H_4 + H_2 \rightarrow C_2H_6$

c. $CH_4 + 2\ O_2 \rightarrow CO_2 + 2\ H_2O$

Collect and Organize We are given three reactions and are asked to determine ΔH_{rxn} based on bond energies.

Analyze Average bond energies (BEs) are listed in your text. To determine what bonds exist in the molecules of the products and reactants, you may have to draw the Lewis structure. The ΔH_{rxn} is determined from the formula

$$\Sigma BE_{broken} - \Sigma BE_{made} = \Delta H_{rxn}$$

Solve

a. Written in semi-Lewis structures (just showing bonds, not lone pairs of electrons), this reaction is

H—Cl + H—Cl + F—F → Cl—Cl + H—F + H—F

Therefore, two H—Cl bonds and one F–F bond are broken; two H—F and one Cl—Cl bonds are formed. The bond energies are H—Cl = 431 kJ/mol, F—F = 155 kJ/mol, Cl—Cl = 242 kJ/mol, and H—F = 565 kJ/mol. Using the preceding formula

$$\Delta H = [2(431) + 155] - [242 + 2(565)] = 1017 - 1372 = -355 \text{ kJ/mol}$$

b. Written as semi-Lewis structures, this reaction is

$$\mathrm{H_2C{=}CH_2} + \mathrm{H{-}H} \rightarrow \mathrm{H{-}CH_2{-}CH_2{-}H}$$

The carbon–hydrogen bonds on the first structure are not broken, so one way to do this is to say the bonds broken are C=C (612 kJ/mol) and H—H of H_2 (436 kJ/mol) and bonds formed are one C—C (348 kJ/mol) and two C—H (413 kJ/mol), so

$$\Delta H = [612 + 436] - [348 + 2(413)] = 1048 - 1174 = -126 \text{ kJ/mol}$$

If that is not obvious, all the bonds in the reactants (C=C is 612 kJ/mol, C—H is 413 kJ/mol, and H—H is 436 kJ/mol) can be broken and all the bonds in the products formed (C—C is 348 kJ/mol and C—H is 413 kJ/mol).

$$\Delta H = [612 + 4(413) + 436] - [348 + 6(413)] = -126 \text{ kJ/mol}$$

c. CH_4 consists of four C—H bonds, O_2 has a O=O bond, and the product carbon dioxide is O=C=O and water is H—O—H, so bonds broken are four C—H bonds (413 kJ/mol) and one O=O bond (497 kJ/mol) and bonds formed are two C=O bonds (743 kJ/mol) and two O—H bonds (463 kJ/mol). So

$$\Delta H = [4(413) + 497] - [2(743) + 2(463)] = 2149 - 2412 = -263 \text{ kJ/mol}$$

Think About It Since it takes energy to break bonds, those values are positive. Energy is released when bonds are made; therefore, those values are negative. The formula is set up to introduce the positive and negative values, so use the absolute values in the table when using the formula.

Self-Test

Key Vocabulary Completion Questions

__________ 1. In a molecule, an atom is surrounded by eight electrons

__________ 2. Attachment of positively charged particles to negatively charged particles

__________ 3. Where a double bond can have more than one exactly equivalent position

__________ 4. Two electrons that do not participate in bonding in a molecule

__________ 5. Connection formed by sharing of electrons

__________ 6. Diagram showing how atoms are connected to each other

__________ 7. Unequally shared electrons

__________ 8. Electrons in the outermost shell

__________ 9. Number of unpaired electrons in an oxygen atom

__________ 10. Tendency of an atom to draw electrons to itself in a bond

__________ 11. Energy released when a bond is formed

__________ 12. Number of valence electrons minus number of assigned electrons

__________ 13. Four electrons shared by two atoms

__________ 14. Electrons equally shared between two atoms

__________ 15. Attachment of atoms by a sea of electrons

__________ 16. Molecule with an odd number of valence electrons

__________ 17. Bond order for a triple bond

__________ 18. Atom that commonly forms molecules with a deficient octet

__________ 19. The most electronegative atom of all

__________ 20. Atom that never forms more than one bond

Multiple-Choice Questions

1. How many valence electrons does phosphorus have?
   a. 3
   b. 5
   c. 7
   d. 8

2. When iron combines with chlorine, it will make
   a. an ionic bond
   b. a polar covalent bond
   c. a nonpolar covalent bond
   d. no bond

3. What type of bond forms between two oxygen atoms?
   a. single
   b. double
   c. triple
   d. none

4. How many valence electrons are in F^-?
   a. one
   b. five
   c. seven
   d. eight

5. In the Lewis structure of PCl_3, how many bonds and how many lone pairs are on the central atom? (Note: In this question a double or a triple bond would count as one bond.)
   a. four bonds, no lone pairs
   b. three bonds, no lone pairs
   c. three bonds, one lone pair
   d. four bonds, one lone pair

6. Of the following, which is the most electronegative atom?
   a. Cs
   b. C
   c. H
   d. O

7. Which element will form the most polar bond with fluorine?
   a. Cl
   b. Si
   c. O
   d. S

8. How many valence electrons are in CBr_4?
   a. 32
   b. 40
   c. 8
   d. 28

9. How many valence electrons are needed by OF_2?
   a. 20
   b. 24
   c. 8
   d. 14

10. Which of the following has the shortest bond between atoms?
    a. Cl_2
    b. O_2
    c. N_2
    d. all are the same

11. Which of the following molecules has resonance?
    a. NO_3^-
    b. N_2
    c. NH_3
    d. none

12. How many bonds and how many lone pairs are around the sulfur of SO_3? (Note: a double or triple bond counts as one bond for this question.)
    a. four bonds and no lone pairs
    b. three bonds and one lone pair
    c. three bonds and no lone pairs
    d. two bonds and one lone pair

13. How many electrons are available (total number of valence electrons) in C_2H_6?
    a. 64
    b. 22
    c. 14
    d. 20

14. What kind of bond is between the bromines of Br_2?
    a. single
    b. double
    c. triple
    d. resonance

15. The bonding in a molecule
    a. must be either totally ionic or totally covalent
    b. may be more ionic-like or more covalent-like, depending on the atoms bonding
    c. is the same mixture of ionic-like and covalent-like bonds, regardless of the atoms involved
    d. there is no difference between ionic and covalent bonding

16. Which bond is the most polar?
    a. the bond between F and F
    b. the bond between F and O
    c. the bond between Si and O
    d. the bond between Si and P

17. Why does hydrogen need two electrons instead of eight?
    a. because of hydrogen's metal-like properties
    b. because that is the typical charge for hydrogen
    c. because hydrogen has a noble gas configuration with two electrons
    d. because hydrogen does need eight electrons, just like all the rest

18. The bond length between atoms involved in resonance
    a. changes between bonds every microsecond or so within the molecule
    b. has one preferred position so that one bond is shorter than the other, which is the same for each molecule; however, the shorter bond can only be detected experimentally
    c. has one preferred position so that one bond is shorter than the other, but its position varies between molecules so that the average bond length over a mole is the same for all resonance positions
    d. is exactly the same for all bond lengths

19. How many resonance structures are there of HNO_3?
    a. one (or there is no resonance)
    b. two
    c. three
    d. four or more

20. What is the formal charge on the sulfur of SO_2?
    a. zero
    b. –1
    c. +1
    d. none of these

21. What is the formal charge on nitrogen in NH_4^+?
    a. zero
    b. +1
    c. –1
    d. none of these

22. What is wrong with the following Lewis structure of SO_3?

    :Ö–S̈–Ö:
        |
      :Ö:

    a. it has the wrong number of valence electrons
    b. O breaks the octet rule
    c. the formal charges are too high
    d. nothing, it's perfect

23. Which of the following atoms will not expand its octet?
    a. P
    b. C
    c. S
    d. Br

24. In $HClO_2$, hydrogen is bonded to
    a. chlorine
    b. oxygen
    c. both chlorine and oxygen
    d. both oxygens

25. In molecules with an odd number of electrons, the unpaired electron goes on
    a. the least electronegative atom
    b. the most electronegative atom
    c. the central atom
    d. the atom experimentally determined to be correct, since there is no way to predict it theoretically

26. Which atom might have a deficient octet?
    a. B
    b. P
    c. Br
    d. all of the above

27. For an atom to expand its octet, it must
    a. have a *d* orbital
    b. have a minimum of four valence electrons as an atom
    c. have a maximum of four valence electrons as an atom
    d. be a metalloid

28. What is the formal charge on chlorine in ClF_5?
    a. +5
    b. +1
    c. 0
    d. –1

29. Which molecule will have the highest bond energy?
    a. Cl_2
    b. O_2
    c. N_2
    d. bond energy must be experimentally determined

30. The bonding in Na_2CO_3 could best be described as
    a. ionic
    b. covalent
    c. both ionic and covalent
    d. metallic

Additional Practice Problems

1. Draw the Lewis structures for each of the following:

   a. S b. SiH_4 c. HNO_3 d. NO_2^- e. SO_3^{2-}

2. Rank the following from highest to lowest electronegativity: Cu, C, N, H, P, F.

3. Draw the Lewis structures for each of the following:

   a. HIO_3 b. XeF_2 c. BH_3 d. ClF_3 e. I_3^-

4. Compare the Lewis structures of N_2O, NO_2^-, and O_3^-.
   a. Which of the three exhibit resonance?
   b. Which has the most lone pairs on the central atom?
   c. What is the formal charge on the central atom for each?

5. When the following atoms bond with hydrogen, rank the bond from most polar to least polar. For each bond, which atom more effectively attracts electrons?

   Na, C, P, F, S

Self-Test Answer Key

Key Vocabulary Completion Answers

1. octet rule
2. ionic bond
3. resonance
4. lone pair
5. covalent bond
6. Lewis structure
7. polar bond
8. valence
9. two
10. electronegativity
11. bond enthalpy
12. formal charge
13. double bond
14. nonpolar bond
15. metallic bond
16. free radical
17. three
18. boron
19. fluorine
20. hydrogen

Multiple-Choice Answers

For further information about the topics these questions address, you should consult the text. The appropriate section of the text is listed in parentheses after the explanation of the answer.

1. b. P has five electrons in its *s* and *p* orbitals. (8.6)
2. a. There is a large electronegativity difference between chlorine and iron, appropriate for an ionic bond. It is also the combination of a metal and a nonmetal. (8.3)
3. b. Oxygen atoms have six valence electrons, so each atom is two electrons short of its octet. By each atom sharing two electrons, there are four shared electrons or a double bond. (8.2)
4. d. Fluorine has seven valence electrons. The negative charge adds one more, for a total of eight. Another way of looking at it is that ions tend to form so that the atom will have a noble gas configuration of electrons, or an octet. (8.2)
5. c. The Lewis structure of PCl_3 (shown below) has three single bonds and one lone pair on phosphorus. (8.2)

   :C̈l:
   |
   :C̈l:—P̤—C̈l:

6. d. Oxygen is the closest to fluorine on the periodic table. It is the second most electronegative element. (8.3)
7. b. The higher the difference in electronegativity, the more polar the bond. (8.3)
8. a. Carbon has four valence electrons and bromine has seven, so $4 + 4(7) = 32$. (8.2)
9. b. Each of the three atoms needs eight electrons, so $3(8) = 24$. (8.2)
10. c. Nitrogen has a triple bond, oxygen has a double bond, and chlorine has a single bond. Multiple bonds are shorter and stronger. (8.7)
11. a. There is one double bond in nitrate ion that can equally occupy any of three positions. N_2 has a triple bond and NH_3 has only single bonds. (8.4)

12. c. The Lewis structure of SO_3 is Ö=S—Ö: (8.2 and 8.4)
    |
    :Ö:

13. c. Carbon has four valence electrons, hydrogen has one. Therefore, for the entire molecule: valence electrons = 2(4) + 6(1) = 14. (8.2)

14. a. The Lewis structure of Br_2 is :B̈r—B̈r: (8.2)

15. b. The higher the difference in electronegativity between atoms, the more ionic-like the bond. The actual split between what is an ionic bond and what is a covalent bond is somewhat arbitrary. (8.3)

16. c. The higher the difference in electronegativity, the more polar the bond. Elements have higher electronegativity the closer to fluorine they are on the periodic table. Silicon and oxygen are farthest apart on the periodic table and therefore, have the greatest difference in electronegativity and the most polar bond. (8.3)

17. c. With two electrons, hydrogen will have the noble gas configuration of helium, $1s^2$, which is a full shell for $n = 1$. (8.2)

18. d. Although the method of Lewis structures requires a different Lewis structure for each resonance form, the reality is that the true molecule is an average of the Lewis structures. (8.4)

19. b. The double bond may exist between either of the oxygens that are not bonded to hydrogen. (8.4)

20. c. The Lewis structure is Ö=S̈—Ö: Sulfur normally has six valence electrons. In the Lewis structure it is assigned five electrons. Valence electrons – assigned electrons = 6 – 5 = +1. (8.5)

21. b. The Lewis structure for ammonium ion consists of each hydrogen singly bonded to the nitrogen. Thus nitrogen has four assigned electrons. Formal charge = valence electrons (5) – assigned electrons (4) = +1. (8.5)

22. a. SO_3 has 24 valence electrons; the Lewis structure has 26. (8.2)

23. b. Only atoms with *d* orbitals will expand their octet. That requires the principal quantum number to be 3 or greater. The highest principal quantum number of carbon's electrons is 2. (8.6)

24. b. Hydrogen can only bond once. In oxyacids it bonds to the oxygen, not to the central atom. (8.2)

25. a. Since an unpaired electron makes the atom short of its octet, it would be on the least electronegative atom. Since electronegative atoms attract electrons, the least electronegative is the most likely to be short an electron. (8.6)

26. a. Boron has an electron configuration of $1s^22s^22p^1$. With only three valence electrons, it can form molecules surrounded by only six electrons. (8.6)

27. a. To expand its octet, the atom must have *d* orbitals available. Since *d* orbitals require a minimum principal quantum number of 3, the atom must be in the third period, or higher, of the periodic table. (8.6)

28. c. The structure of ClF_5 has all five fluorine atoms bonded to the chlorine. The chlorine also has one lone pair. So the number of electrons assigned to the chlorine is one for each bond (five) and both electrons in the lone pair (two), for a total of seven. Since chlorine has seven valence electrons, the formal charge is zero. (8.5)

29. c. Nitrogen has a triple bond, which will re-quire more energy to break than the double bond of oxygen or the single bond of chlorine. (8.7)

30. c. Both ionic and covalent. Carbonate ion has covalent bonds between the carbon and oxygen. However, the bond between sodium and carbonate is ionic. (8.1 and 8.3)

Additional Practice Problem Answers

1. a. ·S̈:

   b.
```
        H
        |
     H—Si—H
        |
        H
```

   c.
```
Ö=N—Ö—H    and    :Ö—N—Ö—H
   |                  ||
  :Ö:                :Ö:
```

   d. Ö=N̈—Ö: and :Ö—N̈=Ö

   e.
```
:Ö—S̈—Ö:
    |
   :Ö:
```

2. Highest electronegativity = F, N, P, C, H, Cu = lowest electronegativity.

3. a. :O—I—O—H
   with :O: bonded below I

   Although this structure meets the criteria of the octet rule, a better structure would be to move a lone pair on each oxygen to create a double bond with the iodine (except for the oxygen already bonded to hydrogen). This would lower the formal charge on iodine from +2 to zero and the formal charges on the oxygens from −1 to zero.

   O=I—O—H
   with :O: double-bonded below I

   b. :F—Xe—F:

   Since all elements have a zero formal charge and xenon can expand its octet, this is the final answer.

   c. H—B—H
   with H bonded below B

   d. :F—Cl—F:
   with :F: bonded below Cl

   e. :I—I—I:

4. There are two possible Lewis structures for dinitrogen oxide (N_2O):

   :N≡N—O: N=N=O

   For the nitrite ion (NO_2^-):

   O—N=O

   For the nitrate ion (NO_3^-):

   :O:
   ‖
   :O—N—O:

   a. NO_2^- and NO_3^- exhibit resonance. Nitrite has two resonance structures. Nitrate has three resonance structures.
   b. Nitrite ion has a lone pair on the nitrogen.
   c. For both dinitrogen oxide structures, the formal charge on the central nitrogen is +1. For nitrite ion the formal charge on nitrogen is zero. For nitrate ion the formal charge on nitrogen is +1.

5. The electronegativity of hydrogen is 2.1.
   (1) The electronegativity of fluorine is 4.0.
   $\Delta\chi = 1.9$. (Electrons are attracted to fluorine.)
   (2) The electronegativity of sodium is 1.0.
   $\Delta\chi = 1.1$. (Electrons are attracted to hydrogen.)
   (3) The electronegativity of both carbon and sulfur is 2.5. $\Delta\chi = 0.4$. (Electrons are more attracted to carbon or sulfur.)
   (4) The electronegativity of phosphorus is 2.1.
   $\Delta\chi = 0$. (Electrons are almost equally distributed.)

   Therefore, the ranking of the bonds is: HF > HNa > HC = HS > HP

CHAPTER 9 Molecular Geometry and Bonding Theories

OUTLINE

REVIEW

Chapter Overview

Lewis structures only show which atoms are bonded to which. They do not determine how these atoms are arranged in space (**molecular geometry**). One theory that does predict molecular geometry is **valence shell electron pair repulsion (VSEPR)** theory. In VSEPR theory, each group of electrons is arranged to be as far from the other groups as possible. A group of electrons is a lone pair, a single bond, a double bond, or a triple bond. The number of groups of electrons around an atom is its **steric number (SN)**. The steric number can also be defined as the number of atoms bonded to the central atom plus the number of lone pairs on the central atom.

The overall shape of the molecule is generally determined by the steric number of the central atom. If the steric number is two, the electron groups will be 180° from each other and the shape is **linear**. The shape of a molecule with steric number three is a **planar triangle** with bond angles of 120°. Atoms with a steric number of four form a **tetrahedral** geometry with an angle between bonds of 109.5°. If the steric number is five, the shape is a **trigonal bipyramid**. A trigonal bipyramid has two positions: **axial** positions are 90° from the **equatorial** positions, which are 120° from each other. If the steric number is six, the shape is **octahedral**, where each group of electrons is 90° from the others.

These basic shapes are the orientation of the atoms if there are no lone pairs present. Even with lone pairs, these shapes describe the **electron pair geometry**. However, the molecular geometry of a molecule with lone pairs described only the position of the atoms, which are in the same approximate positions as determined by the steric number. Thus a molecule with a steric number of three that has one lone pair will have a molecular geometry described as **bent**, since the terminal (outside) atoms are at an angle to each other. However, the repulsion generated by a lone pair is greater than that of a bond; thus the angles between the bonds become smaller than those described in the previous paragraph. Lone pair–lone pair repulsion is greater than lone pair–bonding pair repulsion, which is greater than bonding pair–bonding pair repulsion.

For most geometries, all positions are equivalent and the lone pairs can be positioned at any location with the same result. However, a trigonal bipyramid has two types of positions. Lone pairs position themselves in the equatorial position because they are farther from the other groups than in the axial position. In octahedral geometries, although all positions are initially the same, if there is an even number of lone pairs, those pairs will orient themselves opposite (180°) from each other.

Molecular geometries, with and without lone pairs, are summarized in the table below.

steric number	lone pairs	molecular geometry
3	0	trigonal planar
3	1	bent
3	2	linear
4	0	tetrahedral
4	1	trigonal pyramid
4	2	bent
4	3	linear
5	0	trigonal bipyramid
5	1	see-saw
5	2	T-shaped
5	3	linear
6	0	octahedral
6	1	square pyramid
6	2	square planar
6	3	T-shaped
6	4	linear

Atoms of greater electronegativity have a greater attraction on the electrons of the bond, forming a bond dipole. Within a molecule, the sum of these bond dipoles might cancel with the dipoles that are equal and opposite, creating a molecule that is nonpolar overall. If the bond dipoles do not cancel, there will be a net dipole on the overall molecules creating a **permanent dipole moment** and a **polar molecule**. Because electrons are mobile, they are not always distributed equally throughout the molecule, so even nonpolar molecules have temporary dipoles. The nature of these dipoles in a molecule determines many of its chemical and physical properties.

Another theory that predicts molecular geometry is valence bond theory. **Valence bond theory** predicts that atomic orbitals **hybridize** (mix to become all the same) when part of a molecule. The number of atomic orbitals that hybridize will be the same as the number of hybrid orbitals that are created. When the orbitals overlap, they form a bond. The more overlap, the stronger the bond. The shape of a hybrid orbital is generally two spheres on either side of the nucleus, with one side substantially larger than the other. The larger end of a hybrid orbital points toward another nucleus, which also has a hybrid orbital pointing toward the first nucleus. This results in an overlap with a high electron density on the axis connecting the two nuclei. This overlap is called a **sigma (σ) bond**. When unhybrized p orbitals overlap, the electron density is on either side of the axis between nuclei, forming a **pi (π) bond**. A single bond in a Lewis structure is a sigma bond. A double bond is a combination of a sigma and a pi bond and a triple bond is a sigma and two pi bonds (oriented 90° from each other).

The molecular geometry is predicted from the hybridization of the central atom. (All atoms in a molecule have hybrid orbitals, but the shape of the molecule as a whole is determined by the central atom.) One atomic orbital for each lone pair and one for each bond (regardless of whether it is a single, double, or triple bond) will hybridize. If two atomic orbitals hybridize, both orbitals will become ***sp* orbitals** and exist 180° from each other. Thus *sp* hybridization produces a linear shape. (Just like two groups of electrons in VSEPR theory!) The hybridization of three atomic orbitals is sp^2 with a planar triangle shape with 120° separation between orbitals and sp^3 hybridization with tetrahedral geometries results from hybridization of four atomic orbitals. If five atomic orbitals hybridize, the hybridization is sp^3d. There are no more p orbitals to hybridize, thus d orbitals must be used. Also, if there are five hybrid orbitals, the atom must have expanded its octet. Therefore, valence bond theory explains why only atoms with available d orbitals ($n \geq 3$) expand their octet. As expected, sp^3d hybridization has a trigonal bipyramidal geometry and six hybrid orbitals have sp^3d^2 hybridization and an octahedral geometry.

The unhybridized p orbitals that form a pi bond may have overlap that extends over more than two atoms. In this case, the electrons in these bonds are delocalized (spread over three or more atoms). This occurs in molecules with **resonance** structures. The actual hybridization of each atom involved in resonance is sp^2, thus a p-orbital is available at each atom and the pi bond is spread over all the atoms that participate in the resonance.

Another model of bonding is molecular orbital theory. In **molecular orbital (MO) theory**, atomic orbitals combine to make molecular orbitals. The number of orbitals and the energy of the orbitals are conserved. Therefore, if two atomic orbitals combine, two molecular orbitals will be formed. One of these molecular orbitals will be lower in energy (more stable) than the original atomic orbitals. This is a **bonding orbital** and its electron density is located between the two atoms. The other molecular orbital is the **antibonding orbital**. It is as much higher in energy than the original atomic orbitals as the bonding orbital was lower in energy. The electron density of an antibonding orbital is located on the opposite side of the area between the atoms. Antibonding orbitals are designated with a star(*).

When two s orbitals or two p orbitals oriented along the bonding axis combine, the electron density of the bonding orbital is centered along the axis between the atoms. These molecular orbitals are called **sigma (σ) bonds**, just as hybrid orbitals that overlap along the bonding axis are sigma bonds. When p orbitals oriented sideways to each other combine, the electron density of the bonding orbital is located on either side of the bonding axis and, like in valence bond theory, is called a **pi (π) bond**. The energy change for pi bonds is smaller than for sigma bonds.

Electrons fill molecular orbitals in the same way as they fill atomic orbitals, according to the aufbau principle and Hund's rule. That is, two electrons spinning in opposite directions fill the lowest energy orbital first. If two or more orbitals are degenerate (having the same energy), electrons with the same spin direction (parallel) will fill each orbital before doubling up.

A major strength of MO theory is that it explains magnetic properties of molecular compounds. An element or molecule is attracted to a magnetic field if it has an unpaired electron. Such elements and compounds are called **paramagnetic**. An element or compound will be slightly repelled from a magnetic field if all its electrons are paired with an electron of opposite spin. These elements and compounds are called **diamagnetic**. In most Lewis structures, all electrons appear to be paired. However, MO theory shows that some (O_2 being the classic example) actually do have unpaired electrons.

In addition, MO theory is useful for explaining spectroscopic properties of molecules. Similar to atomic spectra, photons are absorbed when they are of an energy that exactly matches the energy difference between orbitals. This is observed for both molecular and atomic orbitals.

Worked Examples

Molecular Shape

Example 1.
What is the molecular geometry of the following?

a. XeF_4 b. PCl_5 c. I_3^- d. CO_2
e. NO_3^- f. $SiCl_4$ g. SF_6

Collect and Organize We are given five molecular formulas and are asked to determine the molecular geometry of each.

Analyze The geometry of a molecule is determined by the arrangement of atoms around the central atom. VSEPR theory predicts that groups of electrons will repel each other. Therefore, these groups will arrange to be as far apart as possible. Of course, with fewer groups, the electrons can be farther apart. A group of electrons is a bond (single, double, or triple doesn't matter) or a lone pair.

Solve The Lewis structures for each of these atoms are shown in the previous chapter (expanded octets and valence bond theory). You need the Lewis structure before you can make predictions about molecular geometry.

a. Xenon tetrafluoride has two lone pairs and four bonds around the central atom. That is a total of six groups. Therefore, the arrangement of electrons is octahedral. However, the lone pairs do not show up in the molecular geometry. Because lone pairs are particularly repulsive, they will get as far from each other as possible. In this example, that is on opposite sides of the octahedral. Therefore, the overall molecular geometry is square planar.

b. Phosphorus pentachloride has five bonds around the central atom. The electrons arrange in a trigonal bipyramid. Since there are no lone pairs, the molecular geometry is also a trigonal bipyramid.

c. Triiodide ion has three lone pairs and two bonds. The arrangement of electrons is a trigonal bipyramid. In a trigonal bipyramid geometry the lone pairs of electrons prefer equatorial positions to axial positions. This is because the 120° angle of the equatorial positions keeps the electrons farther apart than the 90° angle of the axial position. With a lone pair taking each of the three equatorial positions, the remaining atoms have a linear geometry.

d. The carbon of carbon dioxide has two bonds. Its geometry (both electronic and molecular) is linear.

e. Nitrate ion has three bonds. (That this has resonance structures is irrelevant. Each structure gives the same geometry. It must be this way because the true structure is not any one of the three, but the average of them all.) With three bonds, the molecular geometry is a planar triangle.

f. Silicon tetrachloride has four bonds and a tetrahedral molecular geometry.

g. Sulfur hexafluoride has six bonds and an octahedral molecular geometry.

Think About It Prediction of molecular geometry is greatly facilitated using the rules above.

Polar Bonds and Polar Molecules

Example 1.
Are the following molecules polar or nonpolar?

(a) XeF_4 (b) PCl_5 (c) CO_2 (d) SCl_2

Collect and Organize We are given the molecular formulas of four molecules and are asked to determine whether the molecules are polar on nonpolar.

Analyze The key to determining whether dipoles cancel is to consider their direction as well as their magnitude. The direction is determined by the molecular geometry.

SOLVE The Lewis structures of the molecules are shown in the previous chapter.

a. Because the lone pairs are exactly opposite of each other, their effects cancel. The bond dipoles of the xenon–fluorine bond are also exactly opposite each other. All bond dipoles cancel; the molecule is nonpolar.

b. This molecule has a trigonal bipyramid structure. The phosphorus–chlorine bond dipoles are all of the same magnitude. The axial bond dipoles cancel each other. The equatorial bond dipoles also cancel. Although the bond dipoles aren't exactly opposite, the angle of the two opposing bonds is sufficient to cancel that dipole. Thus the whole molecule is nonpolar. Any molecule for which all terminal atoms are the same and with no lone pairs on the central atom is nonpolar.

c. This molecule has a linear structure. The carbon–oxygen double-bond dipoles cancel. The molecule is nonpolar.

d. This molecule has a bent structure, with two lone pairs on one side and the chlorine atoms on the other. Therefore, it is a polar molecule, with the partial negative charge on the side with the lone pairs.

THINK ABOUT IT A good clue is to look at the symmetry of the molecular geometry. Molecules that are not symmetric are polar.

MOLECULAR ORBITAL THEORY

EXAMPLE 1.
What is the molecular orbital (MO) diagram for O_2? Is it paramagnetic or diamagnetic? What is its bond order?

COLLECT AND ORGANIZE We are asked to determine the MO diagram for O_2, decide whether it is paramagnetic or diamagnetic, and determine its bond order.

ANALYZE We must first determine the number of valence electrons. Once this has been determined we can fill the molecular orbitals that result from the combination of atomic orbitals, starting with the lowest energy orbitals first. (A MO diagram for O_2 is shown in your text.) If there are unpaired electrons in the MO diagram, the molecule is paramagnetic. If all electrons are paired, the molecule is diamagnetic. We then use the MO diagram to calculate the bond order using the following relation: Bond order = (1/2)(number of bonding electrons – number of antibonding electrons).

SOLVE Oxygen has six valence electrons. Therefore, O_2 has 12 valence electrons. This is sufficient electrons to fill the σ_s (2 e), σ_s*(2 e), σ_z (2 e), both π (4 e), and one in each π* (2 e) orbital. The unpaired electrons in the π* orbitals make the molecule paramagnetic. Bond order = (8 – 4)/2 = 2.

THINK ABOUT IT Molecules with a higher bond order have shorter, stronger bonds and are more stable than molecules with a lower bond order.

EXAMPLE 2.
What is the MO diagram for C_2^{+}? Is it paramagnetic or diamagnetic? What is its bond order?

COLLECT AND ORGANIZE We are asked to determine the MO diagram of C_2^+, decide whether it is paramagnetic or diamagnetic, and determine its bond order.

ANALYZE We will use the procedure outlined in the previous problem. For this situation we must take into account the positive charge on the molecule. (A MO diagram for C_2 is shown in your text.)

SOLVE Carbon has four valence electrons. The C_2^+ molecule has 4 + 4 – 1 = 7 valence electrons. This is sufficient to fill the σ_s (2 e), σ_s* (2 e), σ_z (2 e) and one electron in one π orbital. The unpaired electron in the π orbital makes the molecule paramagnetic. The bond order is determined to be (5 – 2) / 2 = 1.5

THINK ABOUT IT Bond order need not be an integer value. The bond order of 1.5 determined in this example suggests that the bond between the two carbon atoms in C_2^+ can be described as being between a single and double bond.

EXAMPLE 3.
Consider the molecule SN.

a. Draw the MO diagram.

b. What is the bond order?

c. The highest-energy electron has the character of which atom?

COLLECT AND ORGANIZE We are given a diatomic molecule consisting of sulfur and nitrogen and are asked to draw the MO diagram, determine bond order, and determine the atom whose atomic orbitals lie closest in energy to the molecular orbital that contains the highest energy electron.

ANALYZE The diagram for a heteronuclear diatomic molecule is a skewed version of the diagram for homonuclear diatomic molecules. The energies of the atomic orbitals are lower for the more electronegative atom. The bonding orbitals are slightly lower in energy than the atomic orbital of the more electronegative atom. The antibonding orbitals are slightly higher in energy than the atomic orbital of the less electronegative atom.

SOLVE The molecule has 11 valence electrons. Nitrogen is more electronegative than sulfur, so the energy of its atomic orbitals is lower. The bond order is (8 – 3)/2 = 2.5. The highest-energy electron is in the π* orbital, which is closest in energy to the *p* orbitals of the sulfur atom. Therefore, the highest-energy electron has the character of sulfur.

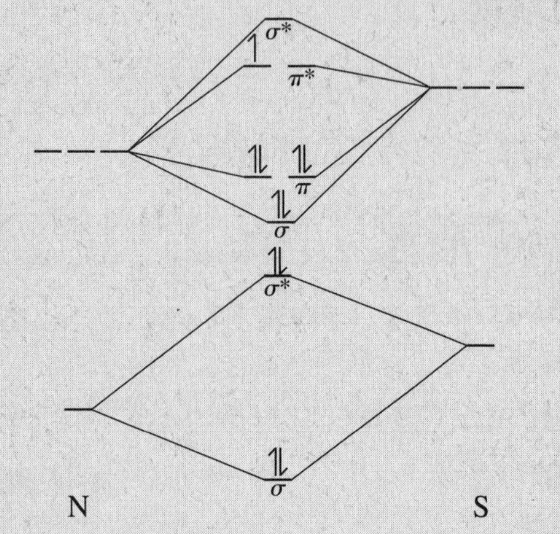

THINK ABOUT IT One difference with heteronuclear molecules is that electrons are not shared equally. Therefore, a bonding (or antibonding) electron is more associated with one atom than the other.

Self-Test

KEY VOCABULARY COMPLETION QUESTIONS

_________ 1. Bonding along the axis connecting two atoms

_________ 2. Shape of sp^3d orbitals

_________ 3. Shape of a molecule whose central atom has three bonds and two lone pairs

_________ 4. Mixing of atomic orbitals so that all the orbitals for bonding are the same

_________ 5. Hybridization of carbon in CO_2

_________ 6. Bonding theory based on hybridization

_________ 7. Shape of a molecule when every bond angle is 120°

_________ 8. Shape of a molecule with four groups of electrons around the central atom

_________ 9. Position of a lone pair in a trigonal bipyramid geometry

_________ 10. Shape of a molecule with one lone pair and five bonds around the central atom

_________ 11. Position in a trigonal bipyramid geometry that is 90° from its nearest neighbor

_________ 12. Molecular geometry of a water molecule

_________ 13. Molecular geometry of PCl_3

_________ 14. Having a partial positive charge on one end and a partial negative charge on the other

_________ 15. Sideways overlap of unhybridized *p* orbitals

_________ 16. Shape when central atom is surrounded by 12 electrons

_________ 17. Molecular geometry of XeF_4

_________ 18. Molecular orbital higher in energy than the atomic orbitals that formed it

_________ 19. Electron pair geometry when an atom is sp^3 hybridized

_________ 20. Substance having unpaired electrons

MULTIPLE-CHOICE QUESTIONS

1. What is the bond order of N_2?
   a. one
   b. one and a half
   c. two
   d. three

2. A π bond is formed from
   a. two *p* orbitals
   b. two *s* orbitals
   c. an *s* and a *p* orbital
   d. two σ orbitals

3. A σ^* orbital's energy
   a. is lower than the σ orbital
   b. is lower than the original *s* orbital
   c. is higher than the original *s* orbital
   d. depends on the atom

4. An advantage of MO theory over Lewis structures is that it
   a. explains the magnetic properties of molecules
   b. shows how atoms are connected to each other
   c. explains the phenomenon of resonance
   d. all of the above
   e. none of the above

5. An advantage of Lewis structures over MO theory is that it
   a. explains the magnetic properties of molecules
   b. shows how atoms are connected to each other
   c. explains the phenomenon of resonance
   d. all of the above
   e. none of the above

6. What is the bond angle between hydrogens in NH_3?
   a. 120°
   b. 109.5°
   c. 107°
   d. 90°

7. How many σ bonds and π bonds are in HCN?
   a. three σ, one π
   b. two σ, no π
   c. two σ, one π
   d. two σ, two π

8. Hybridization of the xenon is XeF_4 is
   a. sp^3
   b. sp^3d^2
   c. sp
   d. sp^3d

9. Cl_2 is a
   a. polar covalent molecule
   b. nonpolar covalent molecule
   c. ionic molecule
   d. metallic molecule

10. If an atom has sp^3 hybridization, it has
    a. four sp^3 orbitals
    b. one s and three p orbitals
    c. three hybridized p orbitals and an unhybridized s orbital
    d. an unhybridized s orbital and three unhybridized p orbitals

11. What is an advantage of valence bond theory over MO theory?
    a. it works well for molecules with an odd number of electrons
    b. it explains and gives a picture of molecular geometry
    c. it explains and gives a picture of double bonds
    d. all of the above

12. What is an advantage of MO theory over valence bond theory?
    a. it works well for molecules with an odd number of electrons
    b. it explains and gives a picture of molecular geometry
    c. it explains and gives a picture of double bonds
    d. all of the above

13. The effect of lone pairs on bond angle is that the bonds
    a. get farther apart
    b. get closer together
    c. remain at the same angle
    d. depend on the identity of the central atom

14. Which end of CCl_2F_2 has a partial negative charge?
    a. carbon
    b. fluorines
    c. chlorines
    d. it is a nonpolar molecule

15. What is the hybridization of XeF_2?
    a. sp
    b. sp^2
    c. sp^3d^2
    d. sp^3d

16. What is the molecular geometry of SF_6?
    a. tetrahedral
    b. octahedral
    c. trigonal bipyramid
    d. square pyramid

17. What is the bond order of PO?
    a. 2.0
    b. 1.5
    c. 0.5
    d. 2.5

18. What is the molecular geometry of $SeCl_2$?
    a. linear
    b. bent
    c. tetrahedral
    d. T-shaped

19. What is the hybridization of fluorine in SF_6?
    a. sp^3d^2
    b. sp^3
    c. sp
    d. terminal atoms don't hybridize

20. The effect of the double bond of the nitrate ion is
    a. to reduce the bond angle of the other oxygens
    b. to make the molecule polar by attracting more electrons to that oxygen
    c. to distort geometry by making one bond shorter than the others
    d. nothing—the ion has resonance, so all of the oxygens are the same

21. The lone pairs on I_3^- are located
    a. all on the axial positions
    b. all on the equatorial positions
    c. two on the axial and one on the equatorial positions
    d. two on the equatorial and one on the axial positions

22. The hybridization of phosphorus in PCl_3 is
    a. sp^2
    b. sp^3
    c. sp^3d
    d. sp^3d^2

23. What does valence bond theory have in common with MO theory?
    a. antibonding orbitals
    b. hybrid atomic orbitals
    c. σ and π bonds
    d. all of the above

24. If the dipole moment is zero, then the molecule must be
    a. homoatomic
    b. heteroatomic
    c. polar
    d. nonpolar

25. The partial negative charge on PCl_3 is
    a. on the phosphorus
    b. on the chlorine
    c. on the lone pair
    d. nowhere because it is a nonpolar molecule

Additional Practice Problems

1. For the following compounds:
   a. HIO_3 b. XeF_2 c. BH_3 d. ClF_3 e. I_3^-
   What is its molecular geometry? What is the hybridization of the central atom? Is the molecule polar or nonpolar?

2. For the following compounds:
   a. CH_3NH_2 b. SiH_4 c. HNO_3 d. NO_2^- e. SO_3^{2-}
   What is its molecular geometry? What is the hybridization of the central atom? Is the molecule polar or nonpolar?

3. Draw the MO diagram of CN.
   a. What is the bond order of the molecule?
   b. Which atom has the unpaired electron?

4. For ClO,
   a. draw the Lewis structure.
   b. draw the MO diagram.
   c. what is the bond order predicted by the Lewis structure?
   d. what is the bond order predicted by the MO diagram?
   e. is there an unpaired electron? If so, which atom is it associated with?

5. Consider Cl_2O.
   a. What is its Lewis structure?
   b. What is the hybridization of each atom?
   c. What is the formal charge of each atom?
   d. Which atom is most likely to expand its octet? Will it? Why or why not?
   e. What is the molecular geometry of the molecule?
   f. Does the molecule have a permanent dipole moment?

6. Using MO theory, compare F_2^+, F_2, and F_2^-.
   a. What is the bond order of each?
   b. Which of the three is the most stable?
   c. Which of the three has the shortest bond?
   d. Which of the three has the weakest bond?
   e. How many unpaired electrons does each have?

Self-Test Answer Key

Key Vocabulary Completion Answers

1. sigma bond
2. trigonal bipyramid
3. T-shaped
4. hybridization
5. *sp*
6. valence bond theory
7. trigonal planar
8. tetrahedral
9. equatorial
10. square pyramid
11. axial
12. bent
13. trigonal pyramid
14. polar
15. pi bond
16. octahedral
17. square planar
18. antibonding
19. tetrahedral
20. paramagnetic

Multiple-Choice Answers

For further information about the topics these questions address, you should consult the text. The appropriate section of the text is listed in parentheses after the explanation of the answer.

1. d. N_2 has 10 valence electrons. Eight are bonding electrons and two are antibonding electrons in the MO diagram of N_2. Bond order = $(8 - 2)/2 = 3$. (9.6)

2. a. *s* orbitals mix to form a σ bond, *p* orbitals oriented along the *z*-axis (the bonding axis) also form a σ bond, but the orbitals oriented along the *x*-axis or the *y*-axis form π bonds with the same *p* orbital of the other atom. (9.6)

3. c. When atomic orbitals combine, the molecular bonding orbital is lower in energy than the atomic orbital, as the bond stabilizes the atoms. However, there is conservation of orbitals and conservation of energy. So there must be another orbital and it must be higher in energy than the atomic orbital, so that the sum of the energy of the molecular orbitals is the same as the energy of the atomic orbitals. (9.6)

4. a. Lewis structures always show electrons moving in pairs, and magnetism comes from unpaired electrons. Yet molecules like O_2 are magnetic. On the other hand, MO theory has molecular orbitals with degenerate energy states that explain magnetic properties. (9.2 and 9.6)

5. b. The whole point of Lewis structures is to show which atoms are sharing electrons and how the atoms are connected. MO theory treats the molecules as a whole, rather than looking at the connections of individual atoms. In addition, although resonance can be determined from a Lewis structure and a Lewis structure shows different resonance forms, in reality, there is only one molecule and one form of that molecule. (9.2 and 9.6)

6. c. The groups of electrons arrange themselves around nitrogen in a tetrahedral orientation, with a bond angle separation of 109.5°. Because one of those groups is a lone pair, which repels the other electrons more strongly than the bond-type groups, the bonds will be a little closer than 109.5°. (9.2)

7. d. The structure of HCN has a single bond between the hydrogen and carbon and a triple bond between carbon and nitrogen. A triple bond is made up of a σ bond and two π bonds. A single bond is made of one σ bond. That gives a total of two σ and two π bonds. (9.4)

8. b. The Lewis structure of XeF_4 has the four bonds of the fluorine to the xenon and two lone pairs. That is, six groups of electrons that are hybridized, so sp^3d^2. (9.4)

9. b. Since it is the bonding of two nonmetals, it is a covalent molecule. Since both atoms in the molecule are the same, the electrons are shared exactly equally. There is no dipole, so it is nonpolar. (9.3)

10. a. Hybridization makes all the orbitals the same. Consequently, all the hybrid orbitals are sp^3 orbitals. With conservation of orbitals you have to end up with the same number of orbitals you start with. (9.4)

11. c. Although molecular geometry can be determined from hybridization, it does not explain it. It does not work for an odd number of electrons. However, that double bonds are an overlap of hybridized orbitals and unhybridized orbitals explains the stronger, but not really double-strength, bonds. (9.4 and 9.6)

12. a. MO theory treats one electron at a time, so it doesn't matter if the electrons are paired. (9.4 and 9.6)

13. b. Since there is no positively charged nucleus associated with the two electrons of a lone pair (unlike the two electrons of a bond), the lone pair is more negative than a bond, repelling the electrons of the bond and pushing the bonds closer together. (9.3)

14. b. Carbon is the central atom (not an end). The fluorines are more electronegative than the chlorines, so the negative electrons tend toward that direction. (9.3)

15. d. The xenon is surrounded by two fluorine bonds and three lone pairs. Therefore, there are five hybrid orbitals. (9.4)

16. b. The sulfur is bonded to six fluorine atoms. There are no lone pairs. Octahedral is the geometry of six groups of electrons. (9.2)

17. d. In the MO diagram there are eight bonding electrons and three antibonding electrons, so the bond order = (8 – 3)/2 = 2.5. (9.6)

18. b. In this molecule selenium has two bonds and two lone pairs. The electrons arrange themselves in a tetrahedral geometry, but the two lone pairs are not "seen" in the molecular geometry, so that is bent. (9.2)

19. b. Each fluorine is surrounded by one bond and three lone pairs. Thus each fluorine has four hybrid orbitals and sp^3 hybridization. (9.4)

20. d. Because the ion has resonance, all oxygens are the same. (9.2)

21. b. There are three lone pairs on the central iodine. All are on the equatorial positions because they are farther apart. (9.2)

22. b. The Lewis structure of PCl_3 has three bonds and one lone pair. Therefore, the hybridization is sp^3. (9.4)

23. c. Antibonding orbitals are characteristic of MO theory; hybridization is characteristic of valence bond theory. However, both have σ and π bonds, which are defined in the same way, by type of overlap. (9.4 and 9.6)

24. d. A dipole moment measures the extent of the polarity of a molecule. A dipole moment of zero is a nonpolar molecule. (9.3)

25. c. The most negative part of any molecule is the lone pairs. (9.3)

Additional Practice Problem Answers

1. Lewis structure for each compound is shown below.

a. Ö—Ï—Ö—H
   ‖
   :O:

b. :F̤̈—Xë—F̤̈:

c. H—B—H
   |
   H

d. :F̤̈—C̈l—F̤̈:
   |
   :F̤:

e. :Ï—Ï—Ï:

a. The molecular geometry of HIO_3 is trigonal pyramid, the hybridization of iodine is sp^3. It is essentially nonpolar, although the hydrogen may induce a slight dipole.
b. The molecular geometry of XeF_2 is linear, the hybridization of xenon is sp^3d, and the molecule is nonpolar.
c. The molecular geometry of BH_3 is trigonal planar (or planar triangle), the hybridization of boron is sp^2, and the molecule is nonpolar.
d. The molecular geometry of ClF_3 is T-shaped, the hybridization of the central chlorine is sp^3d, and the molecule is polar.
e. The molecular geometry of I_3^- is linear, the hybridization of the central iodine is sp^3d. Ions cannot be categorized as polar or nonpolar.

2. The Lewis structures for each are

a.
   H H
   | |
H—C—N:
   | |
   H H

b.
     H
     |
H—Si—H
     |
     H

c. Ö=N—Ö—H and :Ö—N—Ö—H
       | ‖
      :O: :O:

d. Ö=N̈—Ö: and :Ö—N̈=Ö

e. :Ö—S̈—Ö:
       |
      :O:

a. Both carbon and nitrogen can be considered central atoms. The molecular geometry around carbon is tetrahedral. The molecular geometry around nitrogen is trigonal pyramidal. The hybridization of carbon is sp^3. The hybridization of nitrogen is also sp^3. It is the entire molecule that is labeled polar or nonpolar. This molecule is polar.
b. The molecular geometry around the silicon is tetrahedral. Its hybridization is sp^3. The molecule is nonpolar.
c. The molecular geometry around the nitrogen is trigonal planar; its hybrization is sp^2. It is essentially nonpolar, although the hydrogen may induce a slight dipole.
d. The molecular geometry around the nitrogen is trigonal planar; its hybrization is sp^2. It is not appropriate to label an ion as polar or nonpolar. The net (overall) charge determines it properties.
e. The molecular geometry around the sulfur is tetrahedral; its hybridization is sp^3. It is not appropriate to label an ion as polar or nonpolar. The net (overall) charge determines it properties.

3.

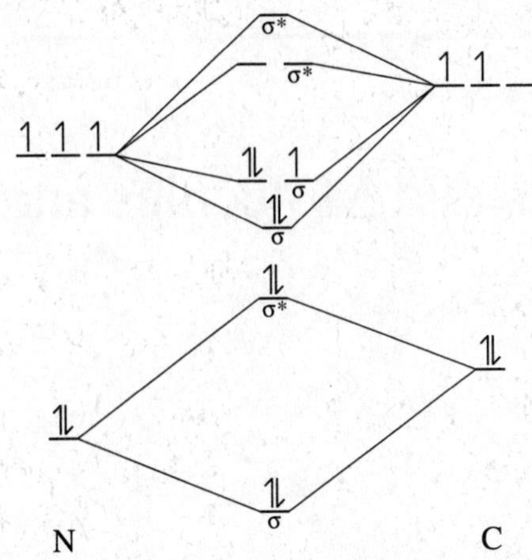

a. The bond order is (7 – 2)/2 = 2.5.

b. The unpaired electron is in the molecular orbital closest to nitrogen, so the unpaired electron is associated with the nitrogen.

4. a. :Ö–C̈l:

b.

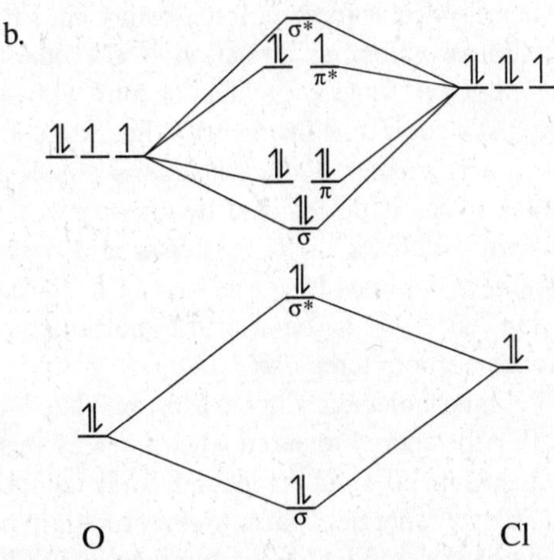

c. The Lewis structure suggests a bond order of one.

d. From the MO diagram, there are 8 bonding electrons and 5 antibonding electrons. (8 – 5)/2 = 1.5. So the bond order is 1.5.

e. Both the Lewis structure and the MO diagram show an unpaired electron associated with chlorine.

5. :C̈l–C̈l–Ö:

a. All atoms satisfy the octet rule, but the middle Cl has a formal charge of +1 and the oxygen has a formal charge of –1, so a better structure would be

:C̈l–C̈l=Ö

b. Using the first Lewis structure, the hybridization of every atom is sp^3. Using the second Lewis structure, the hybridization of both chlorines is sp^3 and the hybridization of oxygen is sp^2.

c. Using the first Lewis structure, the formal charge of the end chlorine is zero, the middle chlorine is +1, and the oxygen is –1. Using the second Lewis structure, the formal charge of every atom is zero.

d. The middle chlorine is most likely to expand its octet. It will do so to reduce the formal charge.

e. Regardless of the Lewis structure, the molecular geometry of this molecule is bent.

f. Yes.

6. F_2^+ has 13 valence electrons, F_2 has 14 valence electrons, and F_2^- has 15 valence electrons. The following MO diagram can be used for any of these molecules or ions by changing the number of electrons. The MO diagram of F_2 is

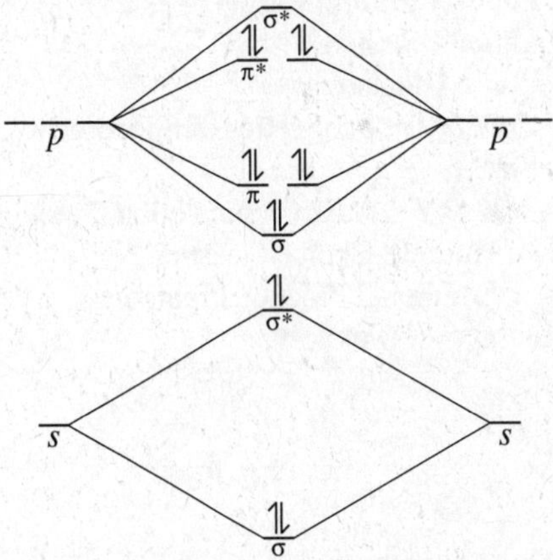

a. Bond order of F_2^+ = 1.5. Bond order of F_2 = 1.0. Bond order of F_2^- = 0.5.

b. F_2^+ is the most stable, since it has the highest bond order and lowest sum of the energies of the electrons.

c. The shortest bond is also that with the highest bond order, F_2^+.

d. The weakest bond has the lowest bond order, F_2^-.

e. F_2^+ has one unpaired electron. F_2 has no unpaired electrons. F_2^- has one unpaired electron.

CHAPTER 10 Forces between Ions and Molecules and Colligative Properties

OUTLINE

REVIEW

Chapter Overview

Liquids and solids are formed by many particles (atoms, ions, or molecules) held closely together. In liquids, the particles can move freely but remain adjacent to the other particles. In solids, not only are the particles adjacent, but the motion of the particles is limited. In gases, the particles move freely and independently; that is, without attaching to other particles.

The particles are held together by **intermolecular forces**, the force of attraction between molecules. All these forces are based on the attraction of a positive charge to a negative charge as expressed by **Coulomb's law**, which states that the energy of attraction is proportional to the product of the charges over the distance of the charges from each other.

The strongest intermolecular force is **ion–ion attraction.** This is the intermolecular force in ionic compounds (salts). The primary factor affecting attraction is the charge on the ion. The attractive force is greater for ions with higher charges. The distance of ions from each other has a smaller effect; attraction is greater if the distance is smaller. The distance between ions is determined by the size of the ion (smaller ions will be closer and have greater attraction) and the arrangement of the ions in space, known as the **lattice structure**. Substances with ion–ion intermolecular forces tend to be solids at room temperature.

The ion–ion intermolecular force is closely related to **lattice energy (U)**, the energy released when free gaseous ions combine to make a solid ionic compound. Ionic compounds with higher lattice energies have higher melting points and tend to be harder and less soluble in water. The lattice energy can be calculated using the **Born–Haber cycle**. The Born–Haber cycle uses the enthalpy of known reactions and Hess's law to determine lattice energies (or if lattice energy is known, one of the other reactions in the cycle can be determined). The reactions in the Born–Haber cycle include the reactions associated with: ionization energy, heat of vaporization, bond energy, heat of formation, and electron affinity. **Electron affinity** is the amount of energy associated with adding a mole of electrons to one mole of gaseous atoms. Values of electron affinity are normally negative (exothermic), but there are no easily determined periodic trends.

Polar molecules have partial charges and are held to each other by **dipole–dipole intermolecular forces**. Because these are only partial charges, unlike the full charges of ions, these forces are much weaker. Thus substances with dipole–dipole intermolecular forces tend to be gases or (occasionally) liquids at room temperature.

When ionic compounds are mixed with polar compounds, they may dissolve. Dissolution will occur if the many small dipoles of the solvent are more attractive than the other ions in the salt. Typically six solvent molecules will surround each ion, creating a **sphere of solvation**. If water is the solvent, it is called the **sphere of hydration**. Thus the ions are said to **solvated** or **hydrated**, as appropriate. The interaction between the ions and the polar solvent are called **ion–dipole intermolecular forces** and, because of the full charge on the ion, tend to be stronger than dipole–dipole forces.

Another special case of dipole–dipole forces is **hydrogen bonding**. Hydrogen bonds occur between molecules that have a hydrogen atom covalently bonded to an oxygen, nitrogen, or fluorine atom. Oxygen, nitrogen, and fluorine are the most electronegative atoms, and therefore, attract most of the electrons in the covalent bond. When the other end of the bond is hydrogen, all the electrons associated with hydrogen have been pulled away, leaving what is nearly a full positive charge. This high charge on hydrogen makes hydrogen bonding a very strong type of dipole–dipole intermolecular force. Substances with hydrogen bonds tend to be liquids at room temperature.

Nonpolar molecules are attracted to each other by temporary dipoles, which are called **London** or **dispersion forces**. The dipoles are created when the electron cloud surrounding the molecule is deformed, either through random motion (**temporary dipole**) or proximity to a permanent dipole (**induced dipole**). With more electrons, either through the size of the atom or number of atoms, the electron cloud is more easily deformed; thus the dipoles are larger and the attraction is greater. Molecules with easily deformed electron clouds are called **polarizable**. Molecules with London forces are normally gases at room temperature.

Forces of attraction can be overcome by increasing temperature and thus the kinetic energy (movement) of the particles. When the temperature is sufficient for the molecules to move freely but not sufficient to escape the proximity of the other molecules, a solid will melt. That temperature when a solid becomes a liquid is called the **melting point**. In the reverse direction, energy is removed so that a liquid becomes a solid. That temperature is the **freezing point**. The value of the melting point and the freezing point is the same. Substances with stronger forces of attraction have higher melting points.

Similarly, kinetic energy allows particles on the surface of a liquid to escape attractive forces of the liquid and become a vapor (gas). The rate at which particles escape depends on the strength of the intermolecular force, the size of the particle, and the temperature and the surface area of the liquid. In a closed container, vapor particles may also be captured and again become liquid (**condensation**). When the number of particles escaping the liquid and the number of particles being captured by the liquid are the same, the system is in **dynamic equilibrium**. The pressure of the gaseous particles in this closed system at equilibrium is called the **vapor pressure**. Vapor pressure is not affected by surface area, since increased surface area increases the rate of capture as well as the rate of escape. Thus substances with stronger intermolecular forces have lower vapor pressure and vapor pressure increases with increasing temperature. When the vapor pressure is the same as the atmospheric pressure, vapor can form throughout the solution rather than just at the surface. The temperature when this occurs is called the **boiling point**. Atmospheric pressure has a dramatic effect on boiling point; therefore, it is convenient to define a **normal boiling point** (the temperature when the vapor pressure is equal to 1 atm) to better compare substances.

Intermolecular forces also determine the solubility of substances. For a substance to dissolve, the solute–solvent attraction must be greater than the solute–solute and solvent–solvent attractions. This often occurs when forces of attraction are similar, so polar solutes dissolve in polar solvents and nonpolar solutes dissolve in nonpolar solvents. When two substances will dissolve in each other in any proportion, they are called **miscible**. Substances that dissolve in water are called **hydrophilic** (water-loving), whereas substances that do not dissolve in water are **hydrophobic** (water-fearing). Some molecules have both hydrophilic and hydrophobic portions. In addition to the effect of intermolecular forces, the solubility of gases is significantly affected by temperature and partial pressure of the gas. Increasing temperature decreases the solubility of gases. Partial pressure is proportional to the solubility of the gas. This relationship is called **Henry's law** and the proportionality constant, which depends on the identity of the gas, is Henry's constant.

The affect of temperature and pressure changes on physical state can be shown in a **phase diagram**. A typical phase diagram has solids at high pressures and low temperatures, gases at low pressures and high temperatures, and liquids in the middle. The physical states are divided by **equilibrium lines**. At a point on the line, both physical states exist. Thus the line dividing liquids and solids defines the melting points, and the line between liquids and gases defines the boiling points. At low pressures and temperatures, solids can become gases without passing through the liquid state. This is **sublimation**. The reverse, when gases become solids, is **deposition**. These lines meet at the **triple point**, the temperature and pressure where all three phases exist. While the melting-point line continues indefinitely, the boiling-point line comes to an end at what is known as the **critical point**. Above this pressure (**critical pressure**) and temperature (**critical temperature**), the liquid and gas states are indistinguishable. This physical state is called a **supercritical fluid**.

Properties of liquids include **surface tension**, the energy to separate molecules in a unit of surface area, and **viscosity**, the resistance to flow. Both of these properties are higher with stronger intermolecular forces and weaker at higher temperatures.

The attraction of a substance to the wall of a container is called *adhesive force* and the attraction of a substance to itself is called *cohesive force*. When the adhesive forces are stronger than the cohesive forces, the substance creeps up the sides of the container, forming a curved surface called a **meniscus**. In capillaries, there is sufficient adhesion that the substance actually climbs the walls of the capillary until gravity overcomes the adhesive and cohesive forces. The rising of a liquid up a capillary tube is called **capillary action**.

Adding a solute affects the properties of a solvent. Vapor pressure (P) is modified by a solute according to **Raoult's law**

$$P_{\text{solution}} = \chi_{\text{solvent}} \cdot P_{\text{solvent}}$$

where χ_{solvent} is the mole fraction of solvent. **Ideal solutions** follow Raoult's law perfectly. However, when the solute–solvent interactions are stronger than solvent–solvent interactions, the vapor pressure is lower than predicted by Raoult's law. When the solute–solvent interactions are weaker than the solvent–solvent interactions, the vapor pressure is higher than predicted by Raoult's law.

The vapor-pressure lowering described by Raoult's law is an example of a **colligative property**, where the number of solute particles rather than the identity of the solute determines the effect on the solvent. Addition of a solute will increase the boiling point of the solvent by ΔT according to

$$\Delta T = K_b m$$

where K_b is a proportionality factor depending on the identity of the solvent and m is the molality of solute particles. **Molality** is defined as moles of solute per kilogram of solvent. Similarly, the solute decreases the freezing point according to

$$\Delta T = K_f m$$

where K_f is the proportionality constant for the solvent's freezing point.

Because colligative properties depend on the number of solute particles, a convenient relationship is the van't Hoff factor. The **van't Hoff factor** is the number of particles per mole of solute. For an ideal strong electrolyte, this could be determined from the formula. For example, since one mole of K_2S produces two moles of K^+ and one mole of S^{2-}, its van't Hoff factor is three. However, at high concentrations, ions may associate with each other in solution, acting as a single particle. Therefore, the actual (measured) van't Hoff value is often less than the predicted value. The effect of ionization on boiling point and freezing point can be accounted for by including the van't Hoff factor in those equations.

Osmosis is a process where solvent moves through a semipermeable membrane from a solution of low concentration to a solution of high concentration until the concentrations are equal. The force driving this process is called **osmotic pressure** (π) and is another example of a colligative property. Osmotic pressure depends on the *molarity* (rather than molality) of the solute particles as well as the temperature. Because even dilute concentrations generate a high osmotic pressure, this property is particularly useful for determining molar mass.

Worked Examples

Ion–Ion Interactions and Lattice Energy

Example 1.
Rank the following ionic compounds according to the strength of their ion–ion forces of attraction:

$$NiCl_2,\ Fe_2O_3,\ MgO,\ KI,\ BeF_2,\ CrCl_3$$

Collect and Organize We are given six ionic compounds and are asked to rank them according to the strength of the forces between their ions.

Analyze Compounds can also be ranked within the category of ion–ion forces. Compounds made of ions with higher charges will be stronger than ions with lower charges. Smaller ions are stronger than larger ions. You can estimate the size from the periodic table. Large ions will be lower (in higher periods). The charge generally has a larger effect on attraction than size does.

Solve First, look at the charges on the ions that make up each compound.

$NiCl_2 = Ni^{2+}$ and Cl^-

$Fe_2O_3 = Fe^{3+}$ and O^{2-}

$MgO = Mg^{2+}$ and O^{2-}

$KI = K^+$ and I^-

$BeF_2 = Be^{2+}$ and F^-

$CrCl_3 = Cr^{3+}$ and Cl^-

Rank first according to the ions with the highest charge. (Multiply charges together if the combination makes it difficult to decide.) The order is

$$Fe_2O_3 > MgO > CrCl_3 > NiCl_2 = BeF_2 > KI$$

Note that MgO is greater than $CrCl_3$, even though +3 > +2. If you multiply the charges, 2(2) = 4, which is greater than 3(1) = 3. The subscripts do not contribute to this decision.

Note also that, based on charge, $NiCl_2$ and BeF_2 are the same. Therefore, you need to go to the second criteria (that of size) to rank these compounds. Beryllium and fluorine are both in the second period of the periodic table. That makes them smaller than the nickel and chlorine (fourth and third, respectively). Smaller ions can be closer together, so they generally have stronger ion–ion forces of attraction.

The final order is

$$Fe_2O_3 > MgO > CrCl_3 > BeF_2 > NiCl_2 > KI$$

Think About It It is generally a good principle to rank the strength of ionic compounds based first on charge, and then on size.

EXAMPLE 2.

Write the chemical reactions associated with the following energy values. (You may want to refer to Chapter 7 for ionization energy reactions and maybe review the naming rules.)

a. the lattice energy of calcium chloride

b. the ionization energy of potassium

c. the second ionization energy of strontium

d. the electron affinity of bromine atom

e. the sublimation of lithium

f. the lattice energy of silver iodide

COLLECT AND ORGANIZE We are given descriptions of energies of several processes and are asked to write the chemical reaction that each describes.

ANALYZE We must know the mechanisms that are described by each energy in order to correctly write the resulting equation.

SOLVE

a. Lattice energy results from the gaseous ions forming a solid. The formula for calcium chloride is $CaCl_2$. This substance is made of the ions Ca^{2+} and Cl^-. Therefore, the reaction is

$$Ca^{2+}(g) + 2\,Cl^-(g) \rightarrow CaCl_2(s)$$

b. Ionization energy is the energy required to remove an ion from an atom in the gaseous state, so

$$K(g) \rightarrow K^+(g) + e^-$$

c. The second ionization energy is the energy required to remove a second electron from the gaseous cation.

$$Sr^+(g) \rightarrow Sr^{2+}(g) + e^-$$

d. Electron affinity is the energy associated with a gaseous atom gaining an electron.

$$Br(g) + e^- \rightarrow Br^-(g)$$

e. Sublimation is a substance going from the solid state to the gaseous state.

$$Li(s) \rightarrow Li(g)$$

f. Lattice energy is the result of gaseous ions creating an ionic solid. The formula for silver iodide is AgI, made of Ag^+ and I^-.

$$Ag^+(g) + I^-(g) \rightarrow AgI(s)$$

THINK ABOUT IT Being able to correctly identify and write the processes given by the above energies represents the first step in the determination of the lattice energy of a compound.

EXAMPLE 3.

When the following reactions are added together, what is the resulting reaction?

$Li(s) \rightarrow Li(g)$

$Li(g) \rightarrow Li^+(g) + e^-$

$Br(g) + e^- \rightarrow Br^-(g)$

$Li^+(g) + Br^-(g) \rightarrow LiBr(s)$

$Br_2(\ell) \rightarrow 2\,Br(g)$

COLLECT AND ORGANIZE We are asked to determine the overall result by combining series of related reactions.

ANALYZE When adding equations together, keep the reactants as reactants and the products as products. Any substance that is exactly the same (including both charge and physical state) on both the product and reactant side can be canceled.

SOLVE Leaving all the reactants together and all the products together, the resulting equation is

$$Li(g) + Br(g) + e^- + Li^+(g) + Br^-(g) + Li(s) + Br_2(\ell) \rightarrow 2\,Br(g) + LiBr(s) + Br^-(g) + Li^+(g) + e^- + Li(g)$$

Canceling all substances that are exactly the same (including stoichiometric coefficients), you get

$$Br(g) + Li(s) + Br_2(\ell) \rightarrow 2\,Br(g) + LiBr(s)$$

Canceling the same substances with different coefficients, you get

$$Li(s) + Br_2(\ell) \rightarrow Br(g) + LiBr(s)$$

THINK ABOUT IT Remember that stoichiometric coefficients are not part of the formula, but denote the amount of the substance. Therefore, if there are two atoms on the reactant side and one on the product side, one atom on each side will cancel, leaving one atom on the reactant side.

EXAMPLE 4.

What is the lattice energy of LiBr? The ionization energy (IE) of lithium is 520 kJ/mol, the vaporization energy (VE) of lithium is 134.7 kJ/mol, the VE for $Br_2(\ell)$ is 15.46 kJ/mol, the energy of atomization (AE) of gaseous bromine ($Br_2 \rightarrow 2\,Br$) is 111.7 kJ/mol, the electron affinity (EA) of bromine is −324 kJ/mol, the formation energy (FE) ($Li(s) + \frac{1}{2}Br_2(\ell) \rightarrow LiBr(s)$) is −351.2 kJ/mol (this is also called the heat of formation, ΔH°_f).

Collect and Organize We are asked to determine the lattice energy of LiBr given the vaporization and ionization energy of lithium, the energies of vaporization, atomization and electron affinity of bromine, and the formation energy of lithium bromide.

Analyze The lattice energy is the energy associated with the chemical reaction of gaseous ions combining to form one mole of a solid ionic compound. This value cannot be determined directly. However, related equations with energy values that can be determined directly can be summed to get the value for lattice energy. This is a specific application of Hess's law, which was studied in Chapter 5.

Solve Writing the equation for each energy value given,

$\text{IE} = \text{Li}(g) \rightarrow \text{Li}^+(g) + e^-$

$\text{VE} = \text{Li}(s) \rightarrow \text{Li}(g)$

$\text{VE} = \text{Br}_2(\ell) \rightarrow \text{Br}_2(g)$

$\text{AE} = \text{Br}_2(g) \rightarrow 2\ \text{Br}(g)$

$\text{EA} = \text{Br}(g) + e^- \rightarrow \text{Br}^-(g)$

$\text{FE} = \text{Li}(s) + \frac{1}{2}\text{Br}_2(\ell) \rightarrow \text{LiBr}(s)$

The equations need to be rearranged to get $\text{Li}^+(g) + \text{Br}^-(g) \rightarrow \text{LiBr}(s)$. A good way to begin is to move the substance to the position needed. This requires reversing the equations for ionization energy and electron affinity and keeping the formation energy the same. Remember that when the equation is reversed, the sign on the energy associated with the reaction (ΔH) is changed.

IE: $\text{Li}^+(g) + e^- \rightarrow \text{Li}(g)\ \Delta H = -520\ \text{kJ}$

EA: $\text{Br}^-(g) \rightarrow \text{Br}(g) + e^-\ \Delta H = -324\ \text{kJ}$

FE: $\text{Li}(s) + \frac{1}{2}\text{Br}_2(\ell) \rightarrow \text{LiBr}(s)\ \Delta H = -351.2\ \text{kJ}$

To get rid of the products in the formation energy equation, reverse the vaporization reactions. In addition, the vaporization energy for bromine should be halved (which also halves the associated energy).

VE: $\text{Li}(g) \rightarrow \text{Li}(s)\ \Delta H = -134.7\ \text{kJ}$

VE: $\frac{1}{2}\text{Br}_2(g) \rightarrow \frac{1}{2}\text{Br}_2(\ell)\ E = -7.73\ \text{kJ}$

Adding these equations gives $\text{Li}^+(g) + \text{Br}^-(g) + \frac{1}{2}\text{Br}_2(g) \rightarrow \text{LiBr}(s) + \text{Br}(g)$

ΔH = sum of values above = −1337 kJ

To get rid of the remaining bromine, you need to add in the reverse of one-half of the atomization equation.

$$\text{Br}(g) \rightarrow \frac{1}{2}\text{Br}_2(g)\ \Delta H = -55.85\ \text{kJ}$$

Adding that in gives the lattice energy of −1395 kJ/mol.

Think About It The key to doing these types of problems is to be able to write the chemical reactions correctly, which must include physical states. The products and reactants can be algebraically added, multiplied, canceled, and so on, to obtain the correct reaction. Whatever is done to the reaction is also done to the associated energy value.

Dispersion Forces

Example 1.
Rank the following from strongest to weakest intermolecular forces.

$$Cl_2,\ CH_4,\ BF_3,\ SCl_2,\ CO_2$$

Collect and Organize We are given five compounds and are asked to rank their intermolecular forces from strongest to weakest.

Analyze While the type of intermolecular force is the major factor that determines the strength of attraction, compounds can also be ranked within the category of London (dispersion) forces. Because London forces are due to a temporary dipole formed by the perturbation of the electron cloud, molecules with bigger electron clouds will be more easily and more greatly perturbed. Since the number of electrons is related to the number of protons, which is related to molar mass, the easiest way to rank compounds with dispersion forces is by molar mass. Nonpolar compounds with higher molar masses will have stronger London forces.

Solve All these compounds, except SCl_2, are nonpolar. SCl_2 has two lone pairs on the central atom of sulfur. Its molecular geometry is bent. Consequently, this compound has dipole–dipole forces of attraction and will be stronger than the other choices.

The other choices are ranked based on molar mass. The molar mass of Cl_2 = 70.91 g/mol, CH_4 = 16.05 g/mol, BF_3 = 67.81 g/mol, CO_2 = 44.01 g/mol. The higher molar mass will be stronger, so

$$SCl_2 > Cl_2 > BF_3 > CO_2 > CH_4$$

Think About It Atoms, ions, and molecules are held together by intermolecular forces.

Polarity and Solubility

Example 1.
Which of the following compounds are likely to dissolve in water?

a. SCl_2 b. O_2 c. NaCl d. CO_2 e. PH_3

COLLECT AND ORGANIZE From the given set of compounds we are asked to determine which are soluble in the polar solvent water.

ANALYZE Water is a polar solvent. Polar molecules and some ionic compounds will dissolve in water.

SOLVE

a. SCl_2 is a polar molecule with a bent geometry. Therefore, it will dissolve in water.

b. O_2 is a nonpolar molecule with a linear geometry and a nonpolar bond between the oxygen atoms. Therefore, it will not be soluble in water.

c. NaCl is an ionic compound. The charges are low (Na^+ and Cl^-), and according to the solubility rules, "group I cations are always soluble." Therefore, NaCl is soluble in water.

d. CO_2 is a linear, nonpolar molecule. Therefore, it is not soluble in water.

e. PH_3 is a trigonal pyramid, polar molecule. It will dissolve in water.

THINK ABOUT IT This is an example of the adage that "like dissolves like."

EXAMPLE 2.
Which of the following will dissolve in C_6H_{14} (hexane)?

a. SCl_2 b. O_2 c. NaCl d. CO_2 e. PH_3

COLLECT AND ORGANIZE From the given set of compounds we are asked to determine which are soluble in the nonpolar solvent hexane.

ANALYZE Hexane is a nonpolar solvent. Hydrocarbons, molecules made solely of hydrogen and carbon, are nearly always nonpolar. Therefore, nonpolar molecules will dissolve in this solvent.

SOLVE In the list, only O_2 and CO_2 are nonpolar, so only these two compounds are soluble in hexane.

THINK ABOUT IT This is another example of "like dissolves like," only in this example the solvent is nonpolar. Therefore, we are required to determine which of the given compounds are nonpolar.

EXAMPLE 3.
How many grams of helium can dissolve in 50.00 mL water at 20°C and 770 torr?

COLLECT AND ORGANIZE We are asked to determine the amount of helium at a given pressure that will dissolve in a given volume of water at a given temperature.

ANALYZE The amount of gas dissolved in a liquid, C, is related to pressure P by Henry's law.

$$C = k_H P$$

The proportionality constant, k_H, depends on the identity of the gas and solvent and the temperature.

SOLVE There are two values for Henry's constant for helium, depending on what unit of pressure is used. Since the pressure units are torr, the value 5.1×10^{-7} mol/kg•mm Hg is appropriate. Recall that 1 mmHg is 1 torr. (The units are also useful for remembering the order of pressure and concentration, since the units must cancel.)

$$C = (5.1 \times 10^{-7} \text{ mol/kg}\cdot\text{mmHg})(770 \text{ mmHg})$$

$$C = 3.9 \times 10^{-4} \text{ mol/kg}$$

Since this is concentration, the kg must refer to the amount of solvent, using the density of water as 1.00 g/mL.

$$50.00 \text{ mL} \times \left(\frac{100 \text{ g}}{1 \text{ mL}}\right) \times \left(\frac{1 \text{ kg}}{1000 \text{ g}}\right) = 0.0500 \text{ kg}$$

To get the amount of helium,

$$3.9 \times 10^{-4} \text{ mol/kg } (0.0500 \text{ kg}) = 1.96 \times 10^{-5} \text{ mol He} \times \left(\frac{4.00 \text{ g}}{1 \text{ mol}}\right) = 7.9 \times 10^{-5} \text{ g}$$

THINK ABOUT IT The amount of gas that dissolves in a liquid is directly proportional to the pressure of the gas above the liquid. The higher the pressure, the higher the concentration of the dissolved gas.

EXAMPLE 4.
What pressure is required to dissolve 0.0045 g O_2 in 1.00 L water at 20°C?

COLLECT AND ORGANIZE We are asked to determine the pressure of O_2 gas that is required for the given mass of O_2 to dissolve in the given volume of water at the given temperature.

ANALYZE After finding the appropriate Henry's law constant, we can use Henry's law to solve the problem.

SOLVE The concentration is

$$0.0045 \text{ g} \times \left(\frac{1 \text{ mol}}{32.00 \text{ g}}\right) = \left(\frac{1.4 \times 10^{-4} \text{ mol}}{1.0 \text{ L}}\right) = (1.4 \times 10^{-4} \text{ mol/L})$$

Henry's constant for oxygen is 1.3×10^{-3} mol/L•atm, so

$$1.4 \times 10^{-4} = (1.3 \times 10^{-3})P$$

$$0.11 \text{ atm} = P$$

THINK ABOUT IT So little gas dissolves in water that it doesn't matter whether it is liters of solvent or liters of solution.

VAPOR PRESSURE

EXAMPLE 1.
Rank the following substances from highest vapor pressure to lowest vapor pressure.

Br_2, NH_3, Ar, PCl_3, PCl_5

COLLECT AND ORGANIZE We are given molecular formulas of five substances and are asked to rank them according to decreasing vapor pressure.

ANALYZE Vapor pressure is determined by the number of molecules that can escape from the surface of a liquid. More molecules can escape if the force of attraction, which holds the molecules together, is weak or if the force pulling them away (kinetic energy measured by temperature) is high.

SOLVE NH_3 must have hydrogen bonds, since the hydrogen must be bonded to the nitrogen. Br_2 and Ar must have dispersion forces, since Br_2 has only nonpolar bonds and Ar has no bonds. You will probably need to draw the Lewis structures for PCl_3 and PCl_5. PCl_3 has three P—Cl bonds and one lone pair. Therefore, it has a trigonal pyramid geometry and is polar. PCl_5 has an expanded octet with five P—Cl bonds and no lone pairs. It is a nonpolar molecule. The nonpolar molecules are ranked by molar mass. The molar masses are Ar = 39.95 g/mol, PCl_5 = 208.24 g/mol, and Br_2 = 159.80 g/mol.

Since the substance with the highest vapor pressure has the weakest intermolecular forces, the nonpolar molecule (dispersion force) with the lowest molar mass will have the highest vapor pressure. The substance with the strongest intermolecular forces will have the lowest vapor pressure, so

$$Ar > Br_2 > PCl_5 > PCl_3 > NH_3$$

THINK ABOUT IT Problems asking you to rank vapor pressure assume all substances are at the same temperature (in the same way that ranking in terms of boiling point assumes that all substances are at the same atmospheric pressure).

PHASE DIAGRAMS

EXAMPLE 1.
Answer the following questions about a hypothetical substance, based on the following phase diagram.

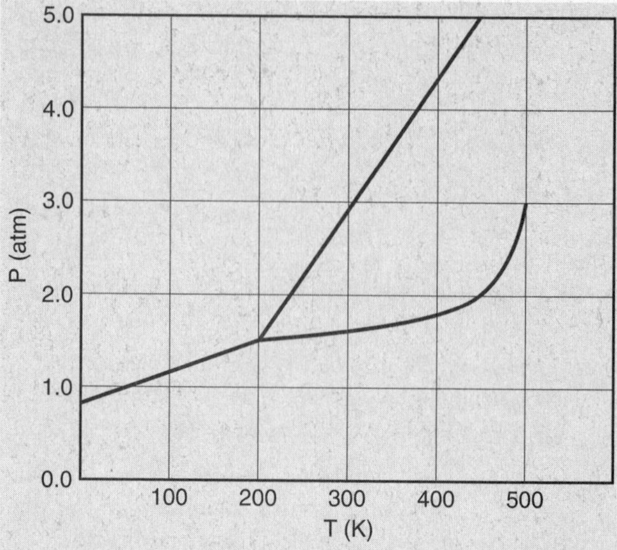

a. What is the triple point?

b. What is the critical point?

c. What is the physical state at 350 K and 2.0 atm?

d. Under what conditions does the substance sublime?

e. What is the physical state at 600 K and 4.0 atm?

f. Under 1.0 atm, at what temperature will the substance boil?

g. At 3.0 atm, at what temperature will the substance melt?

h. At 2.0 atm, at what temperature will the substance boil?

i. What is its physical state at 3.0 atm and 300 K?

COLLECT AND ORGANIZE For a hypothetical substance with the phase diagram shown above, we are asked to answer several questions concerning its physical properties.

ANALYZE A phase diagram is a graph of physical state versus temperature and pressure. The lines dividing the physical state represent conditions where a change of physical state occurs. At a point on the line, both physical states (the ones on either side of the line) exist. The line between the gas and liquid states represents boiling (liquid to gas) or condensation (gas to liquid). The line between the liquid and solid states represents melting (solid to liquid) or freezing (liquid to solid). The line between the solid and gas states represents sublimation (solid to gas) or deposition (gas to solid). The point where all three lines meet is called the triple point. At that point all three phases exist.

The liquid–solid line extends indefinitely. However, the liquid–gas line ends. The point at the end of this line is called the critical point.

SOLVE In the diagram the substance is a solid at low temperature and high pressure, a gas at low pressure and high temperature, and a liquid in between.

a. The triple point is where all three lines meet. In this example that is at 200 K and 1.5 atm.

b. The critical point is where the line between liquid and gas ends, the one with the bit of a curve. In this example the critical point is at 500 K and 3.0 atm.

c. This point is between the two lines. Therefore, the substance is a liquid under these conditions.

d. The sublimation line is the line between solid and gas. In this example the line is at temperatures of less than 200 K. So the primary requirement is that the temperature be less than 200 K. In addition, for a substance to sublime, it must turn from a solid to a gas, so the pressure must be appropriate for a gas. That pressure depends somewhat on the temperature. Any pressure lower than about 0.8 atm will definitely result in sublimation.

e. At 600 K and 4.0 atm the substance is above the critical point (part b). Therefore, the substance acts as a supercritical fluid.

f. The substance will not boil at 1.0 atm. The sublimation line crosses at 1.0 atm. The boiling point line is the line between liquid and gas.

g. At that pressure the substance melts at 300 K. The line between liquid and solid crosses 3.0 atm at that temperature.

h. The substance boils at 450 K. That is the point where the liquid–gas line crosses 2.0 atm.

i. That point is on the solid–liquid line, so both solid and liquid exist at that point. The substance is either freezing or melting.

THINK ABOUT IT Substances with pressures and temperatures above the critical point are in a fourth physical state called a supercritical fluid.

COLLIGATIVE PROPERTIES OF SOLUTIONS

EXAMPLE 1.
What is the molality of a solution made by dissolving 0.1356 g $MgSO_4$ in 200.0 mL of water?

COLLECT AND ORGANIZE We are asked to determine the molality of a solution that is made by dissolving a given mass of solute in a given volume of solvent.

ANALYZE Molality is defined as the moles of solute per kilograms of solvent. We must convert solute from grams to moles using the molar mass of the substance and convert mL of solvent to kg of solvent using the density of the solvent.

SOLVE

$$\text{Molality} = \frac{\text{mol solute}}{\text{kg solvent}}; \text{ solute} = MgSO_4; \text{ solvent} = \text{water.}$$

$$\text{moles of solute} = 0.1356 \text{ g} \times \left(\frac{1 \text{ mol}}{120.367 \text{ g}}\right) = 1.127 \times 10^{-3} \text{ mol}$$

$$\text{kg of solvent} = 200.0 \text{ mL} \times \left(\frac{1.000 \text{ g}}{1 \text{ mL}}\right) \times \left(\frac{1 \text{ kg}}{1000 \text{ g}}\right) = 0.2000 \text{ kg}$$

$$\text{Molality} = \left(\frac{1.127 \times 10^{-3} \text{ mol}}{0.2000 \text{ kg}}\right) = 0.005633\ m \text{ or } 5.633 \times 10^{-3}\ m.$$

If you use a density of water values with fewer than four significant figures, the number of significant figures in the overall answer will be the same as that in the density value used.

THINK ABOUT IT Care must be taken to ensure that one does not confuse *molarity* with *molality*.

EXAMPLE 2.
What is the molality of a solution of 6.44 g of $Cd(NO_3)_2$ in 375.0 g of water?

COLLECT AND ORGANIZE We are asked to determine the molality of a solution that is made by dissolving a given mass of solute in a given mass of solvent.

ANALYZE We must first convert grams of solute to moles of solute using the molar mass of $Cd(NO_3)_2$. Since the solvent is already given as a mass, its density is not necessary; we must simply convert from grams to kilograms before using the definition of molality.

SOLVE

$$\text{Molality} = \frac{\text{mol solute}}{\text{kg solvent}}; \text{ solute} = Cd(NO_3)_2; \text{ solvent} = \text{water.}$$

$$\text{moles of solute} = 6.44 \text{ g} \times \left(\frac{1 \text{ mol}}{236.42 \text{ g}}\right) = 0.0272 \text{ mol } Cd(NO_3)_2$$

$$\text{kg of solvent} = 375.0 \text{ g} \times \left(\frac{1 \text{ kg}}{1000 \text{ g}}\right) = 0.3750 \text{ kg}$$

$$\text{molality} = \frac{0.0272 \text{ mol}}{0.3750 \text{ kg}} = 0.0726\ m$$

THINK ABOUT IT Molality is given the symbol *m* while molarity is given the symbol *M*.

EXAMPLE 3.
How many grams of $Sr(ClO_4)_2$ are required to make a 0.30 *m* aqueous solution using 600 g of water?

COLLECT AND ORGANIZE We are asked how many grams of solute need to be dissolved in 600 g of water in order that the resulting solution has a molality of 0.30 *m*.

ANALYZE We use the definition of molality and the information given to solve for the number of moles of $Sr(ClO_4)_2$ that are required. The final step is to then relate the calculated number of moles to grams using the molar mass of $Sr(ClO_4)_2$.

SOLVE

$$\text{Molality} = \frac{\text{mol solute}}{\text{kg solvent}};\ \text{solute} = Sr(ClO_4)_2;\ \text{solvent} = \text{water.}$$

$$\text{kg solvent} = 600\text{ g water} \times \left(\frac{1\text{ kg}}{1000\text{ g}}\right) = 0.6000\text{ kg}$$

$$\text{mol solute} = \text{molality} \times \text{kg solvent} = 0.30\ m \times 0.6000\text{ kg} = 0.18\text{ mol } Sr(ClO_4)_2$$

$$\text{g solute} = 0.18\text{ mol } Sr(ClO_4)_2 \times \left(\frac{286.52\text{ g}}{1\text{ mol}}\right) = 52\text{ g}$$

or (equally correct!),

$$600\text{ g water} \times \left(\frac{1\text{ kg}}{1000\text{ g}}\right) \times \left(\frac{0.30\text{ mol } Sr(ClO_4)_2}{1\text{ kg water}}\right) \times \left(\frac{286.52\text{ g}}{1\text{ mol } Sr(ClO_4)_2}\right) = 52\text{ g}$$

THINK ABOUT IT Remember that it is necessary to convert the solute to moles.

EXAMPLE 4.
How many grams of $BiCl_3$ are needed to make 500.0 g of a 0.10-*m* aqueous solution?

COLLECT AND ORGANIZE We are asked to calculate the number of grams of $BiCl_3$ that are needed to make 500.0 g of a 0.10-*m* aqueous solution.

ANALYZE We will use the definition of molality to solve the problem. We must recognize that we are given grams of solution, not grams of solvent. However, we can calculate grams of solvent using the relation that g solvent = g solution – g solute.

SOLVE

$$\text{Molality} = \frac{\text{mol solute}}{\text{kg solvent}};\ \text{solute} = BiCl_3;\ \text{solvent} = \text{water. The}$$

molar mass of $BiCl_3$ is 315.34 g/mol. If we let x = g solute, then the molality equation can be rewritten as

$$0.10\ m = \left(\frac{(x/315.34\text{ g/mol})}{[(500.0 - x)/1000]}\right)$$

$$\frac{(50.00\ m - 0.10x\ m)}{1000} = x/315.34\text{ g/mol}$$

$$0.0500\ m - 1 \times 10^{-4}x\ m = x/315.34\text{ g/mol}$$

$$15.767\text{ g} - 0.0315x = x$$

$$15.767\text{ g} = 1.0315x$$

$$15.28\text{ g} = x$$

So, 15 g of $BiCl_3$ is needed (rounded to the appropriate significant figures).

THINK ABOUT IT Make sure you understand whether the question gives a mass of solution or a mass of solvent.

EXAMPLE 5.
What is the molality of a 0.123 *M* HCl(*aq*) solution? The density of the solution is 1.030 g/mL.

COLLECT AND ORGANIZE We are asked to convert from molarity (*M*) to molality (*m*).

ANALYZE We can use the density of the solution to determine the mass of the solution. The definition of molarity (*M*) and the molecular weight of HCl can be used to calculate the mass of solute. This allows for the determination of the mass of solvent that is needed for the molality calculation.

SOLVE

$$\text{Molality} = \frac{\text{mol HCl}}{\text{kg water}};\ \text{molarity} = \frac{\text{mol HCl}}{\text{L solution}}.$$ Since we have a value of molarity given, let's assume that we are working with exactly 1 L of solution. Using the molarity equation

$$M = \frac{\text{mol solute}}{\text{L solution}}$$ or $M \times$ L solution = mol solute, there is 0.123 mol HCl.

$$0.123\text{ mol HCl} \times \left(\frac{36.45\text{ g}}{1\text{ mol}}\right) = 0.0448\text{ g HCl}$$

Using the density,

$$1\text{ L} \times \left(\frac{1000\text{ mL}}{1\text{ L}}\right) \times \left(\frac{1.030\text{ g}}{1\text{ mL}}\right) = 1030\text{ g solution}$$

$$\text{g solvent} = \text{g solution} - \text{g solute} = 1030\text{ g} - 0.0448\text{ g} = 1030\text{ g solvent} \times \left(\frac{1\text{ kg}}{1000\text{ g}}\right) = 1.030\text{ kg}$$

Using the molality equation,

$$\text{molality} = \frac{\text{mol HCl}}{\text{kg solvent}} = \frac{0.123\text{ mol}}{1.030\text{ kg}} = 0.119\ m$$

THINK ABOUT IT Notice how the density of the solution is greater than the density of pure water. This is true because we are added material to the water to form the solution. We know the solvent is water from the (*aq*) description.

EXAMPLE 6.
What is the molarity of a 3.21 *m* KOH(*aq*) solution? (density of solution = 1.163 g/mL)

COLLECT AND ORGANIZE We are asked to convert molality (*m*) to molarity (*M*).

ANALYZE By assuming 1 kg of solvent we can convert from molality to moles of solute. We then use the molar mass of the solute to calculate the grams of solute, which when added to the mass of the solvent will give the mass of the solution. The given density of the solution can then be used to calculate its volume. We then have the information needed to determine the molarity (*M*) of the solution.

SOLVE

$\text{Molarity} = \frac{\text{mol KOH}}{\text{L solution}}$; $\text{molality} = \frac{\text{mol KOH}}{\text{kg water}}$. Since we have a value for molality, let's assume exactly 1 kg (1000 g) of water:

$$\text{mol KOH} = (3.21\ m)(1\ \text{kg}) = 3.21\ \text{mol KOH}$$

$$3.21\ \text{mol KOH} \times \left(\frac{56.10\ \text{g}}{1\ \text{mol}}\right) = 180.1\ \text{g}$$

$$\text{g solution} = \text{g solvent} + \text{g solute} = 1000\ \text{g} + 180.1\ \text{g} = 1180.1\ \text{g}$$

$$1180.1\ \text{g solution} \times \left(\frac{1\ \text{mL}}{1.163\ \text{g}}\right) \times \left(\frac{1\ \text{L}}{1000\ \text{mL}}\right) = 1.015\ \text{L}$$

$$M = \frac{\text{mol KOH}}{\text{L solution}} = \frac{3.21\ \text{mol KOH}}{1.015\ \text{L}} = 3.16\ M$$

THINK ABOUT IT Notice that although the values for molality and molarity are similar, the value for molality will be slightly less. The difference normally becomes larger at higher concentrations.

EXAMPLE 7.
What is the change in boiling point if 3.69 g of K_2SO_4 is added to 100.0 mL of water?

COLLECT AND ORGANIZE We are asked to calculate the boiling point elevation that results when 3.69 g of K_2SO_4 is added to 100.0 mL of water.

ANALYZE The change in boiling point is ΔT, and can be determined using $\Delta T = iK_bm$, where *m* represents the molality of the solute, and *i* is the van't Hoff factor. K_b is the boiling point constant for the solvent, in this case: water.

SOLVE K_b = boiling point constant for water = 0.52°C/*m* and *m* = mol K_2SO_4/kg water.

$$\text{mol } K_2SO_4 = 3.69\ \text{g} \times \left(\frac{1\ \text{mol}}{174.20\ \text{g}}\right) = 0.0212\ \text{mol}$$

When K_2SO_4 dissolves in water it dissociates into two K^+ ions and one SO_4^{2-} ion. Therefore, $i = 3$.

$$\Delta T = iK_bm = (3)(0.52°\text{C}/m)(0.212\ m) = 0.33°\text{C}$$

THINK ABOUT IT The dissolution of solute in pure water will increase its boiling point.

EXAMPLE 8.
What is the boiling point of a solution made by mixing 13.63 g ethanol and 540.00 g water?

COLLECT AND ORGANIZE We are asked to determine the boiling point of a solution made by mixing 13.63 g ethanol and 540.00 g water.

ANALYZE The tricky part of this question has to do with both ethanol and water being listed without defining either as the solvent. Remember that the solvent is the substance in greater quantity. In this case, it is obviously water. Because ethanol is not an electrolyte, one mole of ethanol produces one mole of particles and its van't Hoff factor is 1. Thus, the equation for boiling point elevation ($\Delta T = iK_bm$) can also be written as $\Delta T = K_bm$ for nonelectrolytes.

SOLVE Solute = 13.63 g ethanol (C_2H_5OH) and solvent = 540.00 g water (H_2O).

$$\Delta T = K_bm$$

The value of K_b is the value of the solvent = 0.52°C/*m*. The value of *m* is the value of particles of solute.

$$m = \frac{\text{mol solute}}{\text{kg solvent}} = \frac{\text{mol ethanol}}{\text{kg water}}$$

$$\text{mol ethanol} = 13.63\ \text{g } C_2H_5OH \times \left(\frac{1\ \text{mol}}{46.07\ \text{g}}\right) = 0.2959\ \text{mol}$$

$$\text{kg water} = 540.00\ \text{g} \times \left(\frac{1\ \text{kg}}{1000\ \text{g}}\right) = 0.54000\ \text{kg}$$

$$m = \frac{0.2959\ \text{mol}}{0.54000\ \text{kg}} = 0.5479\ m$$

$$\Delta T = K_bm = (0.52°\text{C}/m)(0.5479\ m) = 0.28°\text{C}$$

Boiling point (bp) of solution = bp of solvent + ΔT = 100°C + 0.28°C = 100.28°C. Recall that the 100°C was *defined* as the boiling point of water; therefore, it is an exact number with infinite significant figures. The significant figure rule for addition is to use the value with the fewest decimal places.

THINK ABOUT IT Don't forget (and it is the most common mistake) that calculating ΔT does not answer the question. We are asked for the new boiling point, not the change in boiling point. Since it is boiling point *elevation*, the change is added to the boiling point of the pure solvent.

EXAMPLE 9.
What is the boiling point of 0.360 *m* $Ca(ClO_4)_2(aq)$?

COLLECT AND ORGANIZE We are asked to determine the boiling point of a 0.360 *m* $Ca(ClO_4)_2$ aqueous solution.

ANALYZE We will calculate the increase in the boiling point of the solvent (water) that results when $Ca(ClO_4)_2$ is added.

SOLVE $\Delta T = iK_bm$, where K_b is the boiling point constant for the solvent, which is water (from "(*aq*)"), and is equal to 0.52°C/*m*. *m* is the molality of the solute. Since $Ca(ClO_4)_2$ is an ionic compound, it dissociates into ions in solution. The ions are one Ca^{2+} and two ClO_4^-, so $i = 3$.

$$\Delta T = (3)(0.52°C/m)(0.360\ m) = 0.56°C$$

The boiling point (bp) of solution = bp solvent + ΔT = 100°C + 0.56°C = 100.56°C.

THINK ABOUT IT It is important to recognize ionic compounds and their component ions in order to predict the value of *i*.

EXAMPLE 10.
What is the boiling point of a solution made by mixing 72.5 g Br_2 with 660.0 g CCl_4?

COLLECT AND ORGANIZE We are asked to determine the boiling point of 660.0 g CCl_4 when 72.5 g of Br_2 has been added.

ANALYZE We can use the expression for boiling point elevation to calculate the change in boiling point and add the result to the normal boiling point of CCl_4. Since bromine is a nonelectrolyte ($i = 1$), the equation for boiling point elevation of nonelectrolytes is used. (The nonaqueous solvent is another tip-off that it will not dissociate.)

SOLVE We have $\Delta T = K_bm$, the solute = Br_2, and the solvent = CCl_4. The solvent determines the K_b. CCl_4 has a K_b = 5.02°C/*m* and a boiling point of 76.8°C. (Found in the appropriate table in Chapter 10.)

$$m = \frac{\text{mol solute}}{\text{kg solvent}} = \frac{\text{mol Br}_2}{\text{kg CCl}_4}$$

$$\text{mol Br}_2 = 72.5\ \text{g} \times \left(\frac{1\ \text{mol}}{159.81\ \text{g}}\right) = 0.454\ \text{mol}$$

$$\text{kg CCl}_4 = 660.0\ \text{g} \times \left(\frac{1\ \text{kg}}{1000\ \text{g}}\right) = 0.6600\ \text{kg}$$

$$m = \frac{0.454\ \text{mol Br}_2}{0.6600\ \text{kg CCl}_4} = 0.688\ m$$

$$\Delta T = K_bm = (5.02°C/m)(0.688\ m) = 3.45°C$$

There are three significant figures this time. The boiling point (bp) of the solution = bp solvent + ΔT = 76.8°C + 3.45°C = 80.3°C. Remember to round off for significant figures to the fewest decimal places.

THINK ABOUT IT Boiling point elevation occurs for solvents other than water.

EXAMPLE 11.
What is the freezing point of 0.544 *m* $KMnO_4(aq)$?

COLLECT AND ORGANIZE We are asked to determine the freezing point of a solution containing an ionic compound.

ANALYZE We will calculate the freezing point depression, $\Delta T = iK_fm$, and subtract from the normal freezing point of the solvent.

SOLVE $\Delta T = iK_fm$ and K_f = 1.86°C/*m* for water. The freezing point of pure water is *exactly* 0°C. The solute, $KMnO_4$, is an ionic compound (metal combined with nonmetal). You might not recognize the anion, but the cation, K^+, should be obvious. Since most ionic compounds consist of only two ions, it is best to assume that the rest of the formula is the anion. That anion, MnO_4^-, must have a charge of –1, so that the compound is electrically neutral. This makes the van't Hoff factor for this compound = 2.

$$\Delta T = iK_fm = (2)(1.86°C/m)(0.544\ m) = 2.02°C$$

The solute lowers the freezing point, so you subtract ΔT from the value of pure solvent. The freezing point of solution = freezing point of solvent – ΔT = 0°C – 2.02°C = –2.02°C.

THINK ABOUT IT Freezing point depression has practical applications. The spreading of salt on roads during winter in northern climates lowers the freezing point of water and prevents the formation of ice.

EXAMPLE 12.
What is the vapor pressure of a mixture of 0.127 mol sugar and 100.0 g water at 25°C?

COLLECT AND ORGANIZE We are asked to determine the vapor pressure of a solution at 25°C that is made by dissolving a given amount of sugar in a given volume of water.

ANALYZE The vapor pressure of a solution is calculated with Raoult's law.

SOLVE Since this refers to solution, where water is the solvent and sugar is the solute, $P_{\text{solution}} = \chi_{\text{solvent}} P_{\text{solvent}}$. P_{solvent} is the vapor pressure of pure water. This value at 25°C is 28.8 torr, according to the table of water vapor pressures in Chapter 6. The mole fraction will be moles water/(moles water + moles sugar). According to the problem, there are 0.127 moles of sugar. Moles of water can be calculated from its molar mass:

$$100.0 \text{ g water} \times \left(\frac{1 \text{ mol}}{18.01 \text{ g}}\right) = 5.552 \text{ mol}$$

So the mole fraction is

$$\chi_{\text{solvent}} = \frac{(5.552)}{(5.552 + 0.127)} = 0.9776$$

The four significant figures are correct, since the addition in the denominator gives a sum with four significant figures. Remember that the mole fraction does not have units.

$$P_{\text{solution}} = \chi_{\text{solvent}} P_{\text{solvent}} = (0.9776)(28.8 \text{ torr}) = 28.2 \text{ torr}$$

The answer is limited to three significant figures by the vapor pressure. The vapor pressure of water is in torr (mmHg), so the final answer has the same units. There is very little change from the vapor pressure of pure water; this is typical. If you do get a very low value for vapor pressure of a solution, check that you have not calculated the mole fraction of solute instead of the mole fraction of solvent.

THINK ABOUT IT Addition of a solute also modifies the vapor pressure of the solvent. This is also a colligative property, because it depends on the number of particles, not the identity of the particles.

EXAMPLE 13.

What is the vapor pressure of a mixture of 3.44 g FeF_2 in 75.00 g water at 20°C?

COLLECT AND ORGANIZE We are asked to determine the vapor pressure of a solution at 20°C that is made by dissolving a given mass of solute in a given mass of water (solvent).

ANALYZE If the solute is an ionic compound in water, you must use the moles of ions rather than the moles of compound, as the moles of solute in the calculation of mole fraction.

SOLVE The vapor pressure of water at 20°C is 17.5 torr. Since FeF_2 is a soluble, ionic compound, when 1 mole of the compound dissolves, it makes 1 mole of Fe^{2+} and 2 moles of F^-. So the mole fraction is

$$\chi_{\text{water}} = \frac{(\text{moles of water})}{(\text{moles of water + moles of ferrous ion + moles of fluoride ion})}$$

$$\text{moles of water} = 75.00 \text{ g water} \times \left(\frac{1 \text{ mol}}{18.01 \text{ g}}\right) = 4.164 \text{ mol}$$

$$\text{moles of } FeF_2 = 3.44 \text{ g} \times \left(\frac{1 \text{ mol } FeF_2}{93.85 \text{ g}}\right) = 0.0367 \text{ mol}$$

$$0.0367 \text{ mol } FeF_2 \times \left(\frac{1 \text{ mol } Fe^{2+}}{1 \text{ mol } FeF_2}\right) = 0.0367 \text{ mol } Fe^{2+}$$

$$0.0367 \text{ mol } FeF_2 \times \left(\frac{2 \text{ mol } F^-}{1 \text{ mol } FeF_2}\right) = 0.0734 \text{ mol } F^-$$

$$X = \frac{4.164}{(4.164 + 0.0367 + 0.0734)} = \frac{4.164}{4.274} = 0.9742$$

$$P_{\text{solution}} = (0.9742)(17.5) = 17.0 \text{ torr}$$

THINK ABOUT IT Recall that colligative properties depend on the number of particles in solution. When an ionic compound dissolves in water, the particles are ions, not the formula unit.

EXAMPLE 14.

What is the osmotic pressure of a solution of 1.455 g of $BeSO_4$ in 500.0 mL aqueous solution at 21°C?

COLLECT AND ORGANIZE We are asked to determine the osmotic pressure resulting from a solution at a temperature of 21°C made by dissolving 1.455 g of $BeSO_4$ in 500.0 mL of water.

ANALYZE Osmotic pressure (π) is determined by the equation $\pi = iMRT$. Since osmotic pressure is a colligative property, it is the molarity of the total number of particles, not the molarity of the solute that is generally used. The van't Hoff factor (i) is used to account for the number of particles. The product iM (where M is the molarity of the solute) is the molarity of particles.

SOLVE We first calculate the number of moles of solute using the given mass of $BeSO_4$ and its molar mass.

$$\text{moles of solute} = \text{moles of } BeSO_4 = 1.455 \text{ g} \times \left(\frac{1 \text{ mol}}{105.07 \text{ g}}\right) = 0.01385 \text{ mol } BeSO_4$$

Since molarity is moles of solute per liter of solution, $M = \frac{0.01385}{0.5000} = 0.02770\ M$. Since we are dealing with a colligative property, we need to determine the moles of particles rather than then the moles of the compound. Since there are 2 moles of ions for every mole of $BeSO_4$, the van't Hoff factor is two. R is the gas constant, 0.0821 L•atm/mol•K. T is temperature in Kelvin, so 21°C + 273.15 = 294 K.

$$\pi = iMRT = (2)(0.02770 \text{ mol/L})(0.0821 \text{ L•atm/ mol•K})(293 \text{ K}) = 1.33 \text{ atm}$$

THINK ABOUT IT *Note:* The temperature limited the value to three significant figures. Although it started with two, it gained one in the conversion to Kelvin. The significant figure rule for addition is to use the value with the fewest decimal places rather than the fewest significant figures.

EXAMPLE 15.
What is the osmotic pressure of a solution of 3.47 mg of PCl_3 in 750.0 mL of ethanol solution at 23°C?

COLLECT AND ORGANIZE We are asked to determine the osmotic pressure resulting from a solution at a temperature of 23°C made by dissolving 3.47 mg of PCl_3 in 750.0 mL of ethanol.

ANALYZE As with the previous example, we will use the expression $\pi = iMRT$.

SOLVE Since molecular compounds do not ionize, the molarity of the compound is the same as the total number of particles. In other words, $i = 1$.

$$\text{molarity} = \frac{\text{moles of solute}}{\text{liters of solution}} = \frac{\text{mol PCl}_3}{\text{L ethanol solution}}$$

$$3.47 \text{ mg PCl}_3 \times \left(\frac{1 \text{ g}}{1000 \text{ mg}}\right) \times \left(\frac{1 \text{ mol}}{137.333 \text{ g}}\right) = 2.53 \times 10^{-5} \text{ mol}$$

$$750.0 \text{ mL} \times \left(\frac{1 \text{ L}}{1000 \text{ mL}}\right) = 0.7500 \text{ L}$$

$$\text{molarity} = \frac{2.53 \times 10^{-5} \text{ mol}}{0.7500 \text{ L}} = 3.37 \times 10^{-5}\, M$$

$$R = 0.0821 \text{ L} \cdot \text{atm/mol} \cdot \text{K}$$

$$T = 23°\text{C} + 273.15 = 296 \text{ K}$$

$$\pi = (1)(3.37 \times 10^{-5} \text{ mol/L})(0.0821 \text{ L} \cdot \text{atm/mol} \cdot \text{K})(296 \text{ K}) = 8.18 \times 10^{-4} \text{ atm}$$

THINK ABOUT IT Osmotic pressure depends upon the concentration of the solute particles. It is independent of the identity of the solute.

EXAMPLE 16.
What is the van't Hoff factor of a 0.010 *M* solution of $CaCl_2$ if the osmotic pressure at 25°C is 0.70 atm?

COLLECT AND ORGANIZE We are given the osmotic pressure of a 0.010 *M* solution of $CaCl_2$ at 25°C and are asked to determine the van't Hoff factor for $CaCl_2$.

ANALYZE We solve the expression for the osmotic pressure, $\pi = iMRT$, for i and using the given parameters.

SOLVE
$\pi = iMRT$

$$0.70 \text{ atm} = i(0.010 \text{ mol/L})(0.0821 \text{ L} \cdot \text{atm/mol} \cdot \text{K})(298 \text{ K})$$

$$i = \frac{0.70}{(0.010 \times 0.0821 \times 298)} = 2.86$$

THINK ABOUT IT For calcium chloride, the predicted van't Hoff factor is 3. Ion pairing causes the true value to be less than the calculated one.

EXAMPLE 17.
What is the molar mass of a nonelectrolyte if 1.50 g in 250.0 mL of solution has an osmotic pressure of 0.521 atm at 15°C?

COLLECT AND ORGANIZE We are asked to determine the molar mass of a solution made up of 1.50 g of nonelectrolyte in 250.0 mL of solution given its osmotic pressure and temperature.

ANALYZE Since the solute is a nonelectrolyte the value of $i = 1$. We can then use $\pi = iMRT$ to solve for the molarity of the solution. We can use the definition of molarity to determine the number of moles of solute and finally determine the molar mass by dividing the given mass of solute by the number of moles of solute.

SOLVE We have $\pi = 0.521$ atm; $R = 0.0821$ L•atm/mol•K; $T = 15 + 273.15 = 288$ K.

$$\frac{0.521}{(0.0821)(288)} = M = 0.0220\, M$$

$$\text{molarity} \times \text{volume} = (0.0220 \text{ mol/L})(0.2500 \text{ L}) = 0.00551 \text{ mol}$$

$$\text{molar mass} = \frac{1.50 \text{ g}}{0.00551 \text{ mol}} = 272 \text{ g/mol}$$

THINK ABOUT IT Because the osmotic pressure is a function of the molarity of the solution, it can be used as a way to determine the molar mass of the solute.

EXAMPLE 18.
What is the molar mass of a nonelectrolyte if 141 mg in 1.00 L of solution has an osmotic pressure of 0.0184 atm at 30°C?

COLLECT AND ORGANIZE We are asked to determine the molar mass of a solution made up of 1.50 g of nonelectrolyte in 250.0 mL of solution given its osmotic pressure and temperature.

ANALYZE As with the previous example, we will use $\pi = iMRT$ to solve for the molarity. Knowing the volume of solution and the molarity we can determine the number of moles of solute, which can then be used to determine the molar mass. Because the solute is a nonelectrolyte, $i = 1$.

SOLVE

$$0.0184 \text{ atm} = (1)M(0.0821 \text{ L}\bullet\text{atm/mol}\bullet\text{K})(303 \text{ K})$$

$$0.000740 \text{ mol/L} = M$$

$$\text{moles of solution} = (0.000740 \text{ mol/L})(1.00 \text{ L}) = 0.000740 \text{ mol}$$

$$\text{mass of solution} = 141 \text{ mg} \times \left(\frac{1 \text{ g}}{1000 \text{ mg}}\right) = 0.141 \text{ g}$$

$$\text{molar mass of solution} = \frac{0.141 \text{ g}}{0.000740 \text{ mol}} = 191 \text{ g/mol}$$

THINK ABOUT IT Because the osmotic pressure is a function of the molarity of the solution, it can be used as a way to determine the molar mass of the solute.

Self-Test

KEY VOCABULARY COMPLETION QUESTIONS

__________ 1. Unit of concentration used in osmotic pressure problems

__________ 2. Moles of particles (ions) per mole of solute

__________ 3. What adding a solute does to the freezing point of a solvent

__________ 4. Moles of solute per kilogram solvent

__________ 5. Force that dissolves salt (NaCl) in water

__________ 6. Resistance to flow

__________ 7. Liquid, solid, and gas all exist at this temperature and pressure

__________ 8. Temperature and pressure above which a supercritical fluid is formed

__________ 9. Curved surface of a liquid in a tube

__________ 10. Solid turning into a liquid

__________ 11. When two liquids dissolve in each other in all proportions

__________ 12. Intermolecular force between molecules containing an O—H bond

__________ 13. Temperature at which the vapor pressure equals the atmospheric pressure

__________ 14. Energy released when gaseous ions produce a mole of ionic solid

__________ 15. Physical state where molecular motion is unrestricted

__________ 16. Attraction due to temporary dipoles

__________ 17. Proportional to mole fraction of the solvent

__________ 18. Attraction between molecules of the same type

__________ 19. Physical state where molecular motion is most limited

__________ 20. Attractive force between oppositely charged particles

__________ 21. Surrounded by water molecules

__________ 22. Attractive force between polar molecules

__________ 23. How easily an electron cloud is deformed

__________ 24. Water fearing

__________ 25. Temperature at which a liquid becomes a solid

__________ 26. Solid becomes a gas without going through a liquid state

__________ 27. Substance cannot be distinguished as a liquid or a gas

__________ 28. Movement of solvent through a membrane

__________ 29. Addition of an electron to a gaseous atom

__________ 30. Relationship between partial pressure and gas solubility

MULTIPLE-CHOICE QUESTIONS

1. What is the strongest intermolecular force between molecules of CH_3OH?
   a. ion–ion
   b. dipole–dipole
   c. hydrogen bonding
   d. London

2. Which compound has the strongest intermolecular force?
   a. KBr
   b. LiF
   c. CrN
   d. MgS

3. Which compound has the strongest intermolecular force?
   a. KI
   b. H_2O
   c. CH_4
   d. HCl
4. What causes the high surface tension of water?
   a. the strong hydrogen bonds
   b. the organization of the lattice
   c. ion–dipole forces
   d. the lack of dispersion forces
5. What is the strongest intermolecular force in CH_3OCH_3?
   a. ion–ion
   b. dipole–dipole
   c. hydrogen bonding
   d. dispersion forces
6. Which compound has the highest melting point?
   a. CsI
   b. NaCl
   c. CF_4
   d. Br_2
7. Which compound has the lowest boiling point?
   a. H_2
   b. Cl_2
   c. CCl_4
   d. CH_2O
8. Which of the following has the lowest vapor pressure?
   a. water
   b. 0.1 *M* NaCl(*aq*)
   c. 0.5 *M* NaCl(*aq*)
   d. all have the same vapor pressure
9. What is the intermolecular force in $MgCl_2(aq)$?
   a. ion–ion
   b. ion–dipole
   c. dipole–dipole
   d. hydrogen bonding
10. In a molecule with only dispersion forces, intermolecular forces are stronger if the molecule
    a. is more polarizable
    b. has lower molar mass
    c. has higher permanent charges
    d. all of the above
11. Intermolecular forces are stronger for compounds with
    a. more highly charged ions
    b. ions closer together in lattice
    c. smaller ions
    d. all of the above
12. Which is most likely to have the strongest intermolecular forces?
    a. solid
    b. liquid
    c. gas
    d. physical state is not related to intermolecular forces
13. Which ionic compound is the most soluble in water?
    a. LiF
    b. MgO
    c. CaS
    d. CsI
14. Which equation represents lattice energy?
    a. $Na(g) \rightarrow Na^+(g) + e^-$
    b. $F_2(g) \rightarrow 2\,F(g)$
    c. $Na(s) + \frac{1}{2}F_2(g) \rightarrow NaF(s)$
    d. $Na^+(g) + F^-(g) \rightarrow NaF(s)$
15. What is the typical ratio of water molecules to ions with ion–dipole forces?
    a. one water to six ions
    b. six water to one ion
    c. one water to one ion
    d. it varies tremendously from ion to ion
16. Which compound is most likely to be miscible in water?
    a. Cl_2
    b. SCl_2
    c. CaS
    d. $CH_3CH_2CH_3$
17. Why are ion–dipole forces stronger than dipole–dipole forces?
    a. larger charges
    b. the distances are closer
    c. there are more of them
    d. ion–dipole forces are not stronger
18. Vapor pressure of a mixture depends on
    a. the identity of the solute
    b. the identity of the solvent
    c. total volume of a mixture
    d. total moles of vapor
19. Which pure liquid has the highest vapor pressure (at a given temperature)?
    a. SBr_2
    b. Br_2
    c. CBr_4
    d. NaBr

20. At higher altitudes, the atmospheric pressure is lower. Consequently, in the mountains, water boils at
    a. higher than 100°C
    b. lower than 100°C
    c. 100°C

21. Increasing the pressure at constant temperature will normally cause
    a. a gas to become a liquid
    b. a solid to become a gas
    c. a liquid to become a gas
    d. a solid to become a supercritical fluid

22. Which physical state exists at the triple point?
    a. solid
    b. liquid
    c. gas
    d. all of the above

23. Under what conditions is a substance likely to sublime?
    a. high pressure and high temperature
    b. high pressure and low temperature
    c. low pressure and low temperature
    d. low pressure and high temperature

24. Under what conditions does a substance become a supercritical fluid?
    a. high pressure and high temperature
    b. high pressure and low temperature
    c. low pressure and low temperature
    d. low pressure and high temperature

25. Why does ice float?
    a. solids are always less dense than liquids
    b. molecules arrange in an open structure to maximize hydrogen bonding
    c. hydrogen bonding in the solid is weaker than hydrogen bonding in the liquid
    d. the intermolecular forces in water are stronger than those in ice

26. How will an increase in temperature affect the viscosity of a substance?
    a. increase
    b. decrease
    c. stay the same
    d. depends on the identity of the substance

27. How will an increase in temperature affect vapor pressure of a substance?
    a. increase
    b. decrease
    c. stay the same
    d. depends on the identity of the substance

28. Which compound has the strongest intermolecular force?
    a. CCl_4
    b. CBr_4
    c. Br_2
    d. Cl_2

29. What is the strongest intermolecular force of PH_3?
    a. hydrogen bond
    b. ion–ion
    c. dipole–dipole
    d. London

30. Which compound will climb the highest in a capillary tube?
    a. the one with the strongest cohesive forces
    b. the one with the strongest adhesive forces
    c. the one with the strongest intermolecular forces
    d. the one with the weakest intermolecular forces

31. What is the molality of CH_3OH in a mixture of 0.10 moles of CH_3OH and 100.0 g of water?
    a. 0.10 *m*
    b. 1.0 *m*
    c. 3.2 *m*
    d. 0.32 *m*
    e. 0.032 *m*

32. Which of the following will have the greatest change in freezing point?
    a. 0.10 *m* $CH_3OH(aq)$
    b 0.10 *m* CH_3OH in benzene
    c. 0.10 *m* CH_3OH in ethanol
    d. 0.10 *m* CH_3OH in carbon tetrachloride
    e. the change in freezing point is the same in all these solutions

33. What is the freezing point of 0.50 *m* NaCl(*aq*)?
    a. 0.93°C
    b. −0.93°C
    c. 0.52°C
    d. −0.52°C
    e. −1.9°C

34. What is the osmotic pressure of 0.10 *M* KI(*aq*) at 20°C?
    a. 4.8 atm
    b. 2.4 atm
    c. 0.33 atm
    d. 0.16 atm
    e. 3.3 atm

35. What is the molecular weight of a nonelectrolyte if 5.0 g dissolved in 100.0 g of water has a freezing point of –0.25°C?
a. 370 g/mol
b. 190 g/mol
c. 37 g/mol
d. 20 g/mol
e. 27 g/mol

36. What additional information do you need to change from units of molarity to units of molality?
a. density of solute
b. density of solvent
c. density of solution
d. none of these
e. all of these

37. What is the boiling point of 0.05 mol of Cl_2 in 200.0 g of benzene?
a. 79.5°C
b. 80.7°C
c. 100.13°C
d. 81.0°C
e. 0.63°C

38. Addition of a solute has the largest effect on which colligative property?
a. osmotic pressure
b. freezing point
c. boiling point
d. the effect is the same for each
e. the effect depends on the type of solute

39. Which of the following would make a 0.1 *m* LiCl(*aq*)?
a. 0.1 mol LiCl and 1 kg solution
b. 0.1 mol LiCl in 1 L of solution
c. 0.1 g LiCl in 1 L of solution
d. 0.1 mol LiCl in 1 kg of water
e. 0.1 g LiCl and 1 mL of water

40. A solution of 0.22 *M* CaI_2 has a density of 1.213 g/mL. How many grams of water are in 1 L of this solution?
a. 1000 g
b. 1213 g
c. 220 g
d. 65 g
e. 1148 g

41. What is the molar mass of a substance if 12.0 g in 100 mL of water make a 1.51 *m* solution? (Density of the solution = 1.045 g/mL.)
a. 76.0 g/mol
b. 7.95 g/mol
c. 79.5 g/mol
d. 12.6 g/mol
e. 760 g/mol

42. If 2.59 g of NH_3 is dissolved in 150 g of ethanol, the freezing point of the solution is
a. –2.0°C
b. –116.6°C
c. –1.9°C
d. –114.4°C
e. 2.0°C

43. Which aqueous solution has the lowest freezing point?
a. 0.20 *m* K_2SO_4
b. 0.10 *m* $(NH_4)_2S$
c. 0.20 *m* FeI_3
d. 0.10 *m* HCl
e. 0.20 *m* C_2H_5OH

44. Which aqueous solution has the lowest boiling point?
a. 0.20 *m* K_2SO_4
b. 0.10 *m* $(NH_4)_2S$
c. 0.20 *m* FeI_3
d. 0.20 *m* HCl
e. 0.10 *m* C_2H_5OH

45. When two solutions with different concentrations are separated by a semipermeable membrane,
a. the solution with the highest concentration of particles will increase in volume.
b. the solution with the lowest concentration of particles will increase in volume.
c. both solutions will increase in volume.
d. both solutions will decrease in volume.
e. the volume of both solutions will remain constant.

46. What van't Hoff factor is predicted for an aqueous solution of sodium carbonate?
a. 0
b. 1
c. 2
d. 3
e. more than 3

47. What is the van't Hoff factor of $CrCl_3$, if 0.15 *m* $CrCl_3$(*aq*) freezes at –0.91°C?
a. 4
b. 3.26
c. 11.7
d. 1
e. 1.7

48. Which gas will have the highest solubility at 20°C?
a. He
b. O_2
c. N_2
d. CO_2

49. What is the van't Hoff factor for a 0.001 *m* $CH_3OH(aq)$?
    a. zero
    b. one
    c. two
    d. three
    e. more than three

50. The solubility of a gas in a solvent increases when
    a. temperature increases
    b. amount of solvent increases
    c. partial pressure of the gas increases
    d. any of the above are true.

ADDITIONAL PRACTICE PROBLEMS

1. What is the freezing point of an aqueous solution of CH_2O with an osmotic pressure of 1.11 atm at 20°C? (The density of the solution is 1.022 g/mL.)

2. Osmotic pressure is useful for determining the molar mass of biological compounds because only a very low concentration is needed to get measurable results. With the large molar masses of biological compounds, the concentrations are very low. What is the molar mass of a biological compound if 1.0 mg in 10.00 mL of solution has an osmotic pressure of 3.71×10^{-4} atm at 30°C?

3. A solution was made by mixing 10.8 g of NaOH with 200.0 mL of water. The density of the resulting solution was 1.065 g/mL. What is the freezing point, boiling point, and osmotic pressure at 25°C of the resulting solution?

4. What is the strongest intermolecular force in each of the following?
   a. CH_3OH b. CH_3CH_3
   c. BCl_3 d. $CaCl_2$
   e. CH_2O

5. Which of the following are highly soluble in water?
   a. CH_3OH b. CH_3CH_3
   c. BCl_3 d. $CaCl_2$
   e. CH_2O

6. Rank the following from strongest to weakest intermolecular forces.
   a. Ne, $CrCl_3$, SCl_2, BaI_2, SiF_4
   b. LiF, CsF, F_2, PH_3, H_2O
   c. CH_3CH_2OH, CH_3CH_3, CH_3OCH_3

7. Rank the following from highest to lowest melting point.
   a. MgO, KBr, $FePO_4$
   b. F_2, NO, Ne
   c. N_2, PH_3, NH_3

8. Rank the following from highest to lowest vapor pressure.
   a. H_2S, H_2O, H_2
   b. $HClO_3$, F_2, NO_2
   c. CH_4, PCl_3, NaCl

9. What is the vapor pressure of the following solutions at 20°C?
   a. 14.0 g SCl_2 in 100 g water
   b. 0.21 *m* $NaCl(aq)$

10. Sketch a phase diagram that has the following four features:
    a. melting point at 1 atm = 300 K
    b. triple point = 175 K and 0.3 atm
    c. boiling point at 0.5 atm = 375 K
    d. critical point = 2 atm and 500 K

Self-Test Answer Key

KEY VOCABULARY COMPLETION ANSWERS

1. molarity
2. van't Hoff factor
3. lowers it
4. molality
5. ion–dipole
6. viscosity
7. triple point
8. critical point
9. meniscus
10. melting
11. miscible
12. hydrogen bonding
13. boiling point
14. lattice energy
15. gas
16. London (or dispersion)
17. vapor pressure
18. intermolecular force
19. solid
20. ion–ion
21. hydrated
22. dipole–dipole

23. polarizability
24. hydrophobic
25. freezing point
26. sublimation
27. supercritical fluid
28. osmosis
29. electron affinity
30. Henry's law

Multiple-Choice Answers

For further information about the topics these questions address, you should consult the text. The appropriate section of the text is listed in parentheses after the explanation of the answer.

1. c. In the Lewis structure, one of the hydrogen atoms is bonded to the oxygen. This makes the molecule capable of hydrogen bonding. (10.2)
2. c. These are all ionic compounds, so all have ion–ion forces of attraction. According to Coulomb's law, the strongest force of attraction will be between molecules with the highest charges that are the closest together. CrN is made up of Cr^{3+} and N^{3-}, the highest charges. Both ions are also rather small. (10.1)
3. a. KI has ion–ion intermolecular forces, stronger than the hydrogen bonding of water, the dispersion force of CH_4, or the dipole–dipole force of HCl. (10.1–10.3)
4. a. The intermolecular force that holds water together is hydrogen bonding. To overcome the surface tension, these forces must be overcome. Hydrogen bonding is one of the stronger forces of attraction, so a lot of energy is needed to overcome that force. (10.8)
5. b. The oxygen in the structure of this compound has two lone pairs and bonded to two carbons. Therefore, the molecular structure is bent and polar. Dipole–dipole forces are the strongest intermolecular force between polar molecules. (10.2)
6. b. Both CsI and NaCl have ion–ion intermolecular forces, which are stronger than the dispersion forces in CF_4 and Br_2. Ion–ion forces are stronger with higher charges (the charges for CsI and NaCl are the same) and smaller distances. Since the ions Na^+ and Cl^- are smaller than Cs^+ and I^-, its ion–ion forces are stronger. (10.1–10.3)
7. a. The compound with the lowest boiling point has the weakest intermolecular forces. CH_2O has dipole–dipole forces; the others have weaker dispersion forces. Dispersion forces increase with molar mass because compounds with higher molar mass are more polarizable. Thus the nonpolar compound with the weakest intermolecular forces will have the lowest molar mass. (10.2–10.3)
8. c. Raoult's law is the $P_{solution} = \chi_{solvent} P_{solvent}$. The solvent is the same for all three choices. In pure water, $\chi = 1$. Mole fraction of solvent will decrease as concentration of solute increases. Consequently, the solution of the highest concentration will also have the lowest vapor pressure. (10.6)
9. b. The solution is a mixture of ions in water. Water has a dipole, magnesium chloride produces the ions, and the interaction of the dipoles with the ions results in the dissolution of magnesium chloride. (10.5)
10. a. Molecules with only dispersion forces do not have a permanent charge (partial or otherwise). Larger (higher molar mass) molecules' electron cloud is more easily deformed (more polarizable), so the temporary dipoles are larger. (10.3)
11. d. Coulomb's law states that the energy of attraction is proportional to charge and inversely proportional to distance. Therefore, higher charges have a higher force of attraction. The distance is smaller with smaller ions that are arranged so that they exist closer together. (10.1)
12. a. Stronger intermolecular forces hold particles close together and the strong attraction can restrict the motion of the particles, making a solid. (10.1–10.3)
13. d. For a compound to be soluble, the ion–dipole forces must overcome the ion–ion forces. Weaker ion–ion forces are easier to overcome. According to Coulomb's law, the forces will be weaker with lower charges and larger ions/distances. (10.5)
14. d. Lattice energy is the energy released when gaseous ions are organized into a solid lattice structure. (10.1)
15. b. The ion–ion forces are stronger than the ion–dipole forces, so there must be more ion–dipole than ion–ion interactions. Since water is the solvent, there is always a lot more water than solute (ions). Because of spacing and orbital arrangements, 6 is the typical number. (10.5)

16. b. When two liquids dissolve in each other they are said to be miscible. Since CaS is an ionic compound, and is solid, it is not miscible. The solubility rules also state that CaS is not water soluble. The substance most likely to be miscible is one with similar attractive forces. Since SCl_2 has a permanent dipole, it is most likely to be water soluble. (10.5)

17. a. All forces of attraction are based on Coulomb's law, where larger charges have larger attraction. Ions have a full charge, whereas dipoles only have a partial charge. Therefore ion–dipole forces are stronger than dipole–dipole forces. (10.5)

18. b. The vapor pressure of the mixture equals $\chi_{solvent}P_{solvent}$. The mole fraction is the ratio of moles of solvent to total moles in the mixture. The vapor pressure of the solvent depends on the temperature and the strength of the intermolecular forces, which depend on the identity of the solvent. (10.6)

19. b. The highest vapor pressure will come from the liquid with the weakest intermolecular forces. The weakest intermolecular forces are dispersion (London) forces. In both CBr_4 and Br_2, dispersion is the strongest type of intermolecular force. Dispersion forces are stronger for substances with a higher molecular weight. (10.3)

20. b. Boiling point is the temperature at which the vapor pressure equals the atmospheric pressure. Water normally boils at 100°C at the normal atmospheric pressure. Since atmospheric pressure is lower in the mountains, water will boil at a lower temperature. (10.7)

21. a. Under pressure, the motion of molecules is restricted, so a gas becomes a liquid. This can also be seen from the phase diagram. (10.7)

22. d. The triple point is the pressure and temperature where all three physical states exist. (10.7)

23. c. Sublimation is when a solid becomes a gas without becoming a liquid. The solid–gas line is normally in the lower left (low pressure, low temperature) corner of the phase diagram. Solids are at a higher pressure, so if the pressure is low the solid will become a gas or sublime. (10.7)

24. a. A substance becomes a supercritical fluid above the critical point, which is at the end of the liquid–gas line. Above the critical point would be at higher pressure and higher temperatures than at the critical point. (10.7)

25. b. Ice, which is solid water, floats because it is less dense than liquid water. Solids have the molecules in fixed positions. The positions of the molecules in water are such that hydrogen bonding is maximized. These positions take up more space (larger volume) than if hydrogen bonding is not maximized. (10.8)

26. b. An increase in temperature increases the kinetic (moving) energy of the molecules. Therefore, molecules at a higher temperature move more freely. Viscosity is resistance to flow, so an increase in temperature decreases viscosity. (10.8)

27. a. An increase in temperature increases the kinetic energy of the molecules. With a higher kinetic energy, more molecules can escape into the vapor phase. Both the higher number of moles and the higher temperature will increase the pressure. (10.6)

28. b. All the choices are nonpolar molecules and therefore, have London intermolecular forces. The strongest London forces are in molecules with the highest molecular weights. (10.3)

29. c. In the Lewis structure of PH_3, the phosphorus has a lone pair in addition to the three P—H bonds. That gives PH_3 a permanent dipole. Phosphorus is not electronegative enough for hydrogen bonding. Therefore, the strongest forces are dipole–dipole forces. (10.2)

30. b. Liquids climb a capillary tube because they are attracted to the walls of the capillary (adhesive forces) more than they are to each other (cohesive forces). Intermolecular forces are another name for cohesive forces. (10.8)

31. b. Molality = moles of solute per kilograms solvent.

$$\frac{0.10 \text{ mol}}{0.10 \text{ kg}} = 1.0\ m \text{ (10.9)}$$

32. d. There is a different freezing point constant for each solvent. That constant is proportional to the change in freezing point. The solvent, carbon tetrachloride, has the largest K_f. (10.9)

33. e. $\Delta T = K_f m$ for freezing point. The K_f for water is 1.86°C/*m*. The molality of *ions* is 1.0 *m*, since each mole of NaCl makes 2 moles of ions. Since the solute decreases the freezing point, the freezing point is $0 - \Delta T$, or $0 - 1.86 = -1.86$°C, –1.9 to two significant figures. (10.9)

34. a. $\pi = iMRT$ for osmotic pressure. $\pi =$

(2)(0.10 mol/L)(0.082 L•atm/mol•K)(293 K) = 4.8 atm.

Take into account that two ions are formed from 1 mole of KI and temperature must be converted to Kelvin. (10.9)

35. a. $\Delta T = K_f m$. In this example, $K_f = 1.86°C/m$ and $\Delta T = 0.25°C$. Therefore, $m = 0.134$ mol/kg. Since there was 100.0 g solvent or 0.1 kg solvent, there are 0.0134 moles. Molar mass $= \frac{5.0 \text{ g}}{0.0134 \text{ mol}} =$ 372 g/mol. (10.9)

36. c. Since molarity is liters of solution and molality is mass of solvent, volume must be changed to mass (or vice versa) using density. To get solvent, the solute must be subtracted as mass, since volume is not additive but mass is. Therefore, you have to change liters of solution to mass of solution, using density. Then grams of solute can be subtracted from grams of solution to get grams, or kilograms of solvent. The molality can be calculated as $\frac{\text{mol solute}}{\text{kg solvent}}$. (10.9)

37. b. The molality of the solution will be $\frac{0.05 \text{ mol}}{0.2 \text{ kg}} =$ 0.25 m. The K_b for benzene is 2.53. Therefore, $\Delta T = K_b\, m = (2.53)(.25) = 0.63°C$. Since it is boiling point elevation, this value is added to the value of the boiling point for pure benzene: 80.1 + 0.63 = 80.7°C. (10.9)

38. a. The proportionality constant between osmotic pressure and concentration is RT, which has a value of about 25 at room temperature. The proportionality constants for boiling point and freezing point are much smaller. In addition, small fractions of an atmosphere are much easier to measure than small fractions of a degree. (10.9)

39. d. Molality is mol solute/kg solvent (10.9)

40. e. Assuming 1 L (1000 mL) of solution, 1000 mL × $\left(\frac{1.213 \text{ g}}{1 \text{ mL}}\right)$ = 1213 g solution. In 1 L there are 0.22 mol CaI_2 × $\left(\frac{293.90 \text{ g}}{1 \text{ mol}}\right)$ = 65 g of CaI_2. 1213 g solution – 65 g solute = 1148 g solvent. (10.9)

41. c. Molar mass is the grams per mol of a substance. The problem specifies that the mass is 12.0 g. The molality is moles of solute per kilogram solvent. The problem specifies that 100 mL of water is the solvent. Using the density of water as 1.00 g/mL, 100 mL × $\left(\frac{1.00 \text{ g}}{1 \text{ mL}}\right) \times \left(\frac{1 \text{ kg}}{1000 \text{ g}}\right)$ = 0.100 kg solvent.

Using the molality, (1.51 m)(0.100 kg) = 0.151 mol solute. Therefore, the molar mass is $\frac{12.0 \text{ g}}{0.151 \text{ mol}} =$ 79.5 g/mol. (10.9)

42. b. Moles of ammonia = 2.59 × $\left(\frac{1 \text{ mol}}{17.03 \text{ g}}\right)$ = 0.152 mol.

Molality of solution = $\frac{\text{mol ammonia}}{\text{kg ethanol}} = \frac{0.152 \text{ mol}}{0.150 \text{ kg}} =$ 1.014 m. $\Delta T = K_f m = (1.99)(1.014) = 2.02°C$. The actual freezing point of ethanol is – 114.6°C. So the freezing point of the solution is –114.6 – 2.02 = –116.6°C. (10.9)

43. c. Addition of a solute lowers the freezing point of a solvent. Since the solvent is the same, the solution with the highest concentration of particles will have the greatest change and therefore, lowest freezing point. The moles of particles can be approximated from the theoretical van't Hoff factor, i, times concentration. For K_2SO_4 and $(NH_4)_2S$, $i = 3$; for FeI_3, $i = 4$; for HCl, $i = 2$; for C_2H_5OH, $i = 1$. Therefore, FeI_3 will produce the most ions, about 0.80 M. (10.9)

44. e. Since the addition of a solute *raises* the boiling point of solvent, the solution with the fewest particles will have the lowest boiling point. The moles of particles can be approximated from the theoretical van't Hoff factor, i, times concentration. For K_2SO_4 and $(NH_4)_2S$, $i = 3$; for FeI_3, $i = 4$; for HCl, $i = 2$; for C_2H_5OH, $i = 1$. Therefore, C_2H_5OH has the fewest particles at about 0.10 M. (10.9)

45. a. Osmosis occurs when solutions of different concentrations are separated by a semipermeable membrane. In osmosis, solvent flows across the membrane to equalize the concentrations. Since it is solvent that flows, it must go into the solution of higher concentration, so that concentration will decrease by increasing its volume. (10.9)

46. d. The formula for sodium carbonate is Na_2CO_3. In water, 1 mole of sodium carbonate will form 2 moles sodium ions and 1 mole of carbonate ion. That is a total of 3 moles of ions per mole of solute and a van't Hoff factor of 3. (10.9)

47. b. The equation for freezing point depression is $\Delta T = iK_f m$. In this example, $K_f = 1.86°C/m$, since water is the solvent; $\Delta T = 0.91$, since the freezing point of pure water is zero. In addition, m is given as 0.15 m. Solving the equation for i, the van't Hoff factor, $0.91 = i(1.86)(0.15)$ and $i = 3.26$. The theoretical value of i is 4, since four ions may be formed. However, in real solutions ions often associate with one another (ion-pairing) leading to a smaller van't Hoff value. (10.9)

48. d. $C = k_H P$, so the gas with the highest k_H will also have the highest solubility. The gas with the highest constant is CO_2. (10.5)

49. b. CH_3OH is a covalent compound and does not form ions in aqueous solutions. One mole of CH_3OH produces one mole of particles, so the van't Hoff factor is one. (10.9)

50. c. According to Henry's law, solubility increases as the partial pressure of the gas increases. An increase in temperature gives the gas molecules more energy so that they escape easier and solubility decreases. Solubility is the ratio of solute to solvent, so the amount of solvent does not matter. (10.5)

Additional Practice Problem Answers

1. $\pi = iMRT$

$1.11 \text{ atm} = M(0.0821 \text{ L·atm/mol·K})(293 \text{ K})$

$M = 0.0461 \text{ mol/L}$

To calculate freezing point, molarity needs to be converted to molality. In 1 L of solution

$$0.0461 \text{ mol} \times \left(\frac{30.03 \text{ g}}{1 \text{ mol}}\right) = 1.384 \text{ g}$$

$$1 \text{ L solution} \times (1000 \text{ mL/L}) \times \left(\frac{1.022 \text{ g}}{1 \text{ mL}}\right) = 1022 \text{ g}$$

$$\text{g solvent} = 1022 \text{ g} - 1.384 \text{ g}$$

$$= 1021 \text{ g solvent} \times \left(\frac{1 \text{ kg}}{1000 \text{ g}}\right) = 1.021 \text{ kg solvent}$$

$$m = \frac{0.0461 \text{ mol}}{1.021 \text{ kg}} = 0.0451\ m$$

$$\Delta T = K_f m = (1.86°\text{C/m})(0.0451\ m) = 0.0840°\text{C}$$

$$\text{freezing point} = 0°\text{C} - 0.0840°\text{C} = -0.0840°\text{C}$$

2. $\pi = iMRT$

$3.71 \times 10^{-4} \text{ atm} = M(0.0821 \text{ L·atm/mol·K})(303 \text{ K})$

$1.49 \times 10^{-5} \text{ mol/L} = M$

$$10.00 \text{ mL} \times \left(\frac{1 \text{ L}}{1000 \text{ mL}}\right) = 0.010 \text{ L}$$

$$1.49 \times 10^{-5} \text{ mol/L} \times (0.010 \text{ L}) = 1.49 \times 10^{-7} \text{ mol}$$

$$1.0 \text{ mg} \times \left(\frac{1 \text{ g}}{1000 \text{ mg}}\right) = 0.0010 \text{ g}$$

$$\text{molar mass} = \frac{\text{g}}{\text{mol}} = \frac{0.0010 \text{ g}}{1.49 \times 10^{-7} \text{ mol}} = 6.7 \times 10^3 \text{ g/mol}$$

3. To determine freezing point and boiling point, you need the molality of the solution $= \frac{\text{mol solute}}{\text{kg solvent}} = \frac{\text{mol NaOH}}{\text{kg water}}$.

$$10.8 \text{ g NaOH} \times \left(\frac{1 \text{ mol}}{40.00 \text{ g}}\right) = 0.270 \text{ mol}$$

$$200.0 \text{ mL water} \times \left(\frac{1.00 \text{ g}}{1 \text{ mL}}\right) \times \left(\frac{1 \text{ kg}}{1000 \text{ g}}\right) = 0.2000 \text{ kg}$$

$$\text{Molality} = \frac{\text{mol}}{\text{kg}} = \frac{0.270}{0.2000} = 1.35\ m.$$

Freezing point:

$\Delta T = iK_f m = (2)(1.86°\text{C}/m)(1.35\ m) = 5.02°\text{C}$

Freezing point of solution = 0°C – 5.02°C = –5.02°C.

For boiling point:

$\Delta T = iK_b m = (0.52°\text{C}/m)(1.35\ m) = 1.40°\text{C}$

Boiling point of solution = 100°C + 1.40°C = 101.40°C.

Osmotic pressure requires molarity:

g solution = g solvent + g solute = 200.0 g + 10.8 g = 210.8 g solution

$$210.8 \text{ g solution} \times \left(\frac{1 \text{ mL}}{1.065 \text{ g}}\right) = 197.9 \text{ mL} \times \left(\frac{1 \text{ L}}{1000 \text{ mL}}\right) = 0.1979 \text{ L}$$

$$\text{Molarity} = \frac{\text{mol}}{\text{L}} = \frac{0.270 \text{ mol}}{0.1979 \text{ L}} = 1.36\ M.$$

$$\pi = iMRT = (2)(1.36 \text{ mol/L})(0.0821 \text{ L·atm/mol·K})(298 \text{ K}) = 66.5 \text{ atm}$$

4. a. Hydrogen bonding. There is a hydrogen bonded to the oxygen in the Lewis structure.
   b. London. Hydrocarbons are normally nonpolar. The molecule is symmetric and overall nonpolar.
   c. London. Because boron has a deficient octet, the chorine atoms arrange themselves in a trigonal planar geometry, and the dipoles on the bonds cancel, creating a nonpolar molecule.
   d. Ion–ion. It is an ionic compound: a metal combined with a nonmetal.
   e. Dipole–dipole. The oxygen is a more negative end than the hydrogens.

5. a. Soluble, both have hydrogen bonds and "like dissolves like." Methanol (CH_3OH) and water are miscible.
   b. Not soluble. Water is polar, CH_3CH_3 is nonpolar.
   c. Not soluble. BCl_3 is nonpolar.
   d. Soluble. An ionic compound, so use the solubility rules to determine solubility. Chlorides are normally soluble. Calcium chloride is not an exception.
   e. Soluble. CH_2O is polar.

6. Generally, strength of intermolecular forces goes from strongest = ion–ion > hydrogen bond > dipole–dipole > London = weakest. Ion–ion forces are stronger, with more highly charged ions and smaller ions. London forces are stronger for molecules with higher molar masses.
   a. BaI_2, $CrCl_3$ have ion–ion, the strongest type. Cr^{3+} is smaller and more highly charged, so it has stronger forces. SCl_2 has dipole–dipole forces, and SiF_4 and Ne both have London forces. So, $CrCl_3 > BaI_2 > SCl_2 > SiF_4 > Ne$.
   b. LiF and CsBr have ion–ion forces, and although all ions have the same charge, the ions of LiF are smaller. F_2 has London forces, PH_3 has dipole–dipole forces, and H_2O has hydrogen bonding. Therefore, $LiF > CsBr > H_2O > PH_3 > F_2$.
   c. The formula implies the structure. CH_3CH_2OH has hydrogen bonding, CH_3CH_3 has only London forces, CH_3OCH_3 has dipole–dipole forces, so $CH_3CH_2OH > CH_3OCH_3 > CH_3CH_3$.

7. The molecules with the strongest intermolecular forces will have the highest melting points.
   a. MgO, KBr, $FePO_4$ all have ion–ion forces. They differ in charge of ions. MgO = Mg^{2+} and O^{2-}, KBr = K^+ and Br^-, $FePO_4$ = Fe^{3+} and PO_4^{3-}. Higher charges have stronger forces, so $FePO_4 > MgO > KBr$.
   b. F_2 and Ne have London forces. Larger molecules with London forces have stronger intermolecular forces. NO has dipole–dipole forces. $NO > F_2 > Ne$.
   c. NH_3 hydrogen bonds. PH_3 has dipole–dipole forces. N_2 has London forces. Therefore, $NH_3 > PH_3 > N_2$.

8. Compounds with weaker intermolecular forces have higher vapor pressures.
   a. H_2O hydrogen bonds. H_2S has dipole–dipole forces. H_2 has London forces, so $H_2 > H_2S > H_2O$.
   b. F_2 has dispersion forces. NO_2 has dipole–dipole forces. $HClO_3$ hydrogen bonds. (Recall that in oxoacids, the hydrogen is bonded to the oxygen.) Therefore, $F_2 > NO_2 > HClO_3$.
   c. NaCl has ion–ion forces. PCl_3 has dipole–dipole forces. CH_4 has dispersion forces. Therefore, $CH_4 > PCl_3 > NaCl$.

9. Since the vapor pressure of water at 20°C is 17.5 torr, the solute will decrease this vapor pressure according to Raoult's law. $P_{solution} = \chi_{solvent}P_{solvent}$. $P_{solvent}$ = 17.5 torr.
   a. $14.0\text{ g } SCl_2 \times \left(\frac{1\text{ mol}}{102.98\text{ g}}\right) = 0.136\text{ mol } SCl_2$

   $100\text{ g } H_2O \times \left(\frac{1\text{ mol}}{18.01\text{ g}}\right) = 5.55\text{ mol}$

   Total moles = 5.69. Water is a solvent, so

   $\chi_{H_2O} = \frac{5.55}{5.69} = 0.975.$

   $P_{solution} = (0.975)(17.5\text{ torr}) = 17.1\text{ torr}$

   b. $0.21\ m\text{ NaCl} = \frac{\text{mol solute}}{\text{kg solvent}}$. For ease, assume 1 kg solvent (water). Then there are 0.21 mol NaCl, but NaCl in water becomes Na^+ and Cl^-, so there are 0.42 mol ions.

   $1\text{ kg water} \times \left(\frac{1000\text{ g}}{1\text{ kg}}\right) \times \left(\frac{1\text{ mol}}{18.01\text{ g}}\right) = 55.52\text{ mol}$

   $\chi H_2O = \frac{55.52}{(55.52 + 0.42)} = 0.993$

   $P_{solution} = (0.993)(17.5\text{ torr}) = 17.4\text{ torr}$

10.

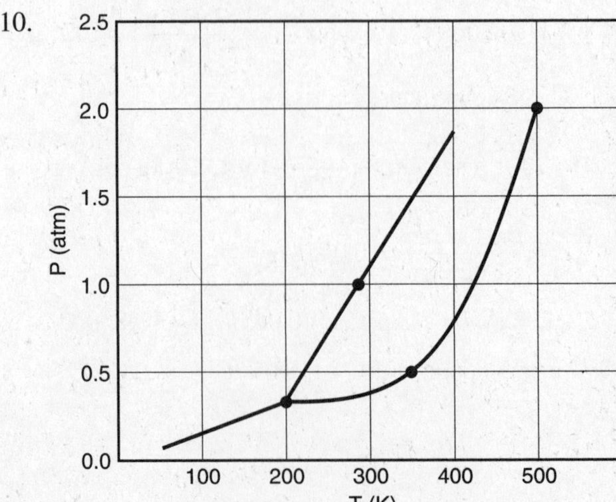

CHAPTER 11 The Chemistry of Solids

OUTLINE

REVIEW

Chapter Overview

Metals are hard, shiny, **malleable** (easily shaped), **ductile** (may be drawn out), thermally conductive (easily transfer heat), and electrically conductive (easily transfer electricity). On the atomic level, the atoms are densely packed such that the electron cloud of one atom can overlap as many as twelve other atoms. This creates the "sea of electrons" described earlier as **metallic bonding**. Because of the substantial overlap, metallic bonds are very strong. However, because the overlap is so diffuse, the individual bonds between atoms are weak.

An alternate way of describing bonding in metals is with band theory. **Band theory** is a variation of molecular orbital theory, but since metals have an enormous number of atoms there are practically an infinite number of molecular orbitals. These orbitals are so close together that their difference in energy is negligible. This stack of orbitals is a band. The band created from overlapping of valence shell orbitals, and containing electrons, is called the **valence band**. Electricity conductivity occurs when the band is only partially full, and electrons can move freely between orbitals. An empty band, higher in energy than the valence band, is the **conduction band**. Because of the existence of the conduction band, even metals with full valence shells are still capable of electrical conductivity. In metals, the **band gap** (E_g), which is the difference in energy between the valence and the conduction bands, is near zero. Thus metals are conductive at room temperature.

Semimetals (or **metalloids**) tend to have the chemical properties of nonmetals and the physical properties of metals. The band gap for semimetals is small, but measurable. Semiconductors are made from metalloids when the valence band is full and the conduction band is empty. Thus a specific minimum value of energy is required to move electrons from the valence band to the conduction band and create a conductive material. Silicon is the most common example of a semiconducting metalloid. The energy needed to induce conduction can be varied dramatically by **doping**, that is adding a similarly sized atom with more or fewer electrons than the intrinsic semiconductor that reduces the band gap. Substances with extra electrons are called **n-type dopants**. Substances with fewer electrons are **p-type dopants**.

Crystalline substances have atoms in specific positions that repeat regularly throughout the substance. Metallic, ionic, and covalent compounds may be crystalline, or have no internal structure, in which case they are called **amorphous**. Amorphous solids are also called glasses.

One way of describing a crystalline structure, or **crystal lattice**, looks at the stacking of layers of the atoms. In each layer, the atoms are arranged hexagonally so that they are as close as possible (layer *a*). The next layer is shifted slightly so that the atoms nestle into the first layer (layer *b*). The third layer can be a repeat of the first layer, thus the stacking is *ababa*... This arrangement is called **hexagonal closest-packed (hcp)**. Alternately, the third layer can be shifted so that it is not directly on top of either the first or second layer

(layer c). In this case, the packing is *abcabcabc...* and is called **cubic closest-packed (ccp)**.

Another way of describing the crystal lattice is with unit cells. A **unit cell** is the smallest repeating volume of a crystal array. Traditionally, each corner of the unit cell is at the center of an atom. Common unit cells include a **simple cubic unit cell**, with one atom at each corner; a **body-centered cubic (bcc) unit cell** with an atom in the center of the cube in addition to an atom at each corner; and a **face-centered cubic (fcc) unit cell**, which has an atom in the center of each face of the cube as well as one at each corner. The **face-centered cubic unit cell** also has a cubic closest-packed lattice structure.

Because the unit cell starts in the center of an atom, only 1/8 of a corner atom is actually contained in the unit cell. For atoms on the face, 1/2 of the atom is in the cell and for atoms on an edge, 1/4 of the atom is within the cell. Of course, atoms in the center of the unit cell are completely contained in the unit cell. The ratio of atoms within a unit cell will be the same as the ratio of the substance overall.

The dimensions of the unit cell are related to the size of the atoms in the cell. For monoatomic simple cubic unit cells, the atoms along the edge are touching so the length is equivalent to two atomic radii. For a face-centered cubic unit cell, the atoms touch in a diagonal along the face. For a body-centered cubic unit cell, the atoms touch in a diagonal from one corner, through the center of the unit cell to the opposite corner. The relationship between the edge length and radii for these unit cells can be determined from the Pythagorean theorem. Density can also be determined from the volume of the cell and the mass of its contents.

Mixtures of metals are called **alloys**. When the metal atoms are of similar radii (within 15%), the added metal will replace the original metal in one of the positions in the crystal lattice. Such materials are called **substitutional alloys**. When one of the metal atoms is substantially smaller than the other, the smaller atom will fit in the holes, or interstices, between atoms that remain in their lattice positions. There are two types of holes. Larger holes are surrounded by six lattice atoms and called **octahedral holes**. The smaller holes, **tetrahedral holes**, are surrounded by only four lattice atoms. Materials where metals fit in the holes within a metal lattice are called **interstitial alloys**.

Nonmetals and covalent compounds can form network covalent or molecular solids. **Network covalent solids** are an extended network of covalent bonds. **Molecular solids** are individual molecules held together by intermolecular forces. When an element has several different arrangements of atoms, the different structures are called **allotropes**. When a compound has the same formula but different structural arrangements, it is a **polymorph**.

Salts (ionic compounds) are commonly crystalline and can also be described using unit cells. For salts, the larger ion generally forms the unit cell and the smaller ion fits in the holes. When an inorganic solid is heated to make it harder and more heat resistant, the resulting substance is called a **ceramic**. The electrical resistance of some salts falls to zero below a certain temperature called the **critical temperature**. When this occurs, the material becomes a **superconductor**. Superconductors repel external magnetic fields, exhibiting the property called the **Meissner effect**. Superconductivity is thought to occur because electrons begin to travel in **Cooper pairs**, where electrons move in a cooperative and, therefore, more efficient fashion.

Crystal structures are determined using **X-ray diffraction (XRD)**. In this study, X-rays of a specific wavelength (λ) impact the solid at different angles (θ). The X-rays diffract from the crystal at different angles. The pattern of the amplitude of the waves versus the angle will depend on the distance between layers according to the Bragg equation. Measuring of different angles and different faces of the crystal can allow scientists to determine the crystal structure.

Worked Examples

The Unit Cell

Example 1.
Describe each of the following unit cells.

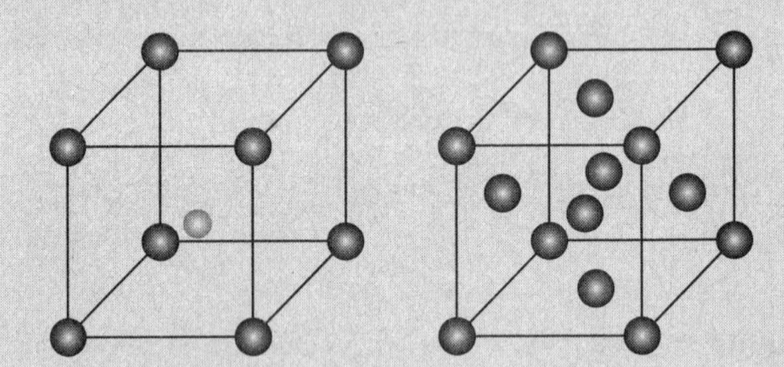

Collect and Organize We are asked to describe the unit cells provided in the figure.

Analyze Define the unit cell by choosing the largest type of atom. (Normally, this is the anion.) Search in one dimension until you find that atom again. Check the same distance in the same direction to see that all the atoms are the same along the second segment of the line as the first. If so, you have found one edge of the unit cell. If not, check both segments together as the potential edge of that unit cell. Return to the original atom and repeat for the other two dimensions.

Solve The unit cell on the left is a simple cubic. Although there is an atom in the center of the unit cell, it is not of the same type as the atoms at the corners. Therefore, it does not count as part of the unit cell description. If the atom in the center was of the same type, it would be a body-centered cubic unit cell.

The unit cell on the right is a face-centered cubic unit cell. It has only one type of atom. It too has an atom in the center, but more important, there is an atom on each face. Face-centered cubic unit cells have an atom in the center as well as on the faces.

Think About It Each unit cell is defined by one type of atom. Consequently, whichever atom you choose when defining that unit cell, it is the *only* atom used in defining the unit cell. (Ignore all other types.)

Example 2.

What is the stoichiometry of the following compound if black dots = X and grey dots = M? (M goes first in the formula.)

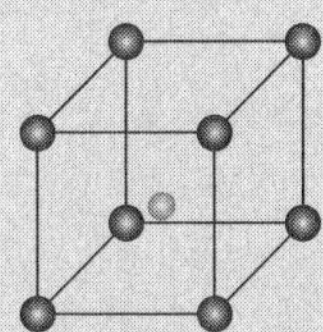

Collect and Organize We are asked to determine the stoichiometry of a compound given its unit cell.

Analyze For ionic and network solids, the stoichiometry is the simplest ratio of particles in an compound. Since the unit cell represents the pattern of the atoms in the compound, the ratio in the unit cell is also the ratio for the entire compound. The way we have drawn unit cells is with the center of the atom at the corners. Consequently, not all of the atom is in the unit cell. The part of the atom in the unit cell is counted according to the following: corner = $\frac{1}{8}$, edge = $\frac{1}{4}$, face = $\frac{1}{2}$, and body = 1.

Solve In this cell there are eight black atoms, one at each corner, but no more. So the total number of X atoms in the cell = $8 \times \frac{1}{8} = 1$. There is one gray dot in the center (body), so the number of M atoms = $1 \times 1 = 1$. So the ratio is 1:1 and the formula = MX.

Think About It Once the number of atoms of each type within a unit cell is determined, the simplest ratio of those atoms can be determined.

Example 3.

What is the stoichiometry of the following compound if black dots = X and gray dots = M? (M goes first in the formula.)

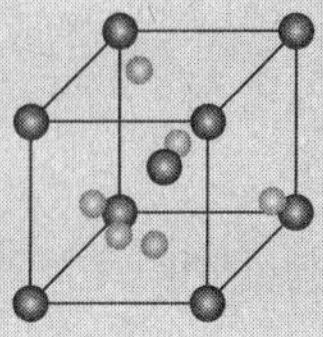

Collect and Organize We are asked to determine the stoichiometry of a compound given its unit cell.

Analyze For ionic and network solids, the stoichiometry is the simplest ratio of particles in an atom. Since the unit cell represents the pattern of the atom, the ratio in the unit cell is also the ratio for the entire compound. The way we have drawn unit cells is with the center of the atom at the corners. Consequently, not all of the atom is in the unit cell. The part of the atom in the unit cell is counted according to the following: corner = $\frac{1}{8}$; edge = $\frac{1}{4}$; face = $\frac{1}{2}$; body = 1.

Solve This is a body-centered cubic unit cell, with X on the corners and in the body, and M on each face. For X, 8 corners ($\times \frac{1}{8}$) = 1 and body (× 1) = 1; total = 2. For M, 6 faces ($\times \frac{1}{2}$) = 3. So the stoichiometry is M_2X_3. (There is no simpler ratio.)

Think About It Once the number of atoms of each type within a unit cell is determined, the simplest ratio of those atoms can be determined.

Example 4.

What is the edge length of a cubic unit cell if iodide ions touch along the face of the unit cell? The radius of an iodide ion is 220 pm.

Collect and Organize We are asked to determine the edge length of a cubic unit cell in which iodide ions touch along the face of the unit cell.

Analyze To determine edge length for unit cell where the ions touch along a face, it is normally assumed that the atoms along the diagonal across the face are touching. The length of this diagonal is determined from the sum of the ions along the diagonal (c). The Pythagorean theorem ($a^2 + b^2 = c^2$) is used to determine the edge length (a). If this is a cubic structure, both edges are the same length ($a = b$) and the formula simplifies to $2a^2 = c^2$.

Solve The corner of the unit cell starts in the center of an iodide ion, so from the corner of the unit cell to the next ion is one radius. The diagonal goes completely through the ion on the face, so that is two radii. The diagonal continues to the center of the ion on the other corner, one more radius, for a total of four. Therefore, the length of the diagonal (c) is 4(220 pm) = 880 pm. Using the Pythagorean theorem, a = 622 pm.

Think About It To determine the edge length from ionic radii, the exact type of lattice is required.

EXAMPLE 5.
What is the edge length of a cubic unit cell where a corner chloride ion ($r = 181$ pm) touches a body-centered cesium ion ($r = 174$ pm) that touches the opposite corner chloride ion?

COLLECT AND ORGANIZE We are asked to determine the edge length for a unit cell in which ions touch through the body of the cell.

ANALYZE To determine edge length for a unit cell where ions touch through the body of the cell, the Pythagorean theorem must be used twice. The sum of the ionic radii corresponds to a diagonal through the center of the unit cell (c). A right triangle can be made from a diagonal through the face (b) and an edge (a), so $a^2 + b^2 = c^2$. The value for the diagonal across the face (b) can be determined by a right triangle along each edge, $2a^2 = b^2$. Putting the two equations together, the edge length can be determined from $a^2 + 2a^2 = 3a^2 = c^2$.

SOLVE The ions are touching along a body diagonal. The body diagonal consists of two radii of chloride ions and two radii of cesium ions. Therefore, its length (c) is 2(181) + 2(174) = 710 pm. Using $3a^2 = c^2$, $a = 410$ pm.

THINK ABOUT IT The volume (size) of a unit cell depends on the ionic radii (size) of the ions and their arrangement (lattice structure). Experimentally the length of an edge can be determined by X-ray diffraction.

EXAMPLE 6.
The cesium chloride unit cell has a chloride ion at each corner and a cesium ion in the center. The ions touch through a body diagonal. What is the density of this solid in g/cm^3? (*Hint:* See the previous examples.)

COLLECT AND ORGANIZE We are asked to determine the density of cesium chloride given its unit cell type.

ANALYZE To determine the volume from edge length, recall that for cubic lattices, each edge length is the same. Consequently, for cubic lattice structures, the volume is simply the edge length cubed.

Density is mass/volume. The density of the unit cell is the same as the density of the compound as a whole. Volume of the unit cell is determined from edge length. The number of particles in the unit cell is determined in the same way as the stoichiometry of the cell. (Do not simplify to the smallest ratio.) The mass can be determined from the atomic weight.

SOLVE The edge length of this unit cell was determined in Example 5 to be 410 pm. To get the volume in cubic centimeters, it is probably more convenient to convert picometers to centimeters rather than cubic picometers to cubic centimeters, so

$$410 \text{ pm} \times \left(\frac{10^{-12}\text{ m}}{1 \text{ pm}}\right) \times \left(\frac{10^{2}\text{ cm}}{1 \text{ m}}\right) = 4.10 \times 10^{-8}\text{ cm}$$

The volume is the edge length cubed: $(4.10 \times 10^{-8}\text{ cm})^3 = 6.8921 \times 10^{-23}\text{ cm}^3 = 6.89 \times 10^{-23}\text{ cm}^3$.

In the unit cell, there are eight chlorides, one on each corner, so that is a total of $8 \times \frac{1}{8}$, or 1 chloride ion. The cesium ion is in the center, so it counts as one atom. The mass, in grams, of one atom of chloride is

$$1 \text{ atom Cl} \times \left(\frac{1 \text{ mol}}{6.022 \times 10^{23}\text{ atoms}}\right) \times \left(\frac{35.453 \text{ g}}{1 \text{ mol}}\right) = 5.887 \times 10^{-23}\text{ g Cl}$$

The four significant figures are due to the value used for Avogadro's number. 1 atom is a counted number; therefore, it has infinite significant figures. The mass in grams of cesium is

$$1 \text{ atom Cs} \times \left(\frac{1 \text{ mol}}{6.022 \times 10^{23}\text{ atoms}}\right) \times \left(\frac{132.9 \text{ g}}{1 \text{ mol}}\right) = 2.200 \times 10^{-22}\text{ g Cs}$$

The total mass of the unit cell is the sum of these two values

$$2.200 \times 10^{-22}\text{ g} + 5.887 \times 10^{-23}\text{ g} = 2.789 \times 10^{-22}\text{ g}$$

The density is mass/volume = 2.789×10^{-23} g/6.8921×10^{-23} cm^3 = 4.05 g/cm^3

THINK ABOUT IT Remember that you are working in atoms, not moles!

PACKING OF SPHERES

EXAMPLE 1.
What kind of hole is the ion represented as an open circle in if it is surrounded as follows?

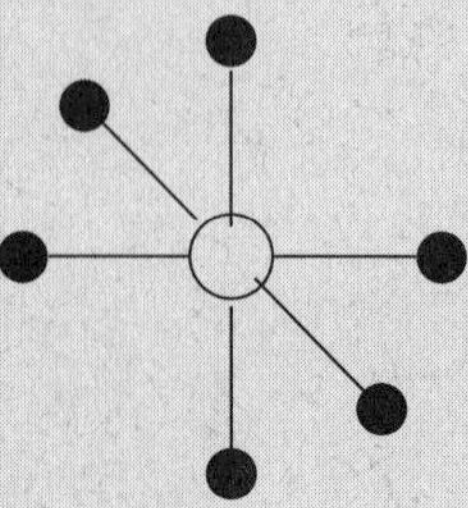

COLLECT AND ORGANIZE We are asked to determine the kind of hole an ion is in based on its neighbors.

ANALYZE The easiest way to determine the type of hole is by looking at how the atoms surround the ion in the hole. Hexagonal and cubic closest-packed have octahedral and tetrahedral holes. Ions in tetrahedral holes are surrounded by ions in a tetrahedral orientation. Ions in octahedral holes are surrounded by ions oriented octahedrally. For simple cubic packing, the holes are cubic holes. The ion is surrounded by eight other ions arranged in a cube.

SOLVE The center ion is surrounded by six other ions. The ions are oriented top, bottom, left, right, front, and back, which is an octahedral geometry. Therefore, the ion is in an octahedral hole.

THINK ABOUT IT The size of the hole depends on the type and size of the atoms making the hole. Larger close-packed atoms leave larger holes. Cubic holes are larger than octahedral holes, which are larger than tetrahedral holes. However, there are more tetrahedral holes than octahedral holes.

EXAMPLE 2.

For the following cations and anions, predict the type of hole the smaller ion will occupy.

a. Fe^{2+} (r = 55 pm) and S^{2-} (r = 184 pm)

b. K^+ (r = 151 pm) and Cl^- (r = 181 pm)

c. Cs^+ (r = 174 pm) and F^- (r = 133 pm)

COLLECT AND ORGANIZE We are given three cation–anion pairs and are asked to determine the type of hole the smaller ion will occupy.

ANALYZE The larger ion will form the packing structure. The ratio of smaller to larger ion determines the types of holes that the smaller ion will fit into.

SOLVE

a. Sulfide is the larger ion, so it forms the packing structure and iron(II) is in the holes. The ratio of smaller/larger = $\frac{55}{184}$ = 0.30. That makes iron(II) small enough to fit in tetrahedral holes.

b. Chloride is the larger ion, so potassium will fit in the holes. The ratio of smaller/larger = $\frac{151}{181}$ = 0.83. So the potassium will only fit in cubic holes.

c. Cesium is the larger ion, so the ratio of smaller/larger = $\frac{133}{174}$ = 0.76. Fluorides are also in cubic holes.

THINK ABOUT IT The ratio of anion to cation size can be used to predict what type of hole the ion will occupy.

Self-Test

KEY VOCABULARY COMPLETION QUESTIONS

__________ 1. Orderly solid structure

__________ 2. Unit cell with particles only at the corners

__________ 3. SiO_2

__________ 4. Solid structure formed by an extended array of covalent bonds

__________ 5. Compound with the same chemical formula but a different internal structure

__________ 6. Solid without a unit cell

__________ 7. Unit cell with particles in each corner and on each face

__________ 8. One metal atom takes the position of another metal atom

__________ 9. Substance that is shiny, malleable, and thermally conductive

__________ 10. Having an electrical resistivity near zero

__________ 11. A “molecular orbital” for an enormous number of metal atoms

__________ 12. Layers where every other layer is different

__________ 13. Smallest repeating volume of a crystal

__________ 14. Unit cell with particles at each corner and one in the center of the cell

__________ 15. Different metal atoms in the holes of a metallic structure

__________ 16. Electrons traveling in tandem to create superconduction

__________ 17. Homogeneous mixture of metals

__________ 18. Layers of abcabcabcabcabc...

__________ 19. Exclusion of a magnetic field by a superconductor

__________ 20. Can be drawn into a wire

MULTIPLE-CHOICE QUESTIONS

1. Lattices are arranged
   a. to accommodate the size of the ions
   b. so that the particles on each corner touch
   c. both a and b
   d. neither a nor b

2. The angle that produces constructive interference of X-rays depends on
   a. distance between each layer of atoms
   b. the intensity of the X-ray
   c. the thickness of the crystal
   d. none of the above; it is empirically determined

3. Compared with visible light, X-rays are
   a. lower frequency
   b. higher frequency
   c. lower intensity
   d. higher intensity

4. If the peaks for an X-ray of 149 pm have a $2\theta = 16.4°$ when $n = 1$, what is the distance between the layers?
   a. 264 pm
   b. 521 pm
   c. 208 pm
   d. 77.7 pm

5. How many atoms are in a unit cell of a body-centered cubic unit cell of iron?
   a. 1
   b. 2
   c. 3
   d. 4 or more

6. Why are cations more likely to be in the holes rather than at lattice points?
   a. they are positive
   b. their electrons are less tightly held
   c. they are smaller
   d. cations are not normally in the holes

7. A face-centered cubic unit cell is the same as
   a. hcp
   b. ccp
   c. bcc
   d. none of the above

8. In a body-centered cubic unit cell of one type of atom, the particles touch each other
   a. along the edge
   b. diagonally across the face
   c. diagonally across the cell
   d. at no point

9. A cubic cell
   a. contains four particles
   b. has the same edge length in all directions
   c. has particles only at the corners
   d. is close packed

10. Nickel metal makes a face-centered cubic unit cell with a density of 8.90 g/cm^3. What is the length of an edge?
   a. 1.51×10^{-7} cm
   b. 4.38×10^{-23} cm
   c. 5.35×10^{-8} cm
   d. 3.53×10^{-8} cm

11. If the layers of a close-packed structure repeat abababab . . . , the structure is
   a. hexagonal closest-packed
   b. cubic closest-packed
   c. simple cubic packed
   d. tetrahedrally packed

12. A cation surrounded by six anions is in
   a. a tetrahedral hole
   b. an octahedral hole
   c. a cubic hole
   d. a hexagonal hole

13. What kind of structures have tetrahedral holes?
   a. hexagonal closest-packed structures
   b. cubic closest-packed structures
   c. face-centered cubic unit cells
   d. all of the above

14. What is an advantage of octahedral holes over tetrahedral holes?
   a. they are larger
   b. there are more of them
   c. they are found in close-packed structures
   d. all of the above

15. What is an advantage of tetrahedral holes over octahedral holes?
   a. they are larger
   b. there are more of them
   c. they are found in close-packed structures
   d. none of the above

16. Which arrangement of spheres most efficiently uses the available space?
   a. cubic closest-packed
   b. hexagonal closest-packed
   c. face-centered cubic cell
   d. all are the same

17. If an anion has a radius of 133 pm and the cation has a radius of 74 pm, then
   a. the anion will fit in a tetrahedral hole
   b. the anion will fit in an octahedral hole
   c. the cation will fit in an octahedral hole
   d. the cation will fit in a tetrahedral hole

18. What type of packing leaves the largest holes?
   a. hexagonal closest-packed
   b. cubic closest-packed
   c. simple cubic packing
   d. all are the same

19. What is the difference between a network solid and an ionic solid?
   a. network solids have specific (connection only between two atoms) covalent bonds
   b. ionic solids have specific covalent bonds
   c. ionic solids have specific ionic bonds
   d. network solids have specific ionic bonds

20. According to band theory, the difference between a conductor and an insulator is
   a. the number of electrons in the bonding orbitals
   b. the number of electrons in the antibonding orbitals
   c. the energy difference between bands
   d. the charge on the ions

21. Which of the following explains the bonding in metals?
    a. many molecular orbitals are so close that they form a band
    b. metal cations that are held together by a sea of electrons
    c. both of the above
    d. neither of the above

22. Which of these is the stoichiometry of a face-centered cubic unit cell with half the tetrahedral holes filled? (Use *X* for the lattice points and *M* for the ion in the holes.)
    a. MX
    b. M_3X
    c. MX_3
    d. MX_3

23. Which is *not* a property of a metal?
    a. shiny
    b. malleable
    c. ductile
    d. insulating

24. Steel is an alloy of iron and
    a. carbon
    b. oxygen
    c. chlorine
    d. all of them

25. An alloy formed when a metal atom is replaced at a lattice point by another metal atom is
    a. intermetallic
    b. interstitial
    c. substitutional
    d. work-hardened

26. Ceramics at room temperature normally have
    a. high thermal conductivity and high electrical conductivity
    b. high thermal conductivity and low electrical conductivity
    c. low thermal conductivity and high electrical conductivity
    d. low thermal conductivity and low electrical conductivity

27. Superconductivity occurs at
    a. high temperatures
    b. low temperatures
    c. room temperature
    d. all temperatures

28. A magnet levitating over a superconductor is an example of
    a. Meissner effect
    b. Cooper pairing
    c. Perovskite behavior
    d. resistivity

Additional Practice Problems

1. If a solid has *XRD* peaks at $2\theta = 29.6, 45.1, 61.5$, and 79.5, when $\lambda = 179$ pm, what is the distance between the layers?

2. Consider the unit cell shown below.

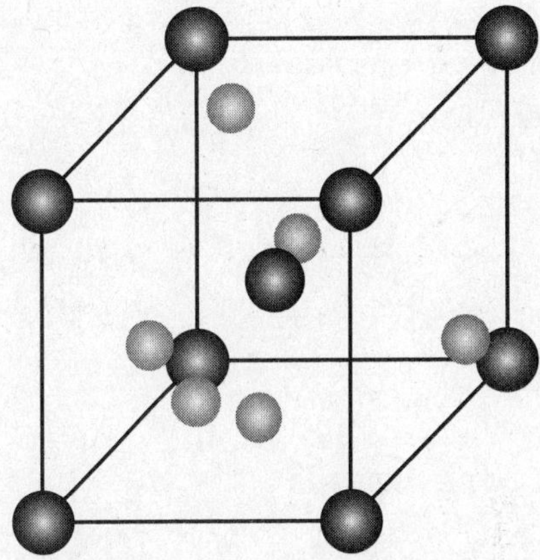

    a. What is the type of unit cell?
    b. What is the formula of the unit cell (anion = black circles = *X*, cation = gray circles = *M*)?
    c. What kind of holes are the cations in?
    d. What is the coordination number of the anions?

3. What is the density (in g/cm^3) of magnesium oxide if it is in a face-centered cubic unit cell with magnesium in half of the tetrahedral holes structure? The size of the magnesium ion is 72 pm. The size of the oxide ion is 140 pm.

4. What type of semiconductor is created when silicon is doped with
    a. aluminum?
    b. antimony?

Self-Test Answer Key

Key Vocabulary Completion Answers

1. crystalline
2. simple cubic
3. silica
4. network covalent
5. polymorph
6. amorphous
7. face-centered cubic
8. substitutional alloy
9. metal

10. superconductor
11. band
12. hexagonal closest-packed
13. unit cell
14. body-centered cubic
15. interstitial alloy
16. Cooper pairs
17. alloy
18. cubic closest-packed
19. Meissner effect
20. ductile

MULTIPLE-CHOICE ANSWERS

For further information about the topics these questions address, you should consult the text. The appropriate section of the text is listed in parentheses after the explanation of the answer.

1. a. Lattice is the arrangement of particles in a crystalline structure. Assuming this is also an ionic structure, the ions must fit (size). (11.7)

2. a. The Bragg equation ($n\lambda = 2d \sin \theta$) defines the angle ($\theta$) of constructive interference. It shows that the angle depends on the wavelength, not the intensity of the X-ray, as well as the distance between layers (d). The symbol n is related to the number of layers the X-ray penetrates, rather than to the thickness of the overall crystal. (11.9)

3. b. X-rays are a more energetic form of light. Therefore, X-rays have higher frequencies and shorter wavelengths than visible light. (11.9)

4. b. In XRD, the distance between layers, d, is determined by the Bragg equation, $n\lambda = 2d \sin \theta$. If $2\theta = 16.40$, then θ must be 8.2°. Substituting in the values, (1)(149 pm) = $2d \sin (8.2)$; $d = \frac{149}{2}(0.143) = 521$ pm. (11.9)

5. b. In a bcc, there is an atom at each of eight corners and one in the center. For each atom in a corner, only one-eighth is in the unit cell. The one in the center is entirely in the unit cell. So $\frac{1}{8}(8) + 1 =$ 2 atoms in the unit cell. (11.3)

6. c. Cations are positive, but charge does not determine their position in the lattice, since the electrostatic attraction that holds an ionic solid together is a general rather than a specific effect. Cations actually hold their electrons more tightly than anions do. This makes their size smaller than the comparable anion. It is their small size that allows them to fit in the holes. (11.7)

7. b. Rotating a face-centered cubic unit cell 45° shows its closest-packed configuration with three alternating layers, a cubic closest-packed (ccp) structure. (11.3)

8. c. If there is only one type of atom, the particles will touch each other. In a body-centered cubic, a particle is in the center of the unit cell. Consequently, the particles on the face will be separated. The particles touch diagonally through the cell. (11.3)

9. b. By definition, a cubic is the same length on all sides. There are many types of cubic cells (face-centered, body-centered, and simple), and some of those could be true for any given type. (11.3)

10. d. In a face-centered cubic, there are four atoms (one on each corner, one on each face). Consequently, the volume is 4 atoms $\times \left(\frac{1 \text{ mol}}{6.022 \times 10^{23} \text{ atoms}}\right) \times \left(\frac{58.70 \text{ g}}{1 \text{ mol}}\right) \times \left(\frac{1 \text{ cm}^3}{8.90 \text{ g}}\right) = 4.38 \times 10^{-23} \text{ cm}^3$. The edge length is the cube root of that value = 3.53×10^{-8} cm. (11.3)

11. a. The definition of a hexagonal closest-packed structure is that it repeats every other row. (11.3)

12. b. An octahedral is made from bonds to six atoms (top, bottom, front, back, left, and right). (11.4)

13. d. Tetrahedral holes are found in both types of closest-packed structures. The face-centered cubic is just a different orientation of the cubic close-packed structure. (11.4)

14. a. Octahedral holes are found in close-packed structures, but so are tetrahedral holes. Octahedral holes are larger, but there are more tetrahedral ones. (11.4)

15. b. Both tetrahedral and octahedral holes are found in close-packed structures. Octahedral holes are larger, but there are more tetrahedral ones. (11.4)

16. d. These are all close-packed structures that use 74% of the space. (11.3)

17. c. The cation is smaller, so it will fit in a hole. The ratio of cation–anion is 0.56, so the cation will fit in an octahedral hole. (11.4)

18. c. While hcp and ccp have the same kind of holes, simple cubic packing produces large holes. (11.4)

19. a. A network solid is an extended network of covalent bonds, each specific between two atoms. An ionic solid is held together by the attraction of positive to negative charges, which is a not specific attraction. (11.6)

20. c. Electrons move freely in a conductor. To do so, they must move into an empty band. If the energy difference between bands is small, the electrons can do so. (11.1)

21. c. Both theories explain the properties of metals. Band theory is similar to molecular orbital theory, and the sea of electrons is more like Lewis structures. (11.1)

22. a. In a face-centered cubic unit cell, there are twice as many tetrahedral holes as lattice points. So if half are filled, the numbers of ions are equal. (11.7)

23. d. Metals are thermal and electrical conductors, not insulators. (11.1)

24. a. Steel is an alloy of iron and carbon, although other metals might also be added. Chlorine and oxygen are not metals and are likely to oxidize iron rather than alloy with it. (11.4)

25. c. A substitutional alloy has metal ions substituting for each other as described here. (11.4)

26. d. Although at low temperatures, some ceramics may be superconductors, most are insulators of both heat and electricity. (11.8)

27. b. Superconductivity occurs at low temperatures when the lack of heat energy minimizes vibration so that the transport of electrons is more efficient. (11.8)

28. a. The Meissner effect is the effect described here. Superconductors normally have a perovskite structure and their low resistivity is attributed to Cooper pairs. (11.8)

Additional Practice Problem Answers

1. We have $n\lambda = 2d \sin \theta$, $\theta = 14.8, 22.55, 30.75$, and 39.75, and $\lambda = 179$ pm. If $n = 1$ for $\theta = 14.8$, then $d = \frac{179}{2 \sin(14.8)} = 350$ pm. If $n = 2$ for $\theta = 22.55$, then $d = \frac{2(179)}{2 \sin(22.55)} = 466$ pm. If $n = 3$ for $\theta = 30.75$, then $d = \frac{3(179)}{2 \sin(30.75)} = 525$ pm.

   This is not working (d should be the same). So maybe $n = 2$ when $\theta = 14.8$? If $n = 2$ for $\theta = 14.8$, then $d = 701$ pm. If $n = 3$ for $\theta = 22.55$, then $d = 700$ pm. If $n = 4$ for $\theta = 30.75$, then $d = 700$ pm. If $n = 5$ for $\theta = 39.75$, then $d = 700$ pm. The average is 700 pm, the distance between layers would be 700 pm.

2. a. Body-centered cubic

   b. We have $X = 8 \times \frac{1}{8} = 1$ (corners) and $1 \times 1 = 1$ (body). So, 2 anions in unit cell. We have $M = 6 \times \frac{1}{2} = 3$ (face). So 3 cations in a unit cell and formula = M_3X_2

   c. octahedral

   d. anion is coordinated to six cations

3. A face-centered cubic unit cell has anions on the corners and each face (corner = $8 \times \frac{1}{8} = 1$ and face = $6 \times \frac{1}{2} = 3$), and therefore, has a total of four oxide ions in the unit cell. There are also four magnesium ions in the unit cell (from stoichiometry of MgO or half-tetrahedral holes). Mass of oxygen:

$$4 \text{ atoms} \times \left(\frac{1 \text{ mol}}{6.022 \times 10^{23} \text{ atoms}}\right) \times \left(\frac{16.00 \text{ g}}{1 \text{ mol}}\right) = 1.063 \times 10^{-22} \text{ g}$$

   Mass of magnesium:

$$4 \text{ atoms} \times \left(\frac{1 \text{ mol}}{6.022 \times 10^{23} \text{ atoms}}\right) \times \left(\frac{24.305 \text{ g}}{1 \text{ mol}}\right) = 1.614 \times 10^{-22} \text{ g}$$

   Mass of unit cell:

   1.063×10^{-22} g + 1.614×10^{-22} g = 2.677×10^{-22} g

   Volume of unit cell: Anions touch across face. There are four radii of oxide ions across the face. The diagonal across the face is 4(140) = 560 pm = c. Using $a^2 + b^2 = c^2$ and $a = b$, $a = 396$ pm is the edge length. Volume = $(\text{edge})^3 = (3.96 \text{ pm})^3 = 6.21 \times 10^{-23} \text{ cm}^3$.

$$\text{Density} = \frac{\text{mass}}{\text{volume}} = \frac{2.677 \times 10^{-22} \text{ g}}{6.21 \times 10^{-23} \text{ cm}^3} = 4.31 \text{ g/cm}^3.$$

4. a. Aluminum has three valence electrons, one less than silicon leaving a "positive hole." Therefore a p-type semiconductor is created.

   b. Antimony (Sb) has five valence electrons, one more than silicon. Therefore, an n-type semiconductor is created.

CHAPTER 12 Organic Chemistry: Fuels and Materials

OUTLINE

REVIEW

Chapter Overview

Because of carbon's propensity to bond to itself, a wide variety of compounds can be made from the arrangement of carbon with itself and other atoms, particularly hydrogen. The study of these compounds, based on carbon, is called **organic chemistry**. If the compound contains only carbon and hydrogen, it is a **hydrocarbon**. However, many organic compounds also contain oxygen or nitrogen atoms in addition to the hydrogen and carbon.

It is possible to arrange the same number of carbons and other atoms in many different ways. Different bonding arrangements of compounds with the same molecular formula are called **structural isomers**. The chemical and physical properties of the compounds depend on this arrangement. Therefore, organic compounds are often represented with condensed structural formulas, formulas that imply the arrangement of atoms as well as the number and each type. When organic chemists are interested in only a small part of the compound, they sometimes use the symbol R to represent the rest of the compound.

Alkanes are the simplest of organic compounds. They are hydrocarbons with only single bonds between the atoms. In general their formula will be C_nH_{2n+2} (where n = number of carbons). Alkanes are also called **saturated hydrocarbons** because they contain the maximum number of hydrogens. **Alkenes** have a carbon–carbon double bond included in their formula. **Hydrogenation** is the process of adding H_2 across the double bond to add a hydrogen to each carbon and change the double bond to a single bond. Each H_2 that can be added is called a **degree of unsaturation**; therefore, alkenes are an example of an unsaturated hydrocarbon. **Alkynes** contain a carbon–carbon triple bond. They also undergo hydrogenation, but two hydrogen molecules can be added across a triple bond. Thus each triple bond has two degrees of unsaturation. Each degree of unsaturation results in a molecule with two fewer hydrogens than the maximum of $2n+2$.

Alkenes have not only structural isomers, but also **geometric isomers**. In structural isomers the atoms are bonded in different orders to different atoms. In geometric isomers, the atoms are bonded in the same way, but are arranged differently in space. Two groups of the same type bonded to each carbon across a double bond may be arranged so that the groups are on the same side of the double bond (***cis* isomer**) or opposite sides of the double bond (***trans* isomer**). The *cis* isomer uses a *Z* prefix in its name to differentiate it from the *trans* (*E*) isomer.

The simplest arrangement of hydrocarbons is for all but the two end carbons to be bonded to only two other carbons. The end carbons are bonded to just one other carbon. These are called **straight-chain hydrocarbons**, and their names reflect this with an *n* prefix. If carbon is bonded to three (or four) carbons, **branched hydrocarbons** are formed. If the chain of carbons is arranged in a circle (there are no "end" carbons), it is a **cyclic compound**. A **homologous series** is a group of organic compounds that differ by a $—CH_2—$ (methylene) group.

Other properties, particularly chemical ones, are determined by a small group of atoms in a particular arrangement. These groups of atoms that are part of an organic molecule are called **functional groups**. Functional groups involving oxygen include **alcohols**, where the oxygen of an OH group is bonded to a carbon, and **ethers**, where there is a carbon–oxygen–carbon sequence of bonds. There are three types of **carbonyls**, where a carbon is doubly bonded to an oxygen. If that carbon (carbonyl carbon) is also bonded to a hydrogen, it is called an **aldehyde**. If the carbonyl carbon is bonded to two other carbons (in addition to its oxygen) the compound is a **ketone**. If the carbonyl carbon is also bound to an alcohol (OH) group, it is a **carboxylic acid**. If the carbonyl carbon is bonded to an additional oxygen that is bonded to another carbon, the compound is called an **ester**.

Functional groups involving nitrogen bonded to carbons and hydrogens are called **amines**. If the nitrogen is bonded to one carbon it is a **primary amine**. If the nitrogen is bonded to two carbons it is a **secondary amine**. If the nitrogen is bonded to three carbons it is a **tertiary amine**.

One particular arrangement of carbons and hydrogens results in an **aromatic compound**. The most common example of this is benzene, which has six carbons in a ring and alternating carbon–carbon double bonds. Because of resonance, the electrons of the double bonds are delocalized throughout the ring. This delocalization of electrons around a ring is one characteristic of aromatic compounds.

Another way of classifying organic compounds is related to the number of carbons in the compound. Small, simple, organic groups often repeat themselves in a formula. The repeating unit is a **monomer**. A compound with a few of these repeating units is called an **oligomer**. When there are many repeating units, the compound is called a **polymer**. Properties generally depend as much on the number of monomer units as the type of monomer units.

Polymers are long chains of repeating units called monomers. If all the monomers of a polymer are the same, the compound is called a **homopolymer**. One type of homopolymer is polyethylene, which has the repeating unit of $\{(CH_2CH_2)\}_n$. Polyethylene can be made from an **addition reaction** of ethylene ($CH_2{=}CH$), where the ethylene groups are added to other ethylene groups. If the ethylene is added in a straight chain, these chains can be packed very close together creating a high density polymer (HDPE). On the other hand, if the chains form branches, they will not pack as well and a low density polymer results (LDPE). Polymers based on ethylene are called **vinyl polymers** and the $CH_2—CH—$ group is a **vinyl group**.

Heteropolymers, made of two or more different monomer groups, can be made with a **condensation reaction**. In a condensation reaction, two molecules are combined to create a larger molecule, and a small molecule (usually water) is also given off.

Properties of polymers are closely related to both their structure and the number of monomers. Because of their large size, there is rarely sufficient kinetic energy for the compound to be a gas. Since the number of monomer units in a sample is rarely constant, the melting and boiling points are ranges rather than specific values. Other properties such as strength and flexibility are also determined by the number of monomers. The way polymers interact with other materials is often a result of its chemical composition.

Organic substances, particularly hydrocarbons, are often used as fuel via combustion reactions. In a **combustion reaction**, the organic fuel reacts with oxygen to produce carbon dioxide and water. The fuels usually come from decayed plant material that is buried, so that it did not decay in the presence of oxygen. The mixture of organic substances produced this way is called **kerogen**. Under higher pressure and temperature, this mixture becomes mostly hydrocarbons and is called **crude oil**.

Crude oil is separated by **fractional distillation**. This process takes advantage of the differences in boiling points of each component of a mixture. Substances with low boiling points will vaporize more easily. Therefore, the vapor of the crude oil at various temperatures can be collected. The vapor collected in each of these temperature ranges is called a **cut** or fraction.

The vapor pressure of a substance depends on the temperature of the liquid and its heat of vaporization. The relationship between these three variables, is expressed with the **Clausius–Clapeyron equation**. The vapor of a mixture is not pure since all volatile substances vaporize to some extent. The partial pressure of each component in the vapor can be determined with Raoult's law. The total pressure of the vapor is the sum of the vapor pressures of each component (Dalton's law). Because substances with lower boiling points also have higher vapor pressures, the vapor is enriched in these substances.

Only **ideal solutions** perfectly follow Raoult's law. In substances for which the solvent–solvent interactions are weaker than the solute–solvent interactions, the solute prevents the solvent from evaporating and the vapor pressure is lower than predicted. However, if the solute–solvent interactions are weaker than solvent–solvent interactions, the vapor pressure is greater than predicted.

Worked Examples

Condensed Formulas

Example 1.
How many hydrogens and carbons are in each of the following structures?

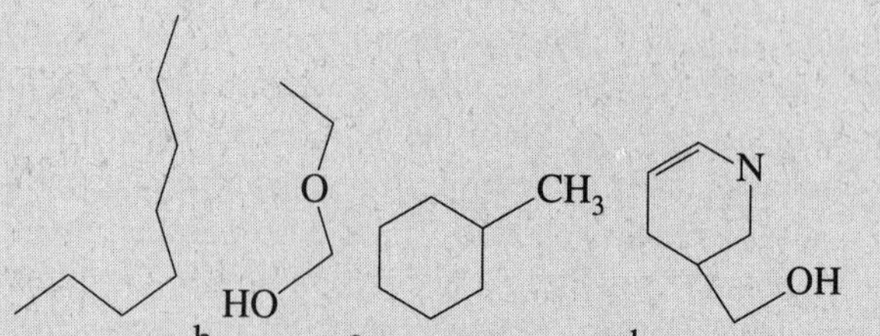

O CH$_3$ N HO OH

a. b. c. d.

Collect and Organize We are given four molecular structures and are asked to determine the number of carbons and hydrogens in each.

Analyze The condensed structures consist of lines representing carbon–carbon bonds. A carbon is assumed at every point where the line segments join unless another element occupies the position. Since four bonds are required for carbon atoms, it is also assumed that carbon–hydrogen bonds make up any missing bonds. The hydrogens are not written.

Solve

a. There are eight places (including each end) where line segments join. Therefore, there are eight carbons. Each of the carbons in the middle of the chain (six of them) needs two more bonds for its total of four. Those bonds will be to hydrogen. The carbons at the end need three more bonds to hydrogen (since it is connected to only one carbon). Therefore, there are 18 hydrogens.

b. There are five places where line segments join. However, at one end an OH replaces the carbon. There is also a place where an oxygen replaces a carbon. Therefore this molecule has only three carbons. The carbon at the end will have three hydrogens. Each of the two carbons in the middle will be bonded to two hydrogens. There is also the hydrogen at the end bonded to the oxygen. That is a total of eight hydrogens.

c. This is a six-membered ring, so it has six carbons. There is also the one at the end of the line, where it says "CH_3." The CH_3 details what is at the end of that line. If it was not present, the CH_3 would still be implied. The line represents the bond between the carbon of the CH_3 and the carbon in the ring. Consequently, this molecule has a total of seven carbons. Except for the carbon bonded to the CH_3, the carbons in the ring are each bonded to two hydrogens. The carbon bonded to the CH_3 needs one more bond, so a bond to a hydrogen is implied. That makes 10 + 1 hydrogens in the ring and three with the CH_3, so there are a total of 14 hydrogens.

d. There is a six-membered ring, with one of the carbon positions replaced with nitrogen; so there are five carbons in the ring. In addition, there are another two line segments that will have an additional carbon where they join. However, the end of this segment has an OH group instead of a carbon. This makes a total of six carbons.

The carbon where the two line segments join must have two hydrogens to complete its four bonds. The carbon where the line segment is joined to the ring needs one more bond. The carbons in the ring on either side of that carbon need two more bonds, so that is a total of four more hydrogens. The carbons on either side of the double bond need one more bond each. The nitrogen will bond three times. Since it is only bonded twice, there is another hydrogen (bonded to the nitrogen) implied. There is also the hydrogen that is part of the OH group. That gives a total of $2 + 1 + 2(2) + 2(1) + 1 + 1 = 11$ hydrogens.

Think About It Structures are condensed by leaving off the lone pairs and grouping the hydrogens with the carbon they are bonded to. Oxygens are also grouped with the carbon they are bonded to.

Structural Isomers

Example 1.
Which of the following pairs are structural isomers?

a.  and $CH_3CH_2CHCH_2$

b. 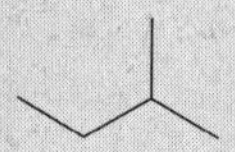 and $CH_3CH_2CH_2CH_2CH_3$

c. $CH_3OCH_2CH_3$ and $CH_3CH_2CH_2OH$ and $(CH_3)_2COH$

d. $CH_2{=}CHCH_2CH_3$ and $CH_3CH_2CH{=}CH_2$

Collect and Organize We are given pairs of molecules and are asked to determine which represent structural isomers.

Analyze Two compounds with the same chemical formula (number and type of atoms) but different bonding patterns (Lewis structures) are called structural isomers.

SOLVE

a. These are the same molecule, one written with line structures and one not. There must be a double bond between the last two carbons to explain the lack of hydrogens.

b. These are structural isomers. The first one branches at the next to last carbon, where it is bonded to three carbons and one hydrogen. In the other structure, there is no carbon bonded to three other carbons. However, both have the same number of carbons and hydrogens, so they are structural isomers.

c. All three are structural isomers, with the general formula C_3H_7O. However, the oxygen in the first structure is bonded to two carbons, whereas the oxygen in the other two structures is bonded to one hydrogen and one carbon. The one at the end has a carbon bonded to three other carbons, whereas the carbons of the middle one are bonded to no more than two other carbons.

d. These are the same molecule. Both have the same number of carbons with a double bond between the first and second carbons. The only difference is that one is flipped over.

THINK ABOUT IT The chemical and physical properties of the compounds depend on the arrangement. Chemical properties, in particular, are often determined by a small group of atoms with a particular arrangement. These groups of atoms are the part of the organic molecule called a functional group.

EXAMPLE 2.
Identify the following as *cis* and *trans* isomers.

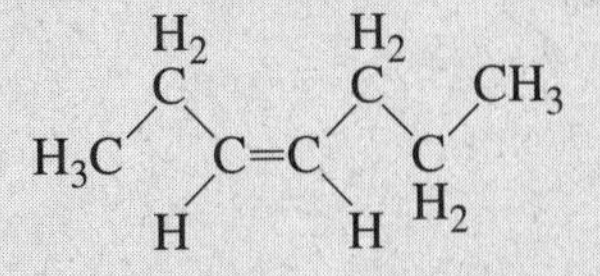

a. $H_3C{-}CH_2{-}CH{=}CH{-}CH_2{-}CH_2{-}CH_3$ (H, H on the bottom of the double bond)

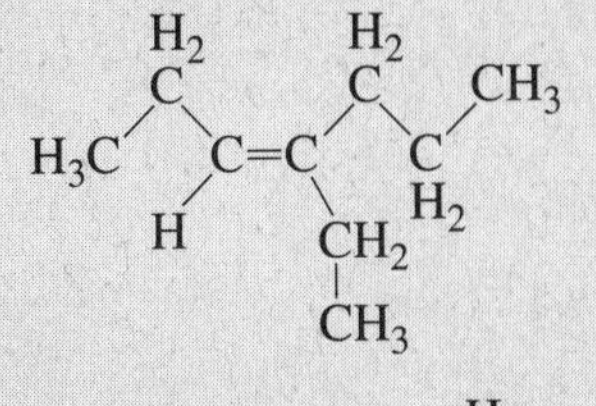

b. $H_3C{-}CH_2{-}CH{=}C(CH_2CH_3){-}CH_2{-}CH_2{-}CH_3$

c. $H_3C{-}C(=O){-}C(CH_2{-}cyclopentyl){=}C(CH_2CH_3){-}CH_2{-}CH_3$

COLLECT AND ORGANIZE We are given four molecular structures and are asked to determine whether they represent the *cis* or *trans* isomer.

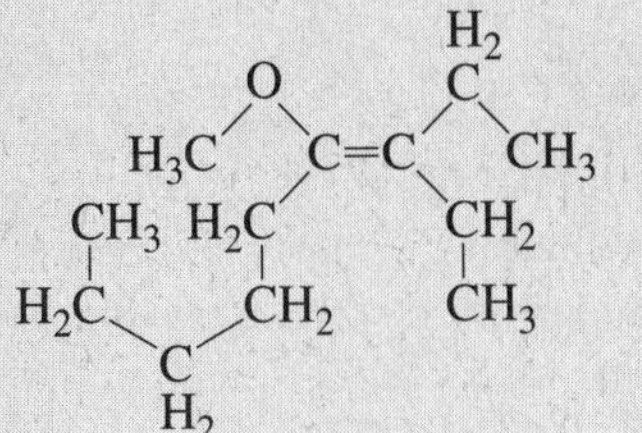

ANALYZE The prefixes *cis* and *trans* are used when one group on the carbons opposite the double bond is the same. If those groups are on the same side, it is a *cis* isomer; if they are on opposite sides, a *trans* isomer.

SOLVE

a. The group that is the same is the H group. It is on the same side (bottom) of the double bond; therefore, it is a *cis* isomer.

b. The group that is the same on either side of the double bond is CH_3CH_2 (*ethyl*). The group is on opposite sides of the double bond; it is therefore, a *trans* isomer.

c. The group that is the same is the ethyl group on the carbon to the right. Because those are the same carbon, this compound does not have a geometric isomer.

THINK ABOUT IT This is only an issue for alkenes. In addition, if the groups attached to one of the carbons in the double bond are the same, there are no geometric isomers.

IDENTIFYING FUNCTIONAL GROUPS

EXAMPLE 1.
Identify the functional groups of the following compounds.

a. 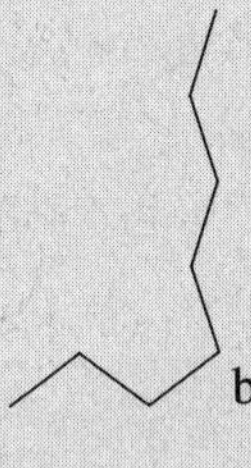 (line structure)

b. 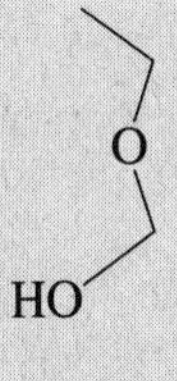 HO (line structure containing O)

c. 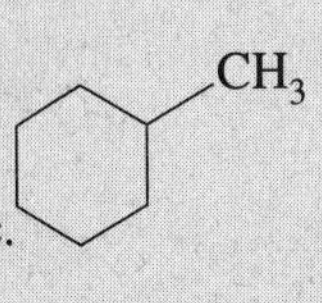 (cyclohexane with CH_3)

d. 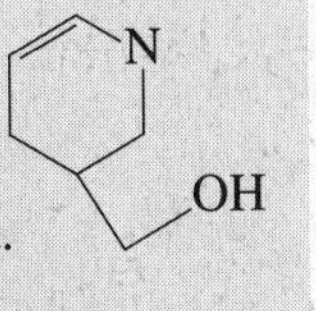 (ring containing N, with OH)

e. CH_3COOH

f. 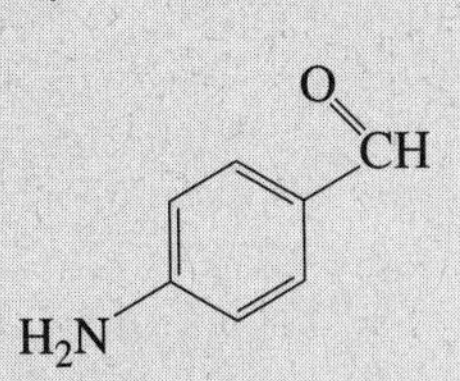 H_2N (benzene ring with CH=O)

g. $H_3C{-}CH_2{-}C(=O){-}CH_2{-}CH_3$

h. (ring containing O, with CH_2OH, OH, OH)

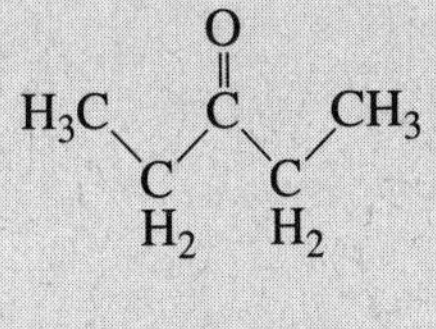

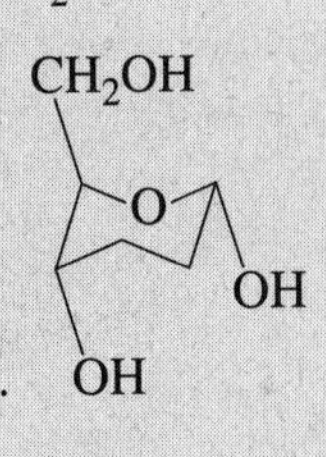

COLLECT AND ORGANIZE We are asked to identify the functional groups contained in the given molecules.

ANALYZE The basic organic structure is carbons and hydrogens, all singly bonded to each other. These are called alkanes. Any variation from that basic structure is called a functional group. Knowing the types of functional groups is basic to an understanding of organic chemistry.

Table 1 FUNCTIONAL GROUPS

Name	Condensed Formula	Description
alkene	$R_2C{=}CR_2$	contains a C=C double bond
alkyne	$RC{\equiv}CR$	contains a C≡C triple bond
alcohol	ROH	contains O singly bonded to a C and an H
ether	ROR	contains O singly bonded to two C's
aldehyde	RCHO	contains C doubly bonded to O and singly to H
ketone	RCOR	contains C doubly bonded to O and singly to two C's
carboxylic acid	RCOOH	contains C doubly bonded to O and singly to O of OH
amine		
primary	RNH_2	contains N singly bonded to one C and two H's
secondary	R_2NH	contains N singly bonded to two C's and one H
tertiary	R_3N	contains N singly bonded to three C's
aromatic	[benzene ring] or [benzene ring with circle]	contains a flat six-membered ring

SOLVE

a. This is all carbons and hydrogens, with no double bonds. It has no functional groups and is an alkane.

b. This has an ether (the oxygen in the middle) and an alcohol (the OH group at the end).

c. Although a cyclic compound, this has no functional groups. It is an alkane.

d. There is a double bond between carbons at the top of the ring, an alkene. There is a nitrogen in the ring, so it must be bonded to two carbons, making it a secondary amine. The OH group at the end is an alcohol.

e. The COOH at the end makes this a carboxylic acid.

f. The hexagonal ring with the alternating double bonds makes this an aromatic. The NH_2 group is a primary amine. The carbon with the double bond to an oxygen and a single bond to a hydrogen is an aldehyde.

g. The carbon in the center with the double bond to the oxygen is a ketone.

h. On the back right side of this molecule is a carbon that is bonded to the oxygen of an ether and the oxygen of an alcohol. There are two more alcohol groups, one with the CH_2OH and the other on the bottom left.

THINK ABOUT IT It is best to memorize the common functional groups that occur in organic molecules.

VOLATILITY

EXAMPLE 1.
In a mixture of 86.0 g C_6H_6 (benzene, $P^\circ_B = 93.96$ torr) and 90.0 g $C_2H_4Cl_2$ (dichloroethane, $P^\circ_D = 224.9$ torr), what is the total vapor pressure?

COLLECT AND ORGANIZE We are asked to find the vapor pressure of a mixture of two volatile substances.

ANALYZE Because the question asked for the vapor pressure of a mixture, Raoult's law is used. However, both components act as solvents. Therefore, Raoult's law is used to find the contribution of each. The total vapor pressure will be the sum of the vapor pressure contributed by each component (Dalton's law).

SOLVE

Raoult's law is $P_A = X_A P^\circ_A$

First, the mole fraction of each component must be determined.

Moles of C_6H_6 (benzene) $= 86.0\text{ g} \times \left(\frac{1\text{ mol}}{78.11\text{ g}}\right) = 1.10$ mol

Moles of $C_2H_4Cl_2$ (dichloroethane) $= 90.0\text{ g} \times \left(\frac{1\text{ mol}}{98.94\text{ g}}\right)$

$= 0.910$ mol

The vapor pressure from C_6H_6 (P_B) is

$$P_B = X_B P^\circ_B = \left(\frac{1.10}{(1.10 + 0.910)}\right)(93.96) = 51.2\text{ torr}$$

The vapor pressure from $C_2H_4Cl_2$ (P_D) is

$$P_D = X_D P^\circ_D = \left(\frac{0.910}{(1.10 + 0.910)}\right)(224.9) = 102\text{ torr}$$

The total pressure is the sum of the partial pressures, so

$$P_{total} = P_B + P_D = 51.2 + 102 = 153\text{ torr}$$

THINK ABOUT IT
The vapor pressure of the mixture falls between the values of each of the pure vapor pressures. This was true even for nonvolatile solutes (whose vapor pressure is zero) that cause the vapor pressure lowering discussed in Chapter 10.

Self-Test

Key Vocabulary Completion Questions

_________ 1. Solution that obeys Raoult's law
_________ 2. Separation based on different volatilities
_________ 3. C=O group
_________ 4. Series of compounds differing by one CH_2 group
_________ 5. ROH
_________ 6. Reaction resulting in a larger molecule and the loss of water
_________ 7. The total pressure equals the sum of the partial pressures
_________ 8. RCOOH
_________ 9. Compound where two carbons are connected by an oxygen
_________ 10. R_2NH
_________ 11. Hydrocarbon where all C—H and C—C bonds are single bonds
_________ 12. Long chains of repeating units
_________ 13. $CH_2CH—$
_________ 14. Functional group in CH_3CH_2CHO
_________ 15. Polymer made of all the same kind of monomers
_________ 16. Molecules with the same chemical formula but different Lewis structures
_________ 17. Molecule with a center carbon singly bonded to one oxygen and doubly bonded to another
_________ 18. Organic compound containing nitrogen
_________ 19. Type of hydrocarbon where carbons are not in a line
_________ 20. Hydrocarbon with a carbon–carbon double bond
_________ 21. The same group on the same side of a double bond
_________ 22. Benzene is an example
_________ 23. Same structural formula, but the groups are oriented differently in space
_________ 24. HCCH, for example
_________ 25. Reaction between an organic compound and oxygen making carbon dioxide and water

Multiple-Choice Questions

1. Which of the following is not a hydrocarbon?
   a. C_2H_2
   b. CH_2O
   c. $CH_3CH_2CH_2CH_2CH_2CH_3$
   d. C_5H_{12}
2. Which of the following is an alkane?
   a. C_2H_2
   b. CH_2O
   c. $CH_3CH_2CH_2CH_2CH_2CH_3$
   d. $CH_2CHCH_2CH_3$
3. How many hydrogens are in a saturated hydrocarbon with eight carbons?
   a. 8
   b. 16
   c. 4
   d. 18
4. One product of a combustion reaction is
   a. gaseous water
   b. hydrocarbon
   c. saturated hydrocarbon
   d. all of the above
5. The real shape of the carbons in a "straight-chain hydrocarbon" is
   a. ———
   b. /\/\/\/\/\
   c. O
   d. []
6. What is the vapor pressure at 20°C of a mixture of 50.0 g water ($P° = 17.5$ torr) and 10.0 g CH_3OH ($P° = 93.3$ torr)?
   a. 15.7 torr
   b. 9.42 torr
   c. 25.1 torr
   d. 30.2 torr
   e. 14.6 torr
7. When a mixture of methanol ($P° = 93.3$ torr @ 20°C) and water ($P° = 17.5$ torr @ 20°C) is distilled, the vapor collected will be
   a. rich in methanol
   b. rich in water
   c. pure methanol
   d. pure water
   e. equally distributed between the two

8. Strong intermolecular forces between solute and solvent will result in a vapor pressure that is
   a. is lower than predicted by Raoult's law
   b. is higher than predicted by Raoult's law
   c. is almost exactly what is predicted by Raoult's law
   d. deviates from Raoult's law, but the direction is not predictable
   e. Raoult's law does not apply to this situation

9. What type of intermolecular forces predominate in alkanes?
   a. hydrogen bonding
   b. dipole–dipole
   c. ion–ion
   d. London
   e. ion–dipole

10. The prefix *iso* means
   a. carbons are arranged in a ring
   b. carbons are arranged in a zig-zag pattern
   c. carbon chain has no branches
   d. carbon chain has more than one branch
   e. carbons are not in a straight chain

11. A characteristic of benzene is that it is
   a. aromatic
   b. planar
   c. delocalized electrons
   d. C_6H_6
   e. all of the above

12. Polystyrene contains which of the following functional groups?
   a. vinyl
   b. ester
   c. aromatic
   d. ether
   e. it is an alkane

13. What is the functional group in

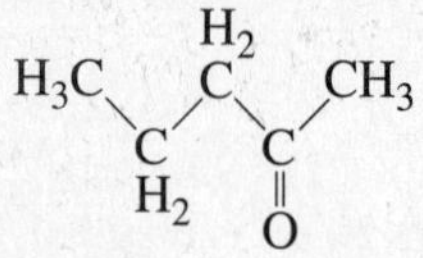

H_3C $\overset{H_2}{C}$ CH_3
$\underset{H_2}{C}$ $C{=}O$

   a. alcohol
   b. ketone
   c. aldehyde
   d. carboxylic acid
   e. amide

14. An amine contains
   a. an OH group
   b. a carbonyl
   c. a ring
   d. ROR
   e. nitrogen

15. A carbonyl is carbon
   a. doubly bonded to an oxygen
   b. bonded to an oxygen and a hydrogen
   c. bonded only to other carbons
   d. singly bonded to oxygen and other carbons
   e. bonded to an OH group

16. The structure of carboxylic acid is carbon doubly bonded to
   a. an oxygen
   b. a hydrogen
   c. an oxygen and singly to hydrogen
   d. an oxygen and singly to a hydroxyl group
   e. a carbon and singly to an oxygen

17. If 1.00 g of an organic molecule contains 0.857 g C and 0.143 g H, how much oxygen does it contain?
   a. none
   b. 0.033 g
   c. 0.533 g
   d. 0.390 g
   e. 1.00 g

18. What is the molecular formula of a compound with an empirical formula of CH_2 and a molar mass of 70.0 g/mol?
   a. C_4H_{22}
   b. C_5H_5
   c. C_5H_{10}
   d. C_6H_{12}
   e. C_7H_{14}

H_3C H
H CH_3

19. 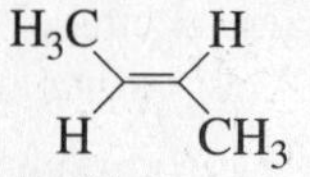 (structure above) is called
   a. *cis*
   b. *trans*
   c. iso
   d. tertiary
   e. secondary

20. In geometric isomers
   a. atoms are bonded to different atoms
   b. the functional groups are different
   c. the arrangement in space of bonds on the same atom is different
   d. the position of the functional groups is different

21. Why doesn't butyne have any *cis* and *trans* isomers?
   a. it is all single bonds
   b. it is a hydrocarbon
   c. there are two hydrogens on one carbon of the triple bond
   d. the triple bond allows only one possible arrangement in space
   e. it does have *cis* and *trans* isomers

22. What type of organic compound is saturated?
    a. cyclic
    b. alkyne
    c. alcohol
    d. aldehyde
    e. aromatic

23. The *ene* suffix means that
    a. there are only single bonds
    b. there is a double bond
    c. there is a triple bond
    d. there is a nitrogen
    e. there is a doubly bonded oxygen

24. How many grams of carbon are in 0.791 g of CO_2?
    a. 0.179 g
    b. 0.431 g
    c. 0.216 g
    d. 0.575 g
    e. 0.791 g

25. A hydroxyl group is
    a. ROR
    b. RHO
    c. ROH
    d. RNH_2
    e. RCOOH

26. Which of the following is an ether?
    a. CH_3OCH_3
    b. CH_3COOH
    c. CH_3CHO
    d. CH_3CH_2OH
    e. CH_3COCH_3

27. Which of the following is an alcohol?
    a. CH_3OCH_3
    b. CH_3COOH
    c. CH_3CHO
    d. CH_3CH_2OH
    e. CH_3COCH_3

28. Which monomer was used to produce Teflon ($\{CF_2CF_2\}$)?
    a. CHF_2CHF_2
    b. CF_2CF_2
    c. CF_3CF_3
    d. CH_2CF_2

29. Which *n*-alkane has the highest boiling point?
    a. butane
    b. ethane
    c. methane
    d. pentane
    e. propane

30. An ideal solution
    a. obeys Raoult's law
    b. has solvent and solute with dissimilar forces
    c. has weak intermolecular forces
    d. all of the above
    e. none of the above

Additional Practice Problems

1. What is the vapor pressure of the following solutions?
   a. 5.00 g CH_3OH (P° = 121.5 torr) and 5.00 g H_2O (P° = 23.76 torr) at 25°C
   b. 10.00 g C_6H_6 (P° = 74.25 torr) and 20.00 g $C_2H_4Cl_2$ (P° = 180.8) at 20°C
   c. 5.00 g CCl_4 (P° = 90.9 torr) and 5.00 g C_6H_6 (P° = 74.25 torr) at 20°C

2. Draw four structural isomers of C_6H_{14}.

3. Classify each of the following as: alkane, alkene, alkyne, aromatic, amine, alcohol, ether, aldehyde, ketone, carboxylic acid, ester, or amide.
   a. $CH_3CH_2CH(CH_3)_2$
   b. $(CH_3)_3N$
   c. CH_3CH_2CHO

   CH_3 CH_3

   d. 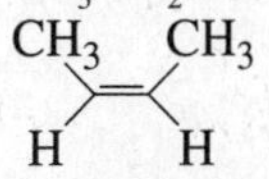 H H
   e. $CH_3(CH_2)_7COOH$

   CH_3
   CH_2
   CH_3

   f.

   CH_3

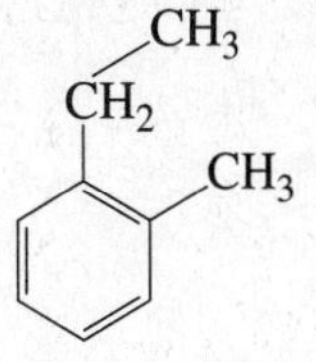

   g. 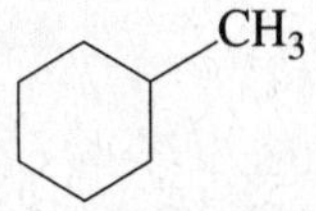
   h. $CH{\equiv}CCH_2CH_2CH_3$

4. Draw four structural isomers of C_5H_{10}. Include the *cis* and *trans* isomers.

5. At 100°C, benzyl alcohol has a vapor pressure of 15.3 torr. At 130°C, benzyl alcohol has a vapor pressure of 63.8 torr. What is the heat of vaporization for benzyl alcohol?

Self-Test Answer Key

Key Vocabulary Completion Answers

1. ideal
2. (fractional) distillation
3. carbonyl

4. homologous series
5. alcohol
6. condensation
7. Dalton's law
8. carboxylic acid
9. ether
10. secondary amine
11. alkane
12. polymer
13. vinyl group
14. aldehyde
15. homopolymer
16. structural isomer
17. ester
18. amine
19. branched
20. alkene
21. *cis*
22. aromatic
23. geometric isomer
24. alkyne
25. combustion

Multiple-Choice Answers

For further information about the topics these questions address, you should consult the text. The appropriate section of the text is listed in parentheses after the explanation of the answer.

1. b. Hydrocarbons are substances made only of hydrogen and carbon. Since CH_2O also contains oxygen, it is not a hydrocarbon. (12.3)

2. c. An alkane is a hydrocarbon consisting only of single bonds. CH_2O is eliminated because it is not a hydrocarbon. C_2H_2 has a triple bond and the first two carbons of $CH_2CHCH_2CH_3$ must be double-bonded, but all the bonds of $CH_3CH_2CH_2CH_2CH_2CH_3$ are single bonds. (12.3)

3. d. In a saturated hydrocarbon, there are $2n + 2$ hydrogens, when $n = 8$, $H = 18$. (12.3)

4. a. A combustion reaction is the reaction of an organic substance (like a hydrocarbon) with oxygen to make the gases carbon dioxide and water. (12.8)

5. b. For carbons in the middle of a straight chain, the carbon is bonded to two hydrogens and two carbons in a tetrahedral geometry. The carbons follow a series of bent shapes, giving an overall zig-zag geometry. (12.2)

6. c. 50.0 g water $\times \left(\frac{1\text{ mol}}{18.01\text{ g}}\right) = 2.77$ mol and 10.0 g methanol (CH_3OH) $\times \left(\frac{1\text{ mol}}{32.04\text{ g}}\right) = 0.312$ mol.

   Therefore, $\chi_{H_2O} = \frac{2.77}{(2.77 + 0.312)} = 0.899$ and

   $\chi_{CH_3OH} = \frac{0.312}{(2.77 + 0.312)} = 0.101.$

   $P_{H_2O} = \chi_{H_2O}P^\circ = (0.899)(17.5) = 15.7$ torr and $P_{CH_3OH} = \chi_{CH_3OH}P^\circ = (0.101)(93.3) = 9.42$ torr. $P_T = 25.1$ torr (12.3)

7. a. Since the vapor pressure of methanol is much higher than that of water, the fraction of methanol in the vapor will be much higher than that in water. However, since the vapor does contain some water, it will not be pure. (12.3)

8. a. The strong intermolecular forces keep the molecules together in the liquid state; therefore, the vapor pressure is lower than would be predicted by Raoult's law (which does apply, since solvent and solute together make a mixture). (12.3)

9. d. Alkanes consist only of carbon and hydrogen atoms, with all the hydrogens on the outside. Consequently, alkanes are always nonpolar. Nonpolar molecules have only London (dispersion) forces. (12.3)

10. e. A straight chain is a "normal," and when it has at least one branch, it is then "iso." Since each carbon has a tetrahedral geometry, a zig-zag is the arrangement of a straight-chain organic compound. (12.2)

11. e. Benzene is a 6-carbon ring with alternating double bonds between the carbons, resulting in a formula of C_6H_6. There is a resonance structure that results in delocalized electrons. The sp^2 geometry of each carbon makes the ring planar. It is a classic example of an aromatic. (12.5)

12. c. Polystyrene is made from an ethylene group with a benzene ring attached to one of the carbons. (12.5)

13. b. A ketone is a carbon doubly bonded to an oxygen and two other carbons, as shown in this structure. (12.7)

14. e. Amines contain nitrogen and are labeled primary, secondary, or tertiary, depending on how many carbons are bonded to the nitrogen. (12.8)

15. a. A carbonyl group is a carbon doubly bonded to an oxygen, which includes both aldehydes and ketones. In an aldehyde, the carbon is also bonded to a carbon and a hydrogen. In a ketone, the carbon additionally bonded to two other carbons. (12.7)

16. d. The structure of a carboxylic acid is $R{-}C(=O){-}OH$, where R represents any length of carbon chain. (12.7)

17. a. Conservation of mass states that mass is neither created nor destroyed. Therefore, the sum of the mass of each element in a compound must equal the total mass of the compound. $C + H + O =$ $1.00\text{ g} = 0.857\text{ g} + 0.143\text{ g} + \text{oxygen}$. Solving the equation, oxygen = zero. (3.6)

18. c. The empirical formula is the simplest ratio of elements in the compound. The molar mass takes into account all the atoms in a molecule. The mass of the empirical formula is 14, the ratio of $\frac{70}{14} = 5$, so there are five times more atoms than in the empirical formula. The molecular formula is then C_5H_{10}. (3.7)

19. b. This compound has geometric (*cis* and *trans*) isomers. Groups that are the same (H) are on opposite sides of the double bond, which makes it a *trans* isomer. (12.4)

20. c. The atoms are all bonded to the same atoms, but geometric isomers are arranged differently in space. Structural isomers bond atoms in different ways. (12.2)

21. d. The *yne* suffix implies a triple bond. That means the carbon attached to the triple bond has only one other place to bond and has a linear geometry. There are no choices about the way these are arranged in space. (12.4)

22. c. A saturated organic has the maximum number of hydrogen atoms ($2n + 2$), where n = number of carbons. Cyclic compounds have only $2n$ hydrogens. Any compound with a double or triple bond has fewer than the maximum hydrogens. Of the choices, only the alcohol consists of all single bonds. (12.2)

23. b. *ene* is the suffix used for double bonds between carbon atoms. (12.4)

24. c. $0.791\text{ g } CO_2 \times \left(\frac{1\text{ mol } CO_2}{44.01\text{ g}}\right) \times \left(\frac{1\text{ mol C}}{1\text{ mol } CO_2}\right) \times$

$$\left(\frac{12.01\text{ g}}{1\text{ mol}}\right) = 0.216\text{ g (3.6)}$$

25. c. A hydroxyl group is the same as an alcohol, ROH. (12.6)

26. a. An ether is two carbons linked by an oxygen. Consequently, it must be in the middle of carbons. In choice e, the link must be through the carbon rather than the oxygen; otherwise, the carbon has insufficient bonds and breaks the octet rule. (12.6)

27. d. An alcohol has a carbon singly bonded to an oxygen and the oxygen bonded to a hydrogen. The hydrogen after the oxygen implies that it is bonded to the oxygen rather than the carbon. The carbon–oxygen must be a single bond, since there are three other atoms that must be bonded to it. (12.6)

28. b. Teflon is an example of a homopolymer, made from an alkene when the double bond becomes a single bond, and the two extra electrons can form bonds with another monomer that does the same thing. (12.4)

29. d. Since the intermolecular force predominating in *n*-alkanes is dispersion (London) forces, molecules with higher molar masses will have higher boiling points. Pentane, with five carbons, has the highest molar mass of the alkanes listed. (12.3 and 10.4)

30. a. An ideal solution, by definition, obeys Raoult's law. This is more likely to occur when the intermolecular forces of the solute and solvent are similar. (12.3)

Additional Practice Problem Answers

1. a. $5.00 \text{ g } CH_3OH \times \left(\frac{1 \text{ mol}}{32.04 \text{ g}}\right) = 0.156 \text{ mol}$ and

$5.00 \text{ g } H_2O \times \left(\frac{1 \text{ mol}}{18.01 \text{ g}}\right) = 0.278 \text{ mol}$. Total moles $= 0.156 + 0.278 = 0.434$ mol. The mole fraction of methanol $= \chi_{\text{methanol}} = \frac{0.156}{0.434} = 0.359$ and $\chi_{\text{water}} = \frac{0.278}{0.434} = 0.641$. The partial pressure of methanol $= P_{\text{methanol}} = \chi_{\text{methanol}} P° = (0.359)(121.5 \text{ torr}) = 43.6$ torr and $P_{\text{water}} = \chi_{\text{water}} P° = (0.641)(23.76 \text{ torr}) = 15.2$ torr. The total vapor pressure $= P_{\text{methanol}} + P_{\text{water}} = 43.6 + 15.2 = 58.8$ torr.

b. $10.00 \text{ g } C_6H_6 \times \left(\frac{1 \text{ mol}}{78.11 \text{ g}}\right) = 0.1280 \text{ mol}$ and

$20.00 \text{ g } CH_3CHCl_2 \times \left(\frac{1 \text{ mol}}{98.96 \text{ g}}\right) = 0.2021 \text{ mol}$.

Total moles $= 0.1280 + 0.2021 = 0.3301$ mol. The mole fraction of benzene $= \chi_{\text{benzene}} = \frac{0.1280}{0.3301} = 0.3878$ and $\chi_{\text{dichloroethane}} = \frac{0.2021}{0.3301} = 0.6122$.

The partial pressure of benzene $= P_{\text{benzene}} = \chi_{\text{benzene}} P° = (0.3878)(74.25 \text{ torr}) = 28.79$ torr and $P_{\text{dichloroethane}} = \chi_{\text{dichloroethane}} P° = (0.6122)(180.8 \text{ torr}) = 110.7$ torr. Total vapor pressure $= P_{\text{benzene}} + P_{\text{dichloroethane}} = 28.79 + 110.7 = 139.5$ torr

c. $5.00 \text{ g } CCl_4 \times \left(\frac{1 \text{ mol}}{153.82 \text{ g}}\right) = 0.0325 \text{ mol}$ and

$5.00 \text{ g } C_6H_6 \times \left(\frac{1 \text{ mol}}{78.11 \text{ g}}\right) = 0.0640 \text{ mol}$.

Total moles $= 0.0325 + 0.0640 = 0.0965$ mol. The mole fraction of carbon tetrachloride $=$ $\chi_{\text{carbon tetrachloride}} = \frac{0.0325}{0.0965} = 0.337$ and $\chi_{\text{benzene}} = \frac{0.0640}{0.0965} = 0.663$. The partial pressures are carbon tetrachloride $= P_{\text{carbon tetrachloride}} = \chi_{\text{carbon tetrachloride}} P° = (0.337)(90.9 \text{ torr}) = 30.6$ torr and $P_{\text{benzene}} = (0.663)(74.25 \text{ torr}) = 49.2$ torr. The total vapor pressure $= P_{\text{carbon tetrachloride}} + P_{\text{benzene}} = 30.6 + 49.2 = 79.8$ torr.

2. $CH_3CH_2CH_2CH_2CH_2CH_3$

$CH_3CHCH_2CH_2CH_3$ with CH_3 on C2; $CH_3CH_2CHCH_2CH_3$ with CH_3 on C3

$CH_3CHCH_2CH_3$ with CH_2CH_3 on C2; $CH_3CCH_2CH_3$ with two CH_3 on C2; $CH_3CHCHCH_3$ with CH_3 on C2 and C3

3. a. alkane (branched)
   b. amine
   c. aldehyde
   d. alkene
   e. carboxylic acid
   f. aromatic
   g. alkane (cycloalkane)
   h. alkyne

4. Cyclopentane (CH_2 ring of five); $H_2C{=}CH{-}CH_2CH_2CH_3$; $CH_3CH{=}CHCH_2CH_3$ (cis: CH_3 and CH_2CH_3 same side, H and H below); $(CH_3)_2C{=}CHCH_3$; $CH_3CH{=}CHCH_2CH_3$ (trans: H / CH_2CH_3 above, CH_3 / H below); $H_2C{=}C(CH_3)CH_2CH_3$; $H_2C{=}CH{-}CH(CH_3)_2$

5. The Clausius–Clapeyron equation is used:

$$\ln\left(\frac{P_1}{P_2}\right) = \left(\frac{\Delta H_{\text{vap}}}{R}\right)\left[\frac{1}{T_1} - \frac{1}{T_2}\right]$$

Note that temperature must be in Kelvin, and vapor pressure can be in any units, provided they are the same for both pressures. We have

$$\ln\frac{15.3}{63.8} = \left(\frac{\Delta H}{8.314}\right)\left[\frac{1}{403} - \frac{1}{373}\right]$$

$\Delta H = 5.9 \times 10^4 \text{ J} = 59 \text{ kJ}$.

CHAPTER 13 Thermodynamics: Spontaneous Processes, Entropy, and Free Energy

OUTLINE

REVIEW

Chapter Overview

A **spontaneous process** is one that proceeds without outside intervention. The driving force that causes a spontaneous reaction is entropy. **Entropy** is a measure of the dispersion of energy, and entropy always increases in a spontaneous process. For example, heat flows from a hot area, where the energy is concentrated, to a cold area, so that it is more dispersed throughout both substances.

Because molecular behavior is ruled by quantum mechanics, each molecule has a variety of specific energy states to choose from. These energy states correspond to differences in energy as molecules move with motions such as **vibration** (atoms moving back and forth along their bonds as if each were a spring), **rotation** (atoms rotating around a bond axis), and **translation** (motion of the entire atom). Each molecule occupies several energy states depending on the positions of its atoms in any moment in time. The distribution of a large number of molecules in these various energy levels is called a **microstate**. Since entropy is the dispersion of energy, when the molecules occupy more energy states, the entropy is high. If most molecules are in just a few energy states, the entropy is low.

There are many conditions that allow molecules to occupy larger number of energy states and increase in entropy. These include a higher volume, higher temperatures, and larger numbers of particles. Entropy also depends on physical state, where gases have the highest entropy and solids the lowest. Entropy is also affected by the complexity of the molecules, where more complex molecules have a higher entropy; the higher number of bonds and atoms increases the number of possible microstates.

Entropy can be determined in several ways. As defined by the **third law of thermodynamics**, when there is only one microstate available, the entropy is zero. Because a zero point can be defined, the **standard molar entropy** (**$S°$**), the entropy of one mole at 298 K and 1 atm, for any substance can be measured. Also, the change in entropy (ΔS) can be calculated from

$$\Delta S_{\text{system}} = S_{\text{final}} - S_{\text{initial}}$$

The **first law of thermodynamics** states that energy is neither created nor destroyed, thus energy must go somewhere. However, the **second law of thermodynamics** states that in a spontaneous change, the entropy of the universe increases. The change in the entropy of the universe ($\Delta S_{\text{universe}}$) is the sum of the change in the entropy of the system (ΔS_{system}) and the change in the entropy of the surroundings ($\Delta S_{\text{surroundings}}$). For a chemical or physical change in a system, ΔS_{system} can be calculated from

$$\Delta S_{\text{system}} = \Sigma n S_{\text{products}} - \Sigma m S_{\text{reactants}}$$

where n is the stoichiometric coefficient for each product and m is the stoichiometric coefficient for each reactant. For isothermal (constant temperature) changes, the change in entropy (ΔS) can be calculated by

$$\Delta S = \frac{q_{rev}}{T} = \frac{n\Delta H}{T}$$

where T is the temperature and q_{rev} is the heat involved in the reversible process, which can be calculated from ΔH, the enthalpy of the process, and n, the moles involved in the process. Since this equation measures entropy from dispersion of heat, it can be used to calculate the entropy of the surroundings as heat moves between the system and surroundings. Because ΔH refers to the system, the change in entropy of the surroundings is actually

$$\Delta S_{surroundings} = \frac{-n\Delta H_{rxn}}{T}$$

A spontaneous change occurs when $\Delta S_{universe} > 0$ (positive). Another, more convenient way to determine spontaneity of a process is to use **free-energy change (ΔG)**. Free energy is the amount of energy available to do useful work, and can be used as a criteria for a spontaneous change. Free-energy change can be calculated from

$$\Delta G = \Delta H - T\Delta S$$

where negative values of free-energy change (exergonic processes) represent spontaneous processes, positive values of free-energy change (endergonic) are nonspontaneous processes and if the value is zero, no net change occurs. In this case, where the forward and reverse processes are the same, the system is in equilibrium.

Worked Examples

Calculating Changes of Entropy in Chemical Reactions

Example 1.
What is the sign of ΔS for the following reactions?

a. $2\ C_8H_{18}(\ell) + 25\ O_2(g) \rightarrow 16\ CO_2(g) + 18\ H_2O(g)$

b. $NaCl(s) \rightarrow Na^+(aq) + Cl^-(aq)$

c. $N_2(g) + 3\ H_2(g) \rightarrow 2\ NH_3(g)$

Collect and Organize We are asked to determine the sign of the change in entropy for the given reactions.

Analyze To qualitatively predict the sign on ΔS for a chemical reaction, consider the following (listed in order of relative importance).

• physical state	gas > aqueous > liquid > solid
• mixtures	mixtures > pure substances
• number of moles of gas	more moles > fewer moles
• bonds in a molecule	more bonds > fewer bonds
	single bonds > double bonds > triple bonds
	less rigid > more rigid

Solve

a. In this example a liquid and a gas become gases. There is also an increase in number of moles (27 to 34). Both of these indicate an increase in energy dispersion or $+\Delta S$.

b. One mole of solid makes 2 moles of a mixture. This is also an increase in energy dispersion and $+\Delta S$.

c. In this case the physical state of each material is a gas, so physical state is not a criterion that can be used. However, the reaction takes 4 moles of gas and creates 2. It also goes from a mixture (two substances) to a pure substance. Therefore, energy dispersion decreases and ΔS is negative.

Think About It Entropy is a measure of the dispersion of energy in a system, which can be related to the physical state. For reactions, we need to compare the entropy of the reactants and the products.

Example 2.
Calculate the value of $\Delta S°$ for the following reactions. Compare to the predictions in Example 1.

a. $2\ C_8H_{18}(\ell) + 25\ O_2(g) \rightarrow 16\ CO_2(g) + 18\ H_2O(g)$

b. $NaCl(s) \rightarrow Na^+(aq) + Cl^-(aq)$

c. $N_2(g) + 3\ H_2(g) \rightarrow 2\ NH_3(g)$

Collect and Organize We are given stoichiometric equations for three reactions and are asked to determine the value of $\Delta S°$ for each.

Analyze The mathematical procedure for calculating ΔS is given as:

$$\Delta S = \Sigma nS_{products} - \Sigma mS_{reactants}$$

Recall that Σ is the summation sign, n is the stoichiometric coefficient of the product, and m is the stoichiometric coefficient of the reactant. This equation assumes that all reactants and products are under the same thermodynamic conditions.

Solve

a. Using either the table within the chapter or the appendix at the end of the text to look up the $S°$ values for each reactant and product, they are $C_8H_{18} = 463.7$, $O_2 = 205.0$, $CO_2 = 213.6$, and $H_2O = 188.7$, all with units J/mol•K. *Note:* Normally two physical states for water are listed. Make sure you pick the one consistent with the reaction. Using the above equation,

$\Delta S = [16(213.6) + 18(188.7)] - [2(483.7) + 25(205.0)] = [\underline{3417}.6 + \underline{3396}.6] - [\underline{967.4} + \underline{5125}.0] = [\underline{6814}.2] - [\underline{6092}.4] = +\underline{721}.8$ J/mol•K

Following significant figures in these problems can be tricky because both multiplication and addition are involved. Remember that it is best to keep as many units as possible and to round at the final answer. In this example, the stoichiometric coefficients are exact and therefore have infinite significant figures. Significant figures are counted in multiplication; decimal places, in addition. The digits that are significant are underlined in each step of the preceding calculation. Therefore the final answer should be $\Delta S = 722$ J/mol•K.

b. The values of $S°$ for each component of the reaction are NaCl = 72.1, Na^+ = 59.0, and Cl^- = 56.5, units being J/mol•K. Substituting into the same equation,

$$\Delta S = [59.0 + 56.5] - [72.1] = [115.5] - [72.1]$$
$$= +43.4 \text{ J/mol•K}$$

Since no multiplication is involved, use the least amount of decimal places to determine the appropriate significant figures.

c. The values of $S°$ for each component of the reaction are N_2 = 191.5, H_2 = 130.6, and NH_3 = 192.3, units being J/mol•K. Substituting into the same equation,

$$\Delta S = [2(192.3)] - [191.5 + 3(130.6)] = -198.7 \text{ J/mol•K}$$

Think About It If there is a symbol °, standard conditions of 25°C (298 K) and 100 Pa (about 1 atm) are indicated. Since tables only list values at standard conditions, they are often assumed, even if not explicitly stated.

Gibbs Free Energy

Example 1.

Without doing any calculations, estimate under what temperature conditions that the following reactions are spontaneous.

a. $NH_4Cl(s) + OH^-(aq) \rightarrow NH_3(g) + H_2O(\ell) + Cl^-(aq)$ (endothermic)

b. $2\ NO(g) + O_2(g) \rightarrow 2\ NO_2(g)$ (exothermic)

c. $CaCO_3(s) \rightarrow CaO(s) + O_2(g)$ (endothermic)

Collect and Organize We are given three reactions and their corresponding signs of enthalpy (endothermic reactions have and $\Delta H > 0$ and exothermic reactions have $\Delta H < 0$) and are asked to determine their spontaneity.

Analyze Spontaneous reactions depend on two forces: enthalpy (ΔH) and entropy (ΔS). If both of these forces decrease the energy of the system ($-\Delta H$ and $+\Delta S$), the reaction will be spontaneous. If both of these forces increase the energy of the system ($+\Delta H$ and $-\Delta S$), the reaction will be nonspontaneous. If one force increases the energy of the system and the other decreases it, spontaneity will be determined by the larger force. Since entropy is linked to temperature, higher temperatures increase the effect of entropy.

Solve

a. Because the reaction is endothermic, ΔH is positive. Entropy increases as the solid becomes a gas and a liquid, so ΔS is positive; since ΔS favors a spontaneous reaction but ΔH does not, to maximize ΔS, this reaction is spontaneous at high temperatures.

b. Because this reaction is exothermic, the sign on ΔH is negative. Going from more moles of gas to fewer decreases entropy, so ΔS is negative. The enthalpy encourages spontaneity; entropy discourages it. To minimize entropy, this reaction will be spontaneous at low temperatures.

c. Because this reaction is endothermic, ΔH is positive. A solid turning into a gas results in an increase in entropy, so ΔS is positive. While ΔS is favorable for a spontaneous reaction, ΔH is not. To maximize the effect of ΔS, high temperatures are required. Therefore, this reaction will be spontaneous at high temperatures.

Think About It Remember that both ΔH and ΔS are needed to make any statement about ΔG, and the effect may be temperature dependent. Also remember that "low" and "high" are relative temperatures. In some situations, 500 K is high; for others it is low.

Example 2.

What is the $\Delta G°$ for the following reaction? (Calculate ΔH and ΔS from values in the appendix.) Is it spontaneous?

$$NH_4Cl(s) + OH^-(aq) \rightarrow NH_3(g) + H_2O(\ell) + Cl^-(aq)$$

Collect and Organize We are given a stoichiometric equation and are asked to determine $\Delta G°$ for the reaction after calculating values of $\Delta H°$ and $\Delta S°$ for the reaction using tabulated values of $\Delta H_f°$ and $\Delta S°$ for the reactants and products.

Analyze The following relations are used in the calculation of ΔG:

$$\Delta H = \Sigma n \Delta H_{products} - \Sigma m \Delta H_{reactants}$$

$$\Delta S = \Sigma n S_{products} - \Sigma m S_{reactants}$$

$$\Delta G = \Delta H - T\Delta S$$

SOLVE Using the appendix to look up values of $S°$ and ΔH_f° for each reactant and product,

$$\Delta H° = [-46.1 + -285.8 + -167.2] - [-314.4 + -230.0]$$
$$= +45.3 \text{ kJ/mol}$$

$$\Delta S° = [192.3 + 69.9 + 56.5] - [94.6 + -10.8]$$
$$= +234.9 \text{ J/mol}\bullet\text{K}$$

Because ΔH and ΔS must have the same units before they can be added,

$$\Delta H° = +45.3 \text{ kJ/mol} \times \left(\frac{1000 \text{ J}}{1 \text{ kJ}}\right) = +45{,}300 \text{ J/mol}$$

$$\Delta G° = +45{,}300 \text{ J/mol} - (298 \text{ K})(+234.9 \text{ J/mol}\bullet\text{K})$$
$$= -2.47 \times 10^4 \text{ J/mol}$$

THINK ABOUT IT The negative sign indicates that the reaction is spontaneous at this temperature.

EXAMPLE 3.
What is the $\Delta G°$ at 25°C and at what temperatures is the following reaction spontaneous?

$$H_2O(\ell) + SO_3(g) \rightarrow H_2SO_4(aq)$$

COLLECT AND ORGANIZE We are given a reaction and are asked to determine $\Delta G°$ and the temperatures at which the given reaction is spontaneous.

ANALYZE We will use tabulated values of ΔH_f° and $S°$ for H_2O, SO_3, and H_2SO_4, and the equations for ΔH and ΔS given above in the expression for the Gibbs free energy: $\Delta G = \Delta H - T\Delta S$.

SOLVE We find that:

$$\Delta H = [-909.3] - [-285.8 + -395.7]$$
$$= [-909.3] - [-681.5] = -227.8 \text{ kJ/mol}$$

$$\Delta S = [20.0] - [69.9 + 256.6] = [20.0] - [257.5]$$
$$= -237.5 \text{ J/mol}\bullet\text{K}$$

$$\Delta G = -227.8 - (298)(-0.2375) = -227.8 + 70.775$$
$$= -157.0 \text{ kJ/mol}$$

To determine the temperatures at which this reaction is spontaneous, set $\Delta G = 0$, and solve for temperature. Not only is $\Delta G = 0$ at equilibrium, but it is the turning point from spontaneous to nonspontaneous reactions.

$$\Delta G = \Delta H - T\Delta S$$

$$0 = -227.8 - T(-0.2375)$$

$$227.8 = 0.2375T$$

$$959.2 \text{ K} = T$$

However, at this point, you have determined the temperature at equilibrium, and the question asks at which temperature the reaction is spontaneous. There are two ways to decide whether it is spontaneous at higher or lower temperatures than the calculated one.

The calculation at 298 K indicates that it is spontaneous at that temperature. Since it is lower than the equilibrium temperature, the reaction is spontaneous at $T < 959.2$ K. Alternately, the ΔS value is negative, which does not favor spontaneity. Consequently, this reaction is spontaneous at low temperatures. Putting that together with the equilibrium temperature, the reaction must be spontaneous at $T < 959.2$ K.

THINK ABOUT IT The most common mistakes using this equation are with units. Typically, the units of ΔH are kJ/mol and of ΔS are J/mol•K. To add these values together, the energy units must be the same. Thus either ΔH must be converted to joules or ΔS must be converted to kJ. (Don't do both. You'll end up as you started!)

EXAMPLE 4.
Using ΔG_f° values, determine ΔG_{rxn}° for the gaseous reaction $2NO + O_2 \rightarrow 2NO_2$. Values of $\Delta G°$ of NO_2 = +51.3 kJ/mol, O_2 = +0 kJ/mol, and NO = +86.6 kJ/mol.

COLLECT AND ORGANIZE
We are asked to determine the change in Gibbs energy for the given reaction, ΔG_{rxn}°, from tabulated values of the Gibbs energy of formation of the products and reactants, ΔG_f°.

ANALYZE We use

$$\Delta G_{rxn}^\circ = \Sigma n\Delta G_{f,\text{products}}^\circ - \Sigma m\Delta G_{f,\text{reactants}}^\circ$$

SOLVE This equation is solved as before:

$$\Delta G_{rxn}^\circ = [2(51.3)] - [2(86.6) + 0] = [102.6] - [173.2]$$
$$= -70.6 \text{ kJ/mol}$$

Significant figures for each value have been underlined except the zero for oxygen, which has infinite significant figures. Like ΔH_f°, ΔG_f° for an element at standard state is defined to be zero.

$$\Delta G° = [2(\underline{51.3})] - [2(\underline{86.6}) + 0] = [\underline{102}.6] - [\underline{173}.2]$$
$$= -\underline{70}.6 \text{ kJ/mol} = -71 \text{ kJ/mol}$$

THINK ABOUT IT Another way of calculating ΔG is from the ΔG_f° values. These are listed in the same appendix as $\Delta H°$ and $S°$. It is an equally valid way of calculating ΔG. Choosing this equation over $\Delta G = \Delta H - T\Delta S$ is appropriate if ΔG_f° values are given. Note that these values refer to formation reactions. You can review formation reactions in Chapter 5.

Self-Test

Key Vocabulary Completion Questions

__________ 1. Reactions with a positive value of ΔG

__________ 2. Measure of energy dispersion

__________ 3. ΔG

__________ 4. Entropy of a perfect crystal at 0 K

__________ 5. Energy is neither created nor destroyed.

__________ 6. Sign of ΔG for a reaction with $+\Delta H$ and $-\Delta S$

__________ 7. Sign of $\Delta S_{universe}$ for a spontaneous reaction

__________ 8. Motion where bond length gets longer or shorter

__________ 9. Sign of ΔG in a spontaneous process

__________ 10. $-\Delta H$

__________ 11. Man whose name is associated with free energy

__________ 12. Process that proceeds without outside intervention

__________ 13. Distribution of molecules in various energy states

__________ 14. Physical state with the highest entropy

__________ 15. Dispersion of the energy of one mole of substance at 298 K and 1 atm

Multiple-Choice Questions

1. According to the second law of thermodynamics, the entropy of the universe
   a. stays constant
   b. increases
   c. decreases
   d. gets closer to zero

2. What is the value of ΔS_{rxn} for the dissolution of sodium sulfate? ($S°_{salt}$ = 149.6 J/mol•K, $S°_{Na}$ = 59.0 J/mol•K, $S°_{sulfate}$ = 20 J/mol•K)
   a. +12 J/mol•K
   b. –12 J/mol•K
   c. +91 J/mol•K
   d. –91 J/mol•K

3. Which of the following has the greatest entropy?
   a. $H_2(g)$
   b. $Ca(s)$
   c. $CH_4(g)$
   d. $NaCl(aq)$

4. Which of the following has the greatest entropy?
   a. $C_6H_6(\ell)$
   b. $NH_4Cl(s)$
   c. $N_2(g)$
   d. $F_2(g)$

5. In the reaction $CaCO_3(s) \rightarrow CaO(s) + O_2(g)$, entropy
   a. increases
   b. decreases
   c. stays the same
   d. more information is needed to predict

6. In the gaseous reaction $2NO_2 \rightarrow N_2O_4$, the sign on ΔS is
   a. positive
   b. negative
   c. zero
   d. impossible to predict

7. Why does benzene (C_6H_6) have a higher entropy than cyclohexane (C_6H_{12})?
   a. cyclohexane has more bonds
   b. benzene is a more rigid molecule
   c. benzene has more double bonds
   d. benzene actually has a lower entropy

8. Is the exothermic, gaseous reaction of $2\ H_2 + O_2 \rightarrow 2\ H_2O$
   a. spontaneous at all temperature
   b. spontaneous at low temperatures
   c. spontaneous at high temperatures
   d. nonspontaneous

9. Is this endothermic reaction $CdSO_4(s) \rightarrow Cd^{2+}(aq) + SO_4^{2-}(aq)$
   a. spontaneous at all temperature
   b. spontaneous at low temperatures
   c. spontaneous at high temperatures
   d. nonspontaneous

10. At what temperature will a system where ΔH = +9.4 kJ/mol and ΔS = +83.9 J/g•mol be at equilibrium?
    a. 112 K
    b. 0.112 K
    c. 8.9 K
    d. 8900 K

11. If a reaction is spontaneous at 298 K, with a positive value of ΔS, then the reaction is
    a. spontaneous at all temperatures
    b. spontaneous at all temperatures higher than 298 K
    c. nonspontaneous at some temperature higher than 298 K
    d. spontaneous only at that temperature

12. If a reaction is nonspontaneous at 298 K with a $-\Delta H$, then the reaction is
    a. nonspontaneous at all temperatures
    b. nonspontaneous at all temperatures higher than 298 K
    c. spontaneous at some temperature higher than 298 K
    d. nonspontaneous only at that temperature

13. The ΔG°_f of $N_2(g)$ is
    a. positive
    b. negative
    c. zero
    d. more information is needed to determine

14. What is ΔG° for the reaction $4Fe + 3O_2 \rightarrow 2Fe_2O_3$.
    (ΔH°_{rxn} = −1648.4 kJ/mol, ΔS°_{rxn} = −549.4 J/mol•K)
    a. −1099.0 kJ/mol
    b. −1484.7 kJ/mol
    c. +1.621 × 10^5 kJ/mol
    d. +1.208 × 10^4 kJ/mol

15. What is the ΔG° for the gaseous reaction $N_2 + 3H_2 \rightarrow 2NH_3$? ($\Delta G^\circ_{f,NH_3}$ = −16.5 kJ/mol)?
    a. −16.5 kJ/mol
    b. +16.5 kJ/mol
    c. −33.0 kJ/mol
    d. more information is needed to determine

Additional Practice Problems

1. Calculate ΔH°, ΔS°, and ΔG° for the reaction $LiOH(s) + HCl(aq) \rightarrow Li^+(aq) + Cl^-(aq) + H_2O(\ell)$.

	LiOH	HCl	Li^+	Cl^-	H_2O
ΔH°_f (kJ/mol)	−484.9	−92.3	−278.5	−167.2	−285.8
S° (J/mol•K)	42.8	186.8	13	56.5	69.9

2. At what temperature will the following gaseous reaction be spontaneous?

$$2NO_2 \rightarrow N_2O_4$$

	NO_2	N_2O_4
ΔH°_f (kJ/mol)	33.2	9.16
S° (J/mol•K)	240.0	304.2

3. What is the ΔG°_{rxn} for $Ag^+(aq) + Cl^-(aq) \rightarrow AgCl(s)$?
   $\Delta G^\circ_{f,Ag}$ = 77.1 kJ/mol, $\Delta G^\circ_{f,Cl}$ = −131.3 kJ/mol; $\Delta G^\circ_{f,AgCl}$ = −109.8 kJ/mol

4. Predict the signs on ΔS and ΔG for the following reactions.
   a. $Zn(s) + 2H^+(aq) \rightarrow Zn^{2+}(aq) + H_2(g)$ (exothermic)
   b. $Br_2(\ell) \rightarrow Br_2(g)$ (endothermic)
   c. $CH_2CH_2(g) + H_2(g) \rightarrow CH_3CH_3(g)$ (exothermic)

5. What is the change in entropy (ΔS) when 100.0 g of gold is melted? (The melting point of gold is 1064°C, and its heat of fusion is 13.1 kJ/mol)

Self-Test Answer Key

Key Vocabulary Completion Answers

1. endergonic (or nonspontaneous)
2. entropy
3. free energy
4. zero
5. first law of thermodynamics
6. positive
7. positive
8. vibration
9. negative
10. exothermic
11. Gibbs
12. spontaneous
13. microstate
14. gas
15. standard molar entropy

MULTIPLE-CHOICE ANSWERS

For further information about the topics these questions address, you should consult the text. The appropriate section of the text is listed in parentheses after the explanation of the answer.

1. b. The second law states that the entropy of the universe increases in a spontaneous reaction. (13.1)

2. b. The dissolution of sodium sulfate is $Na_2SO_4 \rightarrow 2\ Na^+ + SO_4^{2-}$. Using Equation 13.5, $\Delta S = \Sigma nS - \Sigma mS = [2(59.0) + 20] - [149.6] = 138 - 149.6 = -11.6$ J/mol•K. The fact that this turns out to be negative is surprising, but the order is in the organization of water around the ions. (13.2)

3. c. Gases have much more entropy than any other physical state. Molecules with more bonds have more internal motion and therefore, higher entropy. (13.1)

4. d. Gases have more translational motion than liquids and solids. Flourine is more random than nitrogen, because the flourine atoms are connected by a single bond that is less rigid than the triple bond that connects the two nitrogen atoms. (13.1)

5. a. A gas is formed from a solid, making a more random system. Therefore, entropy increases. (13.1)

6. b. Two molecules combine to make one. Therefore, the system becomes less random and ΔS is negative. (13.1)

7. d. Statements a through c are true, but each leads to a higher entropy for cyclohexane than for benzene. (13.1)

8. b. An exothermic reaction has a $-\Delta H$, and going from 3 moles to 2 decreases randomness ($-\Delta S$). In $\Delta G = \Delta H - T\Delta S$, for ΔG to be negative, the ΔS term should be minimized by working at low temperatures. (13.3)

9. c. The dissolving solid increases in entropy ($+\Delta S$). The endothermic reaction indicates a $+\Delta H$. In this case, entropy should be maximized by increasing temperature. (13.3)

10. a. At equilibrium, $\Delta G = 0$. Using the equation $\Delta G = \Delta H - T\Delta S$ requires that ΔH and ΔS have the same energy units. $\Delta H = +9.4$ kJ/mol $= +9400$ J/mol. So $0 = 9400 - T(83.9)$; $9400 = 83.9T$; so $112 = T$. (13.3)

11. b. If ΔS is positive, then increasing temperature will only increase spontaneity. If it is exothermic, the reaction would be spontaneous at all temperatures. However, that information is not given. (13.3)

12. b. A negative value of ΔH tends toward a spontaneous reaction. For a reaction to be nonspontaneous with a negative ΔH, ΔS must be negative as well. Values with negative ΔS values and negative ΔH values are spontaneous at low temperatures. (13.3)

13. c. A formation reaction is when a substance is formed from its elements in their standard state. Since $N_2(g)$ is an element in its standard state, no reaction occurs and ΔG_f° is zero. (13.3)

14. b. To get ΔG from ΔH and ΔS, use the equation $\Delta G = \Delta H - T\Delta S$. The temperature will be 298 K (from the superscript indicating standard conditions). To add ΔH and ΔS, the units must be the same. Converting ΔS to kJ, it is $-0.549.4$ kJ/mol•K. So $\Delta G = -1648.4 - (298)(-0.5494) = -1484.7$ kJ/mol•K. (13.3)

15. c. Like ΔH_f°, ΔG_f° for elements in their standard state is zero. Nitrogen and hydrogen are in their standard states. The ΔG°_{rxn} can be determined from $\Delta G = \Sigma n\Delta G_{products} - \Sigma m\Delta G_{reactants}$. Since the value of ΔG is zero for the reactants, it is only $n\Delta G$ for products. $\Delta G = 2(-16.5) = -33.0$ kJ/mol. (13.3)

ADDITIONAL PRACTICE PROBLEM ANSWERS

1. $\Delta H^\circ = [-278.5 + -167.2 + -285.8] - [-484.9 + -92.3] = -154.3$ kJ/mol

   $\Delta S^\circ = [13 + 56.5 + 69.9] - [42.8 + 186.8]$

   $= -90 \text{ J/mol•K} \times \left(\dfrac{1 \text{ kJ}}{1000 \text{ J}}\right) = -0.090 \text{ kJ/mol•K}$

   $\Delta G = \Delta H - T\Delta S = (-154.3 \text{ kJ/mol}) - (298 \text{ K})(-0.090 \text{ kJ/mol•K}) = -1.3 \times 10^2 \text{ kJ/mol}$

2. $\Delta H = [9.16] - [2(33.2)] = -57.2$ kJ/mol

   $\Delta S = [304.2] - [2(240.0)]$

   $= -175.8 \text{ J/mol•K} \times \left(\dfrac{1 \text{ kJ}}{1000 \text{ J}}\right)$

   $= -0.1758$ kJ/mol•K

   $\Delta G = \Delta H - T\Delta S$

   $0 = -57.2 - T(0.1758 \text{ kJ/mol})$

   $57.2 = 0.1758T$

   $325 = T$

   $325 \text{ K} > T$

3. $\Delta G° = [-109.8] - [77.1 - 131.3] = (-55.6 \text{ kJ/mol})$

4. a. The creation of a gas from a solid means the system increases in entropy. Therefore, ΔS is positive. Since enthalpy and entropy both favor spontaneity, this reaction is spontaneous at all temperatures.

   b. The creation of a gas from a liquid increases entropy. Therefore, ΔS is positive. Entropy favors spontaneity, but enthalpy does not. For spontaneity, this system must maximize entropy. Therefore, this system is spontaneous at high temperatures.

   c. Going from 2 moles of gas to 1 demonstrates a decrease in entropy and a negative ΔS. In this case, enthalpy favors spontaneity and entropy does not. The system is spontaneous when entropy is minimized. Therefore, the system is spontaneous at low temperatures.

5. $100.0 \text{ g} \times \left(\frac{1 \text{ mol}}{196.97 \text{ g}}\right) = 0.5077 \text{ mol Au}$

$$\Delta S = \frac{q_{rev}}{T} = \frac{n\Delta H}{T} = (0.5077 \text{ mol})\frac{(13.1 \text{ kJ/mol})}{1337 \text{ K}}$$

$$= 0.00497 \text{ kJ/K} = 4.97 \text{ J/K}$$

CHAPTER 14 Chemical Kinetics

OUTLINE

REVIEW

Chapter Overview

Kinetics is a term that relates to how fast a reaction occurs. The **rate of reaction** is measured as the change in concentration of a product or reactant ([X]) over a given time (t). The rate of reaction for reactants is negative, since reactants are disappearing, and positive for products, which are appearing. Rate can be measured as **average rate** using the equation

$$\text{rate of } X = \frac{\Delta[\text{X}]}{\Delta t} = \frac{[\text{X}]_{\text{final}} - [\text{X}]_{\text{initial}}}{t_{\text{final}} - t_{\text{initial}}}$$

Rate decreases over time. Therefore, **instantaneous rate**, the rate at any given time, is sometimes used. The instantaneous rate can be determined from a tangent line at the relevant instant of time on a graph of concentration versus time. The instantaneous rate at the start of the reaction (t = 0 s) is called the **initial rate**.

The rates for each product and reactant are related by the stoichiometry of the reaction. For example, a product with a stoichiometric coefficient of 2 appears twice as fast as a product with a stoichiometric coefficient of 1. The relationship between rates of each reactant (A and B) and each product (C and D) depends on the stoichiometric coefficient (lowercase of letter representing products or reactants) according to

$$-\left(\frac{1}{a}\right)\left(\frac{\Delta[\text{A}]}{\Delta t}\right) = -\left(\frac{1}{b}\right)\left(\frac{\Delta[\text{B}]}{\Delta t}\right) = +\left(\frac{1}{c}\right)\left(\frac{\Delta[\text{C}]}{\Delta t}\right) = +\left(\frac{1}{d}\right)\left(\frac{\Delta[\text{D}]}{\Delta t}\right)$$

The relationship between concentration and rate is called the **rate law**. The rate is proportional to the product of the concentration of reactants raised to some exponent. It has the form

$$\text{rate} = k[\text{A}]^m[\text{B}]^n$$

The proportionality constant (k) of this equation is called the **rate constant**. The exponents on the reactant concentration are called the **order**. With the form given, m is the order in A and n is the order in B. The sum of the exponents ($m + n$) is called the **overall order**. The order of the reaction is normally an integer or simple fraction. The order is *not* determined from the stoichiometric coefficients of the reaction.

The rate law can be integrated to get a relationship between time (t) and concentration. For a first-order reaction with a single reactant (rate = k[X]), the **integrated rate law** is

$$\ln[\text{X}] = -kt + \ln[\text{X}]_0$$

where $[\text{X}]_0$ is the initial concentration of X. The integrated rate law for a second-order reaction with a single reactant (rate = $k[\text{X}]^2$) is

$$\frac{1}{[\text{X}]} = kt + \frac{1}{[\text{X}]_0}$$

A reaction that is first-order in two reactants (rate = k[X][Y]) can be expressed as a **pseudo-first-order reaction** if the concentration of one reactant is significantly greater than that of the other. For example, if [Y] is much greater than [X], the change [Y] is negligible. Thus [Y] is essentially constant and the rate law can be expressed as

$$\text{rate} = k'[\text{X}], \text{ where } k' = k[\text{Y}] \quad (14.18)$$

Thus first-order relationships can be used. It is also possible to have a zero-order reaction (rate = k). For zero-order reactions, the integrated rate law is

$$[\text{X}] = -kt + [\text{X}]_0$$

Although there are many other forms of the rate law, those integrated rate laws are too complex to consider at this point.

Another way to express the rate of reaction is with the half-life. **Half-life** is the time required for the reactant concentration to decrease to half its initial value ($[\text{X}] = \frac{1}{2}[\text{X}]_0$). The integrated rate laws can be used to relate the half-life ($t_{1/2}$) to rate constant (k) and initial concentration ($[\text{X}]_0$). For a first-order reaction,

$$t_{1/2} = \frac{\ln 2}{k} = \frac{0.693}{k} \quad (14.12)$$

A first-order reaction is not dependent on concentration of reactant. All nuclear reactions are first-order reactions and the rates of nuclear reactions are commonly designated by the half-life.

The half-life of a second-order reaction is

$$t_{1/2} = \frac{1}{k[\text{X}]_0}$$

and that for a zero-order reaction is

$$t_{1/2} = \frac{[\text{X}]_0}{k}$$

The rate law is determined experimentally, rather than from the chemical reaction. This is because the overall chemical reaction does not necessarily reflect the way in which the reaction occurs. A **mechanism** is the step-by-step sequence by which a chemical reaction occurs. Each of these elementary steps occurs at a specific rate. The overall rate depends on the slowest elementary step of the mechanism. Thus the rate law is determined by the slowest, **rate-determining**, elementary step rather than by the overall reaction.

Reactions occur when bonds are broken and formed. The substance formed during this process, as bonds are breaking and forming, is called an **activated complex**. In some steps, an unstable substance that later undergoes further reaction is formed. This product of one elementary step that is used up in a subsequent step is called an **intermediate**.

Bonds breaking and forming usually occur as a result of a collision. The number of molecules participating in the collision is called the **molecularity** of the step. If only one molecule is involved, the step is **unimolecular**. If two molecules collide, the step is **bimolecular**. In the unlikely event that three molecules collide simultaneously, the step is called **termolecular**.

For bond breakage to occur in the collision, the molecules must have sufficient kinetic energy. The energy required to get a reaction going is called the **activation energy (E_a)**. At higher temperatures, more molecules will have sufficient energy to overcome the activation energy; therefore, the reaction will occur at a faster rate. An **energy profile** diagrams the changes in energy (as ΔH or ΔG) versus the progress of the reaction (reactants to products). The activation energy appears as a hill between the reactants and the products. The substance at the top of the hill is called the **transition state**. The energy difference between the reactants and the transition state is the activation energy. The energy difference between products and reactants is the ΔH (or ΔG) for the reaction.

The relationship between temperature (T) and rate constant (k) is described by the **Arrhenius equation**

$$k = Ae^{-E_a/RT}$$

where E_a is the activation energy, R is the gas constant (8.314 J/mol•K), and A is the frequency factor. The frequency factor is related to how successful the collisions between molecules are. For a collision to result in a reaction, not only must there be sufficient kinetic energy for the bonds to break, but the molecules must collide in the proper orientation. The frequency factor takes orientation into account.

One way to increase the rate of a reaction is to add a catalyst. A **catalyst** increases the rate of reaction without itself being consumed. It does this by lowering the activation energy, often by directing the orientation of the colliding molecules. Catalysts are categorized as **homogeneous**, being in the same phase as the reactants and products, or **heterogeneous**, in a different phase from the reactants and products.

Worked Examples

Reaction Rates

Example 1.
What is the average rate based on the following data?

a. The concentration of reactant changes from 0.50 M to 0.01 M in 1.5 minutes.

b. The concentration of product is 0.056 M after 75 seconds.

c. The $[\text{X}]_0$ = 1.00 M at t = 0 and [X] = 0.34 M at t = 50 s.

COLLECT AND ORGANIZE We are given information regarding concentration at different times and are asked to determine the average rate.

ANALYZE The average rate is determined from initial and final concentration and time data. The tradition is to subtract the final from the initial value.

SOLVE

a. Initial concentration = 0.50 *M*, final concentration = 0.01 *M*, so $\Delta[X] = 0.01 - 0.50 = -0.49\ M$. The time change is 1.5 minutes (or 90 s). So the rate can be either

$$\frac{-0.49\ M}{1.5\ \text{min}} = -0.33\ M/\text{min or } \frac{-0.49\ M}{90\ \text{s}} = 5.4 \times 10^{-3}\ M/\text{s}.$$

b. It is implied that there was no product to start, so the initial concentration = 0 *M*. Thus $\Delta[X] = 0.056\ M$. The change in time is 75 s. So rate = $7.5 \times 10^{-4}\ M/\text{s}$.

c. $[X]_0$ represents initial concentration. Therefore,

$$\text{average rate} = \frac{(0.34 - 1.00)}{(50 - 0)} = -0.013\ M/\text{s}.$$

THINK ABOUT IT The rate of reaction for reactants is negative, since reactants are disappearing, and positive for products, which are appearing.

EXAMPLE 2.
What is the instantaneous rate at 1 minute and 3 minutes, based on the figure below?

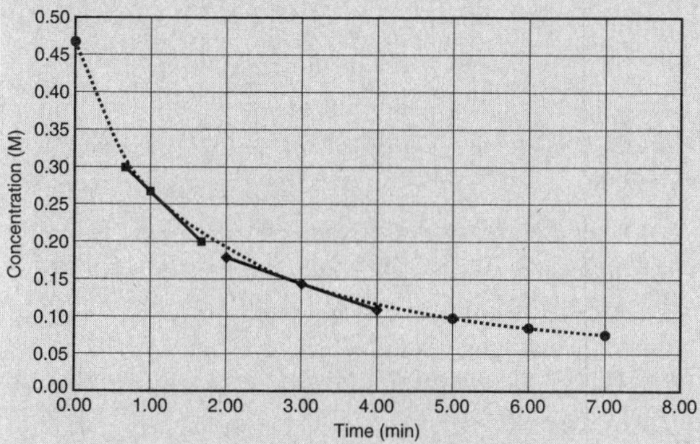

COLLECT AND ORGANIZE We are asked to determine the instantaneous rate at 1 minute and 3 minutes based on the figure above.

ANALYZE To determine the instantaneous rate, draw a line tangent to the curve at the relevant time value. (*Note:* Appropriate tangent lines are shown in the figure.) The instantaneous rate is given by the slope of the tangent line determined using $m = \frac{(y_1 - y_2)}{(x_1 - x_2)}$.

SOLVE For 1 minute your value should be about 0.1 *M*/min. Your tangent line and points chosen to determine the slope may be slightly different. The process is not very exact. For 3 minutes your value should be about 0.03 *M*/min.

THINK ABOUT IT In a graph of concentration versus time, rate is the slope. However, this graph is not usually linear. Consequently, to determine rate from a graph of concentration versus time, a tangent line at the relevant time is used.

EXAMPLE 3.
For the reaction, $2\ H_2 + O_2 \rightarrow 2\ H_2O$, if the rate of disappearance of hydrogen gas is –0.012 *M*/s, what is the rate for the disappearance of oxygen and the appearance of water?

COLLECT AND ORGANIZE We are asked to determine the rate of disappearance of oxygen and the rate of appearance of water given the rate of disappearance of hydrogen gas and the stoichiometric equation relating them.

ANALYZE The rates for each product and reactant are related by the stoichiometry of the reaction. For example, a product with a stoichiometric coefficient of 2 appears twice as fast as a product with a stoichiometric coefficient of 1. The relationship between rates of each reactant (*A* and *B*) and products (*C* and *D*) depends on the stoichiometric coefficient (lowercase of letter representing products or reactants) according to

$$-\left(\frac{1}{a}\right)\left(\frac{\Delta[A]}{\Delta t}\right) = -\left(\frac{1}{b}\right)\left(\frac{\Delta[B]}{\Delta t}\right) = +\left(\frac{1}{c}\right)\left(\frac{\Delta[C]}{\Delta t}\right) = +\left(\frac{1}{d}\right)\left(\frac{\Delta[D]}{\Delta t}\right)$$

SOLVE Using the above equation,

$$\left(-\frac{1}{2}\right)(\text{rate}_{h_2}) = \left(-\frac{1}{1}\right)(\text{rate}_{o_2}) = \left(\frac{1}{2}\right)(\text{rate}_{h_2o}).$$

$$\left(-\frac{1}{2}\right)(-0.012\ M/\text{s}) = (-\text{rate}_{o_2})$$

So the rate of O_2 disappearance = –0.0060 *M*/s and the rate of H_2O appearance = 0.012 *M*/s.

THINK ABOUT IT The greatest difficulty in using this equation is remembering that the entire term $\left(\frac{\Delta[A]}{\Delta t}\right)$ refers to the rate. It is tempting to want to split this into concentration and time terms, but this is not appropriate.

EXAMPLE 4.
For the reaction $P_4 + 6\ Cl_2 \rightarrow 4\ PCl_3$, if the rate of disappearance of chlorine gas is 0.237 *M*/s, what is the rate of the disappearance of phosphorus and the appearance of phosphorus trichloride?

COLLECT AND ORGANIZE We are asked to determine the rate of disappearance of phosphorus and the rate of appearance of phosphorus trichloride given the rate of disappearance of chlorine gas and the stoichiometric equation relating them.

ANALYZE Use the relationship:

$$-\left(\frac{1}{a}\right)\left(\frac{\Delta[A]}{\Delta t}\right) = -\left(\frac{1}{b}\right)\left(\frac{\Delta[B]}{\Delta t}\right) = +\left(\frac{1}{c}\right)\left(\frac{\Delta[C]}{\Delta t}\right) = +\left(\frac{1}{d}\right)\left(\frac{\Delta[D]}{\Delta t}\right)$$

SOLVE We have $-(\text{rate } P_4) = -\left(\frac{1}{6}\right)(\text{rate } Cl_2) = \left(\frac{1}{4}\right)(\text{rate } PCl_3)$.

Using the value given in the problem for the rate of Cl_2 disappearance, $-(\text{rate } P_4) = -\left(\frac{1}{6}\right)(-0.237\ M/\text{s})$. The rate of disappearance of P_4 is $-0.0395\ M/\text{s}$. Likewise the rate disappearance of PCl_3 is $0.158\ M/\text{s}$.

THINK ABOUT IT Since the rate given is for a reactant, it is (by definition) negative.

EFFECT OF CONCENTRATION ON REACTION RATE

EXAMPLE 1.
What is the order of each reactant and overall order for the following reactions?

a. $2\ H_2 + O_2 \rightarrow 2\ H_2O$ (rate = $k[H_2][O_2]$)

b. $2\ P + 3\ Cl_2 \rightarrow 2\ PCl_3$ (rate = $k[Cl_2]^2$)

c. $HCl + NH_3 \rightarrow NH_4Cl$ (rate = $k[HCl][NH_3]^2$)

COLLECT AND ORGANIZE We are asked to determine the order of each reactant and the overall reaction for the given reactions given the rate law for each.

ANALYZE The relationship between concentration and rate is called the rate law. The rate is proportional to the product of the concentration of reactants raised to some exponent. It has the form: rate = $k[A]^m[B]^n$. The exponents on the reactant concentrations are called the order. With the form given, m is the order in A and n is the order in B. The sum of the exponents $(m + n)$ is called the overall order. The order of the reaction is normally an integer or simple fraction.

SOLVE

a. The exponent on both hydrogen and oxygen is an understood 1. Therefore this reaction is first-order in hydrogen, first-order in oxygen, and second-order overall.

b. The exponent on the chlorine concentration is 2, so this reaction is second-order in chlorine. It is zero-order in phosphorus, which is why phosphorus does not appear in the rate law. The reaction is second-order overall (2 + 0 = 2).

c. The exponent on HCl is an understood 1, so this reaction is first-order in HCl. The exponent is 2 for ammonia, so the reaction is second-order in ammonia. The reaction is third-order overall (1 + 2 = 3).

THINK ABOUT IT The proportionality constant (k) of this equation is called the rate constant.

EXAMPLE 2.
What are the units of the rate constant for the following rate laws, assuming the rate is M/s?

a. rate = $k[A]^3$

b. rate = k

c. rate = [R][S]

COLLECT AND ORGANIZE We are asked to determine the units for the given rate constants.

ANALYZE The units on a rate constant depend on the overall order of the reaction. Since one molarity unit is used in the rate, the rate constant must cancel all the other molarity units. It must also account for the time factor portion of the rate. The units are whatever is appropriate for the equation. Any time units can be used, but seconds are usual. Taking this into consideration, the exponent on the molarity unit of the rate constant can be determined from the equation: exponent = 1 – order.

SOLVE

a. The order of the reaction is 3. Using the equation of 1 – order = exponent, we have $M^{1-\text{order}} \cdot \text{time}^{-1}$, $M^{1-3} \cdot \text{s}^{-1}$ or $M^{-2} \cdot \text{s}^{-1}$.

b. The order of the reaction is zero. One minus zero is 1, so the units are $M \cdot \text{s}^{-1}$.

c. The overall reaction order is 2. (One in each R and S.) Using 1 – order, the result is 1 – 2 = –1. So the units are $M^{-1} \cdot \text{s}^{-1}$.

THINK ABOUT IT A useful relationship is that the units on the rate constant are $M^{1-\text{order}} \cdot \text{time}^{-1}$.

EXAMPLE 3.
What is the order of the reaction for the following rate constants?

a. $k = 7.5\ \text{s}^{-1}$

b. $k = 66.2\ M^{-2} \cdot \text{min}^{-1}$

c. $k = 2.53 \times 102\ M^{-1} \cdot \text{s}^{-1}$

COLLECT AND ORGANIZE We are asked to determine the order of the reaction from the units of the given rate constants.

ANALYZE The equation of 1 – order = the exponent on concentration is used.

SOLVE

a. Molarity does not appear in the units. This implies that its exponent is zero, since anything to the zero power is one. 1 – order = 0; so order = 1. The rate law is first-order.

b. The exponent on molarity is –2. 1 – order = –2, so order = 3. The reaction is third-order.

c. The exponent on molarity is –1. 1 – order = –1, so order = 2. This is a second-order reaction.

THINK ABOUT IT
The order of a reaction can be deduced from the units given for its rate constant.

EXAMPLE 4.
For the reaction $A + B \rightarrow C$, what is the rate law and value of rate constant based on the following data.

TABLE 14.E4

Trial	[A] (*M*)	[B] (*M*)	Rate of C (*M*)
1	0.1	0.1	9.0×10^{-4}
2	0.3	0.1	8.1×10^{-3}
3	0.1	0.2	1.8×10^{-3}

COLLECT AND ORGANIZE From the data of rate as a function of reactant concentrations, we are asked to determine the rate law and the value of the rate constant.

ANALYZE The order of the reaction is determined by comparing the rates when reactant concentrations change. The most common method is to set up the rate law with values from one experiment and compare it (by dividing) to the values in the rate law from another experiment (where one reactant concentration has been changed). You may use any of the experiments to determine the value of *k*.

SOLVE First, determine the order for each reactant by comparing two trials where the concentration of only one reactant changes. Using those trials, write the appropriate form of the rate law for each trial. It helps to write the one with the higher concentration values first. The general form is rate = $k[\mathrm{A}]^m[\mathrm{B}]^n$.

Using trials 1 and 2 to determine the order with respect to *A*:

For trial 2 (higher [A]) $8.1 \times 10^{-3} = k[0.3]^m[0.1]^n$.

For trial 1 (lower [A]) $9.0 \times 10^{-4} = k[0.1]^m[0.1]^n$.

Dividing trial 2 by trial 1, the rate constant (*k*) and the concentration of *B* cancel. Therefore,

$$\left(\frac{8.1 \times 10^{-3}}{9.0 \times 10^{-4}}\right) = \frac{[0.3]^m}{[0.1]^m} = \left[\frac{0.3}{0.1}\right]^m$$

$$9 = 3^m$$

There are mathematical ways to solve this problem, but the easiest is by inspection: 3^2 is 9; therefore, $m = 2$.

To determine the order in *B*, consider two trials where the concentration of *B* is the only reactant concentration that changes. This would be trials 1 and 3. Since trial 3 has the higher concentration, it is easier if you write that first. The general form now is: rate = $k[\mathrm{A}]^2[\mathrm{B}]^n$.

For trial 3, $1.8 \times 10^{-3} = k[0.1]^2[0.2]^n$

For trial 1, $9.0 \times 10^{-4} = k[0.1]^2[0.1]^n$.

Dividing trial 3 by trial 2, the rate constant and concentration of *A* cancel. Therefore,

$$\left(\frac{1.8 \times 10^{-3}}{9.0 \times 10^{-4}}\right) = \frac{[0.2]^n}{[0.1]^n}$$

$$2 = 2^n$$

By inspection, $n = 1$. Therefore, the rate = $k[\mathrm{A}]^2[\mathrm{B}]$ (second-order in A, first-order in B, and third-order overall).

To determine the rate constant, use the values from *any* trial and solve for *k*. For trial 1, 9.0×10^{-4} *M*/s = $k[0.1\ M]^2[0.1\ M]$ and $k = 0.9\ M^{-2}\bullet\mathrm{s}^{-1}$. For trial 2, 8.1×10^{-3} *M*/s = $k[0.3\ M]^2[0.1\ M]$ and for trial 3, 1.8×10^{-3} *M*/s = $k[0.1\ M]^2[0.2\ M]$. Any *one* of the trials can be used to determine the rate constant. As you can see, all give the same answer.

THINK ABOUT IT When comparing (by dividing) the values in the rate law between experiments (where one reactant concentration has been changed), if you use the higher concentration values on the top, the numbers are easier to work with.

EXAMPLE 5.
What is the rate law and the value of the rate constant for the reaction $2\,AB_2 + C \rightarrow CB + BA$, based on the data below?

TABLE 14.E5

Trial	[AB_2] (*M*)	[C] (*M*)	Rate of CB (*M*/min)
1	1.0	1.0	1.4×10^{-7}
2	0.5	1.0	3.5×10^{-7}
3	1.0	2.0	5.5×10^{-7}

COLLECT AND ORGANIZE From the data of rate as a function of reactant concentrations, we are asked to determine the rate law and the value of the rate constant.

ANALYZE We will use the procedure outlined in the previous example.

SOLVE The general form of the rate law for this equation is: rate = $k[\mathrm{AB_2}]^m[\mathrm{C}]^n$. Comparing trials 1 and 2, the concentration of *C* stays the same and the concentration of AB_2 changes. Since the concentration is lower in trial 2, trial 1 should be divided by trial 2.

For trial 1, $(1.4 \times 10^{-7}) = k[1.0]^m[1.0]^n$

and for trial 2, $(3.5 \times 10^{-8}) = k[0.5]^m[1.0]^n$,

The rate constant and concentration of C cancel, so

$$2 = 2^m$$

and $m = 2$.

Comparing trials 1 and 3 (when AB_2 is constant). The concentration of C is higher in trial 3.

For trial 3, $5.5 \times 10^{-7} = k[1.0]^2[2.0]^n$

and for trial 1, $1.4 \times 10^{-7} = k[1.0]^2[1.0]^n$.

We have $3.93 = 2^n$.

Since the order is a whole number or simple fraction it makes sense to round to the obvious value of 4. Thus, $n = 2$. Therefore, we have rate $= k[AB_2]^2[C]^2$. To determine the value of k, use the data from any trial. Randomly using trial 1, $1.4 \times 10^{-7} = k[1.0]^2[1.0]^2$ and $k = 1.4 \times 10^{-7}\ M^{-3} \cdot \text{min}^{-1}$.

THINK ABOUT IT Note that the time unit was minutes for this rate constant. That was determined from the rate, given in units of M/min.

EXAMPLE 6.
Determine the rate law, including the value of the rate constant, from the following concentration-versus-time data.

TABLE 14.E1

Time (s)	Concentration (*M*)
0	0.500
10	0.197
20	0.078
30	0.031
40	0.012
50	0.005

COLLECT AND ORGANIZE We are asked to determine the rate law and the value of the rate constant from the given rate-versus-time data.

ANALYZE The rate laws can be integrated to get a relationship between time (t) and concentration. For zero-order reactions, the integrated rate law is $[X] = -kt + [X]_0$. For a first-order reaction with a single reactant (rate $= k[X]$), the integrated rate law is $\ln[X] = -kt + \ln[X]_0$, where $[X]_0$ is the initial concentration of X. The integrated rate law for a second-order reaction with a single reactant (rate $= k[X^2]$) is $1/[X] = kt + 1/[X]_0$. The order of the reaction is determined by graphing $[X]$ versus t, $\ln[X]$ versus t, and $1/[X]$ versus t. The order of the reaction corresponds to the order demonstrating the most linear plot.

SOLVE Determining the natural logarithm and reciprocal for each reactant concentration:

TABLE 14.E1S

Time (s)	Concentration (*M*)	ln [concentration]	1/concentration (1/*M*)
0	0.500	−0.69315	2
10	0.197	−1.62455	5.076142
20	0.078	−2.55105	12.82051
30	0.031	−3.47377	32.25806
40	0.012	−4.42285	83.33333
50	0.005	−5.29832	200

Figure A (below) is the graph of concentration versus time. A line that best fits the data is shown as dotted, and a curve through all points is shown as a solid line. Obviously, the curve is more appropriate to the data. Figure B is the graph of ln [concentration] versus time. A line that best fits the data is shown as dotted, a curve through all points is shown as a solid line. Obviously, the line fits this data well. Figure C (on page 195) is a graph of 1/concentration versus time. A curve better fits this data.

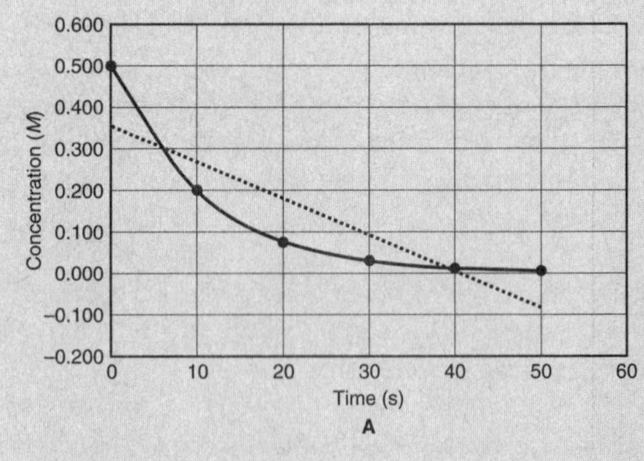

A

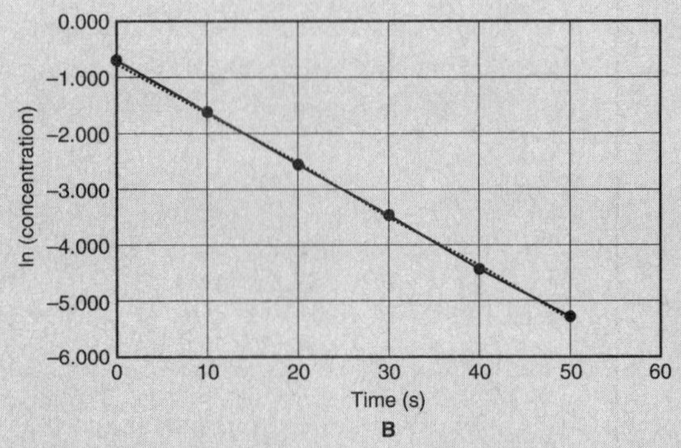

B

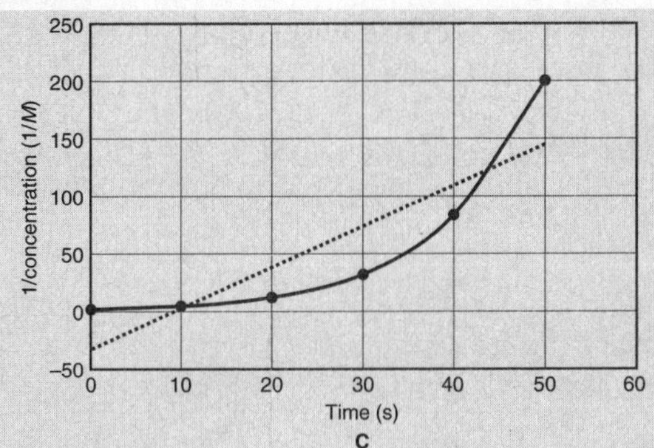

Since Figure B, graphing ln[A], has the line, this is a first-order reaction. Therefore, the rate law is rate = k[A]. From the graph the equation of the line is $y = (-0.0924)x + 0.7004$. The absolute value of the slope is the rate constant. Since the rate law is first-order and the time units are seconds, $k = 0.0924\ s^{-1}$.

Think About It If you use a spreadsheet, check the correlation coefficient (R^2). The closer the value is to 1, the more like a line the data are. It is better to compare all three graphs than to make a decision based on just one.

Example 7.

Determine the rate law, including the value of the rate constant, from the following concentration-versus-time data.

TABLE 14.E2

Time (s)	Concentration (*M*)
0	0.467
1	0.267
3	0.144
5	0.099
6	0.085
7	0.075

Collect and Organize We are asked to determine the rate law and the value of the rate constant from the given rate-versus-time data.

Analyze We will use the graphing procedure described in the previous example.

Solve

Determining the natural logarithm and reciprocal for each reactant concentration:

TABLE 14.E2S

Time (s)	Concentration (*M*)	ln [concentration]	1/concentration (1/*M*)
0	0.467	−0.76143	2.141328
1	0.267	−1.32051	3.745318
3	0.144	−1.93794	6.944444
5	0.099	−2.31264	10.10101
6	0.085	−2.4651	11.76471
7	0.075	−2.59027	13.33333

Graphs of this data are shown below: A for concentration versus time, B for ln [concentration] versus time, and C for 1/concentration versus time. The trend (straight) line is dotted; the best-fit curve is solid. Therefore, the best-fit line is the one with 1/[A] on the y-axis. So this is a second-order reaction. The rate law is rate = $k[A]^2$. The value of the rate constant is the absolute value of the slope of the line in C, which is 1.599 $M^{-1} \cdot min^{-1}$.

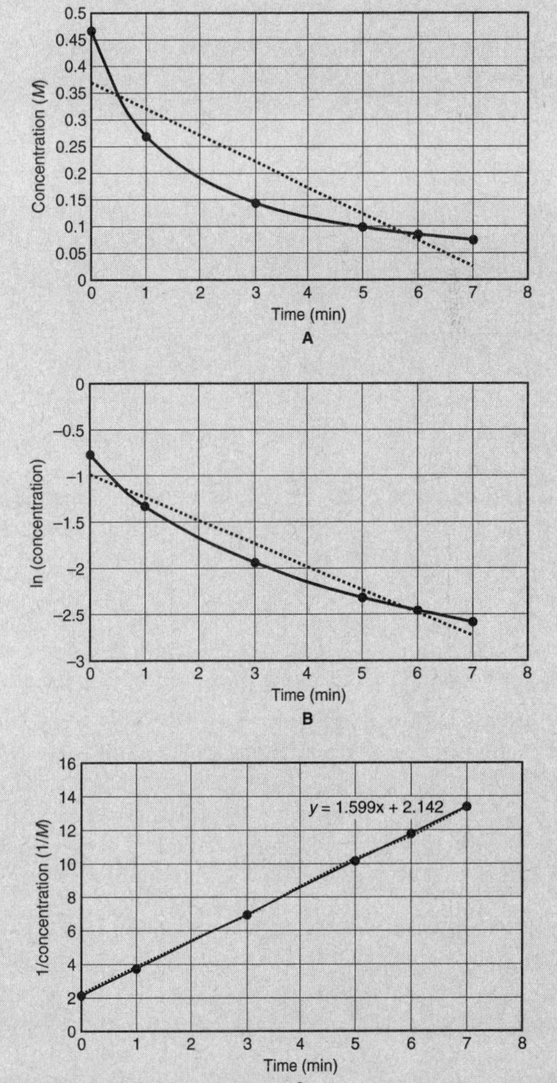

Think About It The units are typical of a second-order reaction. Time is in minutes, since that is how it is given as data.

Example 8.

The reaction of *A* has a rate constant of 1.4×10^{-4} s^{-1}. If 1.0 *M* of reactant reacts for 25 minutes, how much is left?

Collect and Organize We are asked to determine the amount of reactant remaining given its initial amount, the rate constant of the reaction, and the time elapsed.

Analyze From units, this is a first-order reaction and the relevant integrated rate law is $\ln[A] = -kt + \ln[A]_0$. The answer is determined by solving this equation algebraically for the unknown value, in this case [A]. The units of *k* must match the time units.

Solve It is probably easier to change the units of time than to change those of the rate constant. Therefore, $t = 25 \text{ min} \times (60 \text{ s}/1 \text{ min}) = 1500 \text{ s}$ and we have

$$\ln[A] = -(1.4 \times 10^{-4}/\text{s})(1500 \text{ s}) + \ln(1.0)$$

$$\ln[A] = -0.21 + 0$$

$$[A] = 0.81\ M$$

Think About It The units of the rate constant allowed us to quickly determine the order of the reaction and which integrated rate law to use.

Example 9.

If the reactant concentration for a second-order reaction decreases from 0.10 *M* to 0.03 *M* in 1.00 hour, what is the rate constant for the reaction?

Collect and Organize We are asked to determine the rate constant for a second-order reaction given that the reactant concentration decreases from 0.1 *M* to 0.03 *M* in 1 hour.

Analyze This is a second-order reaction (the problem says so), so the relevant integrated rate law is $1/[A] = kt + 1/[A]_0$.

Solve Substituting the concentration values into this equation (and keeping units), $1/(0.03\ M) = kt + 1/(0.10\ M)$, we get $kt = 23.3\ M^{-1}$. The units on the rate constant depend on the value used for time. The easiest thing is to use time as hours. So, $k = 23.3\ M^{-1} \cdot \text{hr}^{-1}$.

If you use time in units of minutes, $t = 60.0$ min and $k = 0.39\ M^{-1} \cdot \text{min}^{-1}$. If you use time in units of seconds, $t = 3600$ s and $k = 6.5 \times 10^{-3}\ M^{-1} \cdot \text{s}^{-1}$. *Any* of these three answers is correct.

Think About It It is essential that the correct integrated rate law expression is used in the calculations.

Example 10.

How long does it take 2.00 *M* of reactant to decrease in concentration to 0.75 *M* if the rate constant is 0.67 *M*/min?

Collect and Organize We are asked to determine the amount of time it takes for 2.00 *M* of reactant to decrease to 0.75 *M* given a rate constant of 0.67 *M*/min.

Analyze The units of rate constant imply that this is a zero-order reaction (rate = *k*). Therefore, the relevant integrated rate law is $[A] = -kt + [A]_0$.

Solve Substituting into the rate law gives

$$(0.75\ M) = -(0.67\ M/\text{min})t + (2.00\ M)$$

The time is 1.9 min.

Think About It In this example, the equation is solved for time. The units of time will be minutes, which is determined from the rate constant.

Example 11.

What is the rate constant for a first-order reaction with a half-life of 300.0 seconds?

Collect and Organize We are asked to determine the rate constant for a first-order reaction with a half-life ($t_{1/2}$) of 300.0 seconds.

Analyze Since this is a first-order reaction, $t_{1/2} = (\ln 2)/k$.

Solve Solving this equation for *k*, $300.0 \text{ s} = (\ln 2)/k$ and $k = 2.310 \times 10^{-3}$ s^{-1}.

Think About It The relationship between half-life and rate constant depends on the order of the reaction. Probably the most important relationship is for a first-order reaction. First-order reactions are the only type of reaction where the concentration of reactant does not affect the half-life.

Example 12.

What is the rate constant for a second-order reaction with a half-life of 35 seconds when the initial concentration is 0.50 *M*?

Collect and Organize Given the initial concentration, the half-life, and the order of the reaction we are asked to determine the rate constant.

Analyze Since this is a second-order reaction, the equation for the half-life is $t_{1/2} = 1/k[A]_0$.

SOLVE Substituting into the equation, 35 s = $1/k(0.50\ M)$ and $k = 0.057\ M^{-1}\bullet s^{-1}$.

THINK ABOUT IT Normally, the integrated rate law is used to determine concentration. The rate constant is determined from the half-life, as in the preceding examples.

EXAMPLE 13.
What is the half-life of a reaction with an initial concentration of 0.300 *M* and a rate constant of $0.00385\ M^{-1}\bullet s^{-1}$?

COLLECT AND ORGANIZE We are asked to determine the half-life of a reaction given its initial concentration and its rate constant.

ANALYZE The units on the rate constant indicate a second-order reaction. Therefore, the equation for half-life is $t_{1/2} = 1/k[A]_0$.

SOLVE Substituting into the above equation gives

$$t_{1/2} = \frac{1}{(0.00385\ M^{-1}\bullet s^{-1})(0.300\ M)} = 866\ s.$$

THINK ABOUT IT Determining half-life from rate constant is a matter of solving for time in the half-life equations rather than *k*.

EXAMPLE 14.
A first-order reaction has a half-life of 0.50 hour. If the initial concentration is 0.80 *M*, what is the concentration after 1.50 hours?

COLLECT AND ORGANIZE We are asked to determine the remaining concentration of reactant given its initial concentration, the order of the reaction, the half-life, and the time elapsed.

ANALYZE We will use the half-life and the time elapsed to determine how many times the concentration has been reduced to one-half.

SOLVE One and one-half hours is three half-lives. A half-life is the time required for the reactant concentration to decrease by half. So after 0.50 hour, the reactant concentration is 0.40 *M*. Another half-life (1.00 hour) decreases that value by one-half to 0.20 *M*. Another half-life (1.50 hour) decreases 0.20 *M* to 0.10 *M*.

THINK ABOUT IT Since the elapsed time is a multiple of the reaction half-life we were able to solve the problem without having to resort to the equations used in the previous examples.

EXAMPLE 15.
A first-order reaction has a half-life of 0.500 hour. If the initial concentration is 1.00 *M*, how much remains after 1.13 hour?

COLLECT AND ORGANIZE We are asked to determine the remaining concentration of reactant given its initial concentration, the order of the reaction, the half-life, and the time elapsed.

ANALYZE Since the reaction is first-order, the appropriate equation for the half life is $t_{1/2} = (\ln 2)/k$.

SOLVE Substituting we get 0.500 hr = $0.6931/k$. So $1.39\ hr^{-1} = k$. Using this rate constant and the integrated rate law for a first-order reaction,

$$\ln[A] = -kt + \ln[A]_0$$

$$\ln[A] = -(1.39\ hr^{-1})(1.13\ hr) + \ln(1.00)$$

$$\ln[A] = -1.57 + 0$$

$$[A] = 0.21\ M$$

THINK ABOUT IT Be sure that the units on *k* and *t* match when using the above equations.

REACTION RATES, TEMPERATURE, AND THE ARRHENIUS EQUATION

EXAMPLE 1.
Label the following energy profile with ΔH, E_a, and transition state.

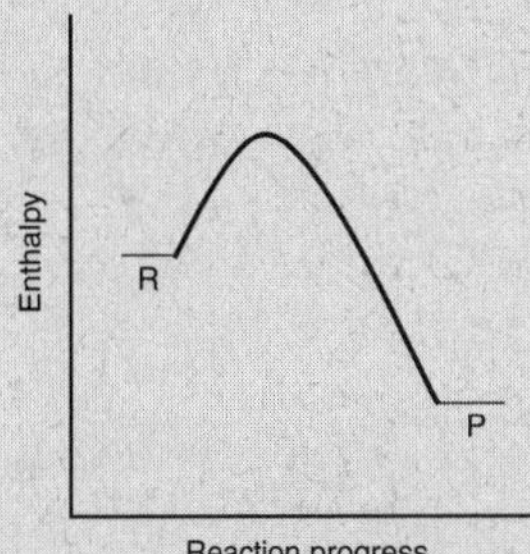

COLLECT AND ORGANIZE We are asked to label the given energy profile with ΔH, E_a, and transition state.

ANALYZE The ΔH is the energy difference between products and reactants. E_a is the energy difference between reactants and transition state. The transition state is the highest energy between reactants and products.

SOLVE

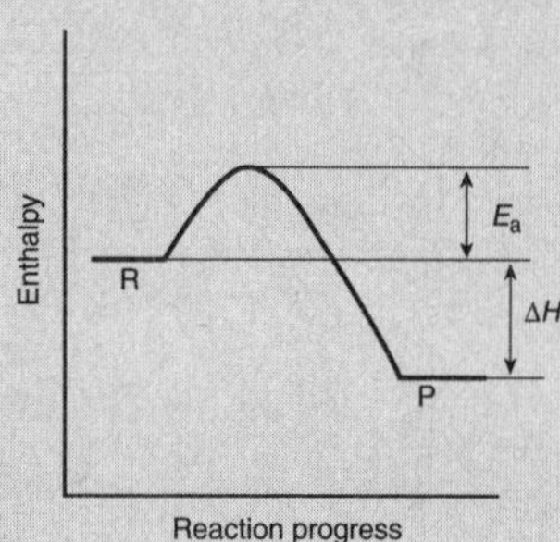

THINK ABOUT IT Energy profiles describe energy changes over the course of a reaction. Energy (as ΔH or ΔG) is graphed on the y-axis; reaction progress (reactants to products) is graphed on the x-axis.

EXAMPLE 2.
What is the activation energy for a reaction with the following temperature and rate constant data?

TABLE 14.E7

Temperature (K)	Rate constant (1/s)
200	0.236
300	0.301
400	0.340
500	0.366
600	0.385
700	0.399
800	0.409

COLLECT AND ORGANIZE We are asked to determine the activation energy from the given data on the temperature dependence of the rate constant.

ANALYZE The linear relationship between rate constant and temperature is $\ln k = \left(\frac{-E_a}{R}\right)\left(\frac{1}{T}\right) + \ln A$. The slope of the plot of $\ln k$ vs. $\left(\frac{1}{T}\right)$ is equal to $\left(\frac{-E_a}{R}\right)$.

SOLVE The data for the graph is

TABLE 14.E7S

Temperature (K)	Rate constant (1/s)	$1/T$	$\ln k$
200	0.236	0.00500	−1.445
300	0.301	0.00333	−1.200
400	0.340	0.00250	−1.078
500	0.366	0.00200	−1.004
600	0.385	0.00167	−0.955
700	0.399	0.00143	−0.920
800	0.409	0.00125	−0.894

When $\ln k$ vs. $\frac{1}{T}$ is graphed, the equation of the line is $\ln k = (-147)\left(\frac{1}{T}\right) - 0.71$. The slope of this line is −147. Therefore, $-147 = \frac{-E_a}{R}$. If R, the gas constant, is given in units of J/mol•K, the activation energy is in units of joules. So, $-147 = \frac{-E_a}{(8.314)}$ and $E_a = 1222$ J.

THINK ABOUT IT The Arrhenius equation relates temperature, rate constant, and activation energy. Solving these equations requires measuring the rate constant at more than one temperature.

REACTION MECHANISMS

EXAMPLE 1.
Consider the following mechanism. Determine the overall reaction and identify reactants, products, intermediates, and catalysts.

$CH_3OH + H^+ \rightarrow CH_3OH_2$ (slow)

$CH_3OH_2 \rightarrow CH_3 + H_2O$ (fast)

$CH_3COOH \rightarrow H^+ + CH_3CO_2^-$ (fast)

$CH_3CO_2^- + CH_3 \rightarrow CH_3CO_2CH_3$ (fast)

COLLECT AND ORGANIZE We are given a series of elementary reactions and are asked to determine the overall reaction and to identify reactants, products, intermediates, and catalysts.

ANALYZE The overall reaction is found by adding all of the elementary reactions and canceling substances that appear on both sides of the equation. Products will be on the right side of the overall reaction. Reactants will be on the left side of the overall reaction. Intermediates and catalysts will not appear in the overall reaction. Whether a substance is an intermediate or product is determined by its position in the first elementary reaction in which it appears. If it first appears as a product, it is an intermediate. If it first appears as a reactant, it is a catalyst.

SOLVE

$CH_3OH + H^+ + CH_3OH_2 + CH_3COOH + CH_3CO_2^- + CH_3 \rightarrow$

$CH_3OH_2 + CH_3 + H_2O + H^+ + CH_3CO_2^- + CH_3CO_2CH_3$

$CH_3OH + CH_3COOH \rightarrow H_2O + CH_3CO_2CH_3$

Consequently, the reactants are CH_3OH and CH_3COOH. The products are H_2O and $CH_3CO_2CH_3$. Intermediates are a product before they are a reactant. Intermediates are CH_3OH_2, CH_3, and $CH_3CO_2^-$. Catalysts are used up, then reformed. H^+ is a catalyst in this reaction.

THINK ABOUT IT The sum of all the steps in a mechanism must result in the overall reaction. The overall mechanism must also be consistent with the rate law. If either of these criteria is not met, the mechanism is wrong.

EXAMPLE 2.
Describe the molecularity of each step in the following mechanism.

$CH_3OH + H^+ \rightarrow CH_3OH_2$	(slow)
$CH_3OH_2 \rightarrow CH_3 + H_2O$	(fast)
$CH_3COOH \rightarrow H^+ + CH_3CO_2^-$	(fast)
$CH_3CO_2^- + CH_3 \rightarrow CH_3CO_2CH_3$	(fast)

COLLECT AND ORGANIZE We are asked to indicate the molecularity of each step in the given mechanism.

ANALYZE The term *molecularity* describes each step of the mechanism in terms of the number of reactants.

SOLVE The first and last steps are bimolecular (two reactants). The two middle steps are unimolecular (one reactant).

THINK ABOUT IT The assumption is that each elementary step is the result of a collision and that the molecularity describes the collision.

EXAMPLE 3.
What rate law is predicted by the following mechanism?

$CH_3OH + H^+ \rightarrow CH_3OH_2$	(slow)
$CH_3OH_2 \rightarrow CH_3 + H_2O$	(fast)
$CH_3COOH \rightarrow H^+ + CH_3CO_2^-$	(fast)
$CH_3CO_2^- + CH_3 \rightarrow CH_3CO_2CH_3$	(fast)

COLLECT AND ORGANIZE We are asked to predict the rate law governing the given mechanism.

ANALYZE Rate is determined from the slowest step in the mechanism. For this reason the rate-determining step is the slowest step. Rate is proportional to the product of the concentration of reactants in the slow step.

SOLVE The reactants are CH_3OH and H^+ in the slow step. Since rate is proportional to the concentration of those substances, the rate law for this mechanism is

$$\text{rate} = k[CH_3OH][H^+].$$

Note that this conforms to the general form of the rate law.

THINK ABOUT IT A mechanism is the step-by-step sequence by which a chemical reaction occurs. Each of these elementary steps occurs at a specific rate. The rate depends on the slowest elementary step of the mechanism.

EXAMPLE 4.
What is the rate law predicted by the following mechanism?

$H_2 \leftrightharpoons 2\,H$	(fast, reversible)
$H + SO_2 \rightarrow HSO_2$	(slow)
$HSO_2 \rightarrow HSO + O$	(fast)
$HSO \rightarrow HS + O$	(fast)
$2\,O \rightarrow O_2$	(fast)
$HS + H \rightarrow H_2S$	(fast)

COLLECT AND ORGANIZE We are asked to predict the rate law based on the given mechanism.

ANALYZE A rate law has the form of rate = $k[A]^m[B]^n$, where *A* and *B* are reactants. Reactants and catalysts appear in the rate law because they can be experimentally controlled. Products occasionally appear in the rate law because they can be experimentally measured and may affect rate. However, intermediates are sort-lived and their concentration is determined by previous reactions, not the experimentalist direction. Therefore, intermediates should not appear in the final version of a rate law.

SOLVE Because each elementary step has its own rate constant, the easiest way to keep track of the rate constants is to use the number of the step as a subscript. (A negative number represents a reverse step.)

The slow step predicts a rate law of: rate = $k_2[H][SO_2]$. However, H is an intermediate. The previous step is reversible. The rate predicted by the forward reaction is: rate = $k_1[H_2]$. The rate law predicted by the reverse of the first step is: rate = $k_{-1}[H]^2$. (Stoichiometric coefficients become exponents, since 2 H really means H + H. Since rate laws are the product of the reactants, rate = $k[H][H] = k[H]^2$.)

For the first step, the rate of the forward reaction equals the rate of the reverse reaction. So, $k_1[H_2] = k_{-1}[H]^2$. Solving for the intermediate, $\left(\frac{k_1[H_2]}{k_{-1}}\right)^{1/2} = [H]$. This expression is used to replace the intermediate in the rate expression of the slow step.

$$\text{rate} = k_2[H][SO_2] = k_2\left(\frac{k_1[H_2]}{k_{-1}}\right)^{1/2}[SO_2]$$

Collecting all the rate constants, the rate law becomes $\text{rate} = k_2\left(\frac{k_1}{k_{-1}}\right)^{1/2}[H_2]^{1/2}[SO_2]$. Fortunately, a constant times a constant, and even the square root of constants, just results in a different constant. Using k for the combined constant, the rate law (finally) is

$$\text{rate} = k[H]^{1/2}[SO_2]$$

THINK ABOUT IT Mechanisms are theoretical explanations for observed experimental results. The rate law determined from experiments with various initial concentrations or the change in concentration with time is correct. If the rate law predicted by the mechanism is not the same as the experimental rate law, the mechanism is wrong. If the rate law predicted by the mechanism matches the rate law, the mechanism *might* be correct.

EXAMPLE 5.

Which mechanism corresponds to the reaction

$$SO_2 + H_2O \rightarrow H_2SO_3$$

if the rate law for the reaction is

$$\text{rate} = k[H_2O]?$$

a. $H_2O \rightarrow H + OH$ (slow)
$SO_2 + OH \rightarrow HSO_2$ (fast)
$HSO_2 + H \rightarrow H_2SO_3$ (fast)

b. $SO_2 + H_2O \rightarrow HSO_2 + OH$ (slow)
$OH + HSO_2 \rightarrow H_2SO_3$ (fast)

c. $H_2O \rightarrow H + OH$ (slow)
$H + SO_2 \rightarrow HSO_2$ (fast)
$HSO_2 + OH \rightarrow H_2SO_3$ (fast)

d. $H_2O \rightarrow H + OH$ (slow)
$OH + SO_2 \rightarrow HSO_2 + O$ (fast)
$O + HSO_2 \rightarrow HSO_3$ (fast)
$HSO_3 + H \rightarrow H_2 + SO_3$ (fast)

COLLECT AND ORGANIZE We are asked to determine which of the given mechanisms corresponds to the given rate law and stoichiometric equation.

ANALYZE The rate law is determined by the slow step, which is the first step for each of these proposed mechanisms. Therefore, the rate law associated with each is easily written.

SOLVE

a. The proposed rate law is: rate = $k[H_2O]$. This is consistent with the experimental rate law, so this mechanism may be correct.

b. The proposed rate law is: rate = $k[H_2O][SO_2]$. This does not match the experimental rate law, so this mechanism is incorrect.

c. The proposed rate law is: rate = $k[H_2O]$. This is consistent with the experimental rate law, so this mechanism may be correct.

d. The proposed rate law is: rate = $k[H_2O]$, but the overall reaction does not correspond to the experimental reaction. Therefore, this mechanism is incorrect.

THINK ABOUT IT Either mechanism *a* or *c* is correct. There is insufficient evidence to determine which. It is also possible that the correct mechanism is something completely different.

Self-Test

KEY VOCABULARY COMPLETION QUESTIONS

__________ 1. Time required for half of the reactant to react

__________ 2. Relationship between rate and concentration

__________ 3. Sequence of making and breaking bonds

__________ 4. Exponents in the rate law

__________ 5. Number of atomic scale reactants in the elementary step

__________ 6. Change in concentration per unit time

__________ 7. Type of reaction where half-life is independent of concentration

__________ 8. Slowest elementary step

__________ 9. Order of reaction with rate law of rate = $k[A]^2$

__________ 10. Step based on collision of two molecules

__________ 11. Midway point in an elementary step

__________ 12. Rate at time = zero

__________ 13. Energy barrier that must be overcome for reaction to proceed

__________ 14. Order of a reaction when rate is constant with concentration

__________ 15. $k = Ae^{-E_a/RT}$

__________ 16. Three-molecule collision

__________ 17. Reaction rate at any given moment

__________ 18. *A* of the Arrhenius equation

__________ 19. Has units of s^{-1} for a first-order reaction

__________ 20. Increases rate of reaction without being produced or consumed

__________ 21. Energy of the activated complex

__________ 22. Rate = k[A][B] but [A] >> [B]

__________ 23. Catalyst in a different phase (physical state) from reactant and products

__________ 24. Produced in one step and consumed in a later one

__________ 25. Single-molecule reactant elementary step

Multiple-Choice Questions

1. Which of the following normally increases the rate of reaction?
   a. larger volumes
   b. higher temperatures
   c. longer times
   d. all of the above

2. What is the rate law for $2\ CO(g) + O_2(g) \rightarrow 2\ CO_2(g)$
   a. rate = $k[CO]^2[O_2]$
   b. rate = $k[CO]$
   c. rate = $k[CO][O_2]$
   d. more information is needed to determine

3. What is the average rate for the reaction $2\ CO(g) + O_2(g) \rightarrow 2\ CO_2(g)$ if the initial concentration is 0.10 *M* CO and its concentration is 0.012 *M* after 2 minutes?
   a. –0.10 *M*/s
   b. –0.044 *M*/s
   c. -7.3×10^{-4} *M*/s
   d. –0.50 *M*/s

4. If the rate is –0.001 *M* CO/s for the reaction $2\ CO(g) + O_2(g) \rightarrow 2\ CO_2(g)$, what is the rate for O_2?
   a. –0.001 *M*/s
   b. –0.002 *M*/s
   c. –0.0005 *M*/s
   d. +0.001 *M*/s

5. If the temperature at which a reaction occurs is increased, then the rate constant will
   a. increase
   b. decrease
   c. stay the same
   d. depend on the order of the reaction

6. In the combustion of gasoline, the pollutant NO is the result of
   a. impurities in the gasoline
   b. incomplete combustion of the gasoline
   c. N_2 in the air
   d. none of the above; NO does not come from combustion

7. Instantaneous rates of reaction are
   a. the same as the average rate of reaction
   b. always increasing over time
   c. normally decreasing over time
   d. normally decreasing linearly with time

8. The stoichiometry of the reaction is used to determine
   a. relative rates
   b. absolute rates
   c. order of the reaction
   d. rate constant

9. Initial rates tend to be fastest because that is when
   a. the rate constant is the highest
   b. the reactant concentration is the highest
   c. the temperature is the highest
   d. initial rates are the slowest

10. Rate depends on concentration, not mass, because
    a. collisions are more likely to occur
    b. collisions are more likely to be effective
    c. collisions are less likely to result in the wrong product
    d. none of the above; rate does depend on mass

11. Which reaction is first-order?
    a. $H_2 + Cl_2 \rightarrow 2\ HCl$
    b. $NH_4^+ \rightarrow H^+ + NH_3$
    c. $S + O_2 \rightarrow SO_2$
    d. The order can only be determined experimentally.

12. If the rate is 1.3×10^{-5} *M*/s for a first-order reaction, what is the rate if the concentration of reactant is doubled?
    a. 1.3×10^{-5} *M*/s
    b. 2.6×10^{-5} *M*/s
    c. 1.3×10^{-10} *M*/s
    d. 5.2×10^{-5} *M*/s

13. If the rate is 0.030 *M*/min for a second-order reaction with one reactant, what is the rate if the concentration of reactant is doubled?
    a. 0.030 *M*/min
    b. 0.060 *M*/min
    c. 0.090 *M*/min
    d. 0.12 *M*/min

14. If the rate law is: rate = $k[Y][X]^2$, the overall order of the reaction is
    a. zero
    b. first
    c. second
    d. third

15. If the rate law is: rate = $k[Y][X]^2$, the units on rate constant are
    a. *M*/s
    b. s^{-1}
    c. $M^{-1} \cdot s^{-1}$
    d. none of the above

16. If the rate law is: rate = $k[Y][X]^2$, the order in *Y* is
    a. zero
    b. first
    c. second
    d. *Y* does not have an order.

17. What are the appropriate units of *k* for a first-order reaction?
    a. *M*/s
    b. s^{-1}
    c. $M^{-1} \cdot s^{-1}$
    d. none of the above

18. If the rate of a second-order, one-reactant reaction is -2.0×10^3 *M*/s when [reactant] = 0.50 *M*, what is the value of *k*?
    a. −4000
    b. +4000
    c. 8000
    d. 2000

19. If the reaction $A + B \rightarrow C$, is second-order overall and second-order in *A*, what is the order in *B*?
    a. zero
    b. first
    c. second
    d. third

20. To determine if a single-reactant reaction has a first-order rate law in one experiment
    a. graph the concentration of reactant versus time
    b. graph the reciprocal of time versus the reactant concentration
    c. graph the natural log of reactant concentration versus time
    d. solve the rate law for *k*

21. The integrated rate law for a second-order reaction is
    a. rate = $k\ [A]^2$
    b. $k = Ae^{-E_a/RT}$
    c. $\frac{1}{[X]} = kt + \frac{1}{[X]_0}$
    d. $[A] = -kt + [A]_0$

22. How can the rate constant for a second-order reaction of *A* be determined?
    a. from the slope of ln [A] versus time
    b. from the *y*-intercept of [A] versus time
    c. from the *y*-intercept of rate versus time
    d. from the slope of $\frac{1}{[A]}$ versus time

23. Half-life is the time when
    a. $[A] = \frac{1}{2}[A]_0$
    b. $[A]_0 = \frac{1}{2}[A]$
    c. $[A] = t$
    d. $[A] = [A]_0$

24. If [A] = 0.725 *M* and $k = 3.6 \times 10^{-3}\ s^{-1}$, the half-life is
    a. 380 s
    b. 0.95 s
    c. 190 s
    d. 265 s

25. If the half-life of a second-order reaction is 1200 s when the reactant concentration is 0.15 *M*, what is the rate constant?
    a. $5.6 \times 10^{-3}\ M^{-1} \cdot s^{-1}$
    b. $180\ M^{-1} \cdot s^{-1}$
    c. $8000\ s^{-1}$
    d. $5.8 \times 10^{-4}\ s^{-1}$

26. If the half-life is independent of reactant concentration, then
    a. the reaction has a low rate constant
    b. the reaction is first-order
    c. the reactant concentration is very low
    d. the rates are very high

27. The following graph represents what type of reaction?

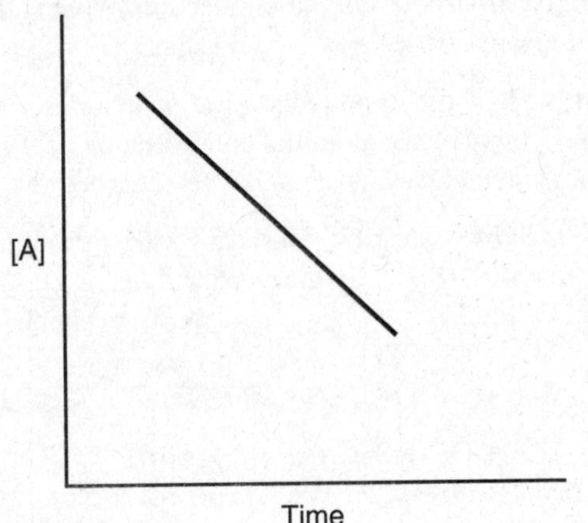

a. first-order
b. second-order
c. zero-order
d. a fast reaction

28. For a reaction to be pseudo–first-order,
a. it must have a high value of rate constant
b. one reactant must be in much higher concentration than the other
c. both of the above
d. neither a nor b

29. If a pseudo–first-order reaction has [A] = 10.0 *M*, [B] = 0.0010 *M*, and rate = 2.0×10^3 *M*/s, then *k′* is
a. 0.2 $M^{-1} \cdot s^{-1}$
b. 20 s^{-1}
c. 2 s^{-1}
d. 0.02 s^{-1}

30. The order of a reaction depends on
a. the overall stoichiometry
b. the fastest elementary step
c. the slowest step in the mechanism
d. the temperature of the reaction

Questions 31–36 refer to the following mechanism

$Cl_2 \leftrightharpoons 2\,Cl$	(fast, both directions)
$I^- + Cl \rightarrow Cl^- + I$	(slow)
$I + I^- \rightarrow I_2^-$	(fast)
$I_2^- + Cl \rightarrow Cl^- + I_2$	(fast)

31. In this reaction, Cl_2 is
a. a reactant
b. a product
c. an intermediate
d. a catalyst

32. In this reaction, Cl is
a. a reactant
b. a product
c. an intermediate
d. a catalyst

33. In this reaction, I is
a. a reactant
b. a product
c. an intermediate
d. a catalyst

34. The molecularity of step 3 is
a. unimolecular
b. bimolecular
c. termolecular
d. none of the above

35. The rate-determining step is
a. the first
b. the second
c. the last
d. none of the above; more information is needed

36. The rate law predicted by the mechanism is
a. rate = $k[Cl_2]$
b. rate = $k[I_2][Cl_2]$
c. rate = $k[I_2]$
d. rate = $k[Cl]^{1/2}[I_2]$

37. Why is a termolecular reaction considered unlikely?
a. because it implies three molecules collide simultaneously
b. because it implies three molecules collide sequentially
c. because it implies three reactants in the overall reaction
d. all of the above

38. What might act as a homogeneous catalyst in a gaseous reaction?
a. Pt(*s*)
b. $Br_2(\ell)$
c. $Cl_2(g)$
d. any of the above

39. Why would increasing temperature increase rate?
a. the molecules would collide more often
b. the molecules would collide with more force
c. a higher fraction of the molecules overcomes the activation energy
d. all of the above

40. Normally, how does a catalyst increase rate?
a. by holding a reactant in the proper orientation
b. by lowering the activation energy
c. by changing the mechanism
d. all of the above

41. Compared with the activation energy of the forward exothermic reaction, the reverse reaction is
a. exothermic with a larger activation energy
b. endothermic with a larger activation energy
c. exothermic with a smaller activation energy
d. endothermic with a smaller activation energy

42. A successful (reaction-producing) collision does not require that
    a. the molecules collide with sufficient force
    b. the molecules collide with proper orientation
    c. the molecules collide termolecularly
    d. the right molecules collide

43. In an energy profile, the transition state
    a. is found at the lowest energy
    b. is found at the highest energy
    c. occurs in the valley (if there is one) between reactants and products
    d. is the difference between products and reactants

44. In an energy profile, the intermediate
    a. is found at the lowest energy
    b. is found at the highest energy
    c. occurs in the valley (if there is one) between reactants and products
    d. is the difference between products and reactants

45. In an energy profile, the slow step will be the step with the
    a. highest-energy transition state
    b. lowest-energy transition state
    c. lowest-energy reactants
    d. lowest-energy products

Additional Practice Problems

1. Given the following concentration and rate data for the reaction: $2A + B_2 \rightarrow 2\,AB$, what is the rate law and the value of the rate constant?

TABLE 14.P1

Trial	[A] (*M*)	[B_2] (*M*)	Rate of AB (*M*/s)
1	0.100	0.100	8.6×10^{-5}
2	0.300	0.100	2.6×10^{-4}
3	0.100	0.200	3.4×10^{-4}

2. For the reaction: $3\,A_2 + B + C \rightarrow BA_2 + CA_2$, what is the rate law and value of the rate constant?

TABLE 14.P2

[A_2] (*M*)	[B] (*M*)	[C] (*M*)	Rate of CA_2 (*M*/s)
0.50	0.50	0.50	2.6×10^{-3}
0.10	0.50	0.50	5.2×10^{-4}
0.50	0.10	0.50	2.6×10^{-3}
0.50	0.50	1.0	2.1×10^{-2}

3. The rate of appearance of $CH_3OH(\ell)$ is 0.167 *M*/s for its formation reaction. What is the rate of disappearance of all the reactants?

4. If the half-life of a second-order reaction is 3.35 minutes for an initial concentration of 0.15 *M*, how much reactant remains after 1.00 hour?

5. Which mechanism is consistent with the rate law, rate = $k[SO_2][O_2]$ for the reaction $2\,SO_2 + O_2 \rightarrow 2\,SO_3$?
    a. $O_2 \Leftrightarrow 2\,O$ (fast, reversible)
       $SO_2 + O \rightarrow SO_3$ (slow)
       $SO_2 + O \rightarrow SO_3$ (same as second step)
    b. $SO_2 + O_2 \rightarrow SO_3 + O$ (slow)
       $SO_2 + O \rightarrow SO_3$ (fast)
    c. $SO_2 + O_2 \Leftrightarrow SO_4$ (fast, reversible)
       $SO_4 \rightarrow SO_3 + O$ (slow)
       $SO_2 + O \rightarrow SO_3$ (fast)

6. From the following rate constant and temperature data, what is the activation energy?

TABLE 14.P6

Temperature (K)	Rate constant (s^{-1})
100	0.608
150	0.637
200	0.652
250	0.661
300	0.667
350	0.671
400	0.675

7. Sketch the energy profile for a one-step, endothermic reaction with a small activation energy (A) and an energy profile for a two-step, exergonic reaction with a large activation energy where the first step is also the slow step (B).

8. For the following mechanism, predict the rate law and identify each substance as reactant, product, intermediate, or catalyst.
   $H_2 + S \leftrightharpoons H + HS$ (fast, reversible)
   $H + HS \rightarrow HS$ (slow)

9. What is the rate law and value of rate constant for a single reactant reaction whose concentration changes according to Table 14.P9?

TABLE 14.P9

Time (s)	Concentration (*M*)
0	1.000
5	0.437
10	0.279
15	0.205
20	0.162
25	0.134
30	0.114
35	0.100
40	0.088
45	0.079
50	0.072

10. What is the rate law and value of rate constant for a single reactant reaction whose concentration changes according to Table 14.P10?

TABLE 14.P10

Time (min)	Concentration (*M*)
0.00	0.150
1.00	0.141
2.00	0.134
3.00	0.126
4.00	0.120
5.00	0.113
6.00	0.107

Self-Test Answer Key

Key Vocabulary Completion Answers

1. half-life
2. rate law
3. mechanism
4. reaction order
5. molecularity
6. reaction rate
7. first-order
8. rate-determining step
9. second
10. bimolecular
11. activated complex
12. initial rate
13. activation energy
14. zero
15. Arrhenius equation
16. termolecular
17. instantaneous rate
18. frequency factor
19. rate constant
20. catalyst
21. transition state
22. pseudo–first-order
23. heterogeneous
24. intermediate
25. unimolecular

Multiple-Choice Answers

For further information about the topics these questions address, you should consult the text. The appropriate section of the text is listed in parentheses after the explanation of the answer.

1. b. An increase in temperature increases the rate constant and thus the rate of reactions. Higher volumes decrease concentration and decrease rate. Rate normally decreases over time because as the reactants are used up, their concentration decrease. (14.2)

2. d. Rate laws are determined from experimental evidence. The overall chemical reaction is insufficient to determine the rate law. (14.3)

3. c. Average rate is $\frac{[X]_{final} - [X]_{initial}}{\Delta t}$, so

$$\frac{(0.012\ M - 0.10\ M)}{120\ s} = 7.3 \times 10^{-4}\ M/s.\ (14.2)$$

4. c. The rate of CO disappearance $\left(\frac{\Delta[CO]}{\Delta t}\right)$ is twice the rate of the disappearance of oxygen $\left(\frac{\Delta[O_2]}{\Delta t}\right)$ according to the stoichiometry. The equation relating rates would be $\left(\frac{1}{2}\right)\frac{\Delta[CO]}{\Delta t} = \frac{\Delta[O_2]}{\Delta t}$, or $\frac{1}{2}$ the rate of CO is equal to the rate of O_2. (14.2)

5. a. According to the Arrhenius equation, rate constant will increase when temperature increases. (14.4)

6. c. The combustion of gasoline is the reaction between hydrocarbons (gasoline) and oxygen in the air. However, since air is 70% N_2, the high temperatures generated by the combustion of the gasoline will also result in some combustion of N_2. (14.1)

7. c. Because concentration decreases with time, so does rate. The decrease in concentration is not normally linear because of the exponential relationship between rate and concentration (order of the reaction). (14.2)

8. a. Because creation or consumption of one substance results in the creation or consumption of another, the relative rates must be related by stoichiometry. However, all the other choices require empirical information. (14.2 and 14.3)

9. b. The reaction, by definition, uses up the reactants. Since, according to the rate law, rate is proportional to concentration, the rate will be highest when concentration is highest (initially). (14.2)

10. a. At higher concentration, molecules are more likely to find each other, so collisions are more likely to occur. More effective collisions also increase rates but are not caused by higher concentrations. (14.5)

11. d. The overall reaction does not give any indication of reaction order. (14.3)

12. b. A first-order reaction with reactant *A* will have a rate law of: rate = k[A]. Therefore, if the concentration is doubled, the rate is doubled. $2(1.3 \times 10^{-5}\ M/s) = 2.6 \times 10^{-5}\ M/s$. (14.3)

13. d. The rate law for a second-order reaction with one reactant is: rate = $k[A]^2$. If the concentration is doubled, the rate will increase by a factor of 4 (2^2). $4(0.030\ M/min) = 0.12\ M/min$. (14.3)

14. d. The overall order is the sum of the exponent on concentration of each reactant. In this case, 1 + 2 = 3. (14.3)

15. d. The units on rate constant can be determined from the formula $M^{1-\text{order}} \cdot s^{-1}$. Since the overall reaction order is 3, the units on rate constant are $M^{-2} \cdot s^{-1}$. (14.3)

16. b. The order in Y is the exponent on the concentration of Y. The exponent is an understood 1. (If the order of Y was zero, as $[Y]^0$, Y would not be included in rate law since $[Y]^0 = 1$.) (14.3)

17. b. Using the formula $M^{1-\text{order}} \cdot \text{time}^{-1}$, 1 – 1 is zero. $M^0 = 1$, so if time is in seconds, the units are just s^{-1}. (14.3)

18. c. The rate law is rate = $k[A]^2$. The sign on rate indicates direction. Rate constants are an absolute value. Solving for k, $2000 = k[0.50]^2$; $2000 = k[0.25]$; $8000 = k$. (14.3)

19. a. The overall order is the sum of the order for each reactant. A + B = order and 2 + B = 2, so B = 0. (14.3)

20. c. In a first-order reaction, a graph of natural log of reactant concentration versus time will be linear. (14.3)

21. c. The integrated rate law for a second-order reaction has a linear relationship between time and the reciprocal of reactant concentration. (14.3)

22. d. The integrated rate law for a second-order reaction is $1/[A] = kt + 1/[A]_0$. Therefore on a graph of 1[A] versus time, the rate constant k will be the slope. (14.3)

23. a. The half-life is the time when the reactant concentration ([A]) is one-half its initial value ($[A]_0$). (14.3)

24. c. Because the units on rate constant are s^{-1}, this is a first-order reaction. The equation for half-life in a first-order reaction is: $\frac{(\ln 2)}{k} = t_{1/2}$. Consequently,

$$\text{the half-life} = \frac{(\ln 2)}{(3.6 \times 10^{-3}\ s^{-1})} = \frac{0.693}{(3.6 \times 10^{-3}\ s^{-1})}$$

= 192.5 s; rounded to two significant figures, it is 1.9×10^2 s. (14.3)

25. a. The formula for the half-life of a second-order reaction is $t_{1/2} = 1/k[A]_0$. Solving the equation for k, $k = 1/t_{1/2}[A]_0 = \frac{1}{(1200)(0.15)} = 5.6 \times 10^{-3}\ M^{-1}\bullet s^{-1}$. (14.3)

26. b. A characteristic of a first-order reaction is that its half-life does not depend on reactant concentration. (14.3)

27. c. The graph is linear, so rate is constant with time. This is only true for a zero-order reaction. (14.3)

28. b. In a pseudo–first-order reaction, the concentration of one reactant is so much higher than that of the other that it is practically a constant. In that case, the reactant with the high concentration becomes part of the rate constant ($k' = k[B]$). (14.3)

29. c. The rate law for the pseudo–first-order reaction is rate $= k[A][B] = k'[B]$, where B is the reactant in lower concentration. Therefore, $2.0 \times 10^{-3} = k'[0.001]$ and $k' = 2$. (14.3)

30. c. The slowest elementary step in the mechanism is called the *rate-determining step*. The timing of all other mechanistic processes depends on this step, either to provide the reactants or to use up the products. (14.5)

31. a. When the steps of the mechanism are added together, Cl_2 does not cancel and it is on the reactant side. Therefore, Cl_2 is a reactant. (14.5)

32. c. Cl is formed in the first step and used up in the second and last steps. Therefore, it is an intermediate. (14.5)

33. c. I is formed in the second step and used up in the next. That makes it an intermediate. (14.5)

34. b. Molecularity refers to the number of reactants in the step. There are two, so the step is bimolecular. (14.5)

35. b. The rate-determining step is always the slow step. The slow step, according to the mechanism, is the second. (14.5)

36. d. The rate law is predicted by the slow step. Rate $= k[Cl][I_2]$. However, intermediates should not appear in the rate law. The intermediate was formed from the previous step, rate $= k_1[Cl_2]$. However, it is also used up in the reverse of that step, rate $= k_{-1}[Cl]^2$. In a fast reversible reaction, the rates of the forward and reverse quickly become the same, so $k_1[Cl_2] = k_{-1}[Cl]^2$. Solving for [Cl], $[Cl] = (k_1k_{-1}[Cl_2])^{1/2}$. Putting that in the rate law for the rate-determining step and combining all rate constants as k, rate $= k[Cl]^{1/2}[I_2]$. (14.5)

37. a. Molecularity describes the number of reactants in an elementary step. If the collisions occurred sequentially, it would be more than one elementary step. The overall reaction reflects all elementary steps. (14.5)

38. c. *Homogeneous* means that it is the same throughout. Therefore, a homogeneous catalyst is in the same physical state as the reactants and products. (14.6)

39. d. At higher temperatures, the molecules have a higher kinetic energy. Because the molecules are traveling at a higher velocity, they will cover more distance and therefore, be more likely to collide. The higher velocity also results in the molecules colliding with more force. A higher average kinetic energy also means that a higher fraction of molecules has more energy than the activation energy. (14.4)

40. d. If a catalyst holds a reactant in the proper orientation, that must be a step in the mechanism, so the mechanism changes. The proper orientation means that the collision does not require as much force and the activation energy is lowered. (14.6)

41. b. In the energy profile of an exothermic reaction, the products are lower in energy than the reactants. If the products and reactants are reversed, the reaction must be endothermic. Since the products were lower in energy, the distance to the transition state (activation energy) must be higher than that for the reactants. (14.4)

42. c. Termolecular collisions are actually unlikely. Unimolecular collisions or bimolecular collisions are just as effective. (14.5)

43. b. The transition state is the highest energy point in the transition between reactants and products. (14.4)

44. c. An intermediate is produced in one step of the mechanism and is used up in the next. Therefore, it will be in a valley as a sort of temporary product. (14.5)

45. a. For the reaction to proceed, it must go through the transition state. The highest-energy transition state will be the most difficult to overcome, so it will be the slowest. (14.4)

ADDITIONAL PRACTICE PROBLEM ANSWERS

1. Comparing trials 1 and 2 and using the general form: rate = $[A]^m[B_2]^n$,

   for trial 2 we have $2.6 \times 10^{-4} = k[0.300]^m[0.100]^n$

   and for trial 1 we have $8.6 \times 10^{-5} = k[0.100]^m[0.100]^n$.

   Dividing, we get $3.02 = 3^m$ or $m = 1$.

   For trial 3, $3.4 \times 10^{-4} = k[0.100][0.200]^n$.

   For trial 1, $8.6 \times 10^{-5} = k[0.100][0.100]^n$.

   Dividing, $3.95 \approx 4 = 2^n$ and $n = 2$. The rate law is rate = $k[A][B_2]^2$.

   Using trial 1 to determine the value of k, $8.5 \times 10^{-5} = k[0.100][0.100]^2$ and $k = 0.086\ M^{-2}\cdot s^{-1}$.

2. The general form of the rate law is rate = $k[A_2]^\ell[B]^m[C]^n$. Comparing trials 1 and 2,

   $2.6 \times 10^{-3} = k[0.50]^\ell[0.50]^m[0.50]^n$

   and $5.2 \times 10^{-4} = k[0.10]^\ell[0.50]^m[0.50]^n$.

   Dividing, $5 = 5^\ell$ and $\ell = 1$.

   The rate law so far is: rate = $[A_2][B]^m[C]^n$ Comparing trials 1 and 3,

   $2.6 \times 10^{-3} = k[0.50][0.50]^m[0.50]^n$

   and $2.6 \times 10^{-3} = k[0.50][0.10]^m[0.50]^n$.

   Dividing, $1 = 5^m$ and $m = 0$.

   The rate law so far is: rate = $[A_2][C]^n$. Comparing trials 1 and 4,

   $2.6 \times 10^{-3} = k[0.50][0.50]^n$

   and $2.1 \times 10^{-2} = k[0.50][1.0]^n$.

   Dividing, $8.08 = 2^n$ and $n = 3$.

   Therefore, the rate law is

   rate = $k[A_2][C]^3$

   Using the first trial to determine the rate constant, $2.6 \times 10^{-3} = k[0.50][0.50]^3$ and $k = 0.042\ M^{-3}\cdot s^{-1}$.

3. A formation reaction produces one mole of substance from its elements at standard state. Therefore, the reaction (without physical states) is

   $$C + 2\,H_2 + \frac{1}{2}O_2 \rightarrow CH_3OH$$

   The relationship between rates is

   $$-\left(\frac{1}{1}\right)(\text{rate C}) = -\left(\frac{1}{2}\right)(\text{rate } H_2) = -\left(\frac{1}{0.5}\right)(\text{rate } O_2)$$
   $$= \left(\frac{1}{1}\right)(\text{rate } CH_3OH)$$

   So, –rate C = (rate CH_3OH) and the rate with respect to carbon is $-1.67\ M/s$.

   $$\text{Rate } CH_3OH = -\left(\frac{1}{2}\right)(\text{rate } H_2) \text{ so (rate } H_2) = (-2)1.67\ M/s$$

   and rate with respect to $H_2 = -3.34\ M/s$.

   $$\text{Rate } CH_3OH = -(2)(\text{rate } O_2) \text{ so rate } O_2 = \left(-\frac{1}{2}\right)1.67\ M/s$$

   and rate with respect to $O_2 = -0.835\ M/s$.

4. For a second-order reaction, $t_{1/2} = \frac{1}{[A]_0}$. We get

   $$3.35 \text{ min} = \frac{1}{k(0.15\ M)}, \text{ so } k = 2.0\ M^{-1}\cdot\text{min}^{-1}.$$

   The integrated rate law for a second-order reaction is

   $$\frac{1}{[A]} = kt + \frac{1}{[A]_0} = (2.0\ M^{-1}\cdot\text{min}^{-1})(60\text{ min}) + \frac{1}{(0.15\ M)}$$

   Thus, $[A] = 0.0079\ M$.

5. All three mechanisms give the correct overall stoichiometry. The rate laws for mechanism *a* are rate = $k_2[SO_2][O]$; rate forward = $k_1[O_2]$; rate reverse = $k_{-1}[O]^2$. The rate forward = rate reverse, so

   $$k_1[O_2] = k_{-1}[O]^2 \text{ and } \left(\frac{k_1}{k_{-1}}[O_2]\right)^{1/2} = [O]. \text{ This means the}$$

   $$\text{overall rate law is: rate} = k_2\left(\frac{k_1}{k_{-1}}\right)^{1/2}[SO_2][O_2]^{1/2} \text{ or}$$

   rate = $k[SO_2][O_2]^{1/2}$.

   The rate law for mechanism *b* is: rate = $k[SO_2][O_2]$. The rate laws for mechanism *c* are: rate = $k_2[SO_4]$; rate forward = $k_1k[SO_2][O_2]$; rate reverse = $k_{-1}[SO_4]$. The rate forward = rate reverse, so

   $$k_1[SO_2][O_2] = k_{-1}[SO_4] \text{ and } \left(\frac{k_1}{k_{-1}}\right)[SO_2][O_2] = [SO_4].$$

   This means the overall rate law is:

   $$\text{rate} = \frac{k_2k_1}{k_{-1}}[SO_2][O_2] \text{ or rate} = k[SO_2][O_2].$$

   Consequently, both *b* and *c* are possible mechanisms.

6. To determine activation energy, graph the following equation:

$$\ln k = \left(\frac{-E_a}{R}\right)\left(\frac{1}{T}\right) + \ln A$$

TABLE 14.A6S

Temperature (K)	Rate constant (1/s)	1/*T*	ln *k*
100	0.608	0.01	−0.49758
150	0.637	0.006667	−0.45099
200	0.652	0.005	−0.42771
250	0.661	0.004	−0.414
300	0.667	0.003333	−0.40497
350	0.671	0.002857	−0.39899
400	0.675	0.0025	−0.39304

From the graph of ln *k* vs $\frac{1}{T}$, the equation of the line is

$$\ln k = (-13.873)\left(\frac{1}{T}\right) - 0.3587$$

The slope $-13.873 = \frac{E_a}{R} = \frac{E_a}{(8.314)}$ and $E_a = 115$ J/mol.

7.

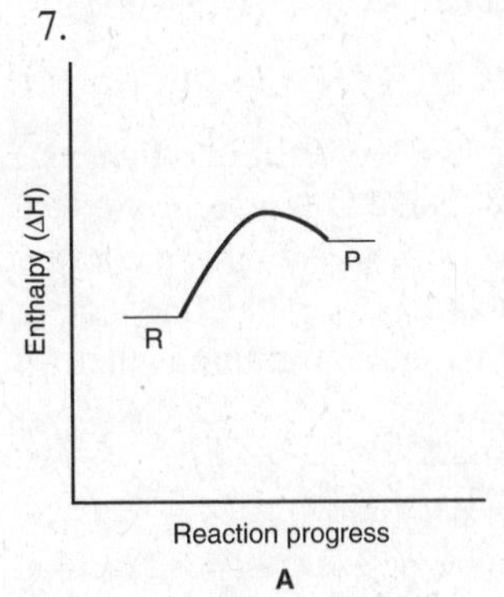

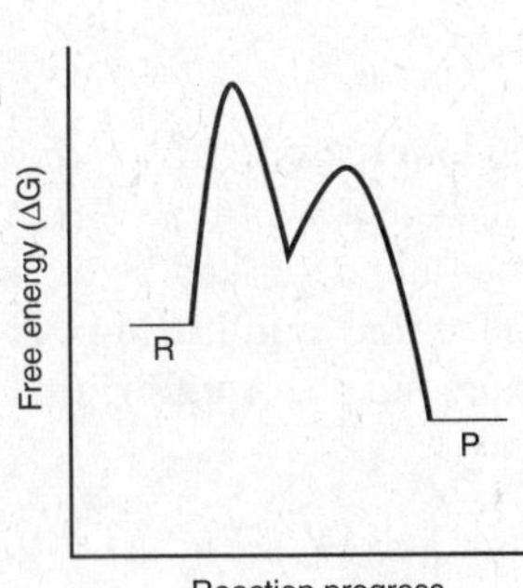

8. The overall reaction is $H_2 + S \rightarrow H_2S$. So the reactants = H_2 and S; the product = H_2S; the intermediates = H and HS; and catalysts = none. The rate law based on the slow step is rate = k_2[H][HS]. The previous step indicates rate forward = $k_1[H_2][S]$, and rate reverse = k_{-1}[H][HS]. Since the reaction is reversible, rate forward = rate reverse. We have

$k_1[H_2][S] = k_{-1}[H][HS]$ and $\frac{k_1}{k_{-1}}[H_2][S] = [H][HS]$.

Substituting into slow step, rate = $\frac{k_2 k_1}{k_{-1}}[H_2][S]$ or

rate = $k[H_2][S]$.

9. The data from the problem yield the following:

TABLE 14.A9S

Time (s)	Concentration (*M*)	ln A	1/A
0	1.000	0.000	1
5	0.437	−0.829	2.29
10	0.279	−1.275	3.58
15	0.205	−1.583	4.87
20	0.162	−1.818	6.16
25	0.134	−2.008	7.45
30	0.114	−2.168	8.74
35	0.100	−2.306	10.03
40	0.088	−2.427	11.32
45	0.079	−2.534	12.61
50	0.072	−2.632	13.9

Make three graphs: ln [A] vs. *t*, 1/[A] vs. *t*; and [A] vs. *t*.

The most linear graph is the graph of 1/[A] which represents a second-order reaction. Therefore the rate law is: rate = $k[A]^2$. The rate constant *k* is the slope of the line from that graph, $k = 0.258\ M^{-1} \cdot s^{-1}$.

10. The processed data are

TABLE 14.A10S

Time (min)	Concentration (*M*)	ln A	1/A
0.00	0.150	−1.89712	6.666667
1.00	0.141	−1.959	7.092199
2.00	0.134	−2.00992	7.462687
3.00	0.126	−2.07147	7.936508
4.00	0.120	−2.12026	8.333333
5.00	0.113	−2.18037	−8.849558
6.00	0.107	−2.23493	9.345794

Make three graphs: ln [A] vs. *t*, 1/[A] vs. *t*; and [A] vs. *t*. The most linear graph is ln [A] vs. *t* so the reaction is first-order and the rate law is: rate = *k*[A]. The slope (absolute value) is the rate constant, so $k = 0.0559\ \text{min}^{-1}$.

CHAPTER 15 Chemical Equilibrium

OUTLINE

REVIEW

Chapter Overview

It is normal in a chemical reaction that not all the reactants become products. The potential energy diagram (Chapter 14) shows that reactions are reversible. As the concentration of products increases, the rate of the reverse reaction increases. At the point where the rate of the forward reaction is the same as the rate of reverse reaction, the concentrations of products and reactants are constant. This point is **chemical equilibrium**. At equilibrium, the concentrations of reactant and product are constant, but not equal. The constant concentration values also do not imply that the reaction stops. Individual molecules of reactants and products react, but the overall amount does not change.

The **law of mass action** states that any reaction mixture eventually reaches a state (equilibrium) in which the ratio of the concentration terms of the products to the reactants, each raised to a power corresponding to the stoichiometric coefficient for that substance in the balanced chemical equation, is a characteristic value for a given temperature. This is sometimes better seen as an equation. For the reaction

$$wA + xB \leftrightharpoons yC + zD$$

the lowercase letters represent stoichiometric coefficients, A and B represent reactants, and C and D represent products. The ratio described by the law of mass action is a constant, called the **equilibrium constant (*K*)**. The mass action expression or **equilibrium constant expression** is then

$$K_c = \frac{[C]^y[D]^z}{[A]^w[B]^x}$$

The c in K_c represents concentration. For gaseous reactions, partial pressures can be used instead of concentration values. If partial pressures are used, the mass action expression is

$$K_p = \frac{(P_C)^y(P_D)^z}{(P_A)^w(P_B)^x}$$

The relationship between K_c and K_p can be derived from the ideal gas law. It is

$$K_p = K_c(RT)^{\Delta n}$$

where Δn is the difference in the number of moles of products (sum of their stoichiometric coefficients) and moles of reactants. Values for equilibrium constants can be determined from equilibrium concentrations of products and reactants. Similarly, reactant and product concentrations at equilibrium can be determined from initial concentration values and the equilibrium constant.

Unlike a gas, the volume of a solid or of a pure liquid depends on its mass. Therefore, the concentration of a solid or pure liquid is a constant. This value is normally combined with the equilibrium constant rather than being included as part of the equilibrium constant expression. Since solvents are generally in much higher concentration than the reacting solutes, the concentration of solvent is not significantly changed by the chemical reaction. Therefore, solvent concentration is also included in the equilibrium constant value rather than in the equilibrium constant expression.

The way the reaction is written affects the value of the equilibrium constant. For example, reversing the reaction reverses the position of the products and reactants in the equilibrium constant expression. Therefore, the equilibrium constant of the reverse reaction is the reciprocal of the equilibrium constant of the forward reaction. Similarly, multiplying the equation by some factor has the effect of raising the value of the equilibrium constant to that factor. In addition, the equilibrium constant for an equation that is the sum of two chemical equations is the product of the equilibrium constants of those two reactions. These relationships do not actually change the chemistry; they are instead artifacts of the math.

While the rate of the forward reaction is equal to the rate of the reverse reaction, equilibrium-constant expressions are not a measurement of rate. The expression is determined from the overall reaction rather than from the rate-determining step. The concentrations are the values at equilibrium but give no information on how long it takes to reach that equilibrium. Catalysts will help the reaction reach equilibrium faster but will not affect the equilibrium concentration. Instead, equilibrium concentrations (and equilibrium constants) are related to thermodynamic parameters like ΔG and ΔH.

Reactions move toward equilibrium from either the products or the reactants. If nonequilibrium concentrations (or pressures) are used in the mass action expression, the value is called the **reaction quotient (Q)**. If the value of Q is smaller than K, the reaction must go in a forward or spontaneous ($-\Delta G$) direction to reach the final value (K). If the value of Q is larger than K, products must react to reach the final value (K). The reaction goes in a reverse or nonspontaneous ($+\Delta G$) direction. The relationship between free energy and the equilibrium constant is

$$\Delta G = \Delta G^\circ + RT \ln Q$$

At equilibrium, the rate is neither forward nor reverse, so ΔG is zero. Consequently, the equilibrium constant can be determined from the free energy at standard state:

$$\Delta G^\circ = -RT \ln K$$

A system at equilibrium can be perturbed by changing conditions. **Le Châtelier's principle** states that if a stress (perturbation) is applied to a system at equilibrium, the equilibrium will adjust to minimize that stress. Consequently, if reactant is added, the reaction must go in a forward reaction to use up that reactant and minimize the stress. Besides changes in concentration, other equilibrium stresses are changes in temperature and volume.

While changes in concentration do not affect the value of the equilibrium constant, a change in temperature does. For exothermic reactions, heat acts as a product, thus an increase in temperature is a stress of too much product and the reaction shifts toward the reactants. With fewer products and more reactants, the value of the equilibrium constant decreases. The reverse is true for endothermic reactions. The relationship between equilibrium constant (K) and temperature (T) is expressed in the Clausius–Clapeyron equation:

$$\ln K = \frac{-\Delta H^\circ}{R}\left(\frac{1}{T}\right) + \frac{\Delta S^\circ}{R}$$

where R is the gas constant, ΔH° is the enthalpy of the reaction, and ΔS° is the entropy of the reaction.

Worked Examples

Equilibrium Constants and Reaction Quotients

Example 1.
What is the equilibrium constant expression (as K_c) for the following gaseous reactions?

a. $C_2H_4(g) + H_2(g) \leftrightharpoons C_2H_6(g)$

b. $Xe + 3\ F_2 \leftrightharpoons XeF_6$

c. $2\ N_2 + O_2 \leftrightharpoons 2\ N_2O$

Collect and Organize From the given stoichiometric equations we are asked to write the equilibrium constant expression.

Analyze The equilibrium constant is written by raising the concentration of each substance to the power of its stoichiometric coefficient. The product of these values for the products is divided by the product of these values for the reactants. Therefore, for the stoichiometric equation

$$wA + xB \leftrightharpoons yC + zD$$

the equilibrium constant would be written as

$$K_c = \frac{[C]^y[D]^z}{[A]^w[B]^x}$$

The square brackets represent the concentration of the substance within the brackets. In instances in which the stoichiometric coefficient of each product and reactant is an understood 1, 1 is the exponent on the concentration of each substance. In instances in which the stoichiometric coefficient is more than 1, the coefficient becomes the exponent.

SOLVE

a. $K_c = \dfrac{[C_2H_6]}{[C_2H_4][H_2]}$

b. $K_c = \dfrac{[XeF_6]}{[Xe][F_2]^3}$

c. $K_c = \dfrac{[N_2O]^2}{[N_2]^2[O_2]}$

THINK ABOUT IT The units on the equilibrium constant will be concentration units raised to some power. The power will vary with the equilibrium constant expression. Normally, if concentrations are in molarity, the equilibrium constant is expressed without units.

EXAMPLE 2.

What is the equilibrium constant expression (as K_c) for the following reactions?

a. $CH_3NH_2(aq) + H_2O(\ell) \leftrightharpoons CH_3NH_3^+(aq) + OH^-(aq)$

b. $H^+(aq) + OH^-(aq) \leftrightharpoons H_2O(\ell)$

c. $Fe_2S_3(s) \leftrightharpoons 2\ Fe^{3+}(aq) + 3\ S^{2-}(aq)$

d. $C(s) + 2\ F_2(g) \leftrightharpoons CF_4(g)$

e. $Hg(\ell) + H_2S(g) \leftrightharpoons HgS(s) + H_2(g)$

COLLECT AND ORGANIZE From the given stoichiometric equations involving reactants and/or products in different physical states, we are asked to write the equilibrium constant expression.

ANALYZE

In heterogeneous equilibria, the products and reactants are not all in the same physical state. The concentrations of solids, solvents, and liquids are constant, and are included in the value of K rather than in the concentration part of the expression. Another (and sometimes more mathematically useful) way of looking at it is that [solid] = [liquid] = [solvent] = 1.

SOLVE

a. "(*aq*)" means the solute is dissolved in the solvent water. Since water is the solvent, it is not included in the equilibrium constant expression:

$$K_c = \frac{[CH_3NH_3^+][OH^-]}{[CH_3NH_2]}$$

b. The reaction occurs in an aqueous solution, so the concentration of water is included in the constant. However, having no value on the top of a fraction is not mathematically appropriate, so a 1 is used:

$$K_c = \frac{1}{[H^+][OH^-]}$$

c. In this expression, the solid on the bottom is included in the value of K instead of the concentration part of the expression. Since it is a more aesthetic expression to not use a 1 on the bottom, the expression is

$$K_c = [Fe^{3+}]^2[S^{2-}]^3$$

d. It is equally appropriate to use the concentration of gases as the concentration of aqueous solutions, so the equilibrium constant expression is

$$K_c = \frac{[CF_4]}{[F_2]^2}$$

e. Mercury metal is a pure liquid, so the equilibrium constant expression is

$$K_c = \frac{[H_2]}{[H_2S]}$$

THINK ABOUT IT A solvent is defined as the substance in greater concentration. Normally, the concentration of solvent is significantly higher than that of solute. Consequently, if the solvent participates in the reaction, very little is used up (or produced) and its concentration is, for all practical purposes, unchanged. Therefore, like solids and liquids, the concentration of the solvent is included in the equilibrium constant expression or defined as 1.

EXAMPLE 3.

If the concentration of all products and reactants is 0.1 M, what will be the direction of the gaseous reaction

$$C_2H_4 + H_2 \leftrightharpoons C_2H_6 \qquad K_c = 0.99$$

COLLECT AND ORGANIZE From the given stoichiometric equation and the corresponding equilibrium constant we are asked to determine the direction of the reaction.

ANALYZE The reaction quotient (Q) has the same form as the equilibrium constant, except that the concentrations are not necessarily equilibrium concentrations. Because a reaction will continue (either forward or backward) until it reaches the equilibrium concentrations, Q can be used to determine the direction of the reaction.

SOLVE

$$Q = \frac{[C_2H_6]}{[C_2H_4][H_2]} = \frac{[0.1]}{[0.1][0.1]} = 10$$

$Q > K$. Since Q is "too high," it can be reduced by using up products and creating reactants. Therefore, the reaction goes in reverse.

THINK ABOUT IT

Because product concentration is proportional to K (or Q), it is easiest to determine reaction direction by considering the required change in product concentration. If Q is greater than K, the product concentration must decrease for the values to match, thus the reaction direction is backward (to decrease products). If $Q < K$, the product concentration must increase for the value of Q to reach K, so the reaction must go forward (create products).

Equilibrium in the Gas Phase and K_p

Example 1.
What is the equilibrium constant expression (as K_p) for the following reactions?

a. $C_2H_4(g) + H_2(g) \leftrightharpoons C_2H_6(g)$

b. $Xe(g) + 3\ F_2(g) \leftrightharpoons XeF_6(g)$

c. $C(s) + 2\ F_2(g) \leftrightharpoons CF_4(g)$

d. $Hg(\ell) + H_2S(g) \leftrightharpoons HgS(s) + H_2(g)$

Collect and Organize We are asked to determine the equilibrium constant for a series of gas phase reactions.

Analyze K_p expressions are written in the same way as K_c expressions except that partial pressures are used instead of the concentration brackets. If the units used for pressure are atmospheres, K_p is usually defined as unit-less.

Solve

a. $K_p = \dfrac{P_{C_2H_6}}{(P_{C_2H_4})(P_{H_2})}$

b. $K_p = \dfrac{P_{XeF_6}}{(P_{Xe})(P_{F_2})^3}$

c. $K_p = \dfrac{P_{CF_4}}{(P_{F_2})^2}$

d. $K_p = \dfrac{P_{H_2}}{P_{H_2S}}$

Think About It Like K_c, solids, liquids, and solvents are not included in the K_p expression.

Example 2.
What is the value of K_p for the following equations?

a. $C_2H_4(g) + H_2(g) \leftrightharpoons C_2H_6(g)$ $\quad K_c = 0.99$ at 500 K

b. $2\ NOCl(g) \leftrightharpoons 2\ NO(g) + Cl_2(g)$
$K_c = 4.4 \times 10^{-6}$ at 250°C

Collect and Organize From the given stoichiometric equations and their corresponding values of K_c, we are asked to determine K_p values for each.

Analyze The relationship between K_c and K_p is based on the ideal gas law. Rearranging the ideal gas law gives $P = MRT$. Substituting that into the K_c and K_p expressions gives $K_p = K_c(RT)^{\Delta n}$. In this equation, Δn = moles of product – moles of reactant. The moles of products are calculated by adding the stoichiometric coefficients of the gaseous products, including the "understood 1's." Similarly, the moles of reactant are the sum of the stoichiometric coefficients of the gaseous reactants. Since solids and liquids are not included in the equilibrium constant expression, they are not included in Δn.

Solve

a. $\Delta n = (1) - (1 + 1) = -1$, so

$$K_p = K_c(RT)^{-1} = (0.99)[(0.08206)(500)]^{-1} = (0.99)[41.03]^{-1} = 0.024$$

b. $\Delta n = (2 + 1) - (2) = 1$, so

$$K_p = K_c(RT)^1 = 4.4 \times 10^{-6}[(0.08206)(523)] = 1.9 \times 10^{-4}$$

Think About It Because this equation comes from the ideal gas law, the value for R is 0.08206 L•atm/mol•K, as it is in the ideal gas law. This is consistent with using units of atmosphere in K_p and units of molarity (mol/L) in K_c. This value also requires that temperature be in units of Kelvin.

Example 3.
What is the value of K_c for the following equations?

a. $Si(s) + 2\ F_2(g) \leftrightharpoons SiF_4(g)$ $\quad K_p = 1.4 \times 10^{82}$ at 1000 K

b. $CO_2(g) + 2\ Cl_2(g) \leftrightharpoons CCl_4(g) + O_2(g)$
$K_p = 6.4 \times 10^{-18}$ at 550°C

Collect and Organize From the given stoichiometric equations and their corresponding values of K_p, we are asked to determine K_c values for each.

Analyze We will follow the procedure outlined in Example 2 but solve for K_c rather than K_p.

Solve

a. $\Delta n = 1 - 2 = -1$. The moles of silicon do not count because it is a solid.

$$1.4 \times 10^{82} = K_c[(0.08206)(1000)]^{-1}$$

$$1.4 \times 10^{82}(82.06) = K_c$$

$$1.1 \times 10^{84} = K_c$$

b. $\Delta n = 2 - 3 = -1$.

$$6.4 \times 10^{-18} = K_c[(0.08206)(823)]^{-1}$$

$$(6.4 \times 10^{-18})(0.08206)(823) = K_c$$

$$4.3 \times 10^{-16} = K_c$$

THINK ABOUT IT Using the ideal gas law and stoichiometry, we can easily convert between K_c and K_p.

EXAMPLE 4.
If the partial pressures are P_{CO_2} = 1.1 atm, P_{Cl_2} = 2.0 atm, P_{CCl_4} = 0.0010 atm, and P_{O_2} = 0.0030 atm in a container, will the gaseous reaction $CO_2 + 2\ Cl_2 \leftrightharpoons CCl_4 + O_2$ go in the forward or reverse direction? ($K_p = 6.4 \times 10^{-18}$.)

COLLECT AND ORGANIZE From the given stoichiometric equation, partial pressures of reactants and products, and equilibrium constant, we are asked to determine the direction of the reaction.

ANALYZE The reaction quotient Q with partial pressures serving as concentrations can be used to determine the direction of the reaction.

SOLVE

$$Q = \frac{(P_{CCl_4})(P_{O_2})}{(P_{CO_2})(P_{Cl_2})^2} = \frac{(0.0010)(0.0030)}{(1.1)(2.0)^2} = \frac{3 \times 10^{-6}}{4.4} = 6.8 \times 10^{-7}$$

Since $Q > K$, the product concentration must decrease. The reaction will go in the reverse direction.

THINK ABOUT IT The system seeks to reach a condition where $Q = K$, at which the reaction is at equilibrium.

K, *Q*, AND Δ*G*

EXAMPLE 1.
Which reactions are spontaneous at standard state?

a. $2\ NOCl \leftrightharpoons 2\ NO + Cl_2$ $K_c = 4.4 \times 10^{-6}$ at 400 K

b. $C_2H_4 + H_2 \leftrightharpoons C_2H_6$ $K_c = 0.99$ at 500 K

COLLECT AND ORGANIZE We are asked to determine which reactions are spontaneous.

ANALYZE We will determine spontaneity by using $\Delta G° = -RT \ln K$ and noting that a negative value of $\Delta G°$ represents a spontaneous reaction, and a positive value is for a nonspontaneous reaction.

SOLVE

a. Since K is a fraction, ln K is negative and $\Delta G°$ will be positive and the reaction is nonspontaneous.

b. Since K is a fraction, $\Delta G°$ will be positive and the reaction is nonspontaneous.

THINK ABOUT IT That the K values given are not at standard state is not relevant, since the equilibrium constant value is related to standard state conditions, not to nonstandard conditions. There is a relationship for nonstandard conditions [$\Delta G = \Delta G° + RT \ln Q$], but since the question does not refer to it, it is not needed.

EXAMPLE 2.
What is the $\Delta G°$ for the following reactions?

a. $2\ NOCl \leftrightharpoons 2\ NO + Cl_2$ $K_c = 4.4 \times 10^{-6}$ at 400 K

b. $C_2H_4 + H_2 \leftrightharpoons C_2H_6$ $K_c = 0.99$ at 500 K

COLLECT AND ORGANIZE We are asked to determine $\Delta G°$ for two reactions at a given temperature given the equilibrium constant of each.

ANALYZE We will use the relation $\Delta G° = -RT \ln K$.

SOLVE

a. $\Delta G° = -(8.314\ \text{J/mol}\bullet\text{K})(400\ \text{K}) \ln (4.4 \times 10^{-6})$
$= -(8.314\ \text{J/mol}\bullet\text{K})(400\ \text{K})(-12.33)$
$= +41,004\ \text{J/mol} = +4.10 \times 10^4\ \text{J/mol}$ or 41.0 kJ/mol

b. $\Delta G° = -(8.314)(500) \ln (0.99) = -(8.314)(500)(-0.01)$
$= 41.57\ \text{J/mol} = 4 \times 10^1\ \text{J/mol}$ or 4×10^{-2} kJ/mol

THINK ABOUT IT The significant figure rule for logs is that the number of significant figures in the number becomes the number of decimal places in the answer (after the log function).

EXAMPLE 3.
What is the equilibrium constant for the following reactions?

a. $C_2H_4 + Cl_2 \leftrightharpoons CH_2ClCH_2Cl$
$\Delta G° = -148$ kJ/mol at 500 K

b. $PbCl_2 \leftrightharpoons Pb + 2\ Cl^-$ $\Delta G° = +28.1$ kJ/mol at 25°C

COLLECT AND ORGANIZE We are asked to calculate the equilibrium constant for two reactions given the temperature at which they were run and their standard free energies.

ANALYZE Again we will use the relation $G° = -RT \ln K$, but in this case we will solve for K. Note that, as usual, $\Delta G°$ values are given in units of kJ. However, since the value of R used is 8.314 J/mol•K, $\Delta G°$ must be converted to J before the equation is solved.

SOLVE

a. $\Delta G° = -RT \ln K$

$-148{,}000 = -(8.314)(500) \ln K$

$35.6 = \ln K$

$3 \times 10^{15} = K$

b. $\Delta G° = -RT \ln K$

$$+28{,}100 = -(8.314)(298) \ln K$$

$$-11.3 = \ln K$$

$$1.19 \times 10^{-5} = K$$

$$1 \times 10^{-5} = K$$

Think About It If you consider the significant figure rule described in the previous example, the reverse is that the number of decimal places in the log term (35.6) becomes the number of significant figures in the antilog.

Le Châtelier's Principle

Example 1.
For the reaction
$CH_3OH(g) + O_2(g) \leftrightharpoons HCOOH(g) + H_2O(g)$,

a. What direction does the equilibrium shift if more oxygen is added?

b. What direction does the equilibrium shift if water is removed?

c. How does the concentration of methanol (CH_3OH) change if more oxygen is added?

d. How does the concentration of methanol change if more water is added?

e. How does the concentration of methanol change when more methanol is added?

Collect and Organize We are asked to consider the effect of changing concentration on the position of equilibrium.

Analyze Le Châtelier's principle says that a system adjusts to minimize stress. The concentration can be changed by adding or removing a substance. If added, the equilibrium shifts to remove the substance. If removed, the equilibrium shifts to replace it.

Solve

a. Because reactant is added, the equilibrium will shift toward the products (right).

b. Because product is removed, the equilibrium will shift toward the products to replace it (right).

c. If more oxygen (reactant) is added, the equilibrium shifts toward products. That uses up the other reactant (methanol); therefore, the concentration of methanol decreases.

d. If water is added, the equilibrium shifts toward the reactants. The creation of more reactants increases the concentration of methanol.

e. When more methanol is added, the concentration of methanol increases. Then the equilibrium shifts to remove some of the extra methanol. Therefore, the concentration of methanol is still higher than it as in the original equilibrium. However, it is less than the total of the methanol from the first equilibrium and the amount added.

Think About It Even if the equilibrium constant is small, the amount of product produced at equilibrium can be increased using Le Châtelier's principle by adding an excess of reactant (usually the less expensive one) or removing the product.

Example 2.
For the reaction $C(s) + 2\ F_2(g) \leftrightharpoons CF_4(g)$,

a. How does the equilibrium shift if carbon tetrafluoride is added?

b. How does the concentration of fluorine change if carbon tetrafluoride is added?

c. How does the concentration of fluorine change if carbon is added?

Collect and Organize From the given stoichiometric equation we are asked to predict the effect of changing the concentrations of reactants and products.

Analyze We will apply Le Châtelier's principle to predict how the reaction is affected by changing concentrations of reactants and products.

Solve

a. As product is added, the equilibrium shifts toward reactants (left).

b. Since the equilibrium shifts toward reactants, the concentration of the reactant fluorine will increase.

c. Carbon is a solid. Adding more will not change its concentration. Therefore, the concentration of fluorine remains the same.

Think About It Adding or removing solid, pure liquid, or solvent does not affect the equilibrium, since the concentration of these substances is constant.

Example 3.
How does an decrease in volume affect the concentration of the first reactant in the following reactions?

a. $C_2H_4(g) + H_2(g) \leftrightharpoons C_2H_6(g)$

b. $Xe(g) + 3\ F_2(g) \leftrightharpoons XeF_6(g)$

c. $C(s) + 2\ F_2(g) \leftrightharpoons CF_4(g)$

d. $H_2S(g) + Hg(\ell) \leftrightharpoons HgS(s) + H_2(g)$

COLLECT AND ORGANIZE For a series of stoichiometric equations, we are asked to determine the change in concentration of the first reactant that occurs when volume is decreased.

ANALYZE Since volume is decreasing, each reaction will shift to the side with fewer moles of gas.

SOLVE

a. There are 2 moles of gas on the reactant side and 1 mole of gas on the product side. The equilibrium will shift toward products (right). This will use up reactants, so the concentration (and partial pressure) of C_2H_4 will decrease.

b. There are 4 moles of gas on the reactant side and 1 mole of gas on the product side, so the equilibrium will shift toward the product (right) side. Therefore, the concentration of xenon will decrease.

c. There are 2 moles of gas on the reactant side and 1 mole of gas on the product side. The equilibrium will shift toward products (right), but since carbon is a solid, its concentration will not change.

d. There is 1 mole of gas on the reactant side and there is 1 mole on the product side. A change in volume will favor neither side, so the concentration of all products and reactants will not change.

THINK ABOUT IT Changing volume affects only gases. Lower volumes favor fewer moles of gas, and higher volumes favor more moles of gas. Only consider moles of gas (not other physical states) in determining equilibrium shifts due to volume changes.

EXAMPLE 4.
How does the concentration of the last product change if the temperature increases?

a. $Fe_2S_3 \leftrightharpoons 2\ Fe^{3+} + 3\ S^{2-}$ $\quad +\Delta H$

b. $Si + 2\ F_2 \leftrightharpoons SiF_4$ $\quad -\Delta H$

COLLECT AND ORGANIZE For the given reactions and their corresponding changes in enthalpy, we are asked to determine the effect of increasing temperature on product concentration.

ANALYZE The easiest way to predict equilibrium shifts with temperature is to consider energy (or heat) as a product or a reactant. It is a product in exothermic reactions ($-\Delta H$) and a reactant in endothermic reactions ($+\Delta H$).

SOLVE

a. Since this reaction is endothermic, energy is a "reactant."

$$\text{energy} + Fe_2S_3 \leftrightharpoons 2\ Fe^{3+} + 3\ S^{2-}$$

So the reaction shifts toward the products, and the concentration of sulfide will increase.

b. Since this reaction is exothermic, energy is a "product."

$$Si + 2\ F_2 \leftrightharpoons SiF_4 + \text{energy}$$

So the reactant shifts toward the reactants, and the concentration of silicon tetrafluoride will decrease.

THINK ABOUT IT Increasing temperature increases the "concentration of energy."

EXAMPLE 5.
How does the concentration of the first reactant change if the temperature decreases?

a. $C_2H_4 + H_2 \leftrightharpoons C_2H_6$ $\quad$ (exothermic)

b. $2\ NOCl \leftrightharpoons 2\ NO + Cl_2$ $\quad$ (endothermic)

COLLECT AND ORGANIZE We are asked to determine the effect of decreasing temperature on the reactant concentration given that the reaction is either exothermic or endothermic.

ANALYZE Consider heat as either a product or a reactant. When the reaction is exothermic, heat will be considered a product. When the reaction is endothermic, heat will be considered a reactant.

SOLVE

a. Since this is an exothermic reaction, energy is a "product."

$$C_2H_4 + H_2 \leftrightharpoons C_2H_6 + \text{heat}$$

At lower temperature, heat is removed. So to replace it, the reaction must shift toward the products. Therefore, the concentration of C_2H_4 decreases.

b. Since this is an endothermic reaction, energy acts as a "reactant."

$$\text{heat} + 2\ NOCl \leftrightharpoons 2\ NO + Cl_2$$

So the decrease in temperature will shift the equilibrium toward reactants. Therefore, the concentration of NOCl will increase.

THINK ABOUT IT Treating heat as part of the stoichiometric equation and using Le Châtelier's principle allows one to predict how temperature changes affect equilibrium.

Calculations Based on *K*

Example 1.
What is the equilibrium constant for the gaseous reaction

$$2\,N_2 + O_2 \leftrightharpoons 2\,N_2O$$

if, at equilibrium, 0.00335 mole N_2, 0.000464 mol O_2, and 0.545 mol N_2O are in a 1.00-L container?

Collect and Organize Given the concentrations of reactants and products at equilibrium, we are asked to determine the equilibrium constant for the given reaction.

Analyze The equilibrium constant is determined by solving the mass action expression. and the concentration of each product and reactant at equilibrium.

Solve The mass action expression for this reaction is $K_c = [N_2O]^2/[N_2]^2[O_2]$. Using the given values in the equation,

$$K_c = \frac{(0.545)^2}{(0.00335)^2(0.000464)} = \frac{0.297025}{5.20724 \times 10^{-9}}$$

$$= 5.7 \times 10^7$$

Think About It The concentrations in the equilibrium constant expression are the concentration values at equilibrium. If the concentrations are known, the equilibrium constant is determined by solving the mass action expression.

Example 2.
What is the concentration for each substance at equilibrium for the following gaseous reaction

$$C_2H_4 + H_2 \leftrightharpoons C_2H_6 \qquad K_c = 0.99$$

if the initial concentration of ethene (C_2H_4) is 0.335 *M* and that of hydrogen is 0.526 *M*?

Collect and Organize Given the initial concentrations of reactants, the stoichiometric relation, and the equilibrium constant for the reaction, we are asked to determine the concentrations of the reactants and products when the reaction reaches equilibrium.

Analyze Use an ICE table in which *I* gives the initial concentrations, *C* the change in concentrations, and *E* the equilibrium concentrations. Initial concentrations are given in the problem. If a substance is not mentioned, the assumption is that its initial concentration is zero. (It makes no sense to list all the things that are not present.) Since the point of the problem is to find out how much substance reacts, x is used to represent this. The reaction shows the relationship between each substance that reacts so in the chance row, each column of the *C* row is the stoichiometric coefficient of that substance times x. The direction of change (+ or –) can be determined using Q. However, if the initial concentration of one of the substances in the *I* line is zero, that substance must increase in concentration. (You can't lose what you don't have). Therefore, every substance on that side of the reaction gets a "+" in the *C* row and every substance on the opposite side gets a "–". The values in the *E* row are found by adding the *I* and *C* rows together. The values in the equilibrium row are used in the equilibrium constant expression.

Solve The equilibrium constant expression for this reaction is $0.99 = \frac{[C_2H_6]}{[C_2H_4][H_2]}$. The resulting ICE table is

	C_2H_4	H_2	C_2H_6
Initial	0.335	0.526	0
Change	$-x$	$-x$	$+x$
Equilibrium	$0.335 - x$	$0.526 - x$	x

The values from the equilibrium line are substituted into the mass action expression:

$$0.99 = \frac{[C_2H_6]}{[C_2H_4][H_2]} = \frac{x}{(0.335 - x)(0.526 - x)}$$

and solved for x.

$$0.99 = \frac{x}{(0.17621 - 0.861x + x^2)}$$

$$0.1744479 - 1.85239x + 0.99x^2 = 0$$

Using the quadratic equation to solve for x,

$$x = \frac{(1.85239 \pm 1.65325)}{1.98}$$

$$x = 0.1006 \text{ or } 1.7705$$

Determine which value for x is correct by calculating equilibrium concentrations:

$$[C_2H_4] = 0.335 - x = 0.335 - 0.1006 = 0.234\ M$$

or

$$[C_2H_4] = 0.335 - x = 0.335 - 1.7705 = -1.435\ M$$

Since a negative concentration is impossible, $[C_2H_4]$ = 0.234 *M*. And

$$[H_2] = 0.526 - x = 0.526 - 0.101 = 0.425\ M$$

$$[C_2H_6] = 0.101\ M$$

Think About It There will be two values obtained for x. One is correct; one is not. The one that is not correct will predict a negative (and thus impossible) concentration for some substance. The equilibrium concentrations can be determined by solving the expression of each *E* column for concentration by substituting in the value for x.

EXAMPLE 3.
What is the equilibrium concentration of silver ion in 1.00 L of solution with 0.010 mol AgCl and 0.010 mol Cl^- in a solution with the equilibrium reaction of

$$AgCl(s) \leftrightharpoons Ag^+(aq) + Cl^-(aq)$$

with an equilibrium constant value of 1.8×10^{-10}?

COLLECT AND ORGANIZE We are asked to determine the equilibrium concentration of Ag^+ ion given initial concentrations and the stoichiometric equation.

ANALYZE The equilibrium constant expression for the reaction is $K = [Ag^+][Cl^-] = 1.8 \times 10^{-10}$. Recall that as a solid, AgCl is not included in the expression. Therefore, it need not be included in the ICE table either. We will then use the ICE table to determine equilibrium concentrations.

SOLVE

	Ag^+	Cl^-
Initial	0	0.010 mol/1.00 L = 0.010 *M*
Change	$+x$	$+x$
Equilibrium	x	$0.010 + x$

Using the equilibrium concentrations in the equilibrium constant expression,

$$1.8 \times 10^{-10} = [x][0.010 + x]$$

Solving for x,

$$1.8 \times 10^{-10} = 0.010x + x^2$$

$$0 = x^2 + 0.010x - 1.8 \times 10^{-10}$$

Using the quadratic equation,

$$x = \frac{-0.010}{2} \pm \frac{[(0.010)^2 - 4(1)(-1.8 \times 10^{-10})]^{1/2}}{2}$$

$x = 0$ if you round off and $x = 1.8 \times 10^{-8}$ if you don't. There is an easier alternative! The value of K as 1.8×10^{-10} implies that the amount of product at equilibrium will be quite small. If you add a small number to a large number, you get the large number back. (This doesn't work for multiplication!) In this reaction, because of the equilibrium constant, the value for x must be small. Therefore, if x is added to or subtracted from a number, you get that number back. Using this assumption, the equilibrium constant expression

$$1.8 \times 10^{-10} = [x][0.010 + x]$$

becomes

$$1.8 \times 10^{-10} = [x][0.010]$$

This is a lot easier to solve: $1.8 \times 10^{-8} = x$. The same answer!

THINK ABOUT IT Generally, even if this assumption turns out not to be valid, this quick calculation is a good way to estimate the answer. (The smaller the value of K, the better the estimate.) You can use it to check your calculation with the quadratic.

EXAMPLE 4.
What are the equilibrium concentrations of all products and reactants for the following aqueous reaction:

$$HSO_4^-(aq) + H_2O(\ell) \leftrightharpoons H_3O^+(aq) + SO_4^{2-}(aq) \quad K = 0.012$$

where the initial concentrations are $[HSO_4^-] = 0.50$ *M*, $[H_3O^+] = 0.020$ *M*, and $[SO_4^{2-}] = 0.060$ *M*?

COLLECT AND ORGANIZE We are asked to determine the concentrations of all products and reactants at equilibrium given their initial concentrations.

ANALYZE The mass action expression for the reaction is $K = \frac{[H_3O^+][SO_4^{2-}]}{[HSO_4^-]}$. Water, as the solvent, is not included.

We will make use of the ICE table to solve the problem.

SOLVE

	HSO_4^-	H_3O^+	SO_4^{2-}
Initial	0.50	0.020	0.060
Change			
Equilibrium			

Since there is no column with a zero, Q is used to determine the reaction direction:

$$Q = \frac{[H_3O^+][SO_4^{2-}]}{[HSO_4^-]} = \frac{(0.020)(0.060)}{0.50} = 0.0024$$

Since $Q < K$, the concentration of products must increase and the reactants decrease. Therefore, the table is

	HSO_4^-	H_3O^+	SO_4^{2-}
Initial	0.50	0.020	0.060
Change	$-x$	$+x$	$+x$
Equilibrium	$0.50 - x$	$0.020 + x$	$0.060 + x$

Using the values in the equilibrium constant expression,

$$K = 0.012 = \frac{(0.020 + x)(0.060 + x)}{(0.50 - x)} = \frac{0.0012 + 0.080x + x^2}{(0.5 - x)}$$

Assuming x is small probably won't work, $K = 0.012$. In addition, all the x's would cancel, so unfortunately, the simplifying assumption won't work here.

$$0.0060 - 0.012x = 0.0012 + 0.080x + x^2$$

$$0 = x^2 + 0.092x - 0.0048$$

$$x = \frac{-0.092}{2} \pm \frac{[8.464 \times 10^{-3} - 4(1)(-0.0048)]^{1/2}}{2} =$$

0.0372 or −0.129

Calculating equilibrium concentrations,

$$[HSO_4^-] = 0.50 - 0.0372 = 0.46\ M$$

or

$$[HSO_4^-] = 0.50 - (-0.129) = 0.63\ M$$

and

$$[H_3O^+] = 0.020 + 0.0372 = 0.057\ M$$

or

$$[H_3O^+] = 0.020 + (-0.129) = -0.109\ M$$

Since this last value is impossible, $[H_3O^+] = 0.057\ M$ and $[HSO_4^-] = 0.046\ M$ are correct, and $[SO_4^{2-}] = 0.060 + 0.0372 = 0.097\ M$.

THINK ABOUT IT Note that the only difference between Q and K is that Q uses the initial line of the table and K uses the equilibrium line.

EXAMPLE 5.
What is the concentration of H^+ if the initial concentration of H_2CO_3 is 0.14 M in the following reaction?

$$H_2CO_3 \rightleftharpoons 2\ H^+ + CO_3^{2-} \qquad K = 2.4 \times 10^{-17}$$

COLLECT AND ORGANIZE Given the stoichiometry, the initial concentration of reactant, and the equilibrium constant, we are asked to determine the equilibrium concentration of the product H^+.

ANALYZE The mass action expression for this reaction is $K = \dfrac{[H^+]^2[CO_3^{2-}]}{[H_2CO_3]}$. As with the previous examples, we will make use of the ICE table to solve for the desired concentrations.

SOLVE The ICE table is

	H_2CO_3	H^+	CO_3^{2-}
Initial	0.14	0	0
Change	$-x$	$+2x$	$+x$
Equilibrium	$0.14 - x$	$2x$	x

Substituting the equilibrium concentrations into the equilibrium constant expression:

$$2.4 \times 10^{-17} = \frac{(2x)^2(x)}{(0.14 - x)}$$

Since the equilibrium constant is very small, it is reasonable to assume that x is small. Therefore $0.14 - x = 0.14$. This is fortunate because there is not an easy way to solve a cubic.

$$2.4 \times 10^{-17} = \frac{4x^3}{0.14}$$

$$3.4 \times 10^{-18} = 4x^3$$

$$9.4 \times 10^{-7} = x$$

Checking the assumption, $0.014 - 9.4 \times 10^{-7} = 0.14$. Yes, x is small! The concentration of H^+ is $2x$. Therefore, $[H^+] = 2x = 2(9.4 \times 10^{-7}) = 1.9 \times 10^{-6}\ M$.

THINK ABOUT IT Note that the change for H^+ is $2x$. This is determined from the stoichiometry, where there are twice as many H^+ ions as carbonate ions. In the change line, use the stoichiometric coefficient times x.

EXAMPLE 6.
If the equilibrium constant of the reaction $2\ NOCl(g) \rightleftharpoons 2\ NO(g) + Cl_2(g)$ is $K_c = 4.4 \times 10^{-6}$ at 250° C, what is the equilibrium constant of

a. $2\ NO + Cl_2 \rightleftharpoons 2\ NOCl$

b. $NOCl \rightleftharpoons NO + \frac{1}{2} Cl_2$

c. $6\ NO + 3\ Cl_2 \rightleftharpoons 6\ NOCl$

COLLECT AND ORGANIZE We are given the equilibrium constant and stoichiometric equation for a reaction and are asked to determine equilibrium constants for variations of the same reaction.

ANALYZE Since reversing the reaction changes the position of the products and reactants, the equilibrium constant is the reciprocal of the original equation. Multiplying by a factor raises the equilibrium constant by the same factor.

SOLVE

a. This reaction is the reverse of the original reaction.

Therefore, $K_{ca} = \dfrac{1}{K_c} = \left(\dfrac{1}{4.4 \times 10^{-6}}\right) = 2.3 \times 10^5$.

b. This reaction is the original reaction multiplied by $\frac{1}{2}$.

Therefore, $K_{cb} = (K_c)^{1/2} = 2.1 \times 10^{-3}$.

c. This reaction is both reversed (compared with the original) and multiplied by 3. Therefore,

$$K_{cc} = \left(\frac{1}{K_c}\right)^3 = \left(\frac{1}{4.4 \times 10^{-6}}\right)^3 = 1.2 \times 10^{16}$$

THINK ABOUT IT The equilibrium constant for the reverse of a reaction is inversely proportional to the equilibrium constant in the forward direction. When the stoichiometry of a reaction is changed by some factor, the effect on the equilibrium constant is to raise it to the power of that same factor.

EXAMPLE 7.
What is the equilibrium constant for the aqueous reaction

$$H_2CO_3 \leftrightharpoons 2\ H^+ + CO_3^{2-}$$

if

$$K_1 = 4.3 \times 10^{-7} \text{ for } H_2CO_3 \leftrightharpoons H^+ + HCO_3^-$$

and

$$K_2 = 5.6 \times 10^{-11} \text{ for } HCO_3^- \leftrightharpoons H^+ + CO_3^{2-}$$

COLLECT AND ORGANIZE We are asked to determine the equilibrium constant for a reaction that can be expressed as the summation of two reactions, each with its own equilibrium constant.

ANALYZE The given reactions must be manipulated in such a way such that when they are added together they produce the desired reaction. The resulting equilibrium constant is then the product of the two reactions being added together.

Solve In this case, when the two reactions are added, bicarbonate (HCO_3^-) cancels, and the desired reaction results. The equilibrium constant is the product of the two reactions that were added together, so $K_c = K_1K_2 = (4.3 \times 10^{-7})(5.6 \times 10^{-11}) = 2.4 \times 10^{-17}$.

THINK ABOUT IT This procedure is very similar to that employed with Hess's law in which ΔH values are obtained for reactions, which are the result of adding component reactions. The difference is that the equilibrium constant for the composite reaction is the product of the equilibrium constants for the component reactions. In the case of Hess's law, the individual enthalpies are summed.

EXAMPLE 8.
What is the equilibrium constant for the reaction (the ions are aqueous)

$$PbCl_2 + 2\ Ag^+ \leftrightharpoons 2\ AgCl + Pb^{2+}$$

if

$$AgCl(s) \leftrightharpoons Ag^+ + Cl^- \qquad K_1 = 1.8 \times 10^{-10}$$

$$PbCl_2(s) \leftrightharpoons Pb^{2+} + 2\ Cl^- \quad K_2 = 1.2 \times 10^{-5}$$

COLLECT AND ORGANIZE As with the previous example, we must manipulate the given reactions such that when they are added together the result is the desired reaction.

ANALYZE To get the desired reaction, the silver chloride reaction must be doubled and reversed and then added to the lead reaction.

SOLVE When the silver reaction is doubled and reversed, it becomes (reaction 3):

$$2\ Ag^+ + 2\ Cl^- \leftrightharpoons AgCl$$

and the equilibrium constant becomes

$$K_3 = \left(\frac{1}{1.8 \times 10^{-10}}\right)^2 = 3.1 \times 10^{19}$$

This equation is added to the lead equation, and the two equilibrium constants are multiplied together.

So, $K_c = K_3K_2 = (3.1 \times 10^{19})(1.2 \times 10^{-5}) = 3.7 \times 10^{14}$.

THINK ABOUT IT Care must be taken to ensure that any multiplication factor and needed reversals of reactions are accounted for in the value of the individual equilibrium constants prior to multiplying them together to get the resulting equilibrium constant.

CHANGING *K* WITH CHANGING TEMPERATURE

EXAMPLE 1.
If the equilibrium constant is 1.5×10^{-8} at 298 K for a reaction with a $\Delta H°$ of +77.2 kJ/mol, what is the equilibrium constant at 400 K?

COLLECT AND ORGANIZE Given $\Delta H°$ for a reaction and the value of its equilibrium constant at a particular temperature, we are asked to determine the value of the equilibrium constant at a different temperature.

ANALYZE The Clausius–Clapeyron equation is used to determine the value of the equilibrium constant with temperature changes. The equation is

$$\ln\left(\frac{K_2}{K_1}\right) = \left(\frac{-\Delta H°}{R}\right)\left[\left(\frac{1}{T_2}\right) - \left(\frac{1}{T_1}\right)\right]$$

SOLVE If the value of equilibrium constant is called K_1, then the values substituted into the equation are $K_1 = 1.5 \times 10^{-8}$, $\Delta H° = +77,\ 200$ J/mol (converted from kJ to match R), $R = 8.314$ J/mol•*K*, $T_1 = 298$ *K*, and $T_2 = 400$ *K*. Substituting these values into the equation,

$$\ln\left[\left(\frac{K_2}{(1.5 \times 10^{-8})}\right)\right] = \left(\frac{-77{,}200}{8.314}\right)\left[\frac{1}{400} - \frac{1}{298}\right] = 7.95$$

$$\frac{K_2}{(1.5 \times 10^{-8})} = 2823$$

$$K_2 = 4.2 \times 10^{-5}$$

THINK ABOUT IT Because the gas constant is used with an energy term (ΔH), the value $R = 8.314$ J/mol•*K* is appropriate. Because *R* uses units of joules, ΔH should also be in units of joules, so that the units will cancel. As usual, temperature must be in units of Kelvin.

Self-Test

Key Vocabulary Completion Questions

__________ 1. Reaction direction favored if $K > 1$

__________ 2. What happens to K if the reaction coefficients are doubled

__________ 3. Rate forward reaction = rate reverse reaction

__________ 4. Adding product to a system at equilibrium, for example

__________ 5. Ratio of product concentration to reactant concentration that is equal to K

__________ 6. Increases rate without affecting equilibrium constant

__________ 7. Has the form of the equilibrium constant expression but uses nonequilibrium concentration values

__________ 8. If a system at equilibrium is subjected to a stress, the system adjusts to minimize the stress

__________ 9. Direction of the reaction if $Q > K$

__________ 10. Type of reaction where an increase in temperature increases equilibrium constant

__________ 11. Effect of a catalyst on activation energy

__________ 12. State where concentrations of reactants and products are constant

__________ 13. Sign of $\Delta G°$ if $K > 1$

__________ 14. Units used in K_p expression

__________ 15. Value of K if $\Delta G°$ is negative

__________ 16. Concentration of a solid

__________ 17. Units used in K_c expression

__________ 18. Relationship between K and temperature

__________ 19. $Q = K$

__________ 20. Product and reactant mixture with various physical states

Multiple-Choice Questions

1. Which of the following does *not* describe a chemical equilibrium?
   a. rate of the forward reaction is the same as the rate of the reverse reaction
   b. the concentration of reactants and products are constant
   c. the reaction stops
   d. $Q = K$

2. If a reaction has an equilibrium constant of 4.6×10^{-6}, then at equilibrium,
   a. there are significantly more products than reactants
   b. there are significantly more reactants than products
   c. the concentrations of reactants and products are about the same
   d. the reaction is too slow to predict equilibrium concentrations

3. What is the mass action expression (K_c) for the reaction $W(s) + O_2(g) \leftrightharpoons WO_2(s)$?
   a. $K = \dfrac{[WO_2]}{[W][O_2]}$
   b. $K = \dfrac{[WO_2] + [O_2]}{[W]}$
   c. $K = \dfrac{[W][O_2]}{[WO_2]}$
   d. $K = \dfrac{1}{[O_2]}$

4. If $K = 9.10 \times 10^3$, then
   a. ΔG is positive
   b. $\Delta G°$ is negative
   c. $Q > 1$
   d. all of the above

5. For the gaseous reaction $N_2O + O_2 \leftrightharpoons N_2 + O_3$, $K_p =$
   a. $\dfrac{[N_2] + [O_3]}{[N_2O] + [O_2]}$
   b. $\dfrac{P_{N_2} \bullet P_{O_3}}{P_{N_2O} \bullet P_{N_2}}$
   c. $\dfrac{[N_2O] + [O_2]}{[N_2] + [O_3]}$
   d. $\dfrac{P_{N_2} \bullet P_{O_3}}{P_{N_2O} \bullet P_{N_2}}$

6. For the endothermic gaseous reaction $CO + Cl_2 \leftrightharpoons COCl_2$, $K_c = 5.0$ at 700 K, if $[Cl_2] = [O_2] = [COCl_2] = 0.10\ M$, then
   a. the reaction will proceed forward until it reaches equilibrium
   b. the reaction will proceed in the reverse direction until it reaches equilibrium
   c. the reaction is at equilibrium
   d. more information is needed to determine

7. For the endothermic gaseous reaction, $CO + Cl_2 \leftrightharpoons COCl_2$, $K_c = 5.0$ at 700 K, what is K_p?
   a. 5.0
   b. 2.9×10^2
   c. 8.7×10^{-2}
   d. 2.9×10^4

8. For the endothermic gaseous reaction, $CO + Cl_2 \leftrightharpoons COCl_2$, $K_c = 5.0$ at 700 K. If chlorine is added to the system at equilibrium,
   a. the equilibrium will shift toward reactants
   b. the equilibrium will shift toward products
   c. the equilibrium will remain unchanged
   d. more information is needed to predict the equilibrium shift

9. For the endothermic gaseous reaction $CO + Cl_2 \leftrightharpoons COCl_2$, $K_c = 5.0$ at 700 K. If the volume of the system at equilibrium is decreased, then
   a. the equilibrium will shift toward reactants
   b. the equilibrium will shift toward products
   c. the equilibrium will remain unchanged
   d. more information is needed to predict the equilibrium shift

10. For the endothermic gaseous reaction, $CO + Cl_2 \leftrightharpoons COCl_2$, $K_c = 5.0$ at 700 K. If the temperature on the system at equilibrium is increased,
   a. the equilibrium will shift toward reactants
   b. the equilibrium will shift toward products
   c. the equilibrium will remain unchanged
   d. the equilibrium shift cannot be predicted without more information

11. For the reaction of $CaCO_3(s) \leftrightharpoons CaO(s) + CO_2(g)$, $\Delta H = +178.3$ kJ/mol. What is the mass action expression for this equilibrium?
   a. $K_c = [CO_2]$
   b. $K_c = \dfrac{[CaCO_3]}{[CaO][CO_2]}$
   c. $K_c = \dfrac{[CaO][CO_2]}{[CaCO_3]}$
   d. $K_c = \dfrac{[CaO]}{[CaCO_3]}$

12. For the reaction of $CaCO_3(s) \leftrightharpoons CaO(s) + CO_2(g)$, $\Delta H = +178.3$ kJ/mol. If, at 350 K, $K_c = 3.5 \times 10^{-20}$, then K_p is
   a. 1.0×10^{-18}
   b. 1.2×10^{-21}
   c. 1.2×10^{-23}
   d. K_p is not appropriate for this equation

13. For the reaction of $CaCO_3(s) \leftrightharpoons CaO(s) + CO_2(g)$, $\Delta H = +178.3$ kJ/mol. If more calcium carbonate was added to the system at equilibrium, the reaction
   a. will shift toward products
   b. will shift toward reactants
   c. will not change
   d. be impossible to predict without more information

14. For the reaction of $CaCO_3(s) \leftrightharpoons CaO(s) + CO_2(g)$, $\Delta H = +178.3$ kJ/mol. If the volume of the system at equilibrium is increased, the reaction will
   a. shift toward products
   b. shift toward reactants
   c. not change
   d. be impossible to predict without more information

15. For the reaction of $CaCO_3(s) \leftrightharpoons CaO(s) + CO_2(g)$, $\Delta H = +178.3$ kJ/mol. If the temperature on the system at equilibrium is decreased, the reaction will
   a. shift toward products
   b. shift toward reactants
   c. not change
   d. be impossible to predict without more information

16. What effect does a catalyst have on the equilibrium concentrations for the reaction

   $$2\ NO + O_2 \leftrightharpoons 2\ NO_2 \qquad K = 4.6 \times 10^{30}$$

   a. increases product concentration
   b. decreases product concentration
   c. depends on the sign of $\Delta H°$
   d. there is no concentration effect

17. For the reaction, $2\ NO + O_2 \leftrightharpoons 2\ NO_2$, $K = 4.6 \times 10^{30}$.

   What is K for $NO + \frac{1}{2}O_2 \leftrightharpoons 2\ NO_2$?

   a. 9.2×10^{30}
   b. 2.3×10^{30}
   c. 2.1×10^{15}
   d. 4.6×10^{30}

18. For the reaction $2\ NO + O_2 \leftrightharpoons 2\ NO_2$, $K = 4.6 \times 10^{30}$. What is the equilibrium constant for $2\ NO_2 \leftrightharpoons 2\ NO + O_2$?
   a. 4.6×10^{30}
   b. 2.2×10^{-31}
   c. 2.1×10^{15}
   d. 9.2×10^{30}

19. For the reaction $2\ NO + O_2 \leftrightharpoons 2\ NO_2$, $K = 4.6 \times 10^{30}$. How much O_2 is left at equilibrium?
   a. a lot
   b. a little
   c. practically none
   d. some

20. For the gaseous reaction $H_2O + SO_3 \leftrightharpoons H_2SO_4$, $K = 1.0$ at 200 K. If $[H_2O] = [SO_3] = [H_2SO_4] = 1.00\ M$, what is the concentration of SO_3 at equilibrium?
   a. greater than 1.00 M
   b. less than 1.00 M
   c. about 1.00 M
   d. more information is needed

21. For the gaseous reaction, $H_2O + SO_3 \leftrightharpoons H_2SO_4$, $K = 1.0$ at 200 K. If more water is added to the system at equilibrium, the concentration of H_2SO_4 will
    a. increase
    b. decrease
    c. stay the same
    d. depend on $\Delta H°$

22. For the gaseous reaction $H_2O + SO_3 \leftrightharpoons H_2SO_4$, $K = 1.0$ at 200 K. If more water is added to the system at equilibrium, the concentration of sulfur trioxide will
    a. increase
    b. decrease
    c. stay the same
    d. depend on $\Delta H°$

23. For the gaseous reaction $H_2O + SO_3 \leftrightharpoons H_2SO_4$, $K = 1.0$ at 200 K. If the volume of the system at equilibrium is increased, the concentration of water will
    a. increase
    b. decrease
    c. stay the same
    d. depend on $\Delta H°$

24. For the gaseous reaction $H_2O + SO_3 \leftrightharpoons H_2SO_4$, $K = 1.0$ at 200 K. If H_2SO_4 is removed from the system at equilibrium, the concentration of sulfur trioxide will
    a. increase
    b. decrease
    c. stay the same
    d. depend on $\Delta H°$

25. For the gaseous reaction $H_2O + SO_3 \leftrightharpoons H_2SO_4$, $K = 1.0$ at 200 K. If H_2SO_4 is removed from the system at equilibrium, the equilibrium constant will
    a. increase
    b. decrease
    c. stay the same
    d. depend on $\Delta H°$

26. For the gaseous reaction $H_2O + SO_3 \leftrightharpoons H_2SO_4$, $K = 1.0$ at 200 K. If the temperature on the system at equilibrium is increased, the concentration of sulfur trioxide will
    a. increase
    b. decrease
    c. stay the same
    d. depend on $\Delta H°$

27. For the gaseous reaction $H_2O + SO_3 \leftrightharpoons H_2SO_4$, $K = 1.0$ at 200 K. If the temperature of this *endothermic* system is increased, K will
    a. increase
    b. decrease
    c. stay the same
    d. be impossible to determine without more information

28. For the gaseous reaction $H_2O + SO_3 \leftrightharpoons H_2SO_4$, $K = 1.0$ at 200 K. How long does it take to reach equilibrium?
    a. 1 second
    b. 200 seconds
    c. 1/200 second
    d. more information is needed

29. For the gaseous reaction, $H_2O + SO_3 \leftrightharpoons H_2SO_4$, $K_c = 1.0$ at 200 K. What is K_p?
    a. 1.0
    b. 16
    c. 0.061
    d. 6.0×10^{-4}

30. If $K_p = 4.7 \times 10^7$ at 400 K, what is K_c for the reaction $A(g) + B(g) + C(g) \leftrightharpoons D(g)$?
    a. 4.7×10^7
    b. 1.5×10^9
    c. 4.4×10^4
    d. 5.1×10^{10}

31. For the reaction $A(g) + B(g) \leftrightharpoons C(g) + D(g)$, $K_c = 3.7 \times 10^{-6}$. If the initial concentrations are $[D] = [C] = 0\ M$ and $[A] = [B] = 0.10\ M$, what is the concentration of A at equilibrium?
    a. $0.10\ M$
    b. $3.7 \times 10^{-6}\ M$
    c. $3.7 \times 10^{-8}\ M$
    d. $1.9 \times 10^{-4}\ M$

32. For $H_2SO_3 \leftrightharpoons H^+ + HSO_3^-$, $K = 1.5 \times 10^{-2}$ and for $HSO_3^- \leftrightharpoons H^+ + HSO_3^{2-}$, $K = 6.3 \times 10^{-8}$, what is K for $H_2SO_3 \leftrightharpoons 2\ H^+ + SO_3^{2-}$?
    a. 1.5×10^{-2}
    b. 9.4×10^{-10}
    c. 2.4×10^6
    d. 4.2×10^{-6}

33. If 0.10 mol of BaF_2 is added to 1 L of water, the equilibrium is $BaF_2(s) \leftrightharpoons Ba^{2+}(aq) + 2\ F^-(aq)$ with $K_c = 1.8 \times 10^{-7}$. What is the concentration of fluoride ion?
    a. $3.5 \times 10^{-3}\ M$
    b. $7.1 \times 10^{-3}\ M$
    c. $1.6 \times 10^{-3}\ M$
    d. $4.5 \times 10^{-3}\ M$

34. What is the equilibrium constant (K_c) for the reaction $A(g) + B(g) \leftrightharpoons C(s) + D(g)$ if, at equilibrium, there is 0.10 mol of each product and reactant in a 2.00 L container?
    a. 1.0
    b. 0.050
    c. 0.10
    d. 20

35. What is the concentration of OH^- in 1.00 L of pure water if $H_2O(\ell) \leftrightharpoons H^+(aq) + OH^-(aq)$ with $K = 1.0 \times 10^{-14}$?
   a. 1.0×10^{-14} *M*
   b. 5.0×10^{-15} *M*
   c. 1.0×10^{-7} *M*
   d. more information is needed to determine

Additional Practice Problems

1. Write the mass action expressions (as K_c) for the following reactions.
   a. $CH_4(g) + 2\,O_2(g) \leftrightharpoons CO_2(g) + 2\,H_2O(g)$
   b. $NH_4Cl(s) \leftrightharpoons NH_3(g) + HCl(g)$
   c. $Br_2(\ell) + H_2(g) \leftrightharpoons 2\,HBr(g)$
   d. $HF(aq) + H_2O(\ell) \leftrightharpoons H_3O^+(aq) + F^-(aq)$
   e. $2\,P(s) + 5\,Cl_2(g) \leftrightharpoons 2\,PCl_5(g)$

2. For the gaseous reaction $H_2 + Cl_2 \leftrightharpoons 2\,HCl$, $K = 2.8 \times 10^{28}$ at 350 K.
   a. What is the equilibrium constant for the reverse reaction?
   b. What is the equilibrium constant for $2\,H_2 + 2\,Cl_2 \leftrightharpoons 4\,HCl$?
   c. What is the equilibrium constant for

   $$H_2O + Cl_2 \leftrightharpoons 2\,HCl + \frac{1}{2}O_2.$$

   $$\text{(Also: } H_2O \leftrightharpoons H_2 + \frac{1}{2}O_2,\ K = 7.6 \times 10^{-35}\text{)}$$

   d. What is the $\Delta G°$ for this reaction?
   e. What is the equilibrium constant for this reaction at 700 K? ($\Delta H° = -184.6$ kJ/mol)
   f. What is the K_p for this reaction?

3. For the reaction $NH_3(aq) + H_2O(\ell) \leftrightharpoons NH_4^+(aq) + OH^-(aq)$, $K = 1.8 \times 10^{-5}$ at 25°C.
   a. What is the equilibrium constant for $2\,NH_4^+(aq) + 2\,OH^-(aq) \leftrightharpoons 2\,NH_3(aq) + 2\,H_2O(\ell)$?
   b. What is the $\Delta G°$ for this reaction?
   c. What is the *K* at 30°C ($\Delta H° = +90$ kJ/mol)?

4. For the endothermic reaction $PbCl_2(s) \leftrightharpoons Pb^{2+}(aq) + 2\,Cl^-(aq)$, $K = 1.2 \times 10^{-5}$, how does the concentration of lead(II) ion change if we
   a. add $PbCl_2$
   b. remove chloride ion
   c. add chloride ion
   d. increase the overall volume
   e. increase the temperature

5. For the exothermic gaseous reaction at equilibrium $SO_2 + 3\,H_2 \leftrightharpoons 2\,H_2O + H_2S$,
   a. What is the mass action expression?
   b. How will the equilibrium shift if volume decreases?
   c. How will the equilibrium shift if temperature increases?
   d. How will the equilibrium shift if steam is added?
   e. How will the concentration of water change if hydrogen is removed?

6. What is the equilibrium concentration of each product and reactant if 1.5 *M* NO_2 and 1.0 *M* SO_2 are allowed to reach equilibrium in the following gaseous reaction?

   $$SO_2 + NO_2 \leftrightharpoons NO + SO_3 \qquad K_c = 1.2$$

7. What is the equilibrium concentration of each product and reactant if the partial pressure of each reactant is 0.50 atm. (All reactants and products are gaseous.)

   $$N_2 + O_2 \leftrightharpoons 2\,NO \qquad K_p = 1.9 \times 10^{-11}$$

8. What is the equilibrium concentration of each product and reactant for the gaseous reaction

   $$N_2O + O_2 \leftrightharpoons N_2 + O_3 \qquad K_c = 8.5 \times 10^{-4}$$

   if the initial concentrations are $[N_2O] = 0.13$ *M*, $[O_2] = 0.012$ *M*, $[N_2] = 0.0091$ *M*, and $[O_3] = 0.014$ *M*.

9. Using the appendix of thermodynamic values in the textbook to determine the value of $\Delta G°$ for the reaction, what is the equilibrium constant for the reaction $2\,CO(g) + O_2(g) \leftrightharpoons 2\,CO_2(g)$ at 700° C?

10. For the reaction $CO(g) + Cl_2(g) \leftrightharpoons COCl_2(g)$, $K_c = 5.1$ at 700 K.
    a. If the concentration of each reactant and product is 1.00 *M*, what is the reaction direction?
    b. What is the concentration of each product and reactant at equilibrium?

Self-Test Answer Key

Key Vocabulary Completion Answers

1. forward
2. squared
3. equilibrium
4. stress
5. mass action expression
6. catalyst
7. reaction quotient

8. Le Châtelier's principle
9. reverse
10. endothermic
11. decrease
12. equilibrium
13. negative
14. atmospheres
15. greater than one
16. constant
17. molarity
18. Clausius-Clapeyron equation
19. equilibrium
20. heterogeneous

Multiple-Choice Answers

For further information about the topics these questions address, you should consult the text. The appropriate section of the text is listed in parentheses after the explanation of the answer.

1. c. At equilibrium, while the concentrations of reactants and products are constant, the reaction continues in both directions. When a reactant is consumed, one is also created. (15.1)

2. b. The equilibrium constant is the ratio of products to reactants (products on the top of the ratio). Consequently, a small fraction, such as 4.6×10^{-6}, reflects a higher concentration of reactants than products. (15.2)

3. d. Solids are not included as part of the equilibrium constant expression. Because the concentration of a solid is constant, it is included in the equilibrium constant rather than the concentration part of the expression. (15.6)

4. b. Equilibrium constant is related to the standard free energy, $\Delta G°$, by the equation $\Delta G° = -RT \ln K$. Thus large values of K result in negative values of $\Delta G°$ and fractional values of K in positive $\Delta G°$. ΔG and Q do not refer to equilibrium conditions. (15.5)

5. b. K_p expressions use partial pressure rather than concentration, but are still the product of the partial pressures rather than the sum. Mass action expressions are always products divided by reactants. (15.2)

6. b. The reaction quotient predicts reaction direction.

$$Q = \frac{[COCl_2]}{[CO][Cl_2]} = \frac{(0.1)}{(0.1)(0.1)} = 10$$

Product concentration must decrease until the equilibrium constant is reached. Thus the reaction proceeds in the reverse direction. (15.4)

7. c. $K_p = K_c(RT)^{\Delta n}$, where Δn = products – reactants = $1 - 2 = -1$. Therefore, $K_p = (5.0)[(0.08206)(700)]^{-1} = 8.7 \times 10^{-2}$. (15.2)

8. b. Le Châtelier's principle states that equilibrium will shift to minimize the stress. Since the stress is added reactant, the system will create products to minimize the stress. (15.7)

9. b. Le Châtelier's principle states that equilibrium will shift to minimize the stress. The stress of a decreased volume is reduced by minimizing the moles of gas. Since the product side has fewer moles of gas, the reaction will shift toward the products. (15.7)

10. b. Le Châtelier's principle states that equilibrium will shift to minimize the stress. Since the reaction is endothermic, energy acts as a reactant. Increasing the temperature is like increasing the reactants. To minimize the reactants, more products are created. (15.7)

11. a. Solids are included in the constant rather than in the mass action expression because the concentration of a solid is constant. (15.6)

12. a. Since all compounds in the mass action expression are gases, K_p is appropriate. $K_p = K_c(RT)^{\Delta n}$. Δn = products – reactants = $1 - 0 = 1$. Therefore, $K_p = (3.5 \times 10^{-20})[(0.08206)(350)] = 1.0 \times 10^{-18}$. (15.2)

13. c. Because the substance added is a solid, the concentration of calcium carbonate does not change. Therefore, the equilibrium does not shift. (15.7)

14. a. Le Châtelier's principle states that the system will adjust to minimize stress. In the case of increased volume, the system will shift to create more moles of gas. There are zero moles of gas on the reactant side and one on the product side; therefore, the reactant will shift toward the product side. (15.7)

15. b. Le Châtelier's principle states that the system will adjust to minimize stress. Since the reaction is endothermic ($+\Delta H$), energy acts as a reactant. With a decrease in temperature, energy is taken away and the system shifts toward the reactants to replace it. (15.7)

16. d. A catalyst increases the rate of reaction but does not affect the final concentration or energies of products and reactions. However, equilibrium is reached at a faster rate. (15.7)

17. c. The reaction $2\ NO + O_2 \leftrightharpoons 2\ NO_2$ has been multiplied by $\frac{1}{2}$. The equilibrium constant is raised to the power of whatever factor the equation was multiplied by. $K = (4.6 \times 10^{30})^{1/2} = 2.1 \times 10^{15}$. (15. 3)

18. b. The reaction has been reversed. The equilibrium constant is the reciprocal of the original reaction.

$$K = \frac{1}{4.6 \times 10^{30}} = 2.2 \times 10^{-31}. \ (15.3)$$

19. c. Since the equilibrium constant is very large, almost no reactant (like O_2) remains. (15.2)

20. c. $Q = \frac{[H_2SO_4]}{[H_2O][SO_3]} = 1.00$. Since $Q = K$, the reaction is at equilibrium and the concentrations are constant. (15.4)

21. a. The addition of reactant will shift the equilibrium toward the products. Therefore, the concentration of H_2SO_4 will increase. (15.7)

22. b. The addition of reactant will shift the equilibrium toward the products. Therefore, the concentration of other reactants (like SO_3) will decrease. (15.7)

23. a. An increase in volume will shift the equilibrium to the side with more moles of gas. In this problem, the reactant side has more moles of gas. Therefore, the concentration of water, a reactant, will increase. (15.7)

24. b. If product is removed, the equilibrium will shift to replace product; thus reactants are used up and the concentration of reactants decreases. (15.7)

25. c. The equilibrium constant does not change, since it shows the relationship between concentrations. (15.7)

26. d. The equilibrium shift depends on whether the system is endothermic or exothermic ($\Delta H°$). (15.7 and 15.9)

27. a. For an endothermic reaction, an increase in temperature will shift the equilibrium toward the products. Therefore, the equilibrium constant must change to express an increase in products (with the corresponding decrease in reactants). Therefore, K increases. (15.9)

28. d. Equilibrium constant does not provide any information about rate of reaction. (15.1)

29. c. $K_p = K_c(RT)^{\Delta n}$. Δn = moles product – moles reactant $= 1 - 2 = -1$. So $K_p = (1)[(0.08206)(200)]^{-1} = 0.061$. (15.2)

30. d. $K_p = K_c(RT)^{\Delta n}$. Δn = moles product – moles reactant $= 1 - 3 = -2$. So

$$4.7 \times 10^7 = K_c[(0.08206)(400)]^{-2}$$

$$4.7 \times 10^7[(0.08206)(400)]^2 = K_c$$

$$5.1 \times 10^{10} = K_c \ (15.2)$$

31. d. If x is the amount by which the concentrations change, C and D increase by x and A and B decrease by x. Thus the equilibrium concentrations are $[D] = [C] = x$ and $[A] = [B] = 0.10 - x$. The low value of K implies that x will be small, so the concentration of A and B will not change significantly. $[A] = [B] = 0.10$. Using these values in the equilibrium constant expression:

$$3.7 \times 10^{-6} = \frac{x^2}{(0.10)^2}. \ x = 1.9 \times 10^{-4}. \ (15.8)$$

32. b. The two reactions are added together to get the final reaction. When adding reactions, the equilibrium constants are multiplied. Therefore, $K_1K_2 = K_{total} = (1.5 \times 10^{-2})(6.3 \times 10^{-8}) = 9.4 \times 10^{-10}$. (15.3)

33. b. The equilibrium constant expression is $K = [Ba^{2+}][F^-]^2$. If the amount of barium ion gained is x, then the amount of fluoride gained is $2x$. Solving the expression for x,

$$1.8 \times 10^{-7} = x(2x)^2 = 4x^3$$

$$3.5 \times 10^{-3} = x$$

but, $[F^-] = 2x = 7.1 \times 10^{-3}\ M$. (15.8)

34. d. Since it is a 2.00-L container, the concentration of A, B, and D is $\frac{0.10 \text{ mol}}{2.00 \text{ L}} = 0.050\ M$. The concentration of C is constant because it is a solid. The mass action expression is $K = \frac{[D]}{[A][B]}$, so

$$K = \frac{(0.05)}{(0.05)(0.05)} = 20. \ (15.2)$$

35. c. The equilibrium constant expression is $K = [H^+][OH^-]$. The amount of OH^- will be the same as the amount of H^+ produced. Therefore, $K = x \cdot x = x^2 = 1.0 \times 10^{-14}$. Thus $[OH^-] = 1.0 \times 10^{-7}\ M$. (15.8)

ADDITIONAL PRACTICE PROBLEM ANSWERS

1. a. $K = \dfrac{[CO_2][H_2O]^2}{[CH_4][O_2]^2}$

   b. $K = [NH_3][HCl]$

   c. $K = \dfrac{[HBr]^2}{[H_2]}$

   d. $K = \dfrac{[H_3O^+][F^-]}{[HF]}$

   e. $K = \dfrac{[PCl_5]^2}{[Cl_2]^5}$

2. a. the reverse $= \dfrac{1}{K} = \dfrac{1}{(2.8 \times 10^{28})} = 3.6 \times 10^{-29}$

   b. times 2 $= K^2 = (2.8 \times 10^{28})^2 = 7.8 \times 10^{56}$

   c. add equations, then $K = K_1K_2 =$
   $(2.8 \times 10^{28})(7.6 \times 10^{-35}) = 2.1 \times 10^{-6}$

   d. $\Delta G° = -RT \ln K = -(8.314)(350) \ln(2.8 \times 10^{28})$
   $= -1.9 \times 10^5 \text{ J} = -190 \text{ kJ}$

   e. $\ln\left(\dfrac{K_1}{K_2}\right) = -\left(\dfrac{\Delta H°}{R}\right)\left(\dfrac{1}{T_2} - \dfrac{1}{T_1}\right)$

   $\ln\left[\dfrac{K_2}{(2.8 \times 10^{28})}\right] = \left(\dfrac{184{,}600}{8.314}\right)\left(\dfrac{1}{700} - \dfrac{1}{350}\right)$

   $\ln\left[\dfrac{K_2}{(2.8 \times 10^{28})}\right] = -31.7$

   $\dfrac{K_2}{(2.8 \times 10^{28})} = 1.73 \times 10^{-14}$

   $K_2 = 4.8 \times 10^{-14}$

3. $NH_3 + H_2O \leftrightharpoons NH_4^+ + OH^- \qquad K = 1.8 \times 10^{-5}$

   a. reverse and multiply by 2

   $K_a = \left(\dfrac{1}{K}\right)^2 = \left(\dfrac{1}{1.8 \times 10^{-5}}\right)^2 = 3.1 \times 10^9$

   b. $\Delta G° = -RT \ln K$
   $\Delta G° = -(8.314 \text{ J/mol}\cdot\text{K})(298 \text{ K}) \ln (1.8 \times 10^{-5})$
   $\Delta G° = +27{,}068 \text{ J/mol} = +2.71 \times 10^4 \text{ J}$

   c. $\ln\left(\dfrac{K_2}{K_1}\right) = -\left(\dfrac{\Delta H°}{R}\right)\left[\left(\dfrac{1}{T_2} - \dfrac{1}{T_1}\right)\right]$

   $\ln\left[\dfrac{K_2}{(1.8 \times 10^{-5})}\right] = -\left(\dfrac{90{,}000 \text{ J/mol}}{8.314 \text{ J/mol}\cdot\text{K}}\right)\left[\dfrac{1}{303} - \dfrac{1}{298}\right]$

   $\dfrac{K_2}{(1.8 \times 10^{-5})} = 0.5994$

   $K_2 = 3.3 \times 10^{-5}$

4. Energy + $PbCl_2(s) \leftrightharpoons Pb^{2+}(aq) + 2\, Cl^-(aq)$

   a. Adding $PbCl_2$ causes no change in $[Pb^{2+}]$.

   b. Removing Cl^- creates more $[Pb^{2+}]$.

   c. Adding Cl^- results in less $[Pb^{2+}]$.

   d. increasing volume does not affect the concentration of $[Pb^{2+}]$.

   e. Increasing temperature increases $[Pb^{2+}]$.

5. $SO_2 + 3\, H_2 \leftrightharpoons 2\, H_2O + H_2S$ + energy

   a. $K = \dfrac{[H_2O]^2[H_2S]}{[SO_2][H_2]^3}$

   b. Decreasing volume shifts the equilibrium toward the products.

   c. Increasing temperature shifts the equilibrium toward the reactants.

   d. Adding steam (H_2O) shifts the equilibrium toward the reactants.

   e. The concentration decreases.

6. $SO_2 + NO_2 \leftrightharpoons NO + SO_3 \qquad K_c = 1.2$

	SO_2	NO_2	NO	SO_3
Initial	1.0	1.5	0	0
Change	$-x$	$-x$	$+x$	$+x$
Equilibrium	$1.0 - x$	$1.5 - x$	x	x

   $1.2 = \dfrac{(x)(x)}{(1.0 - x)(1.5 - x)} = \dfrac{x^2}{1.5 - 2.5x + x^2}$

   $1.8 - 3.0x + 1.2x^2 = x^2$

   $x = \dfrac{3.0}{2(0.2)} \pm \dfrac{[9 - (4)(0.2)(1.8)]^{1/2}}{2(0.2)}$

   $x = 0.63$ or 14.4 (14.4 cannot be correct because it would result in negative reactant concentrations.)

   $[SO_2] = 1.0 - 0.63 = 0.4\ M$

   $[NO_2] = 1.5 - 0.63 = 0.9\ M$

   $[NO] = 0.6\ M$

   $[SO_3] = 0.6\ M$

7. $N_2 + O_2 \rightleftharpoons 2\,NO \qquad K_p = 1.9 \times 10^{-11}$

	N_2	O_3	NO
Initial	0.5	0.5	0
Change	$-x$	$-x$	$+2x$
Equilibrium	$0.5 - x$	$0.5 - x$	$2x$

$$1.9 \times 10^{-11} = \frac{(2x)^2}{(0.5-x)(0.5-x)} = \frac{4x^2}{(0.25 - x + x^2)}$$

$$4.75 \times 10^{-12} - 1.9 \times 10^{-11}x + 1.9 \times 10^{-11}x^2 = 4x^2$$

$$0 = 4x^2 + 1.9 \times 10^{-11}x - 4.75 \times 10^{-12}$$

$$x = \frac{-1.9 \times 10^{-11}}{2(4)} \pm \frac{[(3.61 \times 10^{-22} - (4)(4)(-4.75 \times 10^{-12})]^{1/2}}{2(4)}$$

$$x = \frac{-1.9 \times 10^{-11} \pm 8.7 \times 10^{-6}}{8} = 1.1 \times 10^{-6}$$

Or if you assume "x is small," then

$$1.9 \times 10^{-11} = \frac{(2x)^2}{(0.5-x)(0.5-x)}$$

can be simplified to

$$1.9 \times 10^{-11} = \frac{(2x)^2}{(0.5)(0.5)} = \frac{4x^2}{0.25}$$

$4.75 \times 10^{-12} = 4x^2$
$1.1 \times 10^{-6} = x$
$P_{N_2} = P_{O_2} = 0.5 - 1.1 \times 10^{-6} = 0.5$ atm
$P_{NO} = 2(1.1 \times 10^{-6}) = 2.2 \times 10^{-6}$ atm

Note: The equation can also be solved by taking the square root of each side so that

$$1.9 \times 10^{-11} = \frac{(2x)^2}{(0.5-x)(0.5-x)} \text{ becomes}$$

$$4.35 \times 10^{-6} = \frac{2x}{(0.5-x)}$$

8. $N_2O + O_2 \rightleftharpoons N_2 + O_3 \qquad K_c = 8.5 \times 10^{-4}$

$$Q = \frac{[N_2][O_3]}{[N_2O][O_2]} = \frac{(0.0091)(0.014)}{(0.13)(0.012)} = 0.082$$

Since $Q > K$, the reaction goes in reverse, so

	N_2O	O_2	N_2	O_3
Initial	0.13	0.012	0.0091	0.014
Change	$+x$	$+x$	$-x$	$-x$
Equilibrium	$0.13 + x$	$0.12 + x$	$0.0091 - x$	$0.014 - x$

$$8.5 \times 10^{-4} = \frac{(0.0091-x)(0.014-x)}{(0.13+x)\,(0.012+x)}$$

$$8.5 \times 10^{-4} = \frac{(1.275 \times 10^{-4} - 0.0231x + x^2)}{(1.56 \times 10^{-3} + 0.142x + x^2)}$$

$$1.33 \times 10^{-6} + 1.21 \times 10^{-4} + 8.5 \times 10^{-4}x^2 = 1.275 \times 10^{-4} - 0.0231x + x^2$$

$$0 = 1.26 \times 10^{-4} - 0.0232x + 0.999x^2$$

$$x = \frac{0.0232}{2(0.999)} \pm \frac{[(5.38 \times 10^{-4} - (4)(0.999)(1.26 \times 10^{-4})]^{1/2}}{2(0.999)}$$

$$x = \frac{0.0232}{2} \pm \left[\frac{[3.45 \times 10^{-5}]^{1/2}}{2}\right]$$

$x = 0.0145$ or 0.0087

0.0145 won't work, since $0.0091 - x$ is negative. So,

$[N_2O] = 0.13 + 0.0087 = 0.14\ M$
$[N_2] = 0.012 + 0.0087 = 0.021\ M$
$[N_2] = 0.0091 - 0.0087 = 0.0004\ M$
$[N_3] = 0.014 - 0.0087 = 0.0053\ M$

9. $2\,CO + O_2 \rightleftharpoons 2\,CO_2 \qquad T = 700$ K
Using $\Delta G°_f$ values from the appendix
$\Delta G° = 2(-394.4) - [2(-137.2) + 0] =$
-514.4 kJ/mol $= -514{,}400$ J/mol

$$\Delta G° = -RT \ln K$$

$-514{,}400 = -(8.314)(700) \ln K$
$88.39 = \ln K$
$2.4 \times 10^{38} = K$

10. $CO + Cl_2 \rightleftharpoons COCl_2 \qquad K = 5.1 = \dfrac{[COCl_2]}{[CO][Cl_2]}$

$Q = \dfrac{1}{1 \cdot 1} = 1$; $Q < K$, so reaction goes in the forward direction.

	CO	Cl_2	$COCl_2$
Initial	1.00	1.00	1.00
Change	$-x$	$-x$	$+x$
Equilibrium	$1 - x$	$1 - x$	$1 + x$

$$5.1 = \frac{(1+x)}{(1-x)(1-x)} = \frac{(1+x)}{1 - 2x + x^2}$$

$5.1 - 10.2x + 5.1x^2 = 1 + x$
$5.1 - 9.2x + 5.1x^2 = 0$

$$x = \frac{9.2}{2(5.1)} \pm \frac{[(9.2)^2 - 4(5.1)(5.1)]^{1/2}}{2(5.1)}$$

$$x = \frac{-9.2}{8.2} + \frac{[84.64 - 83.64]^{1/2}}{8.2}$$

$$x = \frac{1}{8.2} = 0.12$$

$[CO] = 1.00 - 0.12 = 0.88\ M$
$[Cl_2] = 1.00 - 0.12 = 0.88\ M$
$[COCl_2] = 1.00 + 0.12 = 1.12\ M$

CHAPTER 16 Equilibrium in the Aqueous Phase

OUTLINE

REVIEW

Chapter Overview

In a **Brønsted–Lowry acid–base reaction**, the transfer of H^+ (also called a proton) occurs in both directions. Therefore, the product of the reaction between an acid and a base is an acid and a base. Thus when an acid reacts in an acid–base reaction, the loss of H^+ creates a base called its conjugate base. Similarly, the gain of a proton by a base creates its conjugate acid. Two substances differing by only one H^+ are a **conjugate acid–base pair**.

Acids can be either strong or weak. A **strong acid** will react completely with water, producing $H_3O^+(aq)$ as one product. **Weak acids** still produce H_3O^+ in their reaction with water, but in an incomplete (equilibrium) reaction. There are seven strong acids: hydrochloric (HCl), hydrobromic (HBr), hydroiodic (HI), nitric (HNO_3), perchloric ($HClO_4$), chloric ($HClO_3$), and sulfuric (H_2SO_4). The easiest way to tell the difference between weak and strong acids is to memorize the strong acids. All other acids are weak.

Stronger acids produce higher concentrations of H_3O^+. The strength of a weak acid depends on the degree of ionization, which is the amount of acid that ionizes in water. The equilibrium constant for a weak acid with water is called the acid-ionization constant, or $\boldsymbol{K_a}$. Since the ions are products of the reaction, larger K_a values represent stronger acids. Strong acids do not have K_as values, since they react completely, rather than react in equilibrium, with water. In addition, all strong acids are the same strength, as they each ionize 100%. Therefore, the product of the ionization, H^+ (or H_3O^+), is the strongest acid that can exist in water. The fact that strong acids are of the same strength is called the **leveling effect** of water.

The relative strength of acids with similar chemical structures can be predicted. For oxoacids, acids containing oxygen as well as hydrogen and another element, substances with more oxygen atoms, but the same otherwise, will be stronger. Similarly, if the substances differ by only the central atom, more electronegative atoms are (generally) stronger acids.

Polyprotic acids have more than one H^+ that is acidic (reacts with a base). For polyprotic acids, reaction of the first proton is K_{a_1}; that of the second is K_{a_2}; and so forth. With polyprotic acids, the second ionization is always weaker than the first ($K_{a_1} > K_{a_2}$). The third ionization (if there is one) will be weaker than the second ($K_{a_2} > K_{a_3}$).

Sulfuric acid is a special case of both a strong acid and a polyprotic acid. The first acidic H^+ is a strong acid and reacts completely with water. The second H^+ reacts in equilibrium with water (a weak acid). However, HSO_4^- is one of the strongest common weak acids.

Bases can also be strong or weak. Like strong acids, strong bases ionize completely in water. The difference is that bases produce OH^- instead of H_3O^+. There are only a few strong bases. Like acids, it is easier to memorize the strong bases and recognize that all other bases are weak. The strong bases are hydroxide salts of a group 1 metal, calcium, strontium, or barium. In water, these salts simply dissolve into their component ions. These bases are strong because they are completely soluble in water and produce OH^-. Other hydroxide salts are not soluble, so they cannot react completely and are, therefore, weak bases. The nitrogen bases react with water by taking an H^+ from the water (the H^+ bonds to the lone pair on the nitrogen), leaving OH^- as a product. The equilibrium constant for this reaction is $\boldsymbol{K_b}$. With a leveling effect parallel to that of acids, the strongest base that can exist in water is hydroxide ion.

When it reacts with acid, water acts as a base. When it reacts with a base, water acts as an acid. Substances that can act as either an acid or a base are called **amphoteric**. Water can also undergo an acid–base reaction with itself, called **autoionization**.

$$H_2O + H_2O \leftrightharpoons H_3O^+ + OH^-$$

Since water is also the solvent in this reaction, the equilibrium constant expression is

$$K_w = [H_3O^+][OH^-]$$

and the value of the equilibrium constant, K_w, is 1.0×10^{-14} at 25°C (room temperature). Therefore, the concentration of H_3O^+ and of OH^- in pure water at room temperature is 1.0×10^{-7} *M*. Acid–base reactions in water can shift this equilibrium. Therefore, in any aqueous solution, if you know the concentration of either H_3O^+ or OH^-, the other can be determined from the K_w expression.

An alternate way of expressing concentration of H^+ is with pH, which is the negative log of the molarity of H^+. Therefore,

$$\mathbf{pH} = -\log[H_3O^+]$$

Similarly, the concentration of hydroxide ion can be expressed as pOH

$$\mathbf{pOH} = -\log[OH^-]$$

Other values can also be written in this format, where pX = –logX. Thus, K_a, K_b, and K_w values can also be written as pK_a, pK_b, and pK_w, where $pK = -\log K$. The advantage of expressing value in a pX format is that scientific notation is not required. However, it is important to recognize that each time pX changes by a factor of one, the value of X changes by a factor of 10. In addition, higher values of pX actually indicate *lower* values of X.

Using these relationships, the K_w expression can be rewritten as

$$pH + pOH = pK_w = 14.00$$

The relative strengths of acid–base conjugate pairs is also related by K_w. There is an inverse relationship between acid and base strength. That is, the weaker the acid, the stronger its conjugate base, and vice versa. This inverse relationship between the strength of conjugate acid–base pairs is quantitatively expressed as

$$K_aK_b = K_w = 1.0 \times 10^{-14} \quad \text{or} \quad pK_a + pK_b = 14.00$$

Since the conjugate base has lost an H^+ ion, it normally has a negative charge. Consequently, most anions act as bases. Exceptions are Cl^-, Br^-, I^-, NO_3^-, ClO_4^-, ClO_3^-, and SO_4^{2-}. These conjugate bases of the strong acids are such weak bases that they will not react with water. The K_b values of an anionic base can be determined from the K_a of its conjugate acid using the $K_aK_b = K_w$ relationship.

Similarly, most cations act as acids. This is most obvious with the nitrogen bases, where the hydrogen bonded to the nitrogen creates the conjugate acid of the nitrogen base. Less obviously, metal ions can also act as bases. The cations associated with the strong bases (group I metal ions, Ca^{2+}, Sr^{2+}, and Ba^{2+}) do not react with water. The K_a of cationic acids can be determined from the K_b of its conjugate base using the $K_aK_b = K_w$ relationship.

The addition of an ion taking part in a reaction causes a shift in the position of the equilibrium according to Le Châtelier's principle. In any ionic equilibrium, the reaction that produces an ion is suppressed when another source of the same ion is added to the system. This is called the **common-ion effect**.

One important example of the common-ion effect is a mixture of an acid and its conjugate base (or a base and its conjugate acid), which forms a buffer solution. **Buffers** resist pH change with the addition of water, small amounts of acid, or small amounts of base. Using Le Châtelier's principle, it can be seen that the addition of conjugate base, a product of the equilibrium, makes the amount of H_3O^+ produced by the reaction much smaller than the acid alone. Because the reaction only occurs to a small extent, the concentration of acid and base is not significantly affected by the equilibrium. Thus the K_a expression can then be rearranged to the **Henderson–Hasselbalch equation**:

$$pH = pK_a + \log\left(\frac{[\text{base}]}{[\text{acid}]}\right)$$

With this equation it is easy to observe that diluting the solution would dilute the acid and base concentrations by the same factor, and not change the pH. The presence of base ensures that a small amount of additional acid is neutralized. Likewise, a small amount of additional base is neutralized by the acid. The amount of acid (or base) that neutralizes a buffer without significant pH change is called the **buffer capacity**. Higher concentrations of acid will neutralize larger amounts of base, as higher concentrations of base neutralize more acid. Therefore, buffer capacity is maximized with high concentrations of acid and base. Since the acid and base concentrations are part of a log term, large changes in concentration are required to change pH significantly. Since both need to be present in significant concentration (so that both acid and base can be neutralized), the most efficient buffers are designed so that the acid and base concentrations are about equal and the pH ≈ pK_a. Thus the pH range over which the buffer is effective is approximately: pH = $pK_a \pm 1$.

Acid–base **indicators** are a special type of weak acid or weak base. What makes these substances special is that the acid is one color and its conjugate base is a different color. Since the concentration of acid or base is controlled by pH, the color of an indicator solution is pH dependent. The Henderson–Hasselbalch equation shows that when pH = pK_a, the acid and base concentrations are equal. Thus the color will be a mixture of both the acid and base forms. (For example, if an indicator has a yellow acid form and a blue basic form, the solution will be green when pH = pK_a.) If there is at least 10 times higher acid concentration than base, the acid form and color predominate (pH < pK_a – 1.0). Likewise, the basic form and color predominate at pH > pK_a + 1.0.

By using a strong acid or a strong base as a reactant, an acid–base reaction can be forced to completion. In a complete reaction, stoichiometry can be used to relate reactants and products to each other. A **titration** is an experiment where the volume of reactant solution (**titrant**) needed to react exactly with some amount of another reactant is measured. This volume is called the **equivalence point** and can be determined stoichiometrically. Calculations involving titrations were studied in Chapter 4.

The progress of an acid–base titration can be followed with a **titration curve**, a graph of pH versus volume of titrant. The pH of the reaction mixture is measured as titrant is added to the other reactant. Before titrant is added, the pH depends on the reactant. After some titrant is added, the concentration of reactant is reduced (due to both reaction and dilution) and the conjugate acid or base of the reactant is created. Consequently, if the reactant is a weak acid or a weak base, a buffer is created. Half way to the equivalence point, the moles of acid and moles of base are equal, and the Henderson–Hasselbalch equation shows that pH = pK_a. Even if the reactant is a weak base, the pH = pK_a of its conjugate acid. At the equivalence point, only the conjugate acid or base is present in the reaction mixture. Since the product of the reaction of a weak base is its conjugate acid, the pH at the equivalence point is weakly acidic. Likewise, the pH of a weak acid at the equivalence point is slightly basic. After the equivalence point, more titrant is added, but it does not react. Therefore, the pH will be determined by the titrant.

Worked Examples

Acids and Bases: A Molecular View

Example 1.
Identify the acid and base in the following reactions.

a. $NH_3 + HNO_2 \rightleftharpoons NH_4^+ + NO_2^-$

b. $CH_3COOH + OH^- \rightleftharpoons H_2O + CH_3CO_2^-$

c. $H_2O + HSO_4^- \rightleftharpoons H_3O^+ + SO_4^{2-}$

d. $H_2O + HS^- \rightleftharpoons OH^- + H_2S$

e. $CO_3^{2-} + HS^- \rightleftharpoons HCO_3^- + S^{2-}$

Collect and Organize We are asked to identify which of the reactants is the acid and which is the base.

Analyze Brønsted–Lowry acids and bases are defined based on the reaction. The substance that loses an H^+ is the acid; the substance that gains it is the base.

Solve

a. NH_3 gains an H^+ to become NH_4^+, making it the base. HNO_2 loses the H^+ to become NO_2^-, making it the acid.

b. CH_3COOH loses the proton on the end, making it the acid. OH^- gains an H^+ to become water, making it the base.

c. H_2O gains H^+ to become H_3O^+, so it is the base. HSO_4^- loses its H^+ to become SO_4^{2-}, so it is the acid.

d. In this case H_2O is the acid, losing H^+ to be OH^-. (It depends on the reaction!) HS^- is the base, gaining H^+ to be H_2S.

e. HS^- is the acid. It loses H^+ to become S^{2-}. CO_3^{2-} is the base; it gains H^+.

Think About It The Brønsted–Lowry definition of acids and bases is one of three classification systems. The other two are the Arrhenius definition in which an acid is a substance that produces H^+ in water and a base is a substance that produces OH^- in water, and the Lewis definition in which the acid acts as an electron acceptor and the base acts as an electron donor.

Example 2.
What is the conjugate acid of the following?

a. H_2O b. OH^- c. $(CH_3)_3N$ d. S^{2-}

Collect and Organize We are asked to determine the conjugate acids of the given bases.

ANALYZE A conjugate acid–base pair differs by one H^+. Adding an H^+ creates the conjugate acid.

SOLVE

a. Since water is normally involved in acid–base reactions, its conjugate acid and base are important. Adding H^+ creates H_3O^+.

b. The conjugate acid is H_2O. (This is common enough that you should recognize it.)

c. Since the H^+ bonds with the nitrogen of the amine, it is customary to add it next to the nitrogen: $(CH_3)_3NH^+$.

d. Adding only one H^+, the answer is HS^-.

THINK ABOUT IT Common mistakes are to add (or subtract) more than one H^+ and to not change the charge along with the formula.

EXAMPLE 3.
What is the conjugate base of the following?

a. H_2O b. H_3O^+ c. CH_3COOH d. H_2S

COLLECT AND ORGANIZE We are asked to determine the conjugate base of the given acids.

ANALYZE Removing an H^+ from the acid creates the conjugate base.

SOLVE

a. Removing H^+ leaves OH^-.

b. Removing H^+ leaves H_2O.

c. With organic acids, the acidic proton (H^+) is the one at the end of the COOH. Therefore, the conjugate base is $CH_3CO_2^-$.

d. Only one H^+ is removed from H_2S, leaving HS^-.

THINK ABOUT IT In a Brønsted–Lowry acid–base equilibrium reaction, the transfer of H^+ occurs in both directions. Therefore, the product of the reaction between an acid and a base is an acid and a base. Thus when an acid reacts in an acid–base reaction, the loss of H^+ creates a base called its **conjugate base**. Similarly, the gain of a proton by a base creates its **conjugate acid**. Two substances differing by only one H^+ are a **conjugate acid–base pair**.

pH

EXAMPLE 1.
What is the hydroxide ion concentration if $[H^+] = 3.8 \times 10^{-4}\ M$?

COLLECT AND ORGANIZE Given the hydrogen ion concentration, we are asked to calculate the hydroxide ion concentration.

ANALYZE Because of the autoionization of water, the concentration of H_3O^+ and the concentration of OH^- are inversely related in any aqueous solution. Either of the following equations can be used.

$$K_w = 1.0 \times 10^{-14} = [H_3O^+][OH^-] \text{ at } 25°C$$

$$pK_w = 14.00 = pH + pOH$$

SOLVE

$$K_w = 1.0 \times 10^{-14} = [H_3O^+][OH^-]$$

$$1.0 \times 10^{-14} = [3.8 \times 10^{-4}][OH^-]$$

$$2.6 \times 10^{-11}\ M = [OH^-]$$

THINK ABOUT IT It would have been equally correct to use the expression $pK_w = 14.00 = pH + pOH$.

EXAMPLE 2.
What is the H_3O^+ concentration if $[OH^-] = 0.011\ M$?

COLLECT AND ORGANIZE We are given the hydroxide ion concentration and are asked to calculate the hydrogen ion concentration.

ANALYZE We will use the expression $K_w = 1.0 \times 10^{-14} = [H_3O^+][OH^-]$ and the given hydroxide concentration to solve for the hydrogen ion concentration.

SOLVE

$$K_w = 1.0 \times 10^{-14} = [H_3O^+][OH^-]$$

$$1.0 \times 10^{-14} = [H_3O^+][0.011]$$

$$9.1 \times 10^{-13}\ M = [H_3O^+]$$

THINK ABOUT IT H^+ and H_3O^+ are equivalent formulas for the hydrogen ion.

EXAMPLE 3.
At 0°C, $K_w = 1.14 \times 10^{-15}$. What is the $[H_3O^+]$ in pure water?

COLLECT AND ORGANIZE We are given the value of K_w at 0°C for pure water and are asked to determine what the $[H_3O^+]$ is at this temperature.

ANALYZE The equation is the same, but the constant (K_w) is different:

$$K_w = 1.14 \times 10^{-15} = [H_3O^+][OH^-]$$

Because water is the only source of H_3O^+ and OH^-, $[H_3O^+] = [OH^-]$.

SOLVE

$$1.14 \times 10^{-15} = [H_3O^+][OH^-] = x^2$$

$$3.38 \times 10^{-8}\ M = x = [H_3O^+]$$

THINK ABOUT IT The extent of dissociation of water into its component ions is temperature dependent (the higher the temperature the greater the dissociation). This can be seen in the larger value of K_w at higher temperature.

EXAMPLE 4.
What is the pH of solutions with the following concentrations of H_3O^+?

a. 0.023 *M* b. $5.74 \times 10^{-6}\ M$

COLLECT AND ORGANIZE We are asked to determine the pH of two solutions given the concentration of H_3O^+ in each.

ANALYZE To calculate the pH given the hydrogen ion concentration we use the expression $pH = -\log[H_3O^+]$.

SOLVE

a. 1.63 = pH. The concentration has two significant figures, so the answer is reported to two decimal places.

b. 5.241 = pH. The concentration has three significant figures, so the answer is reported to three decimal places.

THINK ABOUT IT The number of significant figures in the regular number ($[H_3O^+]$ in this case) becomes the number of decimal places in the log term (pH). For the log term, all digits to the right of the decimal are counted, regardless of whether the digit is zero. None of the digits to the left of the decimal count as significant figures.

EXAMPLE 5.
What is the concentration of H_3O^+ for the following pH values?

a. 8.1 b. 9.13

COLLECT AND ORGANIZE We are given the values of pH for two solutions and are asked to determine the concentration of H_3O^+ in each.

ANALYZE In order to determine the concentration of H_3O^+ from pH we need to rearrange the equation $pH = -\log[H_3O^+]$ to solve for $[H_3O^+]$. This is done by entering the negative of the pH into your calculator and using the 10^x key (usually "shift log").

SOLVE

a. $8 \times 10^{-9} = [H_3O^+]$. Since there is one digit to the right of the decimal, the concentration has one significant figure.

b. $7.4 \times 10^{-10} = [H_3O^+]$. Since there are two digits to the right of the decimal, the concentration value has two significant figures.

THINK ABOUT IT Given either the concentration of H_3O^+ ions or the pH we can easily convert to the other using the definition of pH.

EXAMPLE 6.
What is the pOH of a solution with $[OH^-] = 5.7 \times 10^{-10}\ M$?

COLLECT AND ORGANIZE We are asked to determine the pOH of a solution given its concentration of OH^-.

ANALYZE We will use the expression $pOH = -\log[OH^-]$.

SOLVE The pOH has two decimal places to match the significant figures in the concentration: pOH = 9.24.

THINK ABOUT IT pOH is determined in the same manner as pH, with the concentration of OH^- instead of H_3O^+.

EXAMPLE 7.
What is the hydroxide ion concentration in a solution with pOH = 4.7?

COLLECT AND ORGANIZE We are asked to determine the concentration of hydroxide ions in a solution given its pOH.

ANALYZE In order to determine the concentration of OH^- from pOH we need to rearrange the equation $pOH = -\log[OH^-]$ to solve for $[OH^-]$. This is done by entering the negative of the pOH into your calculator and using the 10^x key.

SOLVE The concentration has one significant figure: $[OH^-] = 2 \times 10^{-5}\ M$.

THINK ABOUT IT Because of the autoionization of water, hydroxide ion concentration can be expressed as pH.

EXAMPLE 8.
What is the hydroxide ion concentration in a solution with pH = 5.82?

COLLECT AND ORGANIZE Given the pH of a solution, we are asked to determine the concentration of *hydroxide* ions.

ANALYZE We can rewrite the K_w expression as pH + pOH = pK_w = 14.00. When the pOH is determined, the formula pOH = –log{OH⁻] can be used to determine the hydroxide ion concentration.

SOLVE Using the given value of pH = 5.82, we solve for pOH: pOH = 14.00 – 5.82 = 8.18. Using the antilog (10^x) of pOH, the concentration is determined to be 6.6×10^{-9} *M* = [OH^-].

THINK ABOUT IT In any aqueous solution, if you know the concentration of either H_3O^+ or OH^- (or conversely the pH or pOH) the other can be determined from the K_w expression.

EXAMPLE 9.
What is the pH if the hydroxide ion concentration is 2.4×10^{-4} *M*?

COLLECT AND ORGANIZE We are given the concentration of *hydroxide ions* in a solution and are asked to determine the pH.

ANALYZE After converting the given hydroxide concentration to pOH, we can then use the expression pH + pOH = 14.00 to solve for the pH.

SOLVE If the [OH^-] = 2.4×10^{-4} *M*, pOH = 3.62. Since pH + pOH = 14.00, 14.00 – 3.62 = 10.38.

THINK ABOUT IT It is important to recognize that each time pX changes by 1, it represents a change in concentration by a factor of 10 and that lower pX numbers represent higher values. Therefore, if solution A has a pH of 2 and solution B has a pH of 5, solution A has an H^+ concentration 1000 times greater than that of solution B.

EXAMPLE 10.
What are the pH and pOH of the following solutions?

a. 0.034 *M* HCl   b. 9.26 *M* HNO_3

COLLECT AND ORGANIZE We are given the concentrations of two solutions of strong acids and are asked to determine the pH and pOH for both.

ANALYZE Strong acids completely ionize in water. As a result, the concentration of the acid is the concentration of H_3O^+.

SOLVE

a. $HCl + H_2O \rightarrow H_3O^+ + Cl^-$

$$\left(\frac{0.034 \text{ mol HCl}}{\text{L}}\right) \times \left(\frac{1 \text{ mol } H_3O^+}{1 \text{ mol HCl}}\right) = 0.034\ M\ H_3O^+$$

$$\text{pH} = 1.47$$

$$\text{pOH} = 14.00 - 1.47 = 12.53$$

b. $HNO_3 + H_2O \rightarrow H_3O^+ + NO_3^-$

$$[H_3O^+] = 9.26\ M$$

$$\text{pH} = -0.967$$

$$\text{pOH} = 14.00 - (-0.967) = 14.97$$

THINK ABOUT IT Despite persistent rumors, the pH can be less than zero, and will be if the concentration of H_3O^+ is greater than 1.

EXAMPLE 11.
What are the pH and degree of ionization for 0.22 *M* HNO_2?

COLLECT AND ORGANIZE We are asked to determine the pH and degree of ionization for a solution of a weak acid of given concentration.

ANALYZE Most weak acids ionize to a very small extent. Under those circumstances, the ionization of the weak acid will not change the concentration of the acid. If the change in concentration is called *x*, you can assume that any *x* added to or subtracted from a number is zero. The general formula for the equilibrium reaction is

$$HA + H_2O \leftrightharpoons H_3O^+ + A^-$$

SOLVE

$$HNO_2 + H_2O \leftrightharpoons H_3O^+ + NO_2^-$$

$$K_a = \frac{[H_3O^+][NO_2^-]}{[HNO_2]} = 4.5 \times 10^{-4}$$

The K_a value is from the appendix at the back of the textbook. The ICE table can be used to determine the amount of each substance at equilibrium.

	HNO_2	H_3O^+	NO_2^-
Initial	0.22	0	0
Change	$-x$	$+x$	$+x$
Equilibrium	$0.22 - x$	x	x

Using the equilibrium values in the equilibrium constant expression,

$$4.5 \times 10^{-4} = \frac{(x)(x)}{(0.22 - x)}$$

Assuming x is small, the equation becomes

$$4.5 \times 10^{-4} = \frac{x^2}{0.22}$$

Solving for x, $9.9 \times 10^{-3} = x$ and pH = 2.00. The degree of ionization (percent ionization) is the amount of acid that is ionized (the change box in the acid column without the sign) divided by the initial amount of acid times 100:

$$\text{percent ionized} = \frac{x}{[\text{acid}]} \times 100 = \frac{(9.9 \times 10^{-3})}{(0.22)} \times 100 = 4.5\%$$

Think About It The assumption that the ionization of a weak acid will not change the concentration of the acid is usually safe to make if K_a is less than 10^{-5}.

Example 12.

What are the pH and percent ionization of 0.56 *M* HOI?

Collect and Organize We are asked to determine the pH and extent of ionization for a given concentration of a weak acid.

Analyze We will use the same procedure outlined in Example 11.

Solve

$$HOI + H_2O \rightleftharpoons H_3O^+ + OI^-$$

$$K_a = \frac{[H_3O^+][OI^-]}{[HOI]} = 2.3 \times 10^{-11}$$

	HOI	H_3O^+	OI^-
Initial	0.56	0	0
Change	$-x$	$+x$	$+x$
Equilibrium	$0.56 - x$	x	x

$$K_a = 2.3 \times 10^{-11} = \frac{(x)(x)}{(0.56 - x)}$$

Assuming x is small,

$$2.3 \times 10^{-11} = \frac{x^2}{0.56}$$

$$3.6 \times 10^{-6} = x$$

Checking the assumption, $0.56 - 3.6 \times 10^{-6} = 0.56$. So,

$$[H_3O^+] = x = 3.6 \times 10^{-6}\ M$$

$$\text{pH} = 5.44$$

Think About It It is always a good idea to verify the assumption that x is small enough to ignore when considering the change in concentration of the acid.

Example 13.

The pH of a 0.11 *M* solution of a weak acid is 4.51. What is its K_a?

Collect and Organize Given the concentration and pH of a weak acid, we are asked to calculate its K_a.

Analyze The general reaction of a weak acid in water is

$$HA + H_2O \rightleftharpoons H_3O^+ + A^-$$

So the K_a expression is $K_a = \frac{[H_3O^+][A^-]}{[HA]}$.

Because acid-base reactions are very fast, whenever pH is given, it is the *equilibrium* concentration.

Solve Using the ICE table,

	HA	H_3O^+	A^-
Initial	0.11	0	0
Change	$-x$	$+x$	$+x$
Equilibrium	$0.11 - x$	x	x

At equilibrium, the pH is 4.51. From pH, the $[H_3O^+]$ can be calculated:

$$[H_3O^+] = 10^{-4.51} = 3.1 \times 10^{-5}$$

According to the ICE table, $[H_3O^+] = x$. Using that value for x in the K_a expression,

$$K_a = \frac{[H_3O^+][A^-]}{[HA]} = \frac{(x)(x)}{(0.11 - x)}$$

$$= \frac{(3.1 \times 10^{-5})^2}{(0.11 - 3.1 \times 10^{-5})}$$

$$= 8.7 \times 10^{-9}$$

Think About It Weak acids undergo ionization to a very small degree. This can be seen by the small value obtained for K_a.

Example 14.

What is the K_a of a weak acid that is 1.49% ionized in water at a concentration of 0.19 *M*?

Collect and Organize We are asked to determine the K_a of a weak acid given its concentration and degree of ionization.

Analyze After setting up and solving the ICE table we will use the equation for percent ionization given by

$$\frac{x}{(\text{initial concentration HA})} \times 100 = \text{percent ionized}$$

to solve for x, which will be substituted into the expression to solve for K_a.

SOLVE The K_a expression is $K_a = \dfrac{[H_3O^+][A^-]}{[HA]}$. Using the ICE table,

	HA	H_3O^+	A^-
Initial	0.19	0	0
Change	$-x$	$+x$	$+x$
Equilibrium	$0.19 - x$	x	x

Solving for x,

$$\frac{x}{(0.19)} \times 100 = 1.49\%$$

$$x = 2.8 \times 10^{-3}$$

Using that value for x in the K_a expression,

$$K_a = \frac{[H_3O^+][A^-]}{[HA]} = \frac{(x)(x)}{(0.19 - x)}$$

$$= \frac{(2.8 \times 10^{-3})^2}{(0.19 - 0.0028)} = 4.2 \times 10^{-5}$$

THINK ABOUT IT The extent of ionization is minimal (only 1.49%) as we would expect for a weak acid.

EXAMPLE 15.
What is the pH of 0.034 *M* H_2SO_4?

COLLECT AND ORGANIZE We are asked to determine the pH of a strong polyprotic acid.

ANALYZE Polyprotic acids can donate more than one proton. These acids must donate one proton (H^+) at a time. The equilibrium constant for the first acidic H^+ is K_{a_1}. The equilibrium constant for the second acidic H^+ is always smaller than K_{a_1} and is called K_{a_2}. If there is a third acidic proton, its equilibrium constant is K_{a_3}. Because these reactions happen in sequence, they can be solved in the same sequence.

SOLVE From the first reaction,

$$H_2SO_4 + H_2O \rightarrow HSO_4^- + H_3O^+$$

So, $[H_3O^+] = [HSO_4^-] = 0.034$ *M*.

From the second reaction,

$$HSO_4^- + H_2O \leftrightharpoons SO_4^{2-} + H_3O^+$$

$$K_a = \frac{[H_3O^+][SO_4^{2-}]}{[HSO_4^-]} = 0.012$$

Using the ICE table, the answers from the first reaction become the initial values for the second equilibrium.

	HSO_4^-	H_3O^+	SO_4^{2-}
Initial	0.034	0.034	0
Change	$-x$	$+x$	$+x$
Equilibrium	$0.034 - x$	$0.034 + x$	x

Because the initial concentration of sulfate was zero, it must increase in concentration. (You can't lose what you don't have.) If the equilibrium shifts to produce sulfate, it must also produce H_3O^+. Consequently, its change is also positive. Using the equilibrium line in the K_a expression,

$$0.012 = \frac{(0.034 + x)(x)}{(0.034 - x)}$$

Assuming x is small (it cancels as both $+x$ and $-x$),

$$0.012 = \frac{(0.034)(x)}{(0.034)}$$

$$0.012 = x$$

Obviously, this is *not* small ($0.034 - 0.012 = 0.022$). Using the quadratic instead,

$$4.08 \times 10^{-4} - 0.012x = 0.034x + x^2$$

$$0 = x^2 + 0.046x - 4.08 \times 10^{-4}$$

$$x = \frac{-0.046}{2} \pm \frac{[(0.046)^2 - 4(1)(-4.08 \times 10^{-4})]^{1/2}}{2}$$

$$x = \frac{-0.046}{2} \pm \frac{[3.748 \times 10^{-3}]^{1/2}}{2} = \frac{-0.046}{2} \pm \frac{[0.06122]}{2}$$

The negative value is impossible, so $x = 0.0076$. Recall from the ICE table that

$$[H_3O^+] = 0.034 + 0.0076 = 0.0416 = 0.042\ M$$

$$pH = 1.38$$

THINK ABOUT IT Sulfuric acid is a special case of a polyprotic acid because the first acidic proton ionizes in a complete reaction while the second acidic proton only partially reacts with water but has a rather high K_a.

EXAMPLE 16.
What is the equilibrium concentration of each species in an aqueous solution of 0.57 *M* $H_2C_2O_4$?

COLLECT AND ORGANIZE Given the concentration of a solution of the weak polyprotic acid $H_2C_2O_4$, we are asked to determine the concentrations of each species present at equilibrium.

ANALYZE In the case of weak polyprotic acids each equilibrium is solved sequentially, with the equilibrium concentrations of the first reaction becoming the initial concentrations of the next reaction.

SOLVE The first equilibrium is

$$H_2C_2O_4 + H_2O \leftrightharpoons HC_2O_4^- + H_3O^+$$

$$K_{a_1} = 5.9 \times 10^{-2} = \frac{[HC_2O_4^-][H_3O^+]}{[H_2C_2O_4]}$$

Using the ICE table,

	$H_2C_2O_4$	H_3O^+	$HC_2O_4^-$
Initial	0.57	0	0
Change	$-x$	$+x$	$+x$
Equilibrium	$0.57 - x$	$+x$	$+x$

$$5.9 \times 10^{-2} = \frac{(x)(x)}{(0.57 - x)}$$

Assuming x is small,

$$5.9 \times 10^{-2} = \frac{x^2}{(0.57)}$$

$$0.18 = x$$

Obviously, x is not small $(0.57 - 0.18 = 0.39)$. Using the quadratic,

$$0.059 = \frac{x^2}{(0.57 - x)}$$

$$0.03363 - 0.059x = x^2$$

$$x = \frac{-0.059}{2} \pm \frac{[(0.059)^2 - 4(1)(-0.03363)]^{1/2}}{2}$$

$$= \frac{-0.059 \pm [0.3715]}{2}$$

Discarding the negative value, $x = 0.16$. So,

$$[H_2C_2O_4] = 0.57 - 0.16 = 0.41\ M$$

$$[H_3O^+] = [HC_2O_4^-] = 0.16\ M$$

For the second equilibrium,

$$HC_2O_4^- + H_2O \leftrightharpoons H_3O^+ + C_2O_4^{2-}$$

$$K_{a_2} = 6.4 \times 10^{-5} = \frac{[H_3O^+][C_2O_4^{2-}]}{[HC_2O_4^-]}$$

Using the ICE table,

	$HC_2O_4^-$	H_3O^+	$C_2O_4^{2-}$
Initial	0.16	0.16	0
Change	$-y$	$+y$	$+y$
Equilibrium	$0.16 - y$	$0.16 + y$	y

$$6.4 \times 10^{-5} = \frac{(0.16 + y)(y)}{(0.16 - y)}$$

Assuming y is small,

$$6.4 \times 10^{-5} = \frac{(0.16)(y)}{(0.16)}$$

$$6.4 \times 10^{-5} = y$$

Checking the assumption, $0.16 - 6.4 \times 10^{-5} = 0.16$, so the assumption is good. Therefore $[H_3O^+] = [HC_2O_4^-] = 0.16\ M$ and $[C_2O_4^{2-}] = 6.4 \times 10^{-5}\ M$ and pH = 0.80. Note that the pH is the same regardless of whether or not the second ionization is considered.

THINK ABOUT IT If the question asks for pH of a weak polyprotic acid, only the first equilibrium needs to be solved. The subsequent equilibria will not significantly change the concentration of H_3O^+.

EXAMPLE 17.
What is the pOH and pH of 0.0902 *M* KOH?

COLLECT AND ORGANIZE We are asked to determine the pOH and pH of a given concentration of a strong base.

ANALYZE Strong bases completely ionize in water. As a result, we can use stoichiometry to determine the number of hydroxide ions produced and therefore, the resulting pOH and pH.

SOLVE

$$KOH \rightarrow K^+ + OH^-$$

$$\frac{0.0902 \text{ mol KOH}}{L} \times \left(\frac{1 \text{ mol OH}^-}{1 \text{ mol KOH}}\right) = \frac{0.0902 \text{ mol OH}^-}{L}$$

$$pOH = -\log[0.0902] = 1.045$$

$$pH = 14.00 - 1.045 = 12.95$$

THINK ABOUT IT The strong bases are the group 1 hydroxides (LiOH, NaOH, KOH, RbOH, CsOH) and the lower group 2 hydroxides [$Ca(OH)_2$, $Sr(OH)_2$, $Ba(OH)_2$].

EXAMPLE 18.
What is the pOH and pH of 0.15 *M* $Ba(OH)_2$?

COLLECT AND ORGANIZE We are asked to determine the pOH and pH of a given concentration of a strong base.

ANALYZE $Ba(OH)_2$ is another example of a strong base, so we will use the procedure outlined in the previous example to solve the problem.

SOLVE

$$Ba(OH)_2 \rightarrow Ba^{2+} + 2\ OH^-$$

$$\frac{0.15 \text{ mol Ba(OH)}_2}{L} \times \left(\frac{2 \text{ mol OH}^-}{1 \text{ mol Ba(OH)}_2}\right) = 0.30\ M\ OH^-$$

$$pOH = -\log[0.30] = 0.52$$

$$pH = 14.00 - 0.52 = 13.48$$

THINK ABOUT IT Since we are dealing with a strong base we would expect to see a high pH.

EXAMPLE 19.
What is the pOH and pH of 0.17 *M* NH_3?

COLLECT AND ORGANIZE We are asked to determine the pOH and pH of a given concentration of a weak base.

ANALYZE The general formula for the equilibrium reaction is

$$B + H_2O \rightarrow BH^+ + OH^-$$

$$K_b = \frac{[BH^+][OH^-]}{[B]}$$

SOLVE The reaction of NH_3 in water is

$$NH_3 + H_2O \leftrightharpoons NH_4^+ + OH^-$$

$$K_b = \frac{[NH_4^+][OH^-]}{[NH_3]} = 1.8 \times 10^{-5}$$

Using the ICE table,

	NH_3	NH_4^+	OH^-
Initial	0.17	0	0
Change	$-x$	$+x$	$+x$
Equilibrium	$0.17 - x$	x	x

Using these values in the K_b equation,

$$1.8 \times 10^{-5} = \frac{(x)(x)}{(0.17 - x)}$$

Assuming x is small,

$$1.8 \times 10^{-5} = \frac{x^2}{0.17}$$

$$1.7 \times 10^{-3} = x$$

$$x = [OH^-] = 1.7 \times 10^{-3}$$

$$pOH = 2.77$$

$$pH = 11.23$$

THINK ABOUT IT The nitrogen bases react with water in an equilibrium reaction, accepting an H^+ from the water. The equilibrium constant for these reactions is K_b.

EXAMPLE 20.
What is the pOH and pH of a 0.35 *M* $C_6H_5NH_2$ solution?

COLLECT AND ORGANIZE We are asked to determine the pOH and pH of a given concentration of a weak base.

ANALYZE Since we are considering a weak base we will use the procedure outlined in the previous example.

SOLVE

$$C_6H_5NH_2 + H_2O \leftrightharpoons C_6H_5NH_3^+ + OH^-$$

$$K_b = 4.3 \times 10^{-10} = \frac{[C_6H_5NH_3^+][OH^-]}{[C_6H_5NH_2]}$$

Using the ICE table,

	$C_6H_5NH_2$	$C_6H_5NH_3^+$	OH^-
Initial	0.35	0	0
Change	$-x$	$+x$	$+x$
Equilibrium	$0.35 - x$	x	x

$$4.3 \times 10^{-10} = \frac{(x)(x)}{(0.35 - x)}$$

Assuming x is small,

$$4.3 \times 10^{-10} = \frac{x^2}{0.35}$$

$$1.2 \times 10^{-5} = x = [OH^-]$$

Checking the assumption, $0.35 - 1.2 \times 10^{-5} = 0.35$; it's good.

$$pOH = 4.91$$

$$pH = 9.09$$

THINK ABOUT IT As expected, the pH obtained for the solution of a weak base in the current example is lower than that obtained in the case of the strong base in Example 17.

THE PH OF SOLUTIONS OF ACIDIC AND BASIC SALTS

EXAMPLE 1.
What is the pH of a 0.029 *M* $(NH_4)_2SO_4$ solution?

COLLECT AND ORGANIZE We are asked to determine the pH of a given concentration of a soluble salt.

ANALYZE Soluble salts dissolve completely into their component ions in water. The concentrations of the ions are determined from the stoichiometry.

SOLVE This is an ammonium salt. All ammonium salts are soluble. Consequently, in water it reacts as

$$(NH_4)_2SO_4 \rightarrow 2\ NH_4^+ + SO_4^{2-}$$

Therefore, the ion concentrations are $[NH_4^+] = 0.058\ M$ and $[SO_4^{2-}] = 0.029\ M$. The sulfate does not react further in water (see the section on basic ions). Ammonium ion is the conjugate acid of ammonia. Its reaction in water is

$$NH_4^+ + H_2O \leftrightharpoons NH_3 + H_3O^+$$

$$K_a = \frac{[NH_3][H_3O^+]}{[NH_4^+]}$$

The value for K_a is determined from the K_b for ammonia. (Since K_a and K_b must be a conjugate acid–base pair, the substance to look for in the table will be a product of the equilibrium reaction.)

$$K_b \text{ of } NH_3 = 1.8 \times 10^{-5}$$

$$K_aK_b = K_W$$

$$K_a(1.8 \times 10^{-5}) = 1.0 \times 10^{-14}$$

$$K_a = 5.6 \times 10^{-10}$$

Using the ICE table to determine equilibrium concentrations, the initial values are determined from the stoichiometry of the ionization. (Always do complete reactions before equilibrium reactions.)

	NH_4^+	H_3O^+	NH_3
Initial	0.058	0	0
Change	$-x$	$+x$	$+x$
Equilibrium	$0.058 - x$	x	x

$$5.6 \times 10^{-10} = \frac{(x)(x)}{(0.058 - x)}$$

Assuming x is small,

$$5.6 \times 10^{-10} = \frac{x^2}{0.058}$$

$$5.7 \times 10^{-6} = x$$

Checking the assumption, $0.057 - 5.6 \times 10^{-6} = 0.057$, we find the assumption is good.

$$[H^+] = x = 5.7 \times 10^{-6}\ M$$

$$pH = 5.24$$

Think About It Solubility can be determined based on the solubility rules. Once a salt is dissolved in water, the ions may (or may not) react with water as an acid or a base.

Example 2.
What is the pH of 0.19 M $CH_3CH_2NH_3Br$?

Collect and Organize We are asked to calculate the pH of a solution containing a nitrogen base cation.

Analyze Since we are dealing with a nitrogen base cation we will use the procedure outlined in the previous example.

Solve

$$CH_3CH_2NH_3Br \rightarrow CH_3CH_2NH_3^+ + Br^-$$

$$[CH_3CH_2NH_3^+] = [Br^-] = 0.19\ M$$

Bromide ion does not react further with water (see the section on basic ions). $CH_3CH_2NH_3^+$ is the conjugate acid of ethyl amine, $CH_3CH_2NH_2$. So its reaction with water is

$$CH_3CH_2NH_3^+ + H_2O \leftrightharpoons CH_3CH_2NH_2 + H_3O^+$$

$$K_a = \frac{[CH_3CH_2NH_2][H_3O^+]}{[CH_3CH_2NH_3^+]}$$

The K_b for $CH_3CH_2NH_2$ is 6.4×10^{-4}. Therefore, the K_a is given by

$$K_aK_b = K_W = 1.0 \times 10^{-14}$$

$$K_a(6.4 \times 10^{-4}) = 1.0 \times 10^{-14}$$

$$K_a = 1.6 \times 10^{-11}$$

Using the ICE table,

	$CH_3CH_2NH_3^+$	$CH_3CH_2NH_2$	H_3O^+
Initial	0.19	0	0
Change	$-x$	$+x$	$+x$
Equilibrium	$0.19 - x$	x	x

Using these values in the K_a expression,

$$1.6 \times 10^{-11} = \frac{(x)(x)}{(0.19 - x)}$$

Assuming x is small,

$$1.6 \times 10^{-11} = \frac{x^2}{0.19}$$

$$1.7 \times 10^{-6} = x$$

Checking the assumption, $0.19 - 1.7 \times 10^{-6} = 0.19$. The assumption is good.

$$[H_3O^+] = x = 1.7 \times 10^{-6}\ M$$

$$pH = 5.76$$

Think About It Some of the nitrogen base cations are not obvious. However, the anions associated with them usually are. Recognizing that there are normally only two ions in a salt is usually sufficient to be able to write the reaction.

Example 3.
What is the pH of a 0.36 M NaF?

Collect and Organize We are asked to determine the pH of a given concentration of a solution containing an anion, F^-, which is a conjugate base of a weak acid.

Analyze The general equilibrium reaction for a basic anion is

$$A^- + H_2O \rightarrow HA + OH^-$$

SOLVE Since sodium salts are always soluble,

$$NaF \rightarrow Na^+ + F^-$$

$$[Na^+] = [F^-] = 0.36\ M$$

Sodium, as the cation of a strong base, does not react further in water. Fluoride ion is the conjugate base of HF. It reacts in water as

$$F^- + H_2O \leftrightharpoons HF + OH^-$$

$$K_b = \frac{[HF][OH^-]}{[F^-]}$$

K_b can be determined from the K_a of the conjugate acid, HF

$$K_aK_b = K_w = 1.0 \times 10^{-14}$$

$$(3.5 \times 10^{-4})\ K_b = 1.0 \times 10^{-14}$$

$$K_b = 2.9 \times 10^{-11}$$

The equilibrium concentrations are determined from the ICE table:

	F^-	HF	OH^-
Initial	0.36	0	0
Change	$-x$	$+x$	$+x$
Equilibrium	$0.36 - x$	x	x

$$2.9 \times 10^{-11} = \frac{(x)(x)}{(0.36 - x)}$$

Assuming x is small,

$$2.9 \times 10^{-11} = \frac{x^2}{0.36}$$

$$3.2 \times 10^{-6} = x$$

$$[OH^-] = x = 3.2 \times 10^{-6}$$

$$pOH = 5.49$$

$$pH = 8.51$$

THINK ABOUT IT Most anions are basic in water. The anions react by accepting an H^+ from water, leaving the water as OH^-. These anions are conjugate bases of weak acids.

EXAMPLE 4.
What is the pH of 0.15 M K_2CO_3?

COLLECT AND ORGANIZE We are asked to determine the pH of a given concentration of a solution containing an anion, CO_3^{2-}, which is a conjugate base of a weak acid.

ANALYZE Since we are asked to consider the equilibrium of a conjugate base of a weak acid we will use the procedure outlined in the previous example.

SOLVE

$$K_2CO_3 \rightarrow 2\ K^+ + CO_3^{2-}$$

Therefore, $[K^+] = 0.30\ M$ and $[CO_3^{2-}] = 0.15\ M$. Potassium, the cation of a strong base, does not react with water. Carbonate does.

$$CO_3^{2-} + H_2O \leftrightharpoons HCO_3^- + OH^-$$

$$K_b = \frac{[HCO_3^-][OH^-]}{[CO_3^{2-}]}$$

$$K_aK_b = K_W = 1.0 \times 10^{-14}$$

$$(5.6 \times 10^{-11})\ K_b = 1.0 \times 10^{-14}$$

$$K_b = 1.8 \times 10^{-4}$$

Using the ICE table to determine equilibrium concentrations,

	CO_3^{2-}	HCO_3^-	OH^-
Initial	0.15	0	0
Change	$-x$	$+x$	$+x$
Equilibrium	$0.15 - x$	x	x

$$1.8 \times 10^{-4} = \frac{(x)(x)}{(0.15 - x)}$$

Assuming x is small,

$$1.8 \times 10^{-4} = \frac{x^2}{0.15}$$

$$5.2 \times 10^{-3} = x$$

Checking the assumption, 0.15 – 0.0052 = 0.144. Not quite true; so, using the quadratic,

$$1.8 \times 10^{-4} = \frac{x^2}{(0.15 - x)}$$

$$2.7 \times 10^{-5} - 1.8 \times 10^{-4}x = x^2$$

$$0 = x^2 + 1.8 \times 10^{-4}x - 2.7 \times 10^{-5}$$

$$x = \frac{-1.8 \times 10^{-4}}{2} \pm \frac{[(1.8 \times 10^{-4})^2 - 4(1)(-2.7 \times 10^{-5})]^{1/2}}{2}$$

$$x = \frac{-1.8 \times 10^{-4}}{2} \pm \frac{[3.24 \times 10^{-8} + 1.08 \times 10^{-4}]^{1/2}}{2}$$

$$x = \frac{-1.8 \times 10^{-4}}{2} \pm \frac{[0.01039]}{2}$$

Ignoring the negative answer,

$$x = 0.0051 = [OH^-]$$

$$pOH = 2.29$$

$$pH = 11.71$$

THINK ABOUT IT As an ion, the K_b for carbonate is not listed. Therefore, the K_b is determined from the K_a of the conjugate acid, HCO_3^-. This is the K_{a_2} for carbonic acid, H_2CO_3. For H_2CO_3, $K_{a_1} = 4.3 \times 10^{-7}$, $K_{a_2} = 5.6 \times 10^{-11}$. K_{a_2} refers to HCO_3^-.

Buffer Solutions and the pH of Natural Waters

Example 1.
What is the pH of a solution of 0.12 *M* HOBr and 0.53 *M* NaOBr?

Collect and Organize We are asked to calculate the pH of the given buffer system.

Analyze We will form an ICE table with the appropriate initial concentrations and use stoichiometry to determine how these concentrations will change.

Solve HOBr is a weak acid, so it will react in equilibrium with water as

$$HOBr + H_2O \leftrightharpoons H_3O^+ + OBr^-$$

and

$$K_a = \frac{[H_3O^+][OBr^-]}{[HOBr]} = 2.0 \times 10^{-9}$$

NaOBr contains always soluble Na^+, so it completely ionizes.

$$NaOBr \rightarrow Na^+ + OBr^-$$

The common ion is OBr^-. From the complete reaction, $[OBr^-] = 0.53\ M$. Since OBr^- is the conjugate base of HOBr, the system is a buffer.

The ICE table is

	HOBr	H_3O^+	OBr^-
Initial	0.12	0	0.53
Change	$-x$	$+x$	$+x$
Equilibrium	$0.12 - x$	x	$0.53 + x$

$$2.0 \times 10^{-9} = \frac{(x)(0.53 + x)}{(0.12 - x)}$$

Assuming x is small,

$$2.0 \times 10^{-9} = \frac{x(0.53)}{(0.12)}$$

$$4.5 \times 10^{-10} = x$$

$$pH = 9.34$$

Think About It Buffers are examples of the common ion effect, since they are mixtures where both substances produce the same ion. One substance reacts completely; the other, in equilibrium. Always do the complete reaction first to determine the initial concentrations for the ICE table. What makes the mixture a buffer is that the concentration of both components of a conjugate acid–base pair is significant.

Example 2.
What is the pH a solution that is 0.13 *M* HCOOH and 0.24 *M* $KHCO_2$?

Collect and Organize We are asked to determine the pH of the given buffer system.

Analyze Because of the common ion effect, changes in concentration will be small. Thus the assumption that "x is small" is more likely to be true. This assumption is made in the Henderson–Hasselbalch equation:

$$pH = pK_a + \log\left(\frac{[base]}{[acid]}\right).$$

Solve Since potassium salts are always soluble,

$$KHCO_2 \rightarrow K^+ + HCO_2^-$$

$$[K^+] = [HCO_2^-] = 0.24\ M$$

And the weak acid reacts as

$$HCOOH + H_2O \leftrightharpoons H_3O^+ + HCO_2^-$$

$$K_a = \frac{[HCO_2^-][OH^-]}{[HCOOH]}$$

In the equilibrium, HCOOH is the acid and HCO_2^- is its conjugate base. The K_a of the acid, HCOOH, is 1.8×10^{-4}. The $pK_a = -\log(1.8 \times 10^{-4}) = 3.74$. So using the Henderson–Hasselbalch equation,

$$pH = pK_a + \log\left(\frac{[base]}{[acid]}\right) = 3.74 + \log\left(\frac{0.24}{0.13}\right) = 4.01$$

Think About It Although the Henderson–Hasselbalch equation was used, the use of K_a is an equally acceptable way to solve the problem.

Example 3.
What is the pH of a buffer made of 0.21 *M* NH_3 and 0.48 *M* NH_4Cl?

Collect and Organize We are asked to determine the pH of the given buffer system.

Analyze After converting K_b for NH_3 into K_a for its conjugate acid NH_4^+, we will use the Henderson–Hasselbalch equation to calculate the pH.

Solve NH_3 is a weak base. Its reaction is

$$NH_3 + H_2O \leftrightharpoons NH_4^+ + OH^-$$

NH_4Cl is a soluble salt. Its reaction is

$$NH_4Cl \rightarrow NH_4^+ + Cl^-$$

The conjugate acid–base pair is NH_3 and NH_4^+, and NH_4^+ is the acid. The K_b for NH_3 is 1.8×10^{-5}. Therefore, the K_a of NH_4^+ can be determined from

$$K_aK_b = 1.0 \times 10^{-14}$$

$$K_a(1.8 \times 10^{-5}) = 1.0 \times 10^{-14}$$

$$K_a = 5.6 \times 10^{-10}$$

$$pK_a = -\log(5.6 \times 10^{-10}) = 9.25$$

Using the Henderson–Hasselbalch equation,

$$pH = pK_a + \log\left(\frac{[\text{base}]}{[\text{acid}]}\right) = 9.25 + \log\left(\frac{0.21}{0.48}\right) = 8.89$$

THINK ABOUT IT The Henderson–Hasselbalch equation can be used for any buffer even if the buffer is based on a weak base rather than a weak acid. Because a conjugate acid–base pair is used, it is always possible to use the pK_a of the acid.

EXAMPLE 4.
What ratio of sodium oxalate ($Na_2C_2O_4$) and sodium hydrogen oxalate ($NaHC_2O_4$) is required for a buffer with a pH of 4.5?

COLLECT AND ORGANIZE We are asked to determine the relative concentrations of a conjugate acid–base pair that will result in a buffer with a pH of 4.5.

ANALYZE After identifying the acid and base of the conjugate acid–base pair, the K_a value of the acid and the desired pH can be used in the Henderson–Hasselbalch equation to solve for the ratio of acid and base needed.

SOLVE

$$Na_2C_2O_4 \rightarrow 2\ Na^+ + C_2O_4^{2-}$$

$$NaHC_2O_4 \rightarrow Na^+ + HC_2O_4^-$$

The conjugate acid–base pair is acid = $HC_2O_4^-$ and base = $C_2O_4^{2-}$. The K_a of the acid is K_{a_2} of oxalic acid ($HC_2O_4^-$) = 6.4×10^{-5}. Consequently, the $pK_a = 4.19$. Using Henderson–Hasselbalch and solving for the acid–base ratio,

$$pH = pK_a + \log\left(\frac{[\text{base}]}{[\text{acid}]}\right)$$

$$4.5 = 4.19 + \log\left(\frac{\text{base}}{\text{acid}}\right)$$

$$0.31 = \log\left(\frac{\text{base}}{\text{acid}}\right)$$

$$2.04 = \frac{\text{base}}{\text{acid}}$$

THINK ABOUT IT You will need twice as much $C_2O_4^{2-}$ as $HC_2O_4^-$ to create this buffer.

EXAMPLE 5.
How many moles of NaOH are required to change the pH of 100 mL of buffer made from 0.10 *M* $NaNO_2$ and 0.22 *M* HNO_2 to pH = 4.00?

COLLECT AND ORGANIZE We are asked to determine the number of moles of NaOH required to change the pH of the given volume of buffer to 4.00.

ANALYZE We can use the Henderson–Hasselbalch equation if we keep in mind that addition of acid or base does two things. First, it reduces the concentration of the substance it reacts with. Second, it increases the amount of the conjugate.

SOLVE For the conjugate acid–base pair of HNO_2/NO_2^-, the pK_a of the acid (HNO_2) is

$$pK_a = -\log(4.5 \times 10^{-4}) = 3.35$$

For the solution to have a pH = 4, the ratio of acid to base must be

$$pH = pK_a + \log\left(\frac{[\text{base}]}{[\text{acid}]}\right)$$

$$4.0 = 3.35 + \log\left(\frac{\text{base}}{\text{acid}}\right)$$

$$0.65 = \log\left(\frac{\text{base}}{\text{acid}}\right)$$

$$4.47 = \frac{\text{base}}{\text{acid}}$$

Adding base will remove HNO_2 and create NO_2^-. For every mole of HNO_2 removed, 1 mole of NO_2^- will be created.

$$HNO_2 + NaOH \rightarrow NaNO_2 + H_2O$$

To work in moles, the moles of NO_2^- are (0.10 mol/L)(0.1 L) = 0.01 mol NO_2^-. The moles of HNO_2 are (0.22 mol/L)(0.1 L) = 0.022 mol HNO_2. Because the volume of both the acid and its conjugate base must be the same (as they are in the same solution), the ratio of base to acid can be expressed in moles instead of molarity. To get the appropriate molar ratio,

$$4.47 = \frac{(0.010 + x)}{(0.022 - x)}$$

$$0.09834 - 4.47x = 0.01 + x$$

$$0.08834 = 5.47x$$

$$0.016 = x$$

THINK ABOUT IT 0.016 mol NaOH will remove 0.016 mol HNO_2 (leaving 0.204 mol) and will add 0.016 mol NO_2^- (making 0.116 mol).

ACID–BASE TITRATIONS

EXAMPLE 1.
What is the equivalence point if 30.0 mL of 0.100 *M* HCl is titrated with 0.150 *M* NaOH?

COLLECT AND ORGANIZE We are asked to determine the equivalence point of a strong acid titrated with a strong base.

ANALYZE The equivalence point is the volume when the reactants are in exact stoichiometric ratio. It requires the use of a balanced chemical equation.

SOLVE The reaction between NaOH and HCl is

$$\text{NaOH} + \text{HCl} \rightarrow \text{NaCl} + \text{H}_2\text{O}$$

Using dimensional analysis, we convert moles of HCl into mL of NaOH:

$$30.0\text{ mL HCl} \times \left(\frac{1\text{ L}}{1000\text{ mL}}\right) \times \left(\frac{0.100\text{ mol HCl}}{1\text{ L}}\right) \times \left(\frac{1\text{ mol NaOH}}{1\text{ mol HCl}}\right) \times \left(\frac{1\text{ L}}{0.150\text{ mol NaOH}}\right) \times \left(\frac{1000\text{ mL}}{1\text{ L}}\right) = 20.0\text{ mL}$$

THINK ABOUT IT By using a strong acid or a strong base as a reactant, an acid–base reaction can be forced to completion. In a complete reaction, stoichiometry can be used to relate reactants and products to each other.

EXAMPLE 2.
What is the equivalence point when 50.0 mL of 0.16 *M* NH_3 is titrated with 0.25 *M* H_2SO_4?

COLLECT AND ORGANIZE We are asked to determine the equivalence point when a weak base is titrated with a strong acid.

ANALYZE The method used in Example 1 can be used regardless of physical state. However, if both reactants are in solution, an alternative procedure might be faster. This procedure is based on $M_AV_A = \left(\frac{1}{X}\right)M_BV_B$, where *M* is molarity, *V* is volume, A is acid, and B is base. $\frac{1}{X}$ represents the ratio of the stoichiometric coefficient of the acid to the stoichiometric coefficient of the base.

SOLVE

$$2\,\text{NH}_3 + \text{H}_2\text{SO}_4 \rightarrow (\text{NH}_4)_2\text{SO}_4$$

$$\frac{1}{X} = \frac{\text{acid}}{\text{base}} = \frac{1}{2}$$

$$M_AV_A = \left(\frac{1}{X}\right)M_BV_B$$

$$(0.25\ M)\ V_A = \left(\frac{1}{2}\right)(0.16\ M)(50.0\text{ mL})$$

$$V_A = 16\text{ mL}$$

THINK ABOUT IT Either the procedure outlined in the current example or that of the previous example can be used to arrive at the correct result.

EXAMPLE 3.
Sketch the titration curve for the titration of 25.0 mL of 0.10 *M* HCl with 0.10 *M* NaOH.

COLLECT AND ORGANIZE We are asked to sketch the titration curve that results from the given titration of a strong acid with a strong base. The phrase "sketch" implies a qualitative answer.

ANALYZE The four regions of a titration curve can be predicted by considering the titration reaction. They are

Region 1. before any titrant is added

Region 2. before the equivalence point

Region 3. at the equivalence point

Region 4. after the equivalence point

SOLVE HCl is the reactant; NaOH is the titrant. The reaction is

$$\text{HCl} + \text{NaOH} \rightarrow \text{H}_2\text{O} + \text{NaCl}$$

Region 1. At volume zero, before any titrant has been added, HCl is present in the reaction solution. It is a strong acid, so the pH will be very low.

Region 2. HCl is still present, so the pH is still very low, but higher than the starting point.

Region 3. HCl is all used up; all the NaOH added has also reacted. The only substances present are water and sodium chloride. Since NaCl does not affect pH, the pH is the pH of water, 7.

Region 4. The NaOH added does not have anything to react with, so the pH is characteristic of a strong base, very high. It increases as more NaOH is added.

THINK ABOUT IT Don't forget what is "obvious" overall: whenever base is added (base is the titrant), pH will increase. Whenever acid is added (acid is the titrant), pH will decrease.

EXAMPLE 4.
Sketch the titration curve for the titration of 30.0 mL of 0.10 *M* HF with 0.20 *M* KOH.

COLLECT AND ORGANIZE We are asked to sketch the titration curve that results from the given titration of a weak acid with a strong base. The phrase "sketch" implies a qualitative answer.

ANALYZE The four regions of a titration curve can be predicted by considering the titration reaction. They are listed in the previous example.

SOLVE HF is the reactant; KOH is the titrant. The reaction is

$$HF + KOH \rightarrow KF + H_2O \text{ (or } K^+ + F^- + H_2O)$$

Region 1. At volume zero, only HF is present. It is a weak acid, so the pH will be less than 7.00 but higher than what it would be for a strong acid.

Region 2. Some of the HF has reacted with the KOH, making the conjugate base F^-. The presence of both acid and base creates a buffer solution. Thus the pH in this region is higher than that in Region 1, but relatively constant (slightly increasing) and centered around 3.45, the pK_a of HF.

Region 3. All the HF is used up and all the KOH has reacted, but F^- is present. F^- is a weak base, so the pH will be slightly greater than 7.

Region 4. F^- is still present, but now OH^- is present too. Consequently, pH will be controlled by the strong base and will be very high.

THINK ABOUT IT If the reactant is a weak acid or base, a buffer solution is formed and the area before the equivalence point is nearly horizontal. The most useful point is halfway to the equivalence point. Since half of the reactant has reacted, forming its conjugate, the concentrations of acid and base are equal. At that volume, pH = pK_a.

EXAMPLE 5.
Sketch the titration curve for 20.0 mL of 0.15 *M* NH_3 with 0.10 *M* HCl.

COLLECT AND ORGANIZE We are asked to sketch the titration curve that results from the given titration of a weak base with a strong acid. The phrase "sketch" implies a qualitative answer.

ANALYZE The four regions of a titration curve can be predicted by considering the titration reaction. They are the same as before.

SOLVE NH_3 is the reactant; HCl is the titrant. The reaction is

$$NH_3 + HCl \rightarrow NH_4^+ + Cl^-$$

Region 1. Only NH_3 is present. It is a weak base, so the pH will be somewhat more than 7.

Region 2. Some NH_3 has reacted, but some remains. Some NH_4^+ has been created. Consequently, there is a buffer solution. The pH is decreasing, but slightly. The pH is near the pK_a of NH_4^+ (the conjugate acid) = 9.25.

Region 3. Only NH_4^+ is present. It is a weak acid, so the pH is slightly less than 7.

Region 4. Not only weak acid, NH_4^+, but strong acid HCl is present. Consequently, the pH is very low.

THINK ABOUT IT Again we have a situation in which the reactant is a weak acid or base. As a result a buffer solution is formed and the area before the equivalence point is nearly horizontal.

EXAMPLE 6.
What is the pH at 0.0 mL, 20.0 mL, and 30.0 mL in the titration of 25.0 mL of 0.10 *M* HCl with 0.10 *M* NaOH.

COLLECT AND ORGANIZE We are asked to determine the pH at several stages of the titration of strong acid with a strong base.

ANALYZE We must first determine the equivalence point and the initial pH of the solution. We will then use a combination of stoichiometry and dimensional analysis to first calculate the number of moles of HCl that remain after a given amount of titrant is added and the resulting pH.

SOLVE Using $M_AV_A = \left(\frac{1}{X}\right)M_BV_B$, where $\frac{1}{X} = 1$, the volume at the equivalence point is 25.0 mL. HCl is the reactant; NaOH is the titrant. The reaction is

$$HCl + NaOH \rightarrow H_2O + NaCl$$

The initial concentration of HCl is 0.10 *M*. Since HCl is a strong acid, the concentration of $[H_3O^+]$ = 0.10 *M*. Therefore, pH = 1.00.

20.0 mL is before the equivalence point (Region 2).
Moles of acid to start: (0.025 L)(0.10 mol//L) = 0.025 mol HCl
Moles base added: (0.020 L)(0.10 mol/L) = 0.020 mol NaOH
Moles of acid remaining: 0.025 – 0.020 = 0.005 mol HCl
Total volume of solution: 25.00 mL + 20.00 mL = 45.00 mL
New concentration HCl: 0.005 mol/0.045 L = 0.011 *M*
pH at 20.00 mL of NaOH: –log(0.011) = 1.95

30 mL is 5.0 mL after the equivalence point. The moles of NaOH in the reaction solution are (0.0050 L)(0.10 mol/L) = 0.00050 mol NaOH. It is in a total volume of 25.0 mL + 30.0 mL = 55.0 mL or 0.0550 L. We have [NaOH] = 0.00050 mol/0.0550 L = 0.0091 *M*, pOH = 2.04. and pH = 11.96.

Think About It By calculating the pH at different stages of the titration we are determining a more quantitative picture of the titration curve.

Example 7.

In the titration of 30.0 mL of 0.10 *M* HF with 0.20 *M* KOH, what is the pH at 10.0 mL, 15.0 mL, and 20.0 mL?

Collect and Organize

We are asked to determine the pH at several stages of the titration of weak acid with a strong base.

Analyze Since this is a buffer, the Henderson–Hasselbalch equation can be used in the region before the equivalence point (Region 2).

Solve This is the same titration as in Example 4. Using $M_A V_A = \left(\frac{1}{X}\right) M_B V_B$, where $\frac{1}{X} = 1$, the volume at the equivalence point is 15.0 mL.

10.0 mL is in Region 2, before the equivalence point, so the Henderson-Hasselbalch equation can be used. Since the volume is the same for both the acid and base, moles can be used instead of concentrations in the ratio. The initial moles of HF are (0.0300 L)(0.10 mol HF/L) = 0.0030 mol HF. The moles of KOH added are (0.010 L)(0.20 mol KOH/L) = 0.0020 mol KOH. The moles of F^- created are 0.0020 mol KOH × (1 mol F^-/1 mol KOH) = 0.0020 mol F^- made (base). The moles of acid reacted are 0.0020 mol KOH × (1 mol HF/1 mol KOH) = 0.0020 mol HF used up. The moles of acid left over are 0.0030 mol – 0.0020 = 0.0010 mol HF left (acid).

$$\text{pH} = \text{p}K_a + \log\left(\frac{\text{base}}{\text{acid}}\right) = 3.46 + \log\left(\frac{0.0020}{0.0010}\right)$$

$$= 3.46 + 0.30 = 3.76$$

15.0 mL is Region 3, the equivalence point. Only F^- is present. Since all the HF has been converted to F^-, there is 0.0030 mol F^-. The total volume is 15.0 mL + 30.0 mL = 45.0 mL = 0.045 L. Consequently, $[F^-]$ = 0.067 *M*. Since F^- is a weak base, its reaction with water is

$$F^- + H_2O \leftrightharpoons HF + OH^-$$

$$K_b = \frac{[HF][OH^-]}{[F^-]} = \frac{K_W}{K_a} = 2.9 \times 10^{-11}$$

Using this equation and the ICE table, concentration of OH^- is 1.4×10^{-6}. So pOH = 5.86 and pH = 8.14.

20 mL is 5.0 mL after the equivalence point. Consequently, the moles of excess base are (0.0050 L)(0.20 mol KOH/L) = 0.0010 mol. The total volume is 20.0 mL + 30.0 mL = 50.0 mL or 0.0500 L solution. Since 0.0010 mol KOH/0.0500 L = 0.020 *M* KOH, $[OH^-]$ = 0.020 *M*, pOH = 1.70, and pH = 12.30.

Think About It Because both acid and conjugate base are in the same volume of solution, the volume cancels in the acid–base ratio and the mole ratio can be used directly.

Acid–Base Indicators

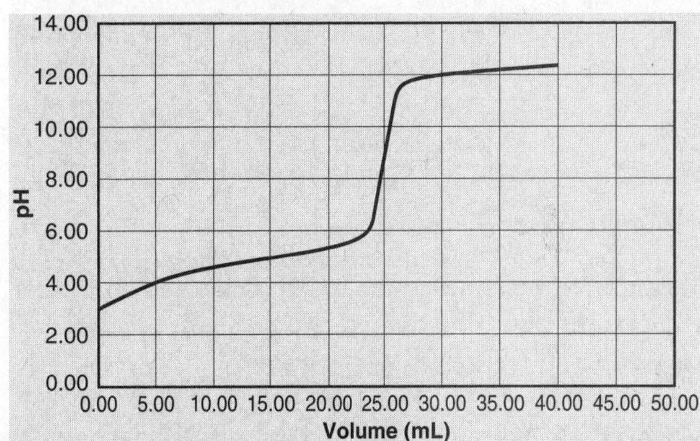

Example 1.

Refer to the figure above. The indicator phenol red has a pK_a of about 7.5. Its acid color is yellow; its base color is red. What color will the indicator be at 10, 25, and 30 mL? Would this be a good indicator for this titration?

Collect and Organize We are asked to determine the color of the indicator at different stages of the titration of a weak acid with a strong base in order to determine whether it can be used for this particular titration.

ANALYZE We must determine if a color change will occur based on the pH values before, near, and after the equivalence point. At pH values about 1 pH unit higher than the pK_a, the basic form and color predominate. Similarly, the acid color will be seen at any pH lower than 1 pH unit below the pK_a.

SOLVE At 10 mL, the pH is 4.4. This is quite acidic; the color will be yellow. At 25 mL, the pH is 8.4. This is just within the ±1 of the pK_a, so it will be orange, likely, a reddish-orange. At 30 mL, the pH is 12.0. This is very basic; the color will be red. Since it changes color at the equivalence point (it doesn't have to hit exactly), this would be a reasonable choice.

THINK ABOUT IT Indicators are simply weak acids or bases where the acid and base forms are different colors. At pH values about 1 pH unit higher than the pK_a, the basic form and color predominate. Similarly, the acid color will be seen at any pH lower than 1 pH unit below the pK_a.

EXAMPLE 2.
Refer again to the titration curve shown in Example 1. The indicator bromocresol green has a pK_a of about 4.6. It is yellow in its acidic form and blue in its basic form. What color will the indicator be at 10, 25, and 30 mL? Would this be a good indicator for this titration?

COLLECT AND ORGANIZE We are asked to determine the color of the indicator at different stages of the titration of a weak acid with a strong base.

ANALYZE We must determine if a color change will occur based on the pH values before, near, and after the equivalence point.

SOLVE At 10 mL, the pH is 4.4. This is near its pK_a, so the color is green. At 25 mL, the pH is 8.4. This is far above the pK_a, so the color is blue. At 30 mL, the pH is 12.30. This is even more above the pK_a, so the color is still blue. (No difference between 15 and 30 mL.) This would be a poor choice of an indicator, since the color change is well before the equivalence point.

THINK ABOUT IT Since the pH change is dramatic at the equivalence point, indicators are chosen so that the pH of the equivalence point is near the pK_a of the indicator. The color change is how the equivalence point is indicated.

Self-Test

KEY VOCABULARY COMPLETION QUESTIONS

__________ 1. Resists pH changes

__________ 2. What NH_4^+ is to NH_3

__________ 3. Reaction of a substance with itself to make ions

__________ 4. H^+ donor

__________ 5. Acid–base theory based on transfer of H^+ ions

__________ 6. Measure of resistance to pH change

__________ 7. $\frac{[H^+][A^-]}{[HA]}$

__________ 8. Reacts incompletely with water to make H_3O^+

__________ 9. One H^+ short of an acid

__________ 10. Logarithmic concentration of H^+

__________ 11. Volume when all reactants have reacted and none are left

__________ 12. Completely ionized to H^+ and an anion

__________ 13. Extent to which an acid reacts with water

__________ 14. Substances that differ by an H^+

__________ 15. Acts as either an acid or a base

__________ 16. Experiment where the amount of one reactant is determined by the amount of another reactant

__________ 17. Can donate more than one H^+

__________ 18. An equation to calculate the pH of a buffer

__________ 19. Acid and conjugate base are different colors

__________ 20. Adding an ion that is the product of another equilibrium reaction

__________ 21. Equilibrium constant for the autoionization of water

__________ 22. Type of solution produced by adding Na_2CO_3 to water

__________ 23. Graph of pH versus titrant volume

__________ 24. $-\log[OH^-]$

__________ 25. All strong acids have the same strength in water

MULTIPLE-CHOICE QUESTIONS

1. A substance that produces H^+ (or H_3O^+) in water is
   a. Arrhenius acid
   b. Brønsted-Lowry acid
   c. Lewis acid
   d. all of the above
   e. none of the above

2. What is the acid in the following reaction?

$$HS^- + HC_2O_4^- \leftrightharpoons H_2S + C_2O_4^{2-}$$

   a. HS^-
   b. $HC_2O_4^-$
   c. $C_2O_4^{2-}$
   d. all substances in the reaction are acids
   e. there are no acids in the reaction

3. Which of the following is a strong base?
   a. NH_3
   b. H_2O
   c. Cl^-
   d. KOH
   e. all are

4. What is the pH of 0.213 *M* HNO_3?
   a. 0.672
   b. 2.009
   c. 7.00
   d. 11.991
   e. 13.328

5. What is the pH of 0.014 *M* $Ba(OH)_2$?
   a. 1.55
   b. 1.85
   c. 11.85
   d. 12.15
   e. 12.45

6. What is the pH of 1.91 *M* HCl?
   a. −0.28
   b. 0.28
   c. 1.91
   d. 13.72
   e. 7.00

7. NO_2 is
   a. strongly acidic
   b. weakly acidic
   c. neutral
   d. slightly basic
   e. strongly basic

8. What is the pH of 0.12 *M* $LiCH_3CO_2$?
   a. 0.92
   b. 2.83
   c. 7.00
   d. 8.91
   e. 11.16

9. The degree of ionization is larger for
   a. larger values of K_a
   b. smaller values of K_a
   c. ionization is independent of K_a

10. Which of the following produce an acidic solution in water?
   a. Na_2SO_4
   b. NH_3
   c. CH_3NH_3Cl
   d. all of the above
   e. none of the above

11. Which of the following produces a basic solution in water?
   a. NH_3
   b. $Ca(OH)_2$
   c. Na_2CO_3
   d. all of the above
   e. none of the above

12. Which of the following is a strong acid?
   a. HF
   b. H_2S
   c. $HClO_2$
   d. all of the above
   e. none of the above

13. Which of the following can be used to make a buffer?
   a. HCl/NaCl
   b. HF
   c. NH_3/NH_4^+
   d. NaOH/NaCl
   e. any of the above

14. What characterizes a buffer?
   a. resists a change in pH with addition of a small amount of acid
   b. resists a change in pH with the addition of a small amount of base
   c. resists a change in pH with the addition of water
   d. all of the above
   e. none of the above

15. Dilution of a weak acid will
   a increase the percent of acid ionized
   b. decrease the percent of acid ionized
   c. have no effect on the percent of acid ionized
   d. increase the percent ionized if $K_a > 10^{-5}$
   e. increase the percent ionized if $K_a > 10^{-5}$

16. In an aqueous solution of $HClO_4$, the base is
   a. $HClO_4$
   b. H_2O
   c. ClO_4^-
   d. H_3O^+
   e. both H_2O and ClO_4^-

17. The products of a Brønsted–Lowry acid–base reaction are
    a. salt and water
    b. complex ion
    c. an acid and a base
    d. water and carbon dioxide
    e. H_3O^+ and OH^-

18. What is the conjugate base of H_2S?
    a. HS^-
    b. S^{2-}
    c. OH^-
    d. H_3S^+
    e. H^+

19. What is the conjugate acid of OH^-?
    a. H_3O^+
    b. H^+
    c. H_2O
    d. O^{2-}
    e. none exist

20. What is a base in the following reaction

$$H_2PO_4^- + HCO_3^- \leftrightharpoons HPO_4^{2-} + H_2CO_3$$

    a. $H_2PO_4^-$
    b. HCO_3^-
    c. both $H_2PO_4^-$ and H_2CO_3
    d. both HCO_3^- and HPO_4^{2-}
    e. no bases exist in this reaction

21. Which acid is the strongest in water?
    a. $HClO_4$
    b. $HClO_3$
    c. HCl
    d. HBr
    e. all are the same

22. Which 0.10 *M* aqueous solution will have the highest pH?
    a. NH_3
    b. LiOH
    c. $Sr(OH)_2$
    d. $Fe(OH)_3$
    e. all are the same

23. HF is a stronger acid than CH_3COOH. Therefore,
    a. CH_3COOH is a stronger base than HF
    b. F^- is a stronger base than $CH_3CO_2^-$
    c. $CH_3CO_2^-$ is a stronger base then F^-
    d. HF is a stronger base than CH_3COOH
    e. base information is not related to acid information

24. Concentrated HCl is about 12 *M*. What is the pH of this solution?
    a. 1.08
    b. –1.08
    c. 12.92
    d. 0.00
    e. 0.12

25. Which is the weakest acid?
    a. $HBrO_4$
    b. $HBrO_3$
    c. $HBrO_2$
    d. HBrO
    e. all are the same

26. Which is the strongest acid?
    a. HClO
    b. HBrO
    c. HIO
    d. all are the same
    e. these are not acids

27. In a polyprotic acid, pH depends on the first ionization because
    a. K_{a_2} is much smaller than K_{a_1}
    b. the common ion effect decreases the H^+ gain in the second reaction
    c. both of the above
    d. neither of the above
    e. every ionization is important to pH

28. If the pH = 5.7, then the hydroxide ion concentration is
    a. 8.3 *M*
    b. 5×10^{-9} *M*
    c. 2×10^{-6} *M*
    d. 5×10^{-5} *M*
    e. 2×10^{-4} *M*

29. Compared to a solution with pH = 5, the concentration of H^+ in a solution with pH = 8
    a. is 1000 times more acidic
    b. is 1000 times more basic
    c. is 30 times more acidic
    d. is 30 times less acidic
    e. there is no H^+ in a solution of pH = 8

30. If the $[OH^-] = 0.066$ *M*, then pH is
    a. 1.18
    b. 12.82
    c. 6.6
    d. –1.18
    e. 7.40

31. The pH of an aqueous solution of a weak acid
    a. will increase if K_a increases
    b. will decrease if K_a increases
    c. is not related to K_a
    d. will only be affected by K_a if $K_a > 10^{-5}$
    e. will only be affected by K_a if $K_a < 10^{-5}$

32. A stronger base
    a. has a high pH and a low K_b
    b. has a low pH and a low K_b
    c. has a high pH and a high K_b
    d. has a low pH and a high K_b
    e. has no K_b

33. Addition of a base to an aqueous solution will
 a. raise pH and lower pOH
 b. raise both the pH and pOH
 c. lower both the pH and pOH
 d. lower the pH and raise the pOH
 e. affect pOH only

34. An anion is likely to be
 a. an acid
 b. a base
 c. neither

35. What is the K_b of HCO_3^-?
 a. 4.3×10^{-7}
 b. 4.7×10^{-11}
 c. 2.3×10^{-8}
 d. 2.1×10^{-4}
 e. HCO_3^- does not have a K_b

36. What is the pH of a solution of 0.10 *M* HOCl and 0.10 *M* NaOCl?
 a. 7.45
 b. 4.23
 c. 6.54
 d. 3.77
 e. 10.22

37. Buffer capacity is higher if
 a. only the acid concentration is very high
 b. only the base concentration is very high
 c. the acid and base are both dilute
 d. the acid and base are both concentrated
 e. concentration does not affect buffer capacity

38. For optimal buffers
 a. pH = K_a
 b. pH = pK_a
 c. pH = pK_b
 d. any of the above work equally well
 e. pH is not related to the equilibrium constant

39. For optimal buffers
 a. the concentration of acid is much higher than the concentration of base
 b. the concentration of base is much higher than the concentration of acid
 c. the concentration of acid is equal to the concentration of base
 d. only acid with no base
 e. only base with no acid

40. The indicator methyl orange has a pK_a of about 3.5. Its acid form is red; its basic form is yellow. What color is the indicator at pH = 7?
 a. red
 b. yellow
 c. orange
 d. colorless
 e. green

41. The K_a of CH_3COOH is 1.8×10^{-5}; therefore, 9.25 is
 a. the pK_a of CH_3COOH
 b. the pK_b of CH_3COOH
 c. the pK_a of $CH_3CO_2^-$
 d. the pK_b of $CH_3CO_2^-$
 e. the pK_W

42. At the equivalence point of the titration of CH_3NH_2 with HCl,
 a. pH equals 7
 b. pH is slightly greater than 7
 c. pH is slightly less than 7
 d. pH is much greater than 7
 e. pH is much less than 7

43. If 30.0 mL of 0.20 *M* HF is titrated with 0.10 *M* NaOH, the pH at 30.0 mL is
 a. 7.00
 b. 3.45
 c. 2.23
 d. 2.31
 e. 8.14

44. If NH_3 is titrated with HCl, the pH at halfway to the equivalence point is
 a. 4.74
 b. 7.00
 c. 9.25
 d. less than 2
 e. more than 10

45. Most hydroxides are not strong bases because
 a. they are insoluble
 b. they contain only weak bases
 c. the associated cations are so acidic that they cancel the base of the hydroxide
 d. they form complex ions instead
 e. All hydroxides are strong bases

Additional Practice Problems

1. What is the conjugate acid of
 a. H_2O b. HCO_3^- c. F^- d. PO_4^{3-} e. NH_3

2. What is the conjugate base of
 a. H_2O b. HCO_3^- c. HNO_2
 d. $H_2PO_4^-$ e. CH_3COOH

3. What is the pH and degree of ionization for 0.057 *M* HOCl?

4. What is the pH for 0.20 *M* piperidine ($C_5H_{11}N$); $K_b = 1.3 \times 10^{-3}$?

5. What is the pH of 0.045 *M* KF?

6. What is the pH of 0.135 *M* $(NH_4)_2SO_4$?

7. What is the pH of 0.161 *M* HNO_3?

8. What is the pH of an aqueous solution of 0.14 *M* C_6H_5COOH and 0.27 *M* $NaC_6H_5CO_2$?

9. What is the pH of diprotic 0.25 *M* tartaric acid (K_{a_1} = 1.0×10^{-3}, $K_{a_2} = 4.6 \times 10^{-5}$)?

10. For the titration of 25.0 mL of 0.15 *M* dimethyl amine [$(CH_3)_2NH$, $K_b = 5.4 \times 10^{-4}$] with 0.10 *M* HCl,
   a. What is the equivalence point?
   b. What is the pH at halfway to the equivalence point?
   c. What is the pH at the equivalence point?
   d. Sketch the titration curve.

Self-Test Answer Key

Key Vocabulary Completion Answers

1. buffer
2. conjugate acid
3. autoionization
4. acid
5. Brønsted–Lowry
6. buffer capacity
7. acid dissociation constant (K_a)
8. weak acid
9. conjugate base
10. pH
11. equivalence point
12. strong acid
13. strength
14. conjugate acid–base pair
15. amphoteric
16. titration
17. polyprotic acid
18. Henderson–Hasselbalch
19. indicator
20. common ion effect
21. $K_w (1.0 \times 10^{-14})$
22. basic
23. titration curve
24. pOH
25. leveling effect

Multiple-Choice Answers

For further information about the topics these questions address, you should consult the text. The appropriate section of the text is listed in parentheses after the explanation of the answer.

1. d. This is the definition of an Arrhenius acid. Brønsted–Lowry and Lewis acids include all Arrhenius acids. (16.1 and 16.2)

2. b. Brønsted–Lowry acids transfer a proton to a base. $HC_2O_4^-$ donates an H^+ to HS^-. (16.2)

3. d. Strong bases include the group 1 hydroxides (like KOH) and calcium, strontium, and barium hydroxide. (16.2)

4. a. HNO_3 is a strong acid, so the concentration of H^+ is 0.213 *M*. pH = $-\log[H^+] = -\log[0.213] = 0.672$. (16.3)

5. e. $Ba(OH)_2$ is a strong base, so the concentration of OH^- is twice the concentration of barium hydroxide, 0.028 *M*. pOH = $-\log[OH^-] = -\log[0.028] = 1.55$. pH = 14.00 – pOH = 12.45. (16.3)

6. a. pH = $-\log[H^+] = -\log[1.91] = -0.28$. Despite common chemistry myths, pH does *not* have to be between 0 and 14! (16.3)

7. b. NO_2 partially reacts with water to create HNO_3. Although HNO_3 is a strong acid (and completely reacts), it was formed from an equilibrium reaction, so all of NO_2 does not react to become H^+, so it is not a strong acid. It does form some H^+, so it is weakly acidic, as are all nonmetal oxides. (16.6)

8. d. Acetate ion ($CH_3CO_2^-$) is the conjugate base of acetic acid. $CH_3CO_2^- + H_2O \rightleftharpoons CH_3COOH + OH^-$.

   Its $K_b = \frac{K_w}{K_a} = 5.6 \times 10^{-10} = \frac{[CH_3COOH][OH^-]}{[CH_3CO_2^-]} =$

   $\frac{x^2}{0.12 - x}$ (from the ICE table). Solving the equation

   $x = [OH^-] = 8.2 \times 10^{-6}$ *M*. pOH = 5.09. So pH = 8.91. (16.7)

9. a. Ions are the product of the hydrolysis of a weak acid. Consequently, higher K_a values lead to higher ion concentration, which leads to a higher degree of ionization. (16.4)

10. c. CH_3NH_3Cl is the salt of a weak base consisting of the ions $CH_3NH_3^+$ and Cl^-. $CH_3NH_3^+$ reacts with water to produce H^+. (16.7)

11. d. Ammonia is a weak nitrogen base, calcium hydroxide is a strong hydroxide base, and carbonate of sodium carbonate is the conjugate base of carbonic acid. (16.1, 16.2, and 16.7)

12. e. The seven strong acids are HCl, HBr, HI, $HClO_4$, $HClO_3$, HNO_3, H_2SO_4. None of these acids are choices. (16.2)

13. c. A buffer is made from a conjugate acid–base pair. Both species of the pair must act as either an acid or a base to react with small amounts of acid or base. NaCl is neither an acid nor a base. (16.9)

14 d. Because a buffer contains a conjugate acid–base pair, the base can react with the additional acid, and the acid can react with the additional base and dilution with water will not change the acid–base ratio, which (along with the identity of the acid–base pair) determines the pH. (16.9)

15 a. The percent ionization increases with more dilute acids, since the ions are more separated and less likely to recombine. (16.4)

16. b. Perchloric acid completely reacts with the solvent water, so water is acting as a base. Because perchloric acid is a strong acid, perchlorate ion is too weak to be considered a base. (The reverse reaction does not occur.) (16.7)

17. c. In a Brønsted–Lowry reaction, an H^+ is transferred. In the reverse, an H^+ is transferred. Consequently, the product of the reaction is an acid and a base. (16.2)

18. a. A conjugate base has one (only one!) H^+ less than its conjugate acid. (16.2)

19. c. A conjugate acid has one (only one!) more H^+ than its conjugate base. (16.2)

20. d. In a Brønsted–Lowry acid–base reaction, a base accepts a H^+. In the forward reaction, that is HCO_3^-. In the reverse reaction, it is HPO_4^{2-}. (16.2)

21. e. All the acids listed are strong acids and ionize to the same extent (100%) in water. Therefore, they all have the same strength. (16.2)

22. c. The highest hydroxide ion concentration will produce the highest pH. Lithium hydroxide and strontium hydroxide are strong bases, so they will completely ionize. The stoichiometry of strontium hydroxide produces twice as much hydroxide as lithium hydroxide. Therefore, the strontium hydroxide solution has the highest pH. (16.3)

23. c. The strength of a conjugate base is inversely related to the strength of its conjugate acid. ($K_aK_b = K_w$) F^- and $CH_3CO_2^-$ are the conjugate bases, and the weaker acid (CH_3COOH) will have the stronger conjugate base ($CH_3CO_2^-$). (16.7)

24. b. $pH = -\log[H_3O^+] = -\log[12] = -1.08$. (16.3)

25. d. More oxygen atoms bonded to the central atom increase the acidity of an oxoacid. (16.6)

26. a. More electronegative central atoms increase the acid strength. (16.6)

27. c. It is harder to remove the second H^+ than the first, so the amount of H^+ contributed to the solution is very small. In addition, the H^+ created in the first ionization decreases the small amount of H^+ created in the second ionization even further. (16.5)

28. b. If pH = 5.7, the pOH = 8.3. $8.3 = -\log[OH^-]$. So $[OH^-] = 10^{-8.3} = 5 \times 10^{-9}$ *M*. (16.3)

29. b. Each pH unit reflects a factor of 10 change in H^+ concentration. Higher pH values represent lower H^+ concentrations. Therefore, a 3 pH unit increase results in a 1000-fold concentration decrease. (16.3)

30. b. $pOH = -\log[OH^-] = -\log[0.066] = 1.18$. $pH = 14 - pOH = 12.82$. (16.3)

31. b. K_a is proportional to concentration of $H^+(H_3O^+)$. Higher H^+ concentrations are represented by lower pH values. (16.4)

32. c. The equilibrium between a base and water is expressed with the K_b. K_b is proportional to the hydroxide ion concentration. Stronger bases (by definition) produce higher hydroxide ion concentrations. The higher hydroxide ion concentration leads to a higher value of pH. (16.4)

33. a. When concentration is expressed as pX, lower values represent higher concentrations. A base (by definition) increases the hydroxide ion concentration and decreases pOH. The hydroxide concentration also shifts the water autoionization equilibrium so that H^+ ion concentration decreases, thus pH increases. (16.3)

34. b. Most anions can be formed by the removal of H^+ from an acid. Thus they are conjugate bases. The anions associated with strong acids are exceptions. (16.7)

35. c. The K_b of an anion can be determined from the K_a of its conjugate acid (in this case, H_2CO_3) and the equation $K_aK_b = K_w$, so $K_b = \dfrac{1.0 \times 10^{-14}}{(4.3 \times 10^{-7})} =$ 2.3×10^{-8}. (16.7)

36. a. This is a buffer solution where HOCl is the acid and OCl^- is the base. According the Henderson-Hasselbalch equation $pH = pK_a + \log\left(\dfrac{[\text{base}]}{[\text{acid}]}\right)$. Since the base and acid have the same concentration, $pH = pK_a = 7.45$. (16.9)

37. d. Since buffer capacity refers to the amount of acid or base that can be added without significant pH change, more acid will neutralize more base, and more base will neutralize more acid. Therefore, a high concentration of both will increase buffer capacity. (16.9)

38. b. Since a large difference between acid and base concentrations is required to shift the pH from the pK_a, and buffers are optimal when both concentrations are large, the best buffers have a pH near their pK_a values. (16.9)

39. c. Since buffer capacity refers to the amount of acid or base that can be added without significant pH change, both acid and base are needed. Being in equal concentrations, both can be neutralized. (16.9)

40. b. Within 1.0 pH unit of the pK_a, a significant amount of both the acid and base form, so from pH of 2.5 to 4.5 the indicator is various shades of orange. Since the pH is higher than 4.5, the basic, yellow form predominates. (16.10)

41. d. The equation $K_aK_b = K_w$ refers to a conjugate acid–base pair. So K_a is for CH_3COOH and K_b is for $CH_3CO_2^-$. Using that equation, $K_b = 5.6 \times 10^{-10}$ and the pK_a is 9.25. (16.7)

42. c. At the equivalence point, the weak base CH_3NH_2 has completely reacted, creating the weak acid $CH_3NH_3^+$. There is also no HCl unreacted. Since the solution contains mostly weak acid, the pH is slightly acidic (less than 7). (16.10)

43. b. The equivalence point for this titration is 60.0 mL. At halfway to the equivalence point (30.0 mL), pH = pK_a. (16.10)

44. c. At halfway to the equivalence point, pH = pK_a. This is true for weak bases as well as weak acids. The K_b of NH_3 is 1.8×10^{-5}. Using $K_aK_b = K_W$, K_a of NH_4^+ is 5.6×10^{-10} and pK_a = 9.25. (16.10)

45. a. Most hydroxides are insoluble salts; therefore, they cannot completely ionize to OH^-, which is the definition of a strong base. (16.2)

Additional Practice Problem Answers

1. a. H_3O^+
   b. H_2CO_3
   c. HF
   d. HPO_4^{2-}
   e. NH_4^+

2. a. OH^-
   b. CO_3^{2-}
   c. NO_2^-
   d. HPO_4^{2-}
   e. $CH_3CO_2^-$

3. $HOCl + H_2O \leftrightharpoons H_3O^+ + OCl^-$

$$K_a = 3.5 \times 10^{-8} = \frac{[H_3O^+][OCl^-]}{[HOCl]}$$

	HOCl	N_3O^+	OCl^-
Initial	0.057	0	0
Change	$-x$	$+x$	$+x$
Equilibrium	$0.057 - x$	x	x

So, $K_a = \dfrac{(x)(x)}{(0.057 - x)} = 3.5 \times 10^{-8}$. Assuming x is small,

$$3.5 \times 10^{-8} = \frac{x^2}{0.057}$$

$$4.5 \times 10^{-5} = x$$

Assumption is correct, $0.057 - 4.5 \times 10^{-5} = 0.057$.

$$x = [H_3O^+]$$

$$pH = -\log[H_3O^+] = -\log[4.5 \times 10^{-5}] = 4.35$$

$$\text{percent ionized} = \frac{\text{amount dissociated}}{\text{initial amount}} \times$$

$$\frac{x}{0.057} \times 100 = \frac{4.5 \times 10^{-5}}{0.057} = 0.079\%$$

4. $C_5H_{11}N + H_2O \leftrightharpoons C_5H_{11}NH^+ + OH^-$

$$K_b = \frac{[C_5H_{11}NH^+][OH^-]}{[C_5H_{11}N]} = 1.3 \times 10^{-3}$$

	$C_5H_{11}N$	$C_5H_{11}NH^+$	OH^-
Initial	0.20	0	0
Change	$-x$	$+x$	$+x$
Equilibrium	$0.20 - x$	x	x

$$1.3 \times 10{-}3 = \frac{(x)(x)}{(0.20 - x)}$$

Assuming x is small,

$$1.3 \times 10^{-3} = \frac{x^2}{0.20}$$

$$0.016 = x$$

Assumption is incorrect, $0.20 - 0.016 \neq 0.20$. Not assuming x is small,

$$1.3 \times 10^{-3} = \frac{x^2}{(0.20 - x)}$$

$$2.6 \times 10^{-4} - 1.3 \times 10^{-3}x = x^2$$

$$0 = x^2 + 1.3 \times 10^{-3}x - 2.6 \times 10^{-4}$$

$$x = \frac{-0.0013}{2} \pm \frac{[(0.0013)^2 - 4(1)(-2.6 \times 10^{-4})]^{1/2}}{2}$$

$$x = \frac{-1.3 \times 10^{-3}}{2} \pm \frac{[1.04 \times 10^{-3}]^{1/2}}{2}$$

$$x = 1.55 \times 10^{-2}$$

$$x = [OH^-]$$

$$pOH = 1.81$$

$$pH = 14.00 - pOH = 14.00 - 1.81 = 12.19$$

5. $KF \rightarrow K^+ + F^-$

$F^- + H_2O \leftrightharpoons HF + OH^-$

K_a of HF $= 3.5 \times 10^{-4}$

K_b of $F^- = 2.86 \times 10^{-11}$

	F^-	HF	OH^-
Initial	0.045	0	0
Change	$-x$	$+x$	$+x$
Equilibrium	$0.045 - x$	x	x

$$K_b = \frac{[HF][OH^-]}{[F^-]}$$

$$2.86 \times 10^{-11} = \frac{(x)(x)}{(0.045 - x)}$$

Assuming x is small,

$$2.86 \times 10^{-11} = \frac{x^2}{0.045}$$

$$1.1 \times 10^{-6} = x = [OH^-]$$

Check assumption, $0.045 - 1.1 \times 10^{-6} = 0.045$. The assumption is good.

$$pOH = 5.95$$

$$pH = 8.05$$

6. $(NH_4)_2SO_4 \rightarrow 2\ NH_4^+ + SO_4^{2-}$

$[NH_4^+] = 0.270\ M$

$NH_4^+ + H_2O \leftrightharpoons H_3O^+ + NH_3$

	NH_4^+	H_3O^+	NH_3
Initial	0.270	0	0
Change	$-x$	$+x$	$+x$
Equilibrium	$0.270 - x$	x	x

$$K_a = \frac{[H_3O^+][NH_3]}{[NH_4^+]}$$

$$K_b \text{ of } NH_3 = 1.8 \times 10^{-5}$$

$$K_a \text{ of } NH_4^+ = 5.6 \times 10^{-10}$$

$$5.6 \times 10^{-10} = \frac{(x)(x)}{(0.270 - x)}$$

Assuming x is small,

$$5.6 \times 10^{-10} = \frac{x^2}{0.270}$$

$$1.2 \times 10^{-5} = x$$

Check assumption, $0.270 - 1.2 \times 10^{-5} = 0.270$. The assumption is good.

$$[H_3O^+] = x = 1.2 \times 10^{-5}\ M$$

$$pH = 4.91$$

7. $HNO_3 + H_2O \rightarrow H_3O^+ + NO_3^-$

$[H_3O^+] = 0.161\ M$

$pH = -\log[0.161] = 0.793$

8. $C_6H_5COOH + H_2O \leftrightharpoons H_3O^+ + C_6H_5CO_2^-$

$NaC_6H_5CO_2 \rightarrow Na^+ + C_6H_5CO_2^-$

	C_6H_5COOH	H_3O^+	$C_6H_5CO_2^-$
Initial	0.14	0	0.27
Change	$-x$	$+x$	$+x$
Equilibrium	$0.14 - x$	x	$0.27 + x$

$$K_a = 6.5 \times 10^{-5} = \frac{[H_3O^+][C_6H_5CO_2^-]}{[C_6H_5COOH]}$$

$$6.5 \times 10^{-5} = \frac{(x)(0.27 + x)}{(0.14 - x)}$$

Assuming x is small,

$$6.5 \times 10^{-5} = \frac{x(0.27)}{(0.14)}$$

$$3.4 \times 10^{-5} = x[H_3O^+]$$

$$4.47 = pH$$

Or,

$$pH = pK_a + \log\frac{[C_6H_5CO_2^-]}{[C_6H_5COOH]}$$

$$pH = 4.19 + \log\frac{(0.27)}{(0.14)}$$

$$pH = 4.47$$

9. Since tartaric acid is diprotic, the acid can be represented as H_2A.

$H_2A + H_2O \leftrightharpoons H_3O^+ + HA^-\ K_{a_1} = 1.0 \times 10^{-3}$

$HA^- + H_2O \leftrightharpoons H_3O^+ + A^{2-}\ K_{a_2} = 4.6 \times 10^{-5}$

Since the first reaction is the only one important to pH,

	H_2A	H_3O^+	HA^-
Initial	0.25	0	0
Change	$-x$	$+x$	$+x$
Equilibrium	$0.25 - x$	x	x

$$1.0 \times 10^{-3} = \frac{(x)(x)}{(0.25 - x)}$$

Assuming x is small,

$$1.0 \times 10^{-3} = \frac{x^2}{0.25}$$

$$1.6 \times 10^{-3} = x$$

$$0.25 - 0.016 \neq 0.25$$

So,

$$1.0 \times 10^{-3} = \frac{x^2}{(0.25 - x)}$$

$$2.5 \times 10^{-4} - 1.0 \times 10^{-3}x = x^2$$

$$0 = x^2 + 1.0 \times 10^{-3}x - 2.5 \times 10^{-4}$$

$$x = \frac{-0.0010}{2} \pm \frac{[(0.001)^2 - (4)(1)(-2.5 \times 10^{-4})]^{1/2}}{2}$$

$$x = 1.5 \times 10^{-2} = [H_3O^+]$$

$$pH = 1.80$$

10. a. $25.0\text{ mL} \times \left(\frac{1\text{ L}}{1000\text{ mL}}\right) \times \left(\frac{0.15\text{ mol amine}}{1\text{ L}}\right) \times \left(\frac{1\text{ mol HCl}}{1\text{ mol amine}}\right) \times \left(\frac{1\text{ L}}{0.10\text{ mol HCl}}\right) \times \left(\frac{1000\text{ mL}}{1\text{ L}}\right) = 37.5\text{ mL}$

Or,

$$M_AV_A = \left(\frac{1}{X}\right)M_BV_B$$

$$X = 1$$

$$(0.10\ M)(V_A) = (0.15\ M)(25.0\text{ mL})$$

$$V_A = 37.5\text{ mL}$$

b. At halfway ($V = 18.75$ mL),

$$pH = pK_a$$

$$K_b = 5.4 \times 10^{-4}$$

$$K_a = \frac{K_w}{K_b} = \frac{(1.0 \times 10^{-14})}{(5.4 \times 10^{-4})} = 1.85 \times 10^{-11}$$

$$pK_a = 10.73$$

c. At the equivalence point, all moles of amine become conjugate acid:

moles base = $(0.025\text{ L})(0.15\ M) = 3.75 \times 10^{-3}$ mol

moles acid = moles base = 3.75×10^{-3} mol

The total volume = 25.0 mL + 37.5 mL = 62.5 mL. The concentration of acid = $\frac{3.75 \times 10^{-3}\text{ mol}}{0.0625\text{ L}} = 0.060\ M$. The reaction is

$$(CH_3)_2NH_2^+ + H_2O \leftrightharpoons H_3O^+ + (CH_3)_2NH$$

	$(CH_3)_2NH_2^+$	H_3O^+	$(CH_3)_2NH$
Initial	0.060	0	0
Change	$-x$	$+x$	$+x$
Equilibrium	$0.060 - x$	x	x

$$K_a = 1.85 \times 10^{-11} = \frac{[H_3O^+][(CH_3)_2NH]}{[(CH_3)_2NH_2^+]}$$

$$1.85 \times 10^{-11} = \frac{(x)(x)}{(0.060 - x)}$$

Assuming x is small,

$$1.85 \times 10^{-11} = \frac{x^2}{0.060}$$

$$1.05 \times 10^{-6} = x$$

$$5.98 = pH$$

d. See figure.

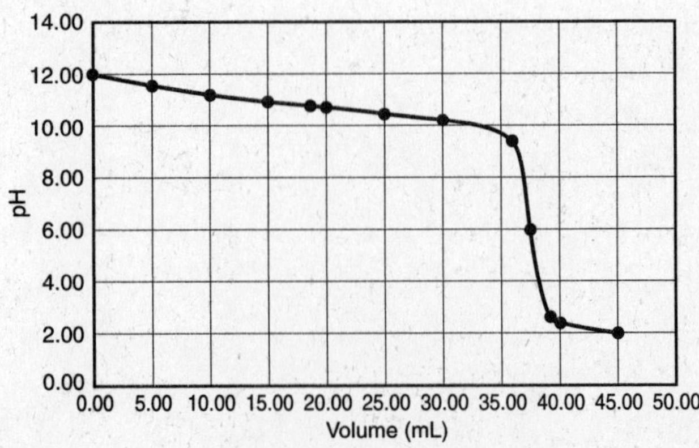

CHAPTER 17 The Colorful Chemistry of Transition Metals

OUTLINE

REVIEW

Chapter Overview

One way to describe the interactions between metal ions and other species in solution is with the Lewis acid–base theory. In this theory, **Lewis bases** are described as electron pair donors and **Lewis acids** are electron pair acceptors. A Lewis base can be easily identified from its Lewis structure as any substance that has a lone pair of electrons. Most anions are examples of Lewis bases. A Lewis acid must have an available location for the lone pair. Typical Lewis acids include H^+, boron compounds (with a deficient octet), and metal ions. When a Lewis acid and Lewis base react, the donated electrons from the base form a covalent bond with the acid. The new bond between the acid and base is called a **coordinate bond**.

Metal ions frequently form coordinate bonds with several Lewis bases at one time, creating a **complex ion**. In complex ions, the Lewis bases are called **ligands**. The ligands directly bonded to the metal form the **inner coordination sphere** of the metal. The number of coordinate bonds that a metal forms is its **coordination number. Coordination compounds** are formed of at least one complex ion. **Counter ions**, which are not bonded to the metal, supply the needed charge to make the compound electrically neutral.

Transition metals commonly form complex ions. Since these metals contain partially occupied *d* orbitals, these orbitals are involved in the bonding. Typical geometric orientations of complex ions include octahedral with d^2sp^3 hybridization (the *d* is first because it represents the lower-energy occupied orbital rather than the higher-energy unoccupied orbital), square planar with dsp^2 hybridization, and tetrahedral with sp^3 hybridization.

The strength of bonding of the ligand to metal can be described with a **formation constant (K_f)**. The reaction it refers to is the metal (M) bonding to *n* ligands (X) to form the complex ion (MX_n). The formation constant expression is $K_f = [MX_n]/[M][X]^n$. Stronger Lewis bases will have higher formation constants.

Hydrated metal ions can act as Brønsted–Lowry acids. Since the metal is bonded to the oxygen of water, the hydrogen of water is more easily released. This can create polyprotic metal acids. Since hydroxide can be a strong ligand, some metals, particularly Cr^{3+}, Al^{3+}, and Zn^{2+}, can bond with four hydroxides, creating a complex ion in strongly basic solution. Thus these ions are soluble in both acidic and basic solutions.

Some ligands bond only once with metal ions, these are **monodentate ligands**. Other Lewis bases, **polydentate** ligands, bond more than once to the metal. For a substance to act as a polydentate ligand, the lone pairs must be sufficiently separated in space so that both lone pairs can coordinate to the metal with the appropriate geometry. Thus water is a monodentate ligand, since both its lone pairs are on the same atom. However, ethylenediamine ($H_2NCH_2CH_2NH_2$) is a bidentate ligand because the lone pair on each nitrogen can both coordinate to the metal. The resulting structure of the metal-nitrogen-carbon-carbon-nitrogen-metal chain of bonds forms a 5-membered ring, which has angles that do not require significant geometric strain. Polydentate ligands often result in 5- or 6-membered rings.

Polydentate ligands are also called **chelates**. Complex ions formed with these ligands tend to be particularly stable. The driving force for this stability is entropy, since fewer ligands can occupy the same number of coordination sites, creating a large number of dissolved species. Ethylenediaminetetraacetic acid (EDTA) is a hexadentate ligand known for its action as a **sequestering agent**. Since EDTA will bond to the metal six times, the formation of a complex ion with EDTA is very strongly favored. Because the metal ion is bound to the EDTA, it is not available to react with other species in the mixture.

When the ligands interact with the *d* orbitals of the metal, the energy levels of the *d* orbitals increase. Due to geometry, all *d* orbitals do not interact equally with the metal ions creating **crystal field splitting**. In an octahedral geometry, the $d_{x^2-y^2}$ and d_{z^2} orbitals interact more strongly with the ligands than the d_{xy}, d_{xz}, and d_{yz} orbitals and are, therefore, higher in energy. This energy difference is called *octahedral crystal field splitting* (Δ_o). In a tetrahedral geometry, the d_{xy}, d_{xz}, and d_{yz} orbitals are higher in energy and the energy difference is the *tetrahedral crystal field splitting* (Δ_t). Square planar geometries also exhibit crystal field splitting. The difference in energy is larger for stronger Lewis bases, higher oxidation state metals, and larger metal ions. The **spectrochemical series** lists ligands in the order of their effect on field strength.

The effect of crystal field splitting can be seen in the different colors of complex ions. The amount of energy required to move an electron from the lower energy *d* orbital to the higher energy *d* orbital is often in the visible range of the spectra. However, the crystal field splitting energy reflects the energy of the absorbed photon, which is not the observed color. The observed color is the complementary color, which is the color of the light with the absorbed photon removed.

Crystal field splitting also affects the magnetic properties of complex ions. Magnetism is determined by the number of unpaired electrons. Electrons will always choose the lowest energy arrangement. It costs electrons energy to be in the higher energy *d* orbital. However, it also costs electrons energy to pair up in an orbital. Therefore, if the crystal field splitting energy is large (**strong field**), the preferential arrangement of electrons is to pair up in the lower-energy *d* orbital, creating a **low-spin complex**. However, if the splitting energy is small (**weak field**), then the electrons will occupy both orbitals creating a **high-spin complex** with more unpaired electrons.

Naming of coordination compounds must account for the metal, ligands, and counter ions. Like other ionic compounds, the name of the cation is first and then the anion. If the cation happens to be a complex ion, each ligand is named in alphabetical order with a prefix to denote how many of that type of ligand. Then the metal is named with its oxidation number in parenthesis as a roman numeral. There are no spaces between any of the ligands, prefixes, and metal names, nor are any letters deleted. If the complex ion is an anion, the ligands are named in the same way, but the root name of the metal ion is given an *ate* ending. Sometimes the Latin root of the metal name is used for anions.

Some complex ions have geometric isomers, also called stereoisomers. These are compounds where the formula and the bonding is the same; however, the geometric arrangement in space is different. With square planar and octahedral geometries, it is possible to have *cis* or *trans* geometric isomers. When two ligands of the same type are arranged opposite each other, the prefix *trans* is added to the name. If the ligands are adjacent, the *cis* prefix is used. When two isomers can be aligned as reflections in a mirror, yet cannot be superimposed on each other, the isomers are called **enantiomers** and each substance is called **chiral**.

Ionic compounds, salts, dissociate into their component ions in water. Soluble salts dissociate completely. However, most salts are not completely soluble in water. These dissociate in an equilibrium reaction to the component ions. The amount of solid that will dissolve in a given quantity of solution is the **solubility** of that salt. The equilibrium constant for the dissociation reaction is called the solubility product (or **solubility product constant**), K_{sp}. Solubility can be determined from K_{sp} values or K_{sp} values can be determined from solubility.

Worked Examples

Complex Ion Equilibria

Example 1.
What is the concentration of metal ion for a 0.13 *M* solution $Ag(CN)_2^-$? ($K_f = 3.0 \times 10^{20}$)

Collect and Organize We are asked to determine the concentration of Ag^+ in a solution of 0.13 *M* solution $Ag(CN)_2^-$ given an equilibrium constant.

Analyze Complex ion problems are solved with the same method as all other equilibrium problems. Formation constants (K_f) refer to the forming of the complex ion from the metal and its ligands.

SOLVE The reaction is

$Ag^+ + 2\ CN^- \rightarrow Ag(CN)_2^-$

$$K_f = \frac{[Ag(CN)_2^-]}{[Ag^+][CN^-]^2}$$

The ICE table

	Ag^+	CN^-	$Ag(CN)_2^-$
Initial	0	0	0.13
Change	$+x$	$+2x$	$-x$
Equilibrium	x	$2x$	$0.13 - x$

Using the K_f formula

$$3.0 \times 10^{20} = \frac{(0.13 - x)}{x(2x)^2}$$

Assuming x is small (after all, K very much favors the complex ion),

$$3.0 \times 10^{20} = \frac{0.13}{4x^3}$$

$1.2 \times 10^{21}x^3 = 0.13$

$x = 4.8 \times 10^{-8}$

Checking the assumption, x *is* much smaller than 0.13.

$$x = [Ag^+] = 4.8 \times 10^{-8}\ M$$

THINK ABOUT IT The equilibrium reaction is the metal ion reacting with ligands creating the complex ion. The equilibrium constant values for this type of reaction, K_f, are usually quite large.

EXAMPLE 2.
What is the concentration of metal ion in a solution of 0.17 M $Pb(OH)_3^-$ at pH = 12.42? ($K_f = 8 \times 10^{13}$)

COLLECT AND ORGANIZE We are asked to determine the concentration of Pb^{2+} in a solution of 0.17 M $Pb(OH)_3^-$ at a given pH given its equilibrium constant.

ANALYZE Using the information provided about hydroxide ion concentration, initial $Pb(OH)_3^-$ concentration, and stoichiometry, we will construct an ICE table to solve for the equilibrium concentration of Pb^{2+}.

SOLVE The reaction is

$Pb^{2+} + 3\ OH^- \leftrightharpoons Pb(OH)_3^-$

$$K_f = \frac{[Pb(OH)_3^-]}{[Pb^{2+}][OH^-]^3} = 8 \times 10^{13}$$

A pH of 12.42 implies a very basic solution. The concentration of OH^- in this solution is 0.026 M. The ICE table is

	Pb^{2+}	OH^-	$Pb(OH)_3^-$
Initial	0	0.026	0.17
Change	$+x$	$+3x$	$-x$
Equilibrium	x	$0.026 + 3x$	$0.17 - x$

Using the K_f formula

$$8 \times 10^{13} = \frac{(0.17 - x)}{(x)(0.026 + 3x)^3}$$

Assume $3x$ is small, and if $3x$ is, so is x. Therefore, the equation can be simplified to

$$8 \times 10^{13} = \frac{0.17}{(x)(0.026)^3}$$

$1.41 \times 10^9 x = 0.17$

$x = 1 \times 10^{-10}$

Well, x is small!

$$x = [Pb^{2+}] = 8 \times 10^{-14}\ M$$

THINK ABOUT IT With such a large equilibrium constant, it is not surprising that at equilibrium only a very small amount of uncomplexed Pb^{2+} remains.

HYDRATED METAL IONS AS ACIDS

EXAMPLE 1.
What is the pH of 0.41 M $AlCl_3$?

COLLECT AND ORGANIZE We are asked to determine the pH of a solution containing a small, highly charged metal ion.

ANALYZE Small, highly charged metal ions are acidic in aqueous solution. The H^+ ions come from the waters coordinated, as ligands or Lewis bases, to the metal ion. The general formula for the equilibrium is

$$M(H_2O)_6^{x+} + H_2O \leftrightharpoons H_3O^+ + M(H_2O)_5(OH)^{(x-1)+}$$

Because it is the lone pair of the oxygen in water that forms the bond with the metal, some prefer to write the formula as

$$M(OH_2)_6^{x+} + H_2O \leftrightharpoons H_3O^+ + M(OH_2)_5(OH)^{(x-1)+}$$

Sometimes it is also convenient to simplify the formula to

$$M^{x+} + H_2O \leftrightharpoons M(OH)^{(x-1)+} + H^+$$

This reaction, whichever way it is written, is essentially the same reaction as for any weak acid. Metal ions also have K_a values (table is found in the text), and the pH of an aqueous solution can be determined in the same way it was for any weak acid.

SOLVE $AlCl_3$ is a soluble salt, so

$AlCl_3 \rightarrow Al^{3+} + 3\ Cl^-$

$[Cl^-] = 1.23\ M$

$[Al^{3+}] = [Al(H_2O)_6{}^{3+}] = 0.41\ M$

Chloride ion does not react with water. However, aluminum ion is acidic, with $K_a = 1.4 \times 10^{-5}$. The reaction of the hydrated ion is

$Al(H_2O)_6{}^{3+} + H_2O \leftrightharpoons H_3O^+ + Al(H_2O)_5\ (OH)^{2+}$

$$K_a = \frac{[H_3O^+][Al(H_2O)_5\ (OH)^{2+}]}{[Al(H_2O)_6{}^{3+}]} = 1.4 \times 10^{-5}$$

Using the ICE table,

	$Al(H_2O)_6{}^{3+}$	$Al(H_2O)_5(OH)^{2+}$	H_3O^+
Initial	0.41	0	0
Change	$-x$	$+x$	$+x$
Equilibrium	$0.41 - x$	x	x

$$1.4 \times 10^{-5} = \frac{(x)(x)}{(0.41 - x)}$$

Assuming x is small,

$$1.4 \times 10^{-5} = \frac{x^2}{0.41}$$

$2.4 \times 10^{-3} = x$

Checking the assumption that x is small, $0.41 - 0.0024 = 0.4076 = 0.41$. The assumption is good.

$x = [H_3O^+] = 2.4 \times 10^{-3}\ M$

pH = 2.62

THINK ABOUT IT Normally, six waters are coordinated with each metal ion. (The exceptions are rarely worth learning.)

SOLUBILITIES OF IONIC COMPOUNDS

EXAMPLE 1.
What is the molar solubility and concentration of each ion in a saturated solution of CaF_2?

COLLECT AND ORGANIZE We are asked to determine the molar solubility and concentration of each ion in a saturated solution of CaF_2.

ANALYZE We will use the solubility product constant K_{sp} and a ICE table to solve for the desired concentrations.

SOLVE

$CaF_2 \leftrightharpoons Ca^{2+} + 2\ F^-$

$K_{sp} = [Ca^{2+}][F^-]^2 = 1.5 \times 10^{-10}$

Recall that the concentration of a solid is included in the constant, not in the concentration part of the mass action expression.

Only compounds that appear in the mass action expression appear in the ICE table,

	Ca^{2+}	F^-
Initial	0	0
Change	$+x$	$+2x$
Equilibrium	x	$2x$

$1.5 \times 10^{-10} = (x)(2x)^2$

$1.5 \times 10^{-10} = 4x^3$

To solve this equation with a nonprogrammable scientific calculator you will need the root button, the $x^{1/y}$ key, usually the shift of the x^y key. This example is solved on a typical nonprogrammable calculator as

$1.5 \times 10^{-10}\ (\div)\ 4 = 3.75 \times 10^{-11}\ (x^{1/y})\ 3 = 3.3 \times 10^{-4}$

$3.3 \times 10^{-4} = x =$ molar solubility (units are mol/L)

$[Ca^{2+}] = 3.3 \times 10^{-4}\ M$

$[F^-] = 6.6 \times 10^{-4}\ M$

THINK ABOUT IT Recall that a saturated solution has the maximum amount of solute dissolved in it. If solid is still present, the solution is saturated. For an equilibrium to occur, both products and reactants will be present; therefore, K_{sp} expressions only apply to saturated solutions.

EXAMPLE 2.
What is the molar solubility and concentration of each ion in a saturated solution of $Al(OH)_3$?

COLLECT AND ORGANIZE We are asked to determine the molar solubility and concentration of each ion in a saturated solution of $Al(OH)_3$.

ANALYZE We will use the solubility product constant K_{sp} and an ICE table to solve for the desired concentrations.

SOLVE

$Al(OH)_3 \leftrightharpoons Al^{3+} + 3\ OH^-$

$K_{sp} = [Al^{3+}][OH^-]^3 = 1.9 \times 10^{-33}$

With the ICE table,

	Al^{3+}	OH^-
Initial	0	0
Change	$+x$	$+3x$
Equilibrium	x	$3x$

$1.9 \times 10^{-33} = (x)(3x)^3$

$1.9 \times 10^{-33} = 27x^4$

For some reason, everyone's first instinct is that $3^3 = 9$, but it's 27. Really!

2.9×10^{-9} mol/L $= x =$ molar solubility

$[Al^{3+}] = 2.9 \times 10^{-9}\ M$

$[OH^-] = 8.7 \times 10^{-9}\ M$

Think About It Compared to the concentrations obtained in the previous example, we see that the degree of ionization of $Al(OH)_3$ is less than the degree of ionization of CaF_2. This is reinforced by the much smaller value of the solubility product constant for $Al(OH)_3$.

Example 3.
What is the concentration of calcium ion in a solution of 0.12 *M* NaF and saturated with CaF_2?

Collect and Organize We are asked to calculate the concentration of calcium ion in a solution containing both an insoluble salt, CaF_2, and a soluble salt, NaF.

Analyze We will use the solubility product constant, an ICE table, and the information provided about the concentration of NaF.

Solve

$NaF \rightarrow Na^+ + F^-$

$[F^-] = 0.12\ M$

$CaF_2 \leftrightharpoons Ca^{2+} + 2\ F^-$

$K_{sp} = [Ca^{2+}][F^-]^2 = 1.5 \times 10^{-10}$

Using the ICE table,

	Ca^{2+}	F^-
Initial	0	0.12
Change	$+x$	$+2x$
Equilibrium	x	$0.12 + 2x$

$$1.5 \times 10^{-10} = (x)(0.12 + 2x)^2$$

Assume $2x$ is small

$1.5 \times 10^{-10} = x(0.12)^2$

$1.5 \times 10^{-10} = x(0.0144)$

$1.5 \times 10^{-8} = x = [Ca^{2+}]$

Think About It Solubility rules and the concentration tip you off that NaF reacts completely. The term "saturated" lets you know that CaF_2 is in equilibrium with its ions.

This is an example of the common ion effect. Fluoride ion is common to both equilibria. The additional fluoride ion will shift the equilibrium toward the solid (Le Châtelier's principle), thus reducing the solubility of the solid.

Example 4.
What is the molar solubility and concentration of aluminum ion in a saturated solution of aluminum hydroxide and 0.0032 *M* calcium hydroxide?

Collect and Organize We are asked to calculate the concentration and molar solubility of aluminum ion in a solution containing both an insoluble salt, aluminum hydroxide, and a slightly soluble salt, calcium hydroxide.

Analyze We will use the solubility product constant, an ICE table, and the information provided about the concentration of calcium hydroxide.

Solve

$Ca(OH)_2 \rightarrow Ca^{2+} + 2\ OH^-$

$[OH^-] = 0.0064\ M$

$Al(OH)_3 \leftrightharpoons Al^{3+} + 3\ OH^-$

$K_{sp} = [Al^{3+}][OH^-]^3 = 1.9 \times 10^{-33}$

Using the ICE table,

	Al^{3+}	OH^-
Initial	0	0.0064
Change	$+x$	$+3x$
Equilibrium	x	$0.0064 + 3x$

$$1.9 \times 10^{-33} = (x)(0.0064 + 3x)^3$$

Assume $3x$ is small.

$1.9 \times 10^{-33} = x(0.0064)^3$

$7.2 \times 10^{-27} = x =$ molar solubility $= [Al^{3+}]$

Think About It The common ion effect refers to a solution that contains both an insoluble salt and a soluble salt with one ion in common. The molar solubility decreases with the addition of a common ion. Compare the answer in this example to the solubility determined in Example 2 of this section.

Crystal Field Theory

Example 1.
Draw the crystal field splitting of the following ions:

a. Iron(II), low spin in an octahedral field

b. Iron(III) in a tetrahedral field

c. High-spin manganese(II) in an octahedral field

d. Platinum(II) in a square planar field

COLLECT AND ORGANIZE We are asked to draw the crystal field splitting of several atoms given the type of splitting pattern observed and whether the electrons prefer a low spin or a high spin orientation.

ANALYZE After determining the number of *d* electrons for the molecule under consideration, it is necessary to then determine how those electrons are distributed among the *d* orbitals. This requires determination of the type of field, (octahedral, tetrahedral, or square planar), which provides information on the splitting pattern of the *d* orbitals and the propensity of the electrons to pair up before seeking higher energy orbitals.

SOLVE

a. First you must determine how many *d* electrons are present in Fe^{2+}.

The electron configuration of Fe is [Ar]$4s^23d^6$. Since higher *n* values are always lost first, the electron configuration of Fe^{2+} = [Ar]$3d^6$. Consequently, iron(II) has six *d* electrons. Since iron(II) is in an octahedral field, the *d* orbitals split as

Since it is a low-spin compound, electrons will fill the bottom row before the top one. Therefore, the crystal field diagram is

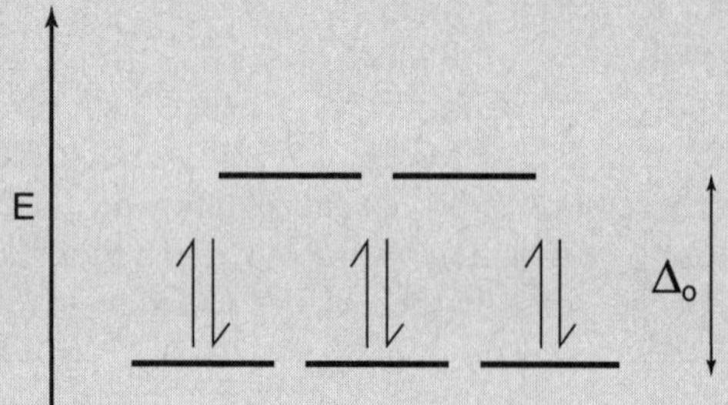

b. For Fe^{3+} the electron configuration is [Ar]$3d^5$. Since this is a tetrahedral field, the splitting is

Since it is a tetrahedral field, it is automatically "high spin." Therefore, the five *d* electrons arrange as

c. Mn^{2+} has an electron configuration of [Ar]$3d^5$. It is high spin and in an octahedral field. Therefore, the diagram looks like

d. Pt^{2+} has an electron configuration of [Xe]$4f^{14}5d^8$. Therefore, it has eight *d* electrons. The square planar splitting diagram is

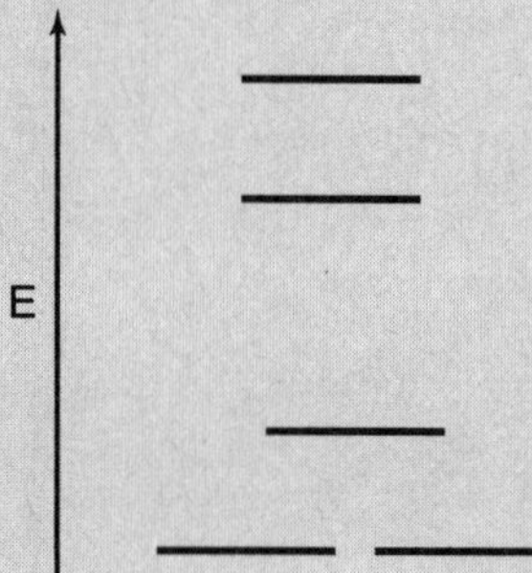

The eight d electrons arrange as if they were "low spin."

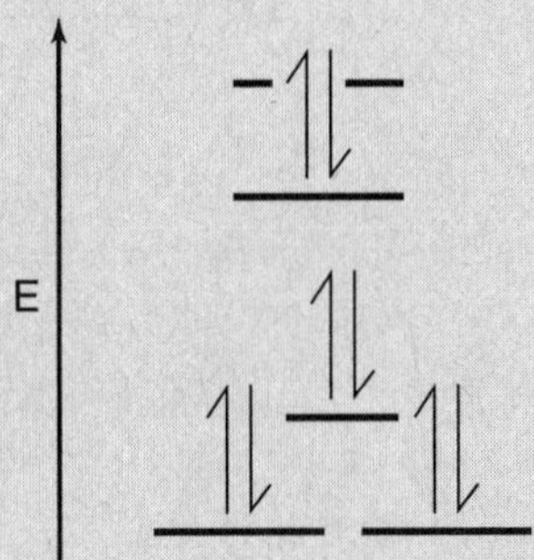

THINK ABOUT IT In an atom, all d orbitals have the same energy because their only difference is the orientation of the orbitals. However, in a molecule, the outermost electrons may interact with electrons from other atoms, if they are oriented correctly.

EXAMPLE 2.
How many unpaired electrons are in each substance listed in the previous example? Which substances are diamagnetic?

COLLECT AND ORGANIZE
We are asked to determine the number of unpaired electrons in each substance listed in Example 1 and to determine which of the substances are diamagnetic.

ANALYZE The number of unpaired electrons determines the magnetic properties of a molecule. A substance without unpaired electrons is "diamagnetic." A substance with unpaired electrons is "paramagnetic."

SOLVE Refer to the preceding splitting diagrams.

a. Low-spin Fe^{2+} (octahedral field) is diamagnetic with no unpaired electrons.

b. Fe^{3+} in a tetrahedral field has five unpaired electrons (paramagnetic).

c. High-spin Mn^{2+} (octahedral field) has five unpaired electrons (paramagnetic).

d. Pt^{2+} in a square planar field is diamagnetic with no unpaired electrons.

THINK ABOUT IT Crystal field splitting will affect the number of unpaired electrons.

NAMING COMPLEX IONS

EXAMPLE 1.
Name the following coordination compounds.

a. $[Fe(H_2O)_4Cl_2]NO_3$

b. $Na[Al(OH)_4]$

c. $[Cu(NH_3)_4]_2[Zn(CN)_6]$

COLLECT AND ORGANIZE We are asked to name several compounds that include complex ions.

ANALYZE Ionic compounds are named by naming first the cation then the anion. If the cation is a complex ion, each ligand should be named (in alphabetical order) with a prefix denoting the number of ligands when there is more than one of that type. Then the metal is named with its charge as a roman numeral in parenthesis after the name. The prefix, ligand names, metal, and charge are strung together without spaces. Common names of ligands are given in the text. The prefixes are the same ones used for covalent molecules (Chapter 2). Anions are named similarly, except that the metal has an "ate" ending and often uses the Latin root of the name.

SOLVE

a. $[Fe(H_2O)_4Cl_2]NO_3$. The two ions are $[Fe(H_2O)_4Cl_2]^+$ and NO_3^-. The charge on the complex ion was determined based on the charge of the nitrate. The entire compound, as with all salts, must be neutral.

   The ligands are water, which is called *aqua* when it acts as a ligand and chloride, which is called *chloro* when it acts as a ligand. The prefix for four is *tetra* and for two is *di*. Since water is neutral and chloride has a −1 charge, the charge on iron must be +3 so that the overall charge of the complex is +1. $(4(0) + 2(-1) + 3 = +1)$. The ligands are listed in alphabetical order (disregard prefixes to determine order); therefore, the name is: tetraaquadichloroiron(III) nitrate.

b. $Na[Al(OH)_4]$. In this case, the two ions are Na^+ and $[Al(OH)_4]^-$. The ligand is *hydroxo*, from the hydroxide ion. This means aluminum has a +3 charge. Normally the roman numeral for aluminum is not included in its name; however, complex ions may always contain the roman numeral. Because the complex ion is the anion, the metal will have an "ate" ending added. The name of this compound is: sodium tetrahydroxoaluminate(III).

c. $[Cu(NH_3)_4]_2[Zn(CN)_6]$. Since zinc always has a +2 charge and it is combined with six cyanide ions (*cyano* ligands), the total charge on the anion must be $+2 + 6(-1) = -4$. Therefore, the charge on the cation must be +2. Since ammonia (*ammine* ligand) is neutral, the charge on the copper must also be +2. The name of this compound is: tetraamminecopper(II) hexacyanozincate(II).

THINK ABOUT IT Names must always provide sufficient information to determine the formula. Complex ion naming is very systematic and does not make assumptions like the charge of the metal or combine double letters into one (i.e., it is tetraamminecopper(II) not tetramminecopper).

EXAMPLE 2.
What is the formula of each of the following compounds?

a. potassium trichlorotribromoferrate(III)

b. hexacarbonylnickel(II) sulfate

c. calcium diaquatetrahydroxoplumbate(II)

COLLECT AND ORGANIZE We are asked to determine the formulas of several ionic compounds containing complex ions.

ANALYZE In the formula, cations are always before anions regardless of whether or not it is the complex ion. The complex ion is grouped in square brackets with the metal first followed by the ligands in any order. The ratio of cation to anion is the simplest ratio that results in a neutral charge.

SOLVE

a. potassium trichlorotribromoferrate(III). The cation is potassium ion, K^+. The anion has three *chloro*, Cl^-, and three *bromo*, Br^-, ligands and iron (note that the symbol for iron comes from its Latin name *ferrum*) has a +3 charge. Therefore, the charge on the complex ion is 3(–1) + 3(–1) + 3 = –3. There must be three potassium ions to neutralize the charge of the anion. Therefore, the formula is: $K_3[FeCl_3Br_3]$.

b. hexacarbonylnickel(II) sulfate. In this case, the cation is a complex ion containing six *carbonyl* (CO) ligands and Ni^{2+}. Therefore, its overall charge is +2. The anion is sulfate, SO_4^{2-}. The formula for the compound is $[Ni(CO)_6]SO_4$.

c. calcium diaquatetrahydroxoplumbate(II). The cation is calcium ion, Ca^{2+}. The anion contains two *aqua* ligands (H_2O), four *hydroxo* (OH^-) ligands, and Pb^{2+} ion (from the Latin *plumbum* for lead). The net charge on the complex ion is –2. Therefore, the formula is: $Ca[Pb(H_2O)_2(OH)_4]$.

THINK ABOUT IT If the chemical symbol refers to the Latin name, the anion will also likely use the Latin version of the name. Lead was often used to make ancient Roman plumbing.

Self-Test

KEY VOCABULARY COMPLETION QUESTIONS

_________ 1. Arrangement of electrons in complex ion that minimizes unpaired electrons

_________ 2. Having the same chemical formula but a different geometric arrangement of atoms

_________ 3. Number of ligands attached to a metal

_________ 4. The ligand in $[Fe(H_2O)_6]^{2+}$

_________ 5. Electron pair donor

_________ 6. Ligand that can bind to a metal in two locations

_________ 7. Prefix for a compound where identical ligands are adjacent rather than across from each other

_________ 8. Compound that has an enantiomer

_________ 9. Energy difference between *d* orbitals in a complex ion

_________ 10. Reducing solubility of a solid by adding a soluble salt containing one of the ions also in the solid.

_________ 11. Metal ions, boron compounds, and H^+ are examples

_________ 12. Ion in a coordination compound that is *not* bonded to the metal

_________ 13. Equilibrium constant for the reaction of a metal with a ligand

_________ 14. Number of times EDTA binds to a metal

_________ 15. Ligand that will bond more than once with a metal

_________ 16. Amount of solid that will dissolve in a given quantity of solution

_________ 17. Another name for a Lewis base in a complex ion

_________ 18. Number of unpaired electrons in tetrahedral $[Ni(NH_3)_4]^{2+}$

_________ 19. Equilibrium constant that refers to the hydrolysis of an insoluble salt

_________ 20. Hybridization of metal in an octahedral complex ion

MULTIPLE-CHOICE QUESTIONS

1. The visible colors of gems are normally due to
   a. absorption of energy making a *d* electron become a *p* electron
   b. absorption of energy making a *p* electron become a *d* electron

c. absorption of energy making an *s* electron become a *d* electron
d. leftover visible light after some was used to promote a *d* electron to a higher-energy *d* electron

2. When *d* orbitals split in an octahedral field, the lower-energy *d* orbitals are
a. *xy*, *xz*, and *yz*
b. $x^2 - y^2$, z^2
c. z^2, *xz*, and *yz*
d. *xy*, $x^2 - y^2$

3. When *d* orbitals split in a tetrahedral field, the orbitals lower in energy are
a. *xy*, *xz*, and *yz*
b. $x^2 - y^2$, z^2
c. z^2, *xz*, and *yz*
d. *xy*, $x^2 - y^2$

4. When a compound absorbs orange light, its color is
a. orange
b. red
c. yellow
d. blue

5. Typically, which is largest?
a. high-spin Δ_o
b. low-spin Δ_o
c. Δ_t
d. all are the same

6. Which ion will not have high- and low-spin compounds?
a. Co^{2+}
b. Mn^{2+}
c. Fe^{3+}
d. Zn^{2+}

7. How many unpaired electrons are in a high-spin Fe^{3+} (octahedral field)?
a. zero
b. one
c. two
d. three or more

8. How many unpaired electrons are in low-spin Fe^{2+} (octahedral field)?
a. zero
b. one
c. two
d. three or more

9. Which *d* orbitals are highest in energy in a square planar compound?
a. *xy*, *xz*, and *yz*
b. z^2
c. $x^2 - y^2$
d. *xy*

10. Which low-spin compound is diamagnetic?
a. Fe^{3+}
b. Mn^{2+}
c. Co^{2+}
d. none are

11. Whether a compound is high spin or low spin depends primarily on
a. the number of *d* electrons
b. the magnitude of Δ
c. the type of metal ion
d. none of the above

12. Which of the following produce an acidic solution in water?
a. Na_2SO_4
b. NH_3
c. $AlCl_3$
d. all of the above
e. none of the above

13. Another name for a Lewis base is
a. conjugate
b. amphoteric
c. ligand
d. hydroxide
e. all of the above

14. In water, most metal ions
a. are insoluble
b. form complex ions with water
c. are acidic
d. both b and c
e. none of the above

15. Which of the following can act as a Lewis base?
a. hydroxide ion
b. chloride ion
c. water
d. none of the above
e. all of the above

16. If a metal has six ligands, its geometry will be
a. tetrahedral
b. square planar
c. trigonal bipyramid
d. octahedral
e. hexagonal

17. High formation constants imply that
a. most of the metal exists freely in solution
b. most of the ligands exist freely in solution
d. both metal and ligands exist freely in solution
e. ligands are bound to most of the metal ions

18. Which of the following is a polydentate ligand?
    a. H_2O
    b. Cl^-
    c. $H_2NCH_2CH_2NH_2$
    d. Al^{3+}

19. Ethylenediaminetetraacetic acid (EDTA) is what type of ligand?
    a. monodentate
    b. bidentate
    c. tridentate
    d. hexadentate
    e. it is not a ligand

20. Polydentate ligands are likely to displace monodentate ligands because
    a. it increases the entropy of the system
    b. it increases the enthalpy of the system
    c. monodentate ligands are stronger bases than polydentate ligands
    d. polydentate ligands are more magnetic than monodentate ligands

21. What is the charge on the metal in $[Fe(H_2O)_4Cl_2]NO_3$?
    a. zero
    b. +1
    c. +2
    d. +3
    e. more than +3

22. When NH_3 acts as a ligand, the name of the ligand is
    a. amine
    b. ammine
    c. ammonia
    d. ammonium
    e. trihydrogennitride

23. A geometric isomer where two ligands of the same type are exactly opposite each other uses the prefix
    a. *cis*
    b. *trans*
    c. di
    d. bi

24. What is the name of $Na_4[Ni(CN)_4Cl_2]$?
    a. sodium nickeltetracyanodichloride
    b. tetrasodium tetracyanodichloronickelate(IV)
    c. sodium dichlorotetracyanonickelate(II)
    d. sodium dichlorotetracyanonickel(IV)

25. What is the charge on the metal of $K_4[Fe(CN)_6]$?
    a. zero
    b. +1
    c. +2
    d. +3
    e. more than +3

26. Which of the following is most soluble in water?
    a. $AlPO_4$ ($K_{sp} = 9.8 \times 10^{-21}$)
    b. $BaSO_4$ ($K_{sp} = 9.1 \times 10^{-11}$)
    c. CuI ($K_{sp} = 1.3 \times 10^{-12}$)
    d. $FeCO_3$ ($K_{sp} = 3.1 \times 10^{-11}$)

27. In which of the following solutions will $AlPO_4$ ($K_{sp} = 9.8 \times 10^{-21}$) be the least soluble?
    a. water alone
    b. 0.10 M $Na_3PO_4(aq)$
    c. 0.10 M $HCl(aq)$
    d. Since they are all aqueous solutions, $AlPO_4$ will be equally soluble in all three.

28. What is the solubility of lead(II) fluoride ($K_{sp} = 3.2 \times 10^{-8}$) in water?
    a. 3.2×10^{-8} mol/L
    b. 1.8×10^{-4} mol/L
    c. 3.2×10^{-3} mol/L
    d. 2.0×10^{-3} mol/L

29. In a saturated solution of $Mg(OH)_2$, the pH will be
    a. very high
    b. slightly less than 7
    c. slightly greater than 7
    d. very low
    e. almost exactly 7, since magnesium hydroxide is insoluble

30. Metal ions produce acidic solutions by
    a. reacting with water so that the OH^- becomes a ligand, leaving the H^+ in solution.
    b. reacting with water so that the H^+ becomes a ligand, leaving OH^- in solution.
    c. forming a complex ion with water and this removal of water affects autoionization of water.
    d. metal ions do not produce acidic solutions.

Additional Practice Problems

1. Name the following coordination compounds when given the formula. Write the formula of the following compounds when given the name. If a substance is an isomer, sketch the appropriate orientation of ligands.
   a. $[Cu(NH_3)_4]SO_4$
   b. $Na_3[FeCl_2Br_3F]$
   c. $K_2[Mn(OH)_4(NH_3)_2]$
   d. tetraaquazinc(II) chloride
   e. *cis*-tetraamminedicyanochromium(III) nitrate
   f. calcium tetrachlorocobaltate(II)

2. What is the solubility of a saturated solution of calcium hydroxide? What is the pH of a saturated solution of calcium hydroxide? What is the solubility of calcium hydroxide if the pH of the solution is 14.00?

3. What is the equilibrium concentration of zinc ion and ammonia in a 0.14 *M* $Zn(NH_3)_4^{2+}$?

4. What is the pH of 0.15 *M* $CoCl_2$?

5. Draw the crystal field splitting diagrams for Ni^{2+} in all its variations. Nickel(II) ion can exist in tetrahedral, a square planar, high-spin octahedral, and low-spin octahedral fields.

Self-Test Answer Key

Key Vocabulary Completion Answers

1. low spin (strong field)
2. stereoisomers
3. coordination number
4. aqua
5. Lewis base
6. bidentate
7. *cis*
8. chiral
9. crystal field splitting
10. common ion effect
11. Lewis acids
12. counter ion
13. formation constant (K_f)
14. six
15. polydentate or chelate
16. solubility
17. ligand
18. two
19. solubility product constant (K_{sp})
20. d^2sp^3

Multiple-Choice Answers

For further information about the topics these questions address, you should consult the text. The appropriate section of the text is listed in parentheses after the explanation of the answer.

1. d. Transmitted light, *not* absorbed, is seen. The light energy absorbed is the energy required to promote the split *d* orbitals from one to the other *d* orbital. (17.8)

2. a. In an octahedral field, orbitals lying between the axes (*xy*, *xz*, *yz*) are lower in energy, and orbitals lying on the axes ($x^2 - y^2$, z^2) are higher in energy. That is because the orbitals on the axes interact with the electrons of the ligands. (17.8)

3. b. In a tetrahedral field, the orbitals lying on the axes ($x^2 - y^2$, z^2) are lower in energy than those between the axes (*xy*, *xz*, *yz*). The tetrahedral field is oriented between the axes. (17.8)

4. d. The color seen is the color transmitted, not the one absorbed. Since yellow and red are absorbed (orange), blue is the color observed. Since blue is leftover, that is the color seen. (17.8)

5. b. The crystal field splitting energy (Δ) is generally larger for octahedral fields than for tetrahedral fields. In a low-spin compound, the energy is so large that electrons will pair up rather than go to a higher energy level as in high-spin compounds. (17.8)

6. d. High- and low-spin compounds differ by their arrangement of electrons in *d* orbitals. Since zinc ion has 10 *d* electrons, there is only one possible arrangement. (17.8)

7. d. Fe^{3+} has five *d* electrons. Since it is a high-spin compound, the electrons will distribute in all five *d* orbitals rather than pair up. Consequently, there are five unpaired electrons. (17.8)

8. a. Fe^{2+} has six *d* electrons. In an octahedral field, three *d* orbitals are lower in energy than the other two. Since it is low spin, the electrons will fill the lower orbitals before going into the higher orbitals. Consequently, all six electrons, all paired, will be in the lower *d* orbitals. (17.8)

9. c. A square planar compound is similar to an octahedral compound. However, only four positions of the six are occupied. These positions would be in the same place as the lobes of the $d_{x^2-y^2}$ orbital, so that orbital is the highest energy. (17.8)

10. d. If it is a low-spin compound, electrons will first fill the three lower-energy *d* orbitals. Iron(III) and manganese(II) both have five electrons, so they all exist in the three lower orbitals, leaving one unpaired. Cobalt(II) has seven unpaired electrons. Six will fill the lower orbitals, with one unpaired electron remaining in the higher energy level. In each case, there is at least one unpaired electron, making all three paramagnetic. (17.9)

11. b. The electrons organize to have the lowest total energy. Since pairing up electrons costs energy, a high-spin compound must use less energy, putting electrons in a higher energy level than pairing up. Consequently, the higher the value of [delta], the less likely it is to be high spin. (17.8)

12. c. Hydrated metal ions, particularly small ions with a high charge (like Al^{3+}) are acidic. (17.4)

13. c. A ligand is a Lewis base, normally used in the context of the base bonding to a metal to make a complex ion. (17.1 and 17.2)

14. d. Metal ions generally form complex ions with water in a Lewis acid–base reaction. The H^+ from one of the waters bonded to the metal can react as an acid. (17.4)

15. e. Any substance with a lone pair of electrons can act as a Lewis base. (17.1)

16. d. Six substances arrange around a central atom in an octahedral geometry. (17.2)

17. e. Formation constants are the ratio of the complex ion over the free metal and ligand. If it is a high value, the substance exists primarily in the complex ion form. (17.3)

18. c. Each nitrogen has a lone pair. The lone pairs are sufficiently separated that both can bond to the metal. (17.5)

19. d. There are six locations that EDTA can bond. It surrounds the metal, which is why it is also a sequestering agent. (17.5 and 17.6)

20. a. Because six ligands are replaced by three (or fewer), there are more species when the polydentate ligand is bound to the metal than if monodentate ligands are bound to the metal. (17.7)

21. d. Nitrate and chloride each have a –1 charge. Water does not have a charge. Fe + 4(0) + 2(–1) + (–1) = 0. Therefore, Fe is +3. (17.10)

22. b. When ammonia acts as a ligand, its name is ammine. (17.10)

23. b. Ligands oriented opposite each other use the prefix *trans*. (17.11)

24. c. The cation is sodium. In ionic compounds, prefixes are not used to designate the number of ions. Because the complex ion is the anion, the metal ends in "ate." The ligands are cyano and chloro and should be in put in alphabetical order. The charge on nickel is +2 to balance the charges. (17.10)

25. c. Each cyano ligand has a –1 charge. Each potassium is +1. So 4(+1) + Fe + 6(–1) = 0; Fe = +2. (17.10)

26. b. Since each compound produces two ions in water, solubility will be equal to the square root of K_{sp}. Therefore, the highest value of K_{sp} will also have the highest solubility. (17.5)

27. b. The common ion, PO_4^{3-}, shifts the equilibrium toward the solid reactant, reducing the solubility. (17.5)

28. d. $K_{sp} = [Pb^{2+}][F^-]^2$. If solubility = x, then $[Pb^{2+}] = x$ and $[F^-] = 2x$. Therefore, $K_{sp} = 4x^3$ and $x = 2.0 \times 10^{-3}$ mol/L. (17.5)

29. c. While classes as insoluble, $Mg(OH)_2$ does produce a small amount of hydroxide ion in solution, thus slightly increasing the pH. (17.5)

30. a. Acidic solutions have an H^+ concentration higher than pure water. Therefore, to create an acidic solution (as metal ions do), H^+ must be created. (17.4)

Additional Practice Problem Answers

1. a. tetraamminecopper(II) sulfate
   b. sodium tribromodichlorofluoroferrate(III)
   c. potassium diamminetetrahydroxomanganate(II)
   e. $[Zn(H_2O)_4]Cl_2$
   f. $[Cr(NH_3)_4(CN)_2]NO_3$ (note: *cis* means the CN ligands are adjacent not across)
   g. $Ca[CoCl_4]$

2. $Ca(OH)_2 \rightleftharpoons Ca^{2+} + 2\ OH^-$

   $K_{sp} = [Ca^{2+}][OH^-]^2 = 4.7 \times 10^{-6}$

   Let x represent the solubility of calcium hydroxide and use that in the ICE table.

	Ca^{2+}	OH^-
Initial	0	0
Change	$+x$	$+2x$
Equilibrium	x	$2x$

   $4.7 \times 10^{-6} = (x)(2x)^2 = 4x^3$

   $x = 1.0 \times 10^{-2}$ mol/L = solubility

   $[OH^-] = 2x = 2.0 \times 10^{-2}\ M$

   pOH = 1.67 and pH = 13.32

   If the pH is 14.00, then the equilibrium concentration of hydroxide ion is 1.00 *M*. Using that value in the K_{sp} expression (with x still representing solubility)

   $4.7 \times 10^{-6} = x(1.00)^2$

   4.7×10^{-6} mol/L = x = solubility

3. $Zn^{2+} + 4\ NH_3 \leftrightharpoons Zn(NH_3)_4^{2+} + OH^-$

$$K_f = 7.8 \times 10^8 = \frac{[Zn(NH_3)_4^{2+}]}{[Zn^{2+}][NH_3]^4}$$

	Zn^{2+}	NH_3	$Zn(NH_3)_4^{2+}$
Initial	0	0	0.14
Change	+*x*	+4*x*	−*x*
Equilibrium	*x*	4*x*	0.14 − *x*

$$2.9 \times 10^9 = \frac{(0.14 - x)}{x(4x)^4}$$

Assuming x is small,

$$2.9 \times 10^9 = \frac{(0.14)}{256x^5}$$

$7.4 \times 10^{11}x^5 = 0.14$

$x = 2.9 \times 10^{-3}$

$[Zn^{2+}] = x = 2.9 \times 10^{-3}\ M$

$[NH_3] = 4x = 1.1 \times 10^{-2}\ M$

4. $CoCl_2 \rightarrow Co^{2+} + 2\ Cl^-$

$Co(H_2O)_6^{2+} \leftrightharpoons Co(H_2O)_5(OH)^+ + H^+$

$$K_a = \frac{[Co(H_2O)_5(OH)^+][H^+]}{[Co(H_2O)_6^{2+}]} = 2 \times 10^{-10}$$

	$Co(H_2O)_6^{2+}$	H^+	$Co(H_2O)_5(OH)^+$
Initial	0.15	0	0
Change	−*x*	+*x*	+*x*
Equilibrium	0.15 − *x*	*x*	*x*

$$2 \times 10^{-10} = \frac{(x)(x)}{(0.15 - x)}$$

Assuming x is small,

$$2 \times 10^{-10} = \frac{x^2}{0.15}$$

$5.5 \times 10^{-6} = x = [H^+]$

$5.3 = pH$

5. The electron configuration of $Ni^{2+} = [Ar]3d^8$.

Tetrahedral

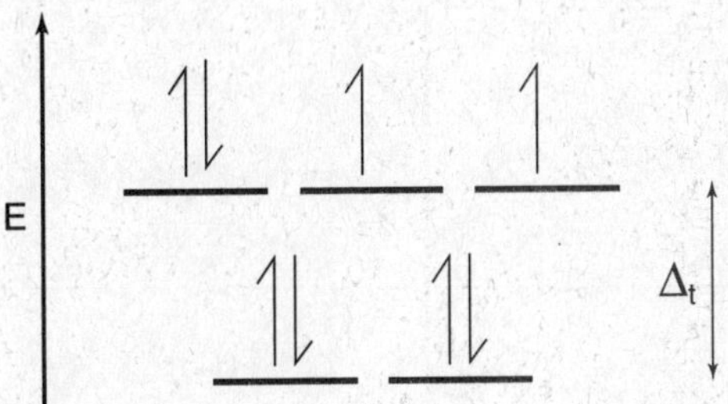

Square Planar

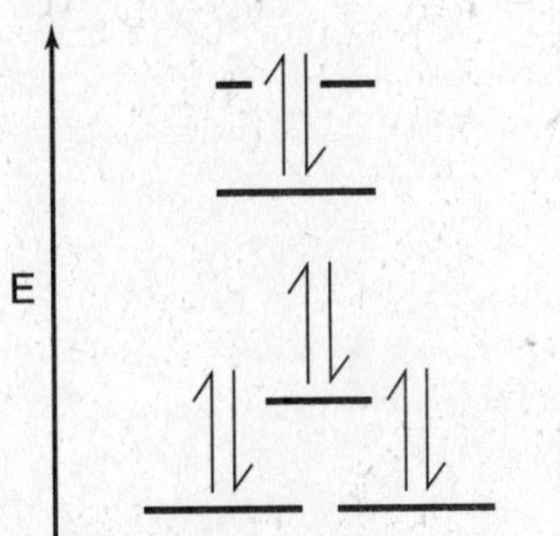

High-spin octahedral

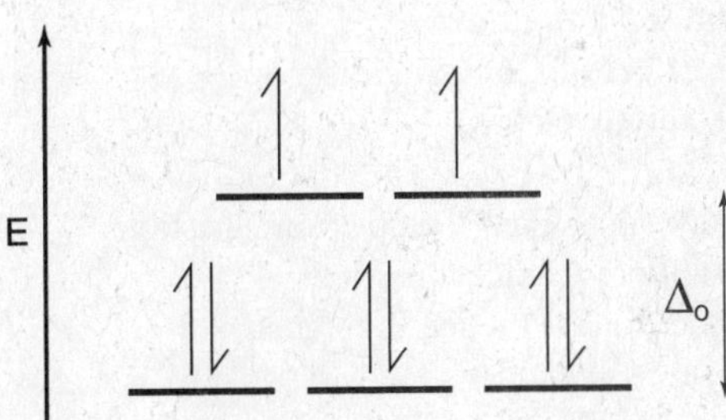

Low-spin octahedral

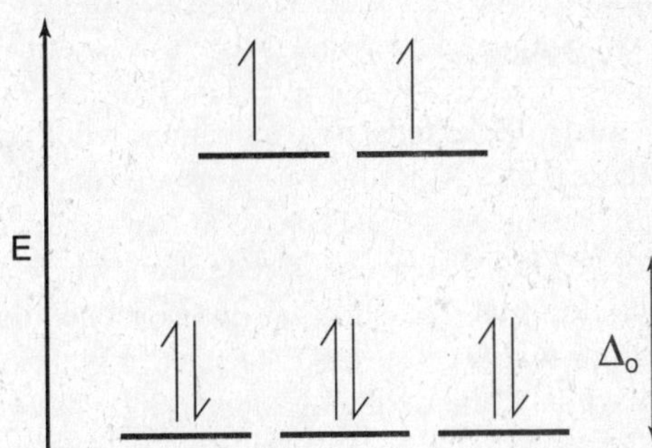

(*Note:* arrangement is the same as high spin for a d^8, but the energy separation is still larger.)

CHAPTER 18 Electrochemistry and Electric Vehicles

OUTLINE

REVIEW

Chapter Overview

Electrochemistry links reduction–oxidation (redox) chemistry to electrical energy. Redox reactions are characterized by the transfer of electrons and consist of two simultaneous half-reactions. One half-reaction is **reduction**, where a substance gains electrons. The other half-reaction is **oxidation**, where a substance loses electrons. The reactant of the reduction half-reaction is the **oxidizing agent**, since reaction of that substance is required for the oxidation half-reaction to occur. Similarly, the reactant of the oxidation half-reaction is the **reducing agent**. When the two half-reactions are combined, the number of electrons should cancel. This may require multiplying one or both half-reactions by some factor before the addition.

By separating the half-reactions in an **electrochemical cell**, the transfer of electrons can convert the chemical potential energy to electrical work. An electrochemical cell consists of a permeable **bridge**, so that ions can migrate and provide electrical neutrality without solution mixing, and **electrodes**, which provide the pathway for electron gain or loss. The electrode where oxidation occurs is the **anode**; the electrode where reduction occurs is the **cathode**. A **cell diagram** is a convenient way to describe an electrochemical cell. It starts on the left with the anode and separates the electrode from the surrounding solution with a vertical line. A double vertical line represents the bridge that separates the two solutions, and it is followed by the solution surrounding the cathode, a vertical line, and then the cathode.

Electromotive force (emf), also called **voltage**, is the measure of how forcefully electrons are pushed through an external circuit between electrodes. When measured in volts, electromotive force is also called **cell potential (E_{cell})**. Cell potential is proportional to the chemical potential energy released in the reaction. Therefore, electrical work is also the free energy (ΔG) of the electrochemical cell. Free energy is related to cell potential (E_{cell}) by **Faraday's constant** ($F = 9.65 \times 10^4$ C/mol) and the moles of electrons transferred (n) in the equation: $\Delta G_{cell} = -n\,FE_{cell}$.

Cell potentials (E_{cell}) depend primarily on the identity of the reaction. Since electrochemical cells separate the half-reactions, it is often convenient to talk about the potential of the half-reaction or **electrode potential**. Unfortunately, since one-half of a reaction does not occur without the other half, it is impossible to measure electrode potentials directly. So that electrode potentials could be tabulated, one half-cell was defined as having an electrode potential of exactly zero volts. This half-cell is called the **standard hydrogen electrode (SHE)**. It refers to the reduction of hydrogen ion (H^+) to hydrogen gas (H_2) using a platinum electrode. In this half-cell, concentration of H^+ is 1 M and the partial pressure of H_2 is 1 atm.

By tradition, tables list potentials for reduction half-reactions at **standard state** (each dissolved substance at a concentration of 1 *M*, each gas at a partial pressure of 1 bar and 25°C) with SHE as the oxidation half-cell. These are called the **standard reduction potentials (E°_{red})**. Since oxidation half-reactions are the reverse of the oxidation reactions, to obtain **standard oxidation potentials (E°_{ox})**, the sign of the reduction potential is reversed. (As with Hess's law for both ΔH and ΔG, when the reaction is reversed the sign is changed.) These tabulated values can be used to determine the **standard cell potential** for any electrochemical cell (**E°_{cell}**) by adding the standard oxidation and reduction potentials. Unlike Hess's law, however, any factor used to multiply the reactions before the addition is not used to calculate the standard cell potential.

To correct for nonstandard conditions, the **Nernst equation** is used to determine the **cell potential (E_{cell})**:

$$E_{cell} = E^\circ_{cell} - \left(\frac{RT}{nF}\right) \ln Q$$

where R is the gas constant (J/mol•K), T is the temperature (K), n is the moles of electrons in the overall reaction (mol), F is Faraday's constant (C/mol), and Q is the reaction quotient (Chapter 15) with all solutions in molarity and the partial pressure of all gases in atmospheres. Because the temperature effect is relatively small and most cells operate at room temperature, a convenient form of the Nernst equation, which assumes a temperature of 25°C is

$$E_{cell} = E^\circ_{cell} - \left(\frac{0.0592}{n}\right) \log Q$$

Spontaneous electrochemical cells, **voltaic cells**, are often used as batteries. The capacity of a battery depends on its potential and the amount of current it produces. **Current** is the rate of electron flow, with units of amperes (A). One **ampere** is one coulomb of charge per second. The power of a battery, a **watt**, is the product of the potential and amperes, 1 watt = 1 V•A = 1 J/s.

Since this power comes from a redox reaction, the battery is no longer useful when the reactants have been used up. Faraday's constant and the stoichiometry of the redox reaction can be used to calculate how long the battery will be useful.

The addition of an electrical current to an electrochemical cell can induce a nonspontaneous redox reaction, this process is called **electrolysis**. The actual potential required for this process is often more than calculated from the electrochemical potential to overcome kinetic as well as thermodynamic factors. This extra energy is called the **overpotential**. Faraday's constant can be used to relate the current to the time required to generate a specific amount of substance using electrolysis.

Worked Examples

Redox Chemistry Revisited

Example 1.
What is the oxidation number for each element in the following compounds:

a. LiH b. H_2 c. HNO_3 d. H_2O_2 e. $KClO_3$

Collect and Organize We are asked to determine the oxidation number for each element in the given compounds.

Analyze The rules for oxidation numbers: (Consider each ion of an ionic compound separately.)

1. The sum of all oxidation numbers is equal to the charge. Therefore,
   a. The oxidation number of a pure element is zero.
   b. The oxidation number of a monoatomic ion is equal to its charge.
2. The oxidation number of hydrogen is
   a. +1 if combined with a nonmetal
   b. –1 if combined with a metal
   c. 0 if combined only with itself
3. The oxidation number of oxygen is –2 (unless that contradicts rule 1 or 2).

Solve

a. Since hydrogen is combined with a metal, its oxidation number is –1. Since the sum of the oxidation numbers must equal the overall charge, the oxidation number of Li must be +1.

b. In this example, hydrogen is a pure element. Therefore, its oxidation number is zero.

c. Here hydrogen is combined with nonmetals, so its oxidation number is +1. Oxygen normally has an oxidation number of –2. Since there is no contradictory information, this must be correct. So that the oxidation numbers add up to the overall charge (zero), the oxidation number of nitrogen is calculated from the equation $1 + N + 3(-2) = 0$. Solving for N, the oxidation number of nitrogen is +5.

d. Since hydrogen is combined with a nonmetal, it has an oxidation number of +1. Oxygen normally has an oxidation number of –2, but if that were true, the sum of the oxidation numbers would not be equal to the charge. Since that rule (rule 1) has priority, the oxidation number of oxygen is –1. [From $2(+1) + 2(x) = 0$, where x is the oxidation number of oxygen.]

e. Since this is an ionic compound, it will be easier to split it into ions before determining the oxidation numbers. The ions that make up $KClO_3$ are K^+ and ClO_3^-. According to rule 1b, the oxidation number of potassium ion will be +1. Using rule 3, the oxidation number of oxygen is −2. Using rule 1, the sum of the oxidation numbers equals the charge, and the oxidation number on chlorine is determined from $Cl + 3(-2) = -1$, so the oxidation number of chlorine is +5.

THINK ABOUT IT The assignment of oxidation numbers to atoms in compounds is sometimes a process of elimination by using the rules provided. This is a review. Oxidation numbers were also covered in Chapter 4.

EXAMPLE 2.
Balance the following reaction in acid:

$$O_2 + H_2CO \rightarrow H_2O + HCO_2H$$

COLLECT AND ORGANIZE We are asked to balance the given redox reaction in acidic solution.

ANALYZE Redox reactions in acidic solutions are balanced by following the steps below.

1. Separate into the skeletal half-reactions.
2. Balance all elements except hydrogen and oxygen.
3. Add H_2O to balance water.
4. Add H^+ to balance hydrogen.
5. Add electrons (e^-) to balance charge.
6. Multiply each half-reaction so that the number of electrons is the same for each half.
7. Add the half-reactions together and cancel appropriately.

SOLVE

1. The elements changing oxidation state are oxygen and carbon. Therefore, the skeletal half-reactions are

$$O_2 \rightarrow H_2O$$
$$H_2CO \rightarrow HCO_2H$$

2. All elements except hydrogen and oxygen are balanced.
3. Balancing oxygen by adding water,

$$O_2 \rightarrow H_2O$$
$$H_2CO + H_2O \rightarrow HCO_2H$$

4. Balance hydrogen by adding H^+:

$$O_2 + 4H^+ \rightarrow 2H_2O$$
$$H_2CO + H_2O \rightarrow HCO_2H + 2H^+$$

5. Balance the charge by adding electrons:

$$O_2 + 4H^+ + 4e^- \rightarrow 2H_2O$$
$$H_2CO + H_2O \rightarrow HCO_2H + 2H^+ + 2e^-$$

6. Multiply the bottom half-reaction by 2 so that both half-reaction have four electrons:

$$O_2 + 4H^+ + 4e^- \rightarrow 2H_2O$$
$$2H_2CO + 2H_2O \rightarrow 2HCO_2H + 4H^+ + 4e^-$$

7. Add the two reactions together and cancel appropriately:

$$O_2 + 4H^+ + 4e^- + 2H_2CO + 2H_2O \rightarrow 2H_2O + 2HCO_2H + 4H^+ + 4e^-$$
$$O_2 + 2H_2CO \rightarrow 2HCO_2H$$

THINK ABOUT IT In an acid solution, use H^+ and H_2O to balance the charges and other atoms. (More examples in Chapter 4.)

EXAMPLE 3.
Balance the following reaction in base:

$$Ni + SO_4^{2-} \rightarrow Ni(OH)_2 + S$$

COLLECT AND ORGANIZE We are asked to balance the given redox reaction in basic solution.

ANALYZE Balancing redox reactions in basic solutions is accomplished in the same way as for acids. However, an additional step is needed to remove the H^+ and make the solution basic. This step is done immediately after balancing the hydrogens, so we'll call it step 4b.

4b. Add the same number of OH^- ions to BOTH sides, as there are H^+ ions. When H^+ and OH^- are on the same side of the equation, replace them with an equal number of waters. (e.g., $2H^+$ and $2OH^-$ becomes $2H_2O$).

SOLVE

1. The elements changing oxidation state are sulfur and nickel. Therefore, the skeletal half-reactions are

$$Ni \rightarrow Ni(OH)_2$$
$$SO_4^{2-} \rightarrow S$$

2. All elements except hydrogen and oxygen are balanced.
3. Balance oxygen by adding water. However, you may recognize that this is just a first step to balancing with hydroxide. In this half-reaction, a little common sense should tell you that the easiest way to balance the nickel reaction is to just balance with hydroxide directly. It is OK to do so!

$$Ni + 2OH^- \rightarrow Ni(OH)_2$$

Always let common sense beat out the rules. (Although the rules will work in this example, they are more steps than necessary.)

$$SO_4^{2-} + \rightarrow S + 4H_2O$$

The rules *are* helpful for the second reaction.

4a. Hydrogens are balanced for the nickel reaction. For the sulfur reaction,

$$8H^+ + SO_4^{2-} + \rightarrow S + 4H_2O$$

4b. To remove H^+ because this is a basic solution, add 8 OH^- (same as the number of H^+ ions) to each side. On the side where there are both H^+ and OH^-, these two ions combine to make water.

$$8OH^- + 8H^+ + SO_4^{2-} + \rightarrow S + 4H_2O + 8OH^-$$

$$8H_2O + SO_4^{2-} + \rightarrow S + 4H_2O + 8OH^-$$

You can simplify a little now or wait until the end.

$$4H_2O + SO_4^{2-} + \rightarrow S + 8OH^-$$

5. Balance the charge by adding electrons:

$$Ni + 2OH^- \rightarrow Ni(OH)_2 + 2e^-$$

$$SO_4^{2-} + 4H_2O + 6e^- \rightarrow S + 8OH^-$$

6. Multiply so that the number of electrons is the same in each half-reaction:

$$3Ni + 6OH^- \rightarrow 3Ni(OH)_2 + 6e^-$$

$$SO_4^{2-} + 4H_2O + 6e^- \rightarrow S + 8OH^-$$

7. Add the reactions together:

$$SO_4^{2-} + 4H_2O + 6e^- + 3Ni + 6OH^- \rightarrow 3Ni(OH)_2 + 6e^- + S + 8OH^-$$

Cancel:

$$SO_4^{2-} + 4H_2O + 3Ni \rightarrow 3Ni(OH)_2 + S + 2OH^-$$

THINK ABOUT IT In a basic solution, use OH^- and H_2O to balance the charges and other atoms. (There is another example of balancing a redox reaction in a basic solution in Chapter 4.)

EXAMPLE 4.

Identify the oxidation and reduction half-reactions in Examples 2 and 3.

COLLECT AND ORGANIZE We are asked to identify the oxidation and reduction half-reactions in the previous two examples.

ANALYZE When electrons are a reactant, the substance is gaining the electrons and therefore, is being reduced. The half-reaction is then referred to as the reduction half-reaction. When electrons are a product, the substance is losing electrons and therefore, is being oxidized. In this case the half-reaction is referred to as the oxidation half-reaction.

SOLVE The balanced half-reactions from the previous examples are

Ex 2. $O_2 + 4H^+ + 4e^- \rightarrow 2H_2O$

$H_2CO + H_2O \rightarrow HCO_2H + 2H^+ + 2e^-$

Ex 3. $Ni + 2OH^- \rightarrow Ni(OH)_2 + 2e^-$

$SO_4^{2-} + 4H_2O + 6e^- \rightarrow S + 8OH^-$

Using these examples, the position of electrons within the half-reaction, oxidation, and reduction can be identified.

Ex 2. $O_2 + 4H^+ + 4e^- \rightarrow 2H_2O$
(reduction; electrons are reactants)

$H_2CO + H_2O \rightarrow HCO_2H + 2H^+ + 2e^-$
(oxidation; electrons are products)

Ex 3. $Ni + 2OH^- \rightarrow Ni(OH)_2 + 2e^-$
(oxidation; electrons are products)

$SO_4^{2-} + 4H_2O + 6e^- \rightarrow S + 8OH^-$
(reduction; electrons are reactants)

THINK ABOUT IT Oxidation and reduction can also be determined from the change in oxidation numbers. If the oxidation number decreases (less positive, more negative), the substance must have gained electrons and been reduced. If the oxidation number increases (more positive, less negative), the substance must have lost electrons and been oxidized.

EXAMPLE 5.

What is the oxidizing and reducing agent in Examples 2 and 3?

COLLECT AND ORGANIZE We are asked to identify the oxidizing and reducing agents in the half-reactions from Examples 2 and 3.

ANALYZE The oxidizing agent causes oxidation by gaining electrons and is reduced in the process. The reducing agent causes reduction by losing electrons and is oxidized in the process.

SOLVE In Example 2 the half-reactions are

$$O_2 + 4H^+ + 4e^- \rightarrow 2H_2O \text{ (reduction)}$$

$$H_2CO + H_2O \rightarrow HCO_2H + 2H^+ + 2e^- \text{ (oxidation)}$$

Therefore, O_2 is the oxidizing agent. It is the reactant containing the element changing oxidation state by undergoing reduction. H_2CO is the reducing agent, the reactant containing the element undergoing oxidation.

In Example 3 the half-reactions are

$$Ni + 2OH^- \rightarrow Ni(OH)_2 + 2e^- \text{ (oxidation)}$$

$$SO_4^{2-} + 4H_2O + 6e^- \rightarrow S + 8OH^- \text{ (reduction)}$$

Ni is losing electrons; it is the reducing agent. Sulfur gains electrons; SO_4^{2-} is the oxidizing agent.

THINK ABOUT IT The reactant in the oxidation reaction is the reducing agent, and the reactant in the reduction reaction is the oxidizing agent. (There are more examples of determining reducing and oxidizing agents in Chapter 4.)

STANDARD POTENTIALS

EXAMPLE 1.
What is the standard cell potential of electrochemical cell with the following half-reactions with the given reduction potentials?

a. $O_2 + 4H^+ + 4e^- \rightarrow 2H_2O$ $\quad E° = 1.229$ V

$H_2CO + H_2O \rightarrow HCO_2H + 2H^+ + 2e^-$ $\quad E° = +0.237$ V

b. $Sn^0 \rightarrow Sn^{2+} + 2e^-$ $\quad E° = -0.141$ V

$2e^- + 2H^+ + H_2O_2 \rightarrow 2H_2O$ $\quad E° = +1.763$ V

c. $Ni + 2OH^- \rightarrow Ni(OH)_2 + 2e^-$ $\quad E° = -0.714$ V

$SO_4^{2-} + 4H_2O + 6e^- \rightarrow S + 8OH^-$ $\quad E° = -0.751$ V

COLLECT AND ORGANIZE We are asked to determine the standard cell potential of the given cells given the standard reduction potentials.

ANALYZE The standard cell potential ($E°_{cell}$) is the sum of the reduction and oxidation potentials of the two half-reactions: $E°_{cell} = E°_{red} + E°_{ox}$.

SOLVE

a. The reduction potential of the oxygen reaction is +1.229 V. The oxidation of carbon has a reduction potential of +0.237 V. To determine its oxidation potential, reverse the sign. So, $E°_{cell} = E°_{red} + E°_{ox} =$ $+1.229 + -0.237 = +0.922$ V. Recall that the significant-figure rule for addition is that the answer has the fewest number of decimal places.

b. The oxidation of tin has a reduction potential of −0.141 V. Reverse the sign to change it to the oxidation potential. The reduction of oxygen has reduction potential of +1.763 V. So, $E°_{cell} = E°_{red} + E°_{ox} =$ $+1.763 + (+0.141) = +1.904$ V.

c. The oxidation of nickel has a reduction potential of −0.714 V; therefore, its oxidation potential is +0.714 V. The reduction of sulfur has a reduction potential of −0.751 V. So, $E°_{cell} = E°_{red} + E°_{ox} =$ $-0.751 + (+.741) = -0.010$ V.

THINK ABOUT IT Reduction potentials are the values that are tabulated, therefore those are the numbers given regardless of whether it is a reduction or oxidation reaction. The values are not modified for multiplication to match the number of electrons.

EXAMPLE 2.
What is the free energy of an electrochemical cell undergoing the following half-reactions under standard conditions?

a. $O_2 + 4H^+ + 4e^- \rightarrow 2H_2O$

$H_2CO + H_2O \rightarrow HCO_2H + 2H^+ + 2e^-$

b. $Sn^0 \rightarrow Sn^{2+} + 2e^-$

$2e^- + 2H^+ + H_2O_2 \rightarrow 2H_2O$

c. $Ni + 2OH^- \rightarrow Ni(OH)_2 + 2e^-$

$SO_4^{2-} + 4H_2O + 6e^- \rightarrow S + 8OH^-$

COLLECT AND ORGANIZE We are asked to determine the free energy of the given cells under standard conditions.

ANALYZE The relationship between ΔG and E is $\Delta G = -nFE_{cell}$, where n = number of electrons in the total reaction, F is Faraday's constant (9.65×10^4 C/mol), and E_{cell} is the cell potential. Under standard conditions, $\Delta G° = -nFE°_{cell}$.

SOLVE The standard cell potentials were determined in Example 1.

a. Before the two half-reactions are combined, each must be multiplied by some factor, so that the number of electrons is the same in each half. The least common multiple of electrons for these reactions is 4. In Example 1, $E°_{cell} = +0.922$ V. Therefore, for the overall reaction, including electrons is

$$4e^- + O_2 + 2H_2CO \rightarrow HCO_2H + 4e^-$$

The standard free energy $\Delta G°$ is

$$\Delta G° = -nFE°_{cell} = -(4 \text{ mol } e^-)(9.65 \times 10^4 \text{ C/mol } e^-)(0.922 \text{ V})$$

$$= -3.56 \times 10^5 \text{ J}$$

b. When the two reactions are combined, and without canceling the electrons, the overall reaction is

$$2e^- + 2H^+ + H_2O_2 + Sn \rightarrow Sn^{2+} + 2H_2O + 2e^-$$

The standard free energy is

$$\Delta G° = -nFE°_{cell} = -(2)(9.65 \times 10^4)(+1.904) = -3.67 \times 10^5 \text{ J}$$

c. For the overall reaction:

$$6e^- + 3\,Ni + 4\,H_2O + SO_4^{2-} \rightarrow 3\,Ni(OH)_2 + S + 2\,OH^- + 6e^-$$

The standard free energy is

$$\Delta G^\circ = -nFE_{cell}^\circ = -(6)(9.65 \times 10^4)(-0.010) = +5.8 \times 10^3 \text{ J}$$

Think About It A positive potential represents a spontaneous reaction and voltaic cell. A negative potential denotes a nonspontaneous reaction and an electrolytic cell.

Example 3.

What is the equilibrium constant at 25°C of the following reactions?

a. $O_2 + 2\,H_2O_2 \rightarrow 2\,HCO_2H$

b. $2\,H^+ + H_2O_2 + Sn \rightarrow Sn^{2+} + 2\,H_2O$

c. $3\,Ni + 4\,H_2O + SO_4^{2-} \rightarrow 3\,Ni(OH)_2 + S + 2\,OH^-$

Collect and Organize We are asked to determine the equilibrium constant from the given reactions.

Analyze As ΔG° is related to the equilibrium constant (K), by $\Delta G^\circ = -\text{RTln K}$. E°_{cell} is related to ΔG° by $\Delta G^\circ = nFE^\circ_{cell}$. Combining the two equation, you get $E^\circ_{cell} = (\text{RT}/nF)\ln \text{K}$. As with ΔG°, it is the standard state only (E°_{cell}) that is related to the equilibrium constant. If you change the natural logarithm to a base 10 log and assume 25°C, the equation becomes: $\log K = nE^\circ/0.059$.

Solve The E°_{cell} values were determined in Example 1. The number of electrons transferred was determined in Example 2. Remember that the number of decimal places in a log term becomes the number of significant figures in the antilog.

a. $E^\circ_{cell} = \left(\dfrac{\text{RT}}{nF}\right)\ln \text{K}$

$$0.922 = \left[\frac{(8.314)(298)}{(4)(9.65 \times 10^4)}\right]\ln \text{K}$$

$$143 = \ln \text{K}$$

Hence, $K = e^{143} = 10^{62}$

b. $E^\circ_{cell} = \left(\dfrac{\text{RT}}{nF}\right)\ln \text{K}$

$$+1.904 = \left[\frac{(8.314)(298)}{(2)(9.65 \times 10^4)}\right]\ln \text{K}$$

$$148 = \ln \text{K}$$

Hence, $K = e^{148} = 10^{64}$

c. $E^\circ_{cell} = \left(\dfrac{\text{RT}}{nF}\right)\ln \text{K}$

$$-0.010 = \left[\frac{(8.314)(298)}{(6)(9.65 \times 10^4)}\right]\ln \text{K}$$

$$-2.33 = \ln \text{K}$$

Hence, $K = e^{-2.33} = 0.1$

Think About It Positive values of standard potential (E°_{cell}) indicate that the forward process is favored, and $K > 1$. Negative values of E°_{cell} indicate reactants are favored, with $K < 1$.

The Effect of Concentration on E_{cell}

Example 1.

What is the potential of an electrochemical cell with the reaction:

$$O_2 + 2\,H_2CO \rightarrow 2\,HCO_2H$$

if $P_{O_2} = 0.200$ atm, $[H_2CO] = 0.0137\ M$ and $[H_2CO_2H] = 0.124\ M$.

Collect and Organize For an electrochemical cell based on the given redox reaction as well as concentrations of reactants and products, we are asked to determine the cell's potential.

Analyze Standard potentials assume that the concentration of every substance in the reaction is 1 M. At other concentrations, it is necessary to use the Nernst equation

$$E_{cell} = E^\circ_{cell} - \left(\frac{0.0592}{n}\right)\log Q$$

to correct for the difference.

Solve The half-reactions are

$$O_2 + 4H^+ + 4e^- \rightarrow 2\,H_2O\ (E^\circ = +1.229\text{ V})$$

$$H_2CO + H_2O \rightarrow HCO_2H + 2H^+ + 2e^-\ (E^\circ = +0.237\text{ V})$$

The Nernst equation with Q for the overall reaction written out is

$$E_{cell} = E^\circ_{cell} - \left(\frac{0.0592}{n}\right)\log\left(\frac{[HCO_2H]^2}{P_{O_2} \cdot [HCO_2H]}\right)$$

The value of $E^\circ_{cell} = 1.229 + -0.237 = 0.922$ V. The number of electrons transferred, (n) is 4. Including those values in the Nernst Equation we have

$$E_{cell} = 0.922 - \left(\frac{0.0592}{4}\right)\log\left(\frac{[0.124]^2}{[0.200][0.0137]^2}\right) = +0.883\text{ V}$$

THINK ABOUT IT This form of the Nernst equation assumes that the temperature is 298 K (room temperature). Q is the reaction quotient, the form of the equilibrium constant at conditions other than equilibrium. In the Nernst equation, the concentration is in the form of partial pressures with units of atmospheres for gases and molarity for aqueous solutes, even if the units are mixed.

EXAMPLE 2.
What is the potential of an electrochemical cell with the reaction

$$2H^+ + H_2O_2 + Sn \rightarrow Sn^{2+} + 2H_2O$$

if $[Sn^{2+}] = 0.100\ M$, $[H_2O_2] = 0.759\ M$, and pH = 3.49.

COLLECT AND ORGANIZE For an electrochemical cell based on the given redox reaction as well as concentrations of reactants and products, we are asked to determine the cells potential.

ANALYZE Since we are dealing with a situation in which concentrations are not 1 M, it will be necessary again to use the Nernst equation.

SOLVE The half-reactions and reduction potentials are

$$Sn^0 \rightarrow Sn^{2+} + 2e^- \ (E° = -0.141\ V)$$

$$2e^- + 2H^+ + H_2O_2 \rightarrow 2H_2O \ (E° = +1.763\ V)$$

and the Nernst equation is

$$E_{cell} = E°_{cell} - \left(\frac{0.0592}{n}\right)\log\left(\frac{[Sn^{2+}]}{[H_2O_2][H^+]^2}\right)$$

The value of $E°_{cell} = +1.763 + (+0.141) = +1.904$ V. The value of $n = 2$. If the pH = 3.49, then $[H^+] = 10^{-3.49} = 3.2 \times 10^{-4}\ M$.

$$E_{cell} = 1.904 - \left(\frac{0.0592}{2}\right)\log\left(\frac{[0.10]}{[0.759][3.2 \times 10^{-4}]^2}\right)$$

$$E_{cell} = 1.904 - \left(\frac{0.0592}{2}\right)\log(6.1)$$

$$E_{cell} = 1.72\ V$$

THINK ABOUT IT Solids, like Sn^0, and solvents, like H_2O, are not included in Q.

EXAMPLE 3.
What is the potential of a cell with the reaction of

$$3\,Ni + 4H_2O + SO_4^{2-} \rightarrow 3\,Ni(OH)_2 + S + 2OH^-$$

if $[SO_4^{2-}] = 0.17\ M$ and pH = 12.66.

COLLECT AND ORGANIZE For an electrochemical cell based on the given redox reaction as well as concentrations of reactants and products, we are asked to determine the cell's potential.

ANALYZE Since we are dealing with a situation in which concentrations are not 1 M, it will be necessary again to use the Nernst equation.

SOLVE

$$E_{cell} = E°_{cell} - \left(\frac{0.0592}{n}\right)\log\left(\frac{[OH^-]^2}{[SO_4^{2-}]}\right)$$

Ni, $Ni(OH)_2$, and S are solids. H_2O is the solvent. The half-reactions and reduction potentials are

$$Ni + 2OH^- \rightarrow Ni(OH)_2 + 2e^- \ (E° = -0.741\ V)$$

$$SO_4^{2-} + 4H_2O + 6e^- \rightarrow S + 8OH^- \ (E°_{cell} = -0.751\ V)$$

The value of $E°_{cell} = -0.751 + (+0.741) = -0.010$ V. The value of $n = 6$. If pH = 12.66, pOH = 14 – 12.66 = 1.34, so $[OH^-] = 0.046\ M$. Using these values in the Nernst equation, we have

$$E_{cell} = -0.010 - \left(\frac{0.0592}{6}\right)\log\left(\frac{[0.046]^2}{[0.17]}\right)$$

$$E_{cell} = -0.010 - \left(\frac{0.0592}{6}\right)\log(0.012)$$

$$E_{cell} = +0.009\ V$$

THINK ABOUT IT Decreasing the pH would result in a higher concentration of hydroxide ion, resulting in an increase in the potential of the cell.

EXAMPLE 4.
At pH = 5.50 and at equilibrium, what is the ratio of $[ClO_3^-]$ to $[ClO_4^-]$? The standard potential for the reduction of perchlorate ion to chlorate ion in acidic solution is +1.266 V.

COLLECT AND ORGANIZE At a pH of 5.50, we are asked to determine the ratio of equilibrium concentrations resulting from the reduction of perchlorate to chlorate.

ANALYZE First, the balanced half-reaction must be determined. Since the question asks about concentration, the Nernst equation is appropriate.

SOLVE The skeletal reaction is $ClO_4^- \rightarrow ClO_3^-$. Using the directions for balancing in acid, the reaction becomes

$$ClO_4^- + 2H^+ + 2e^- \rightarrow ClO_3^- + H_2O$$

With electrons on the reactant side, this is written as a reduction, so the standard potential is +1.266 V for this half-reaction. For this reaction the Nernst equation is

$$E_{cell} = 1.266 - \left(\frac{0.0592}{2}\right)\log\left(\frac{[ClO_3^-]}{[ClO_4^-][H^+]^2}\right)$$

Since the reaction is at equilibrium, $E_{cell} = 0$. At a pH of 5.50, $[H^+] = 3.2 \times 10^{-6}$. Using these values in the equation,

$$0 = 1.266 - \left(\frac{0.0592}{2}\right)\log\left(\frac{[ClO_3^-]}{[ClO_4^-][3.2 \times 10^{-6}]^2}\right)$$

$$1.266 = \left(\frac{0.0592}{2}\right)\log\left(\frac{[ClO_3^-]}{[ClO_4^-][3.2 \times 10^{-6}]^2}\right)$$

Since $\log xy = \log x + \log y$,

$$1.266 = \left(\frac{0.0592}{2}\right)\log\left(\frac{[ClO_3^-]}{[ClO_4^-]}\right) + \left(\frac{0.0592}{2}\right)\log\left(\frac{1}{[3.2 \times 10^{-6}]^2}\right)$$

$$1.266 = \left(\frac{0.0592}{2}\right)\log\left(\frac{[ClO_3^-]}{[ClO_4^-]}\right) + 0.326$$

$$0.940 = \left(\frac{0.0592}{2}\right)\log\left(\frac{[ClO_3^-]}{[ClO_4^-]}\right)$$

$$6 \times 10^{31} = \frac{[ClO_3^-]}{[ClO_4^-]}$$

Think About It At equilibrium, the cell potential is zero. The ratio of various components at equilibrium can then be determined by the standard cell potential and reaction conditions. The Nernst equation is as valid for half-reactions as for total cell reactions.

Electrolytic Cells

Example 1.
How many grams of chromium metal can be made from chromium(III) ion with a current of 0.54 A for 1.00 hour?

Collect and Organize We are asked to determine the amount of chromium metal that is produced from the electrolysis of chromium(III) for 1.00 hour with a current of 0.54 A.

Analyze The half-reaction will be used to relate moles of electrons to moles of substance.

Solve The reaction is that of chromium(III) ion (Cr^{3+}) to its metal (Cr^0), so the balanced half-reaction is

$$Cr^{3+} + 3e^- \rightarrow Cr^0$$

From the reaction, 3 moles of electrons are required for each mole of chromium metal. Faraday's constant relates moles of electrons to the electrical unit coulombs (C). Amps (or amperes) are C/s, so it's a conversion factor, not the starting place. Using dimensional analysis and these relationships,

$$1.00 \text{ hour} \times \left(\frac{60 \text{ min}}{1 \text{ hr}}\right) \times \left(\frac{60 \text{ s}}{1 \text{ min}}\right) \times \left(\frac{0.54 \text{ C}}{1 \text{ s}}\right) \times \left(\frac{1 \text{ mol } e^-}{9.65 \times 10^4 \text{ C}}\right) \times \left(\frac{1 \text{ mol Cr}}{3 \text{ mol } e^-}\right) \times \left(\frac{52.00 \text{ g}}{1 \text{ mol Cr}}\right) = 0.35 \text{ g}$$

Think About It In electrolytic cells, nonspontaneous reactions are forced to occur through the application of current. These cells are used to produce some substance in a process called electrolysis.

Example 2.
What volume of hydrogen gas at STP can be produced from aqueous hydrochloric acid with a constant current of 5.30 A for 90.0 minutes?

Collect and Organize We are asked to determine the volume of hydrogen gas that is produced from the electrolysis of aqueous HCl for 90.0 minutes with a current of 5.30 A.

Analyze For the reaction, first recognize that hydrochloric acid in water is really the ions H^+ and Cl^-. Since hydrogen gas (H_2) is being produced, it must be made from H^+ ions.

Solve The reaction is

$$2H^+ + 2e^- \rightarrow H_2$$

Rather than grams of hydrogen, the question asks for volume of hydrogen. Recall that the volume of 1 mole of gas at STP = 22.4 L. Using dimensional analysis to solve the problem,

$$90.0 \text{ min} \times \left(\frac{60 \text{ s}}{1 \text{ min}}\right) \times \left(\frac{5.30 \text{ C}}{1 \text{ s}}\right) \times \left(\frac{1 \text{ mol } e^-}{9.65 \times 10^4 \text{ C}}\right) \times \left(\frac{1 \text{ mol } H_2}{2 \text{ mol } e^-}\right) \times \left(\frac{22.4 \text{ L}}{1 \text{ mol } H_2}\right) = 3.32 \text{ L}$$

Think About It Hydrochloric acid electrolysis is also of interest for the production of chlorine.

Example 3.
How long does it take to make 5.56 g nickel metal from Ni^{2+} with a constant current of 3.78 A?

Collect and Organize We are asked to determine the time it takes to produce a given quantity of nickel metal from the electrolysis of Ni^{2+} with a constant current of 3.78 A.

Analyze These problems are also solved by dimensional analysis. The same conversion factors are used, only the order of use changes.

SOLVE The reaction of making nickel metal from its ion is

$$Ni^{2+} + 2e^- \rightarrow Ni^0$$

Therefore, 2 moles of electrons create 1 mole of nickel metal. The electrons are being introduced at a rate of 3.78 C/s (amps). The relationship between coulombs (C) and moles of electrons is Faraday's constant (9.65×10^4 C/mol). Using these relationships in dimensional analysis,

$$5.56 \text{ g Ni} \times \left(\frac{1 \text{ mol Ni}}{58.69 \text{ g}}\right) \times \left(\frac{2 \text{ mol } e^-}{1 \text{ mol Ni}}\right) \times \left(\frac{9.65 \times 10^4 \text{ C}}{1 \text{ mol } e^-}\right) \times$$

$$\left(\frac{1 \text{ s}}{3.78 \text{ C}}\right) = 4837 \text{ s} = 4.84 \times 10^3 \text{ s} \times \left(\frac{1 \text{ min}}{60 \text{ s}}\right) = 80.6 \text{ min}$$

THINK ABOUT IT The superscript zero is not required for the formula of a metal but is sometimes convenient.

EXAMPLE 4.
How many minutes are required to make 1.0 g potassium metal from potassium ion at a current of 3.53 A?

COLLECT AND ORGANIZE We are asked to determine the time it takes to produce a given quantity of potassium metal from the electrolysis of potassium ion at a current of 3.53 A.

ANALYZE We will use the dimensional analysis method of the previous example.

SOLVE The reaction of potassium metal from potassium ion is

$$K^+ + e^- \rightarrow K$$

So,

$$1.0 \text{ g K} \times \left(\frac{1 \text{ mol K}}{39.10 \text{ g}}\right) \times \left(\frac{1 \text{ mol } e^-}{1 \text{ mol K}}\right) \times \left(\frac{9.65 \times 10^4 \text{ C}}{1 \text{ mol } e^-}\right) \times$$

$$\left(\frac{1 \text{ s}}{3.53 \text{ C}}\right) \times \left(\frac{1 \text{ min}}{60 \text{ s}}\right) = 12 \text{ min}$$

THINK ABOUT IT The information used for the preceding conversions was the following: mass × (molar mass) × (chem rxn) × (Faraday's constant) × (current) × (unit conversion) = time. The two significant figures were determined by the grams of potassium.

EXAMPLE 5.
What current is required to reduce 0.21 g of zinc ion to zinc metal in 30.0 minutes?

COLLECT AND ORGANIZE We are asked to determine the current that is required to reduce 0.21 g of zinc ion to zinc metal in 30.0 minutes.

ANALYZE We will solve this problem in a manner similar to that in the previous problems, in which we calculated the amount of material produced or the time required, except in this case we will calculate the required current.

SOLVE The reaction is

$$Zn^{2+} + 2e^- \rightarrow Zn$$

The charge on zinc is always +2. Current in amps is C/s. Since time is given in the problem, it just needs to be converted to seconds: $30.0 \text{ min} \times \left(\frac{60 \text{ s}}{1 \text{ min}}\right) = 1.80 \times 10^3$ s.

The coulombs required can be determined from the reaction and Faraday's constant:

$$0.21 \text{ g Zn} \times \left(\frac{1 \text{ mol Zn}}{65.38 \text{ g}}\right) \times \left(\frac{2 \text{ mol } e^-}{1 \text{ mol Zn}}\right) \times \left(\frac{9.65 \times 10^4 \text{ C}}{1 \text{ mol } e^-}\right) =$$

$$619.91 \text{ C} = 6.2 \times 10^2 \text{ C}$$

So the current $= \left(\frac{6.2 \times 10^2 \text{ C}}{1.80 \times 10^3 \text{ s}}\right) = 0.34 \text{ C/s} = 0.34 \text{ A}$.

THINK ABOUT IT In order to produce the same quantity of zinc in a shorter period of time, one would apply a stronger current.

Self-Test

KEY VOCABULARY COMPLETION QUESTIONS

__________ 1. Substance that is reduced

__________ 2. Oxidation and reduction half-reactions are physically separated

__________ 3. Surface where oxidation or reduction occurs

__________ 4. Rate of electron flow

__________ 5. Relationship of chemical reactions to the production or use of electricity

__________ 6. Sum of the standard reduction potential and the standard oxidation potential

__________ 7. Gain of electrons

__________ 8. Unit of electrical potential energy

__________ 9. Spontaneous electrochemical cell

__________ 10. Relationship between cell potential and concentration

__________ 11. Loss of electrons

__________ 12. An electrochemical cell where reactants are continually supplied

__________ 13. Substance that causes another to gain electrons

__________ 14. 25°C and all dissolved substances have a concentration of 1.00 *M*

__________ 15. Surface on which oxidation occurs

__________ 16. Coulomb × volt

__________ 17. Connects the solutions of an electrochemical cell without mixing

__________ 18. SI unit of charge

__________ 19. Surface at which reduction occurs

__________ 20. Ratio of products to reactants when reaction is NOT at equilibrium

__________ 21. A 1.00 *M* solution of H^+ is reduced to H_2 on a Pt electrode

__________ 22. Force that pushes electrons through a wire

__________ 23. Unit of current

__________ 24. 9.65×10^4 C/mol

__________ 25. Value of E_{cell} at equilibrium

__________ 26. A volt × ampere of electrical power

__________ 27. Using electrical energy to produce a substance

__________ 28. Extra electrical energy needed to induce a nonspontaneous reaction

__________ 29. $-nFE$

__________ 30. Moles of electrons required to reduce Fe^{3+} to the metal

Multiple-Choice Questions

1. In the half-reaction where elemental sulfur becomes sulfide ion, the type of reaction is
   a. oxidation
   b. reduction
   c. neutralization
   d. precipitation

2. For $2H^+ + 2e^- \rightarrow H_2$, hydrogen is
   a. oxidized
   b. reduced
   c. neutralized
   d. precipitated

3. For $8H^+ + 5Fe^{2+} + MnO_4^- \rightarrow Mn^{+2} + 5Fe^{3+} + 4H_2O$, the element oxidized is
   a. hydrogen
   b. oxygen
   c. manganese
   d. iron

4. For $8H^+ + 5Fe^{2+} + MnO_4^- \rightarrow Mn^{+2} + 5Fe^{3+} + 4H_2O$, the element reduced is
   a. hydrogen
   b. oxygen
   c. manganese
   d. iron

5. For $8H^+ + 5Fe^{2+} + MnO_4^- \rightarrow Mn^{+2} + 5Fe^{3+} + 4H_2O$, the oxidizing agent is
   a. H^+
   b. Fe^{2+}
   c. MnO_4^-
   d. H_2O

6. For $8H^+ + 5Fe^{2+} + MnO_4^- \rightarrow Mn^{+2} + 5Fe^{3+} + 4H_2O$, the reducing agent is
   a. H^+
   b. Fe^{2+}
   c. MnO_4^-
   d. H_2O

7. For $8H^+ + 5Fe^{2+} + MnO_4^- \rightarrow Mn^{+2} + 5Fe^{3+} + 4H_2O$, the number of electrons transferred is
   a. none
   b. one
   c. two
   d. more than two

8. For the reduction of 1 mole of chromium(III) to the metal, how many electrons are transferred?
   a. 3 moles
   b. 96,500 coulombs
   c. 3 amp
   d. all of the above

9. If $E° = +0.078$ V with four electrons transferred, then $\Delta G°$ is
   a. 3.0×10^4 J
   b. -3.0×165^4 J
   c. -7.8×10^3 J
   d. 7.8×10^{-2} J

10. For $8H^+ + 5Fe^{2+} + MnO_4^- \rightarrow Mn^{+2} + 5Fe^{3+} + 4H_2O$, $E°_{cell}$ is
    a. +2.28 V
    b. +0.74 V
    c. −2.28 V
    d. −2.34 V

11. For a spontaneous reaction,
    a. $E > 0, \Delta G > 0$
    b. $E > 0, \Delta G < 0$
    c. $E < 0, \Delta G > 0$
    d. $E < 0, \Delta G < 0$

12. If $E° > 0$, then the equilibrium constant is
   a. $K > 0$
   b. $K < 0$
   c. $K < 1$
   d. $K > 1$

13. If $E° = +0.77$ for the reduction of iron(III) to iron(II), K for this reaction at 25°C is
   a. 10^{13}
   b. 10^{-13}
   c. 10^{39}
   d. 1.6

14. The potential of the standard hydrogen electrode is
   a. exactly 1 V
   b. exactly –1 V
   c. exactly 0 V
   d. 0.0592 V

15. The potential of the standard hydrogen electrode at pH = 2 (but otherwise standard conditions) is
   a. 0.0 V
   b. –0.059 V
   c. –0.009 V
   d. –0.12 V

16. How long would it take to reduce 1.00 g of Ni^{2+} to its metal at 2.00 A?
   a. 13.7 min
   b. 54.8 min
   c. 27.4 min
   d. 26.8 hr

17. How much tin metal can be made from Sn^{4+} at a current of 3.5 A for 1.00 hr?
   a. 37 g
   b. 0.25 g
   c. 15.5 g
   d. 3.9 g

18. How many coulombs are required to reduce 1.0 mol of Cu^{2+}?
   a. 2
   b. 9.65×10^4
   c. 1.93×10^5
   d. 6.02×10^{23}

19. The Nernst equation corrects for nonstandard
   a. temperature
   b. concentrations
   c. both of the above
   d. neither of the above

20. If $\Delta G° = -1.03$ kJ/mol for a reaction that transfers 3 electrons, then $E° =$
   a. 3.56×10^{-6} V
   b. 2.98×10^5 V
   c. 3.20×10^3 V
   d. 3.56×10^{-3} V

21. What is $E°_{cell}$ for the reaction $2Ag^+ + Zn \rightarrow 2Ag + Zn^{2+}$?
   a. +1.56 V
   b. –0.04 V
   c. +0.04 V
   d. +0.84 V

22. For $E° = +1.78$ V, the cell is
   a. spontaneous, voltaic
   b. spontaneous, electrolytic
   c. nonspontaneous, voltaic
   d. nonspontaneous, electrolytic

23. In a voltaic cell, current flows
   a. from anode to cathode
   b. from cathode to anode
   c. through the salt bridge
   d. all of the above

24. Oxidation potential is
   a. the negative of reduction potential
   b. the reciprocal of the reduction potential
   c. the sum of the reduction potentials
   d. not related to the reduction potential

25. For the half-reaction $K^+ + e^- \rightarrow K$, $E° = -2.936$ V. Therefore,
   a. K^+ is a good oxidizing agent
   b. K^+ is a good reducing agent
   c. K is a good oxidizing agent
   d. K is a good reducing agent

26. The potential for H_2O_2 reducing to water is +1.763 V. Therefore,
   a. H_2O_2 is a good reducing agent
   b. H_2O_2 is a good oxidizing agent
   c. water is a good reducing agent
   d. water is a good oxidizing agent

27. Given the half-reactions:

   $Pb^{4+} + 2e^- \rightarrow Pb^{2+}$ $\quad E°_{red} = +1.210$ V

   $Fe^{3+} + e^- \rightarrow Fe^{2+}$ $\quad E°_{red} = +0.771$ V

   The spontaneous reaction is
   a. $Pb^{4+} + 2\,Fe^{3+} \rightarrow Pb^{2+} + 2\,Fe^{2+}$
   b. $Pb^{4+} + 2\,Fe^{2+} \rightarrow Pb^{2+} + 2\,Fe^{3+}$
   c. $Pb^{2+} + 2\,Fe^{3+} \rightarrow Pb^{4+} + 2\,Fe^{2+}$
   d. $Pb^{2+} + 2\,Fe^{2+} \rightarrow Pb^{4+} + 2\,Fe^{3+}$

28. For $8H^+ + 5Fe^{2+} + MnO_4^- \rightarrow Mn^{+2} + 5Fe^{3+} + 4H_2O$, what will have the biggest effect on the cell potential?
   a. changing the concentration of H^+
   b. changing the concentration of Fe^{2+}
   c. changing the concentration of MnO_4^-
   d. changing the concentration of water

29. In an electrochemical cell
    a. the oxidation and reduction reactions must be physically separated
    b. the oxidation and reduction reactions must be physically mixed
    c. a platinum electrode must be used
    d. it must be connected to a battery

30. Which reaction would occur at the cathode?
    a. $Sn^{4+} + 2e^- \rightarrow Sn^{2+}$
    b. $H_2 \rightarrow 2H^+ + 2e^-$
    c. $Sn^{4+} + H_2 \rightarrow Sn^{2+} + 2H^+$
    d. $Sn^{2+} + 2H^+ \rightarrow Sn^{4+} + H_2$

Additional Practice Problems

1. How many grams of manganese can be made from manganese(II) in 86.0 minutes with a current of 7.62 mA?

2. How many minutes are required to make 5.00 mL of oxygen at 25°C and 752 torr with a constant current of 2.10 A in acidic, aqueous solution?

3. For $HClO_3 + S \rightarrow SO_2 + Cl_2$:
    a. Balance the reaction in acidic, aqueous solution.
    b. Calculate the standard cell potential.
    E°_{red} for the chlorine half-reaction = +1.458 V
    E°_{red} for sulfur half-reaction = +0.450 V
    c. Calculate the standard free energy.
    d. Calculate the equilibrium constant.

4. For $BrO^- + Al \rightarrow Br^- + Al(OH)_4^-$:
    a. Balance the reaction in basic, aqueous solution.
    b. Determine the oxidizing and reducing agents.
    c. Calculate the standard cell potential.
    E°_{red} for the aluminum half-reaction = –2.328 V
    E°_{red} for the bromine half-reaction = +0.766 V
    d. What is the cell potential when pH = 12.31, $[BrO^-] = 0.10\ M$, $[Br^-] = 0.50\ M$, and $[Al(OH)_4^-] = 0.20\ M$.

5. Calculate the cell potential for the following reaction in a solution where pH = 2.73 and the concentration of all other ions is 0.10 *M*:

$$Cr_2O_7^{2-} + Sn \rightarrow Sn^{2+} + Cr^{3+}$$

Self-Test Answer Key

Key Vocabulary Completion Answers

1. oxidizing agent
2. electrochemical cell
3. electrode
4. current
5. electrochemistry
6. standard cell potential (E°_{cell})
7. reduction
8. volt
9. voltaic
10. Nernst equation
11. oxidation
12. fuel cell
13. reducing agent
14. standard conditions
15. anode
16. joule
17. permeable bridge (salt bridge)
18. coulomb
19. cathode
20. reaction quotient (Q)
21. standard hydrogen electrode (SHE)
22. electromotive force (or voltage or potential)
23. ampere
24. Faraday's constant
25. zero
26. watt
27. electrolysis
28. overpotential
29. free energy (ΔG)
30. three

Multiple-Choice Answers

For further information about the topics these questions address, you should consult the text. The appropriate section of the text is listed in parentheses after the explanation of the answer.

1. b. Elemental sulfur has an oxidation number of zero; sulfide ion has an oxidation number (and charge) of –2. This change in oxidation number requires that sulfur gain two electrons. A gain of electrons is reduction. (18.1)

2. b. Since hydrogen gains two electrons, it is reduced. (18.1)

3. d. Loss of electrons is oxidation. Iron loses an electron to change from Fe^{2+} to Fe^{3+}. (18.1)

4. c. Mn goes from an oxidation state of +7 in MnO_4^- to +2 in Mn^{2+}; therefore, it must gain electrons and be reduced. (18.1)

5. c. The oxidizing agent causes another substance to be oxidized. It does this by undergoing reduction. Permanganate ion (MnO_4^-) causes the oxidation of iron and is reduced to Mn^{2+}. (18.1)

6. b. The reducing agent causes another substance to be reduced. It does this by undergoing oxidation. Since Fe^{2+} is oxidized to Fe^{3+}, it is the reducing agent. (18.1)

7. d. The oxidation of Fe^{2+} to Fe^{3+} requires the loss of one electron. Since five irons are oxidized, five electrons are transferred. (18.1)

8. a. The reaction is $Cr^{3+} + 3e^- \rightarrow Cr^0$. The reaction states that 3 moles of electrons are used. One mole of electrons is 96,500 coulombs. Current, in amps, is the rate of electron travel in coulombs per second. (18.8)

9. b. $\Delta G° = -nFE^0 = -(4 \text{ mol})(96{,}500 \text{ C/mol})(+0.078 \text{ V}) = -30{,}108 \text{ C}\bullet\text{V} = -3.0 \times 10^4 \text{ J}$. (18.3)

10. b. The reduction (cathodic) reaction is for manganese; its reduction potential is +1.51 V. The oxidation (anodic) reaction is for iron; its reduction potential is +0.77 V, so its oxidation potential is –0.77 V. $E°_{cell} = E°_{red} + E°_{ox} = 1.51 - 0.77 = 0.74$ V. (18.4)

11. b. Spontaneous reactions have negative values of ΔG and positive values of E. The relationship is $\Delta G = -nFE$. (18.3)

12. d. Equilibrium constants are never negative. The relationship between equilibrium constant is $\log K = \frac{nE°}{0.0592}$. If E is positive, $K > 1$. If E is negative, $K < 1$. (18.4 and 18.6)

13. a. The relationship between equilibrium constant and cell potential at 25°C is $\log K = \frac{nE°}{0.0592}$. In this reaction: $Fe^{3+} + e^- \rightarrow Fe^{2+}$, $n = 1$. Therefore, $\log K = (1)\frac{(0.77)}{0.0592} = 13.00$, so $K = 10^{13}$. (18.6)

14. c. The standard hydrogen electrode is defined as having a potential of exactly zero volts. (18. 5)

15. d. Standard conditions have the concentration of H^+ at 1.0 M. Since this is nonstandard conditions, the Nernst equation is used. $E = E° - 0.0592/n \log Q$. The relevant reaction is $2H^+ + 2e^- \rightarrow H_2$. So $E = 0 - \left(\frac{0.0592}{2}\right)\log\left(\frac{P_{H_2}}{[H^+]^2}\right)$. At pH = 2, $[H^+] = 0.01\ M$, and $P_{H_2} = 1$ atm (standard conditions). Therefore, $E = -\left(\frac{0.0592}{2}\right)\log\left(\frac{1}{10^{-4}}\right) = -0.12$ V. (18.5 and 18.6)

16. c. $1.00 \text{ g Ni}^{2+} \times \left(\frac{1 \text{ mol}}{58.69 \text{ g}}\right) \times \left(\frac{2 \text{ mol } e^-}{1 \text{ mol Ni}^{2+}}\right) \times \left(\frac{9.65 \times 10^4 \text{ C}}{1 \text{ mol } e^-}\right) \times \left(\frac{1 \text{ s}}{2.00 \text{ C}}\right) \times \left(\frac{1 \text{ min}}{60 \text{ s}}\right) =$ 27.4 min. (18.8)

17. d. $Sn^{4+} + 4e^- \rightarrow Sn^0$, so $1.00 \text{ hr} \times \left(\frac{3600 \text{ s}}{1 \text{ hr}}\right) \times \left(\frac{3.5 \text{ C}}{1 \text{ s}}\right) \times \left(\frac{1 \text{ mol } e^-}{9.65 \times 10^4 \text{ C}}\right) \times \left(\frac{1 \text{ mol Sn}}{4 \text{ mol } e^-}\right) \times \left(\frac{118.71 \text{ g}}{1 \text{ mol}}\right) = 3.9$ g. (18.8)

18. c. Two moles of electrons are required to reduce 1 mole of copper(II) ion. Each mole of electrons is 9.65×10^4 C. (18.8)

19. c. The Nernst equation is $E = E° - (RT/nF) \ln Q$. The concentration factor is in the Q, and the temperature is the constant before the logarithm. (18.6)

20. d. $\Delta G° = -nFE°$. So $-\Delta G°/nF = E°$. If E is in volts, ΔG is in joules. Therefore, $-\left(\frac{(-1030)}{[(3)(9.65 \times 10^4)]}\right) = 0.00356$ V. (18.3)

21. a. $E° = E°_{red} + E°_{ox} = +0.80 + 0.76 = +1.56$ V. (18.4)

22. a. Another name for a spontaneous cell is a voltaic cell. Positive cell potentials are spontaneous. (18.3)

23. a. Oxidation occurs at the anode, giving electrons to the wire. Reduction occurs at the cathode, so it is removing electrons from the wire. Therefore, electrons travel from the anode to the cathode. (18.2)

24. a. Since the oxidation reaction is the reverse of the reduction reaction, the oxidation potential (like ΔH, ΔS, and ΔG) is the negative of the reduction potential. (18.4)

25. d. The spontaneous reaction is the reverse reaction, the oxidation reaction. Therefore, K reacts, not K^+. Since K undergoes oxidation, it acts as a reducing agent. (18.1 and 18.4)

26. b. Since the reduction of H_2O_2 is spontaneous, H_2O_2 is a good oxidizing agent. (18.1 and 18.4)

27. b. So that electrons cancel, one of the reactions must be reversed. If the iron reaction is reversed, the cell potential is positive and the reactions are spontaneous. (If the lead reaction were reversed, the cell potential would be negative and the overall reaction would be nonspontaneous.) (18.3 and 18.4)

28. a. The effect of changing concentration on potential is expressed by the Nernst equation. Concentration is part of the Q expression. In the reaction quotient (Q), the concentration of each reactant and product is raised to its stoichiometric coefficient. Therefore, the concentration of substance with the largest stoichiometric coefficient (H^+) will have the largest effect on potential. (18.6)

29. a. To create an electrochemical cell, the electron transfer must occur through a wire. If the reactions are not physically separate, the electron transfer will occur directly (molecule to molecule) rather than through the wire. (18.2)

30. a. Reduction occurs at the cathode. The tin reaction is the reduction reaction. The hydrogen reaction is the oxidation reaction, and the overall cell reaction occurs overall, not at one electrode. (18.2)

Additional Practice Problem Answers

1. $Mn^{2+} + 2e^- \rightarrow Mn$

$$86.0 \text{ min} \times \left(\frac{60 \text{ s}}{1 \text{ min}}\right) \times \left(\frac{7.62 \text{ mC}}{1 \text{ s}}\right) \times \left(\frac{1 \text{ C}}{1000 \text{ mC}}\right) \times$$

$$\left(\frac{1 \text{ mol } e^-}{9.65 \times 10^4 \text{ C}}\right) \times \left(\frac{1 \text{ mol Mn}}{2 \text{ mol } e^-}\right) \times \left(\frac{54.94 \text{ g}}{1 \text{ mol Mn}}\right) = 0.0112 \text{ g}$$

2. The reaction is $H_2O \rightarrow O_2 + 4H^+ + 4e^-$. We have $PV = nRT$, with $R = 0.08206$ L•atm/mol•K, $P = 752$ torr $\times \left(\frac{1 \text{ atm}}{760 \text{ torr}}\right) = 0.989$ atm, $V = 5.00 \text{ mL} \times \left(\frac{1 \text{ L}}{1000 \text{ mL}}\right) =$ 0.00500 L, and $T = 25°\text{C} + 273.15 = 298$ K. Thus,

$(0.989 \text{ atm})(0.00500 \text{ L}) = n(0.08206 \text{ L•atm/mol•K})(298 \text{ K})$

and $n = 0.0202$ mol. Therefore,

$$0.0202 \text{ mol } O_2 \times \left(\frac{4 \text{ mol } e^-}{1 \text{ mol } O_2}\right) \times \left(\frac{9.65 \times 10^4 \text{ C}}{1 \text{ mol } e^-}\right) \times$$

$$\left(\frac{1 \text{ s}}{2.10 \text{ C}}\right) \times \left(\frac{1 \text{ min}}{60 \text{ s}}\right) = 0.619 \text{ min}$$

3. a. The balanced half-reactions are

$$10e^- + 10H^+ + 2HClO_3 \rightarrow Cl_2 + 6H_2O$$

$$S + 2H_2O \rightarrow SO_2 + 4H^+ + 4e^-$$

The total reaction is

$$4HClO_3 + 5S \rightarrow 2Cl_2 + 2H_2O + 5SO_2$$

b. $E°$ for the reduction of $HClO_3 = +1.458$ V. This is the cathodic reaction. $E°$ for the reduction of SO_2 is +0.450 V. This is the anodic reaction. Its oxidation potential is –0.450 V.

$$E°_{cell} = E°_{red} + E°_{ox} = +1.458 + -0.450 = +1.008 \text{ V}$$

c. $\Delta G = -nFE° = -(20)(9.65 \times 10^4)(+1.008) =$ 1.95×10^6 J/mol.

d. $\log K = \frac{nE°_{cell}}{0.0592} = \frac{(20)(1.008)}{0.0592} = 340$ and $K = 10^{340}$.

4. a. The half-reactions are

$$BrO^- + H_2O + 2e^- \rightarrow Br^- + 2OH^-$$

$$Al + 4OH^- \rightarrow Al(OH)_4^- + 3e^-$$

So the total reaction is

$$3BrO^- + 3H_2O + 2Al + 2OH^- \rightarrow 3Br^- + 2Al(OH)_4^-$$

b. The oxidizing agent is the reactant that changes oxidation state in the reduction half-reaction, BrO^-. The reducing agent is the reactant-changing oxidation state in the oxidation half-reaction, Al.

c. The reduction potential for the bromine reaction is +0.766 V. The reduction potential for the aluminum reaction is –2.328 V. Therefore, its oxidation potential is +2.328 V.

$$E°_{cell} = E°_{red} + E°_{ox} = +0.766 + (+2.328) = +3.094 \text{ V}$$

d. pH = 12.31, pOH = 1.69, and $[OH^-] = 0.020$ *M*

The Nernst equation is

$$E = E° - \left(\frac{0.0592}{n}\right)\log\left(\frac{[Br^-]^3[Al(OH)_4^-]^2}{[BrO^-]^3[OH^-]^2}\right)$$

$$= +3.094 - \left(\frac{0.0592}{6}\right)\log\left(\frac{[0.50]^3[0.20]^2}{[0.10]^3[0.020]^2}\right)$$

$$= 3.094 - (0.00987)\log\left(\frac{0.0050}{4.0 \times 10^{-3}}\right)$$

$$= 3.094 - (0.00987)\log(12500)$$
$$= 3.053 \text{ V}$$

5. The balanced half-reactions are

$$Cr_2O_7^{2-} + 14H^+ + 6e^- \rightarrow 2Cr^{3+} + 7H_2O \quad E°_{red} = +1.36 \text{ V}$$

$$Sn \rightarrow Sn^{2+} + 2e^- \quad E°_{red} = -0.141 \text{ V}$$

The total reaction is

$$Cr_2O_7^{2-} + 14H^+ + 3Sn \rightarrow 2Cr^{3+} + 7H_2O + 3Sn^{2+}$$

The standard cell potential is $E°_{cell} = E°_{red} + E°_{ox} =$ $+1.36 + (+0.141) = +1.50$ V. If pH = 2.73, $[H^+] = 1.9 \times 10^{-3}$ *M*. The Nernst equation is

$$E = E° - \left(\frac{0.0592}{n}\right)\log\left(\frac{[Cr^{3+}]^2[Sn^{2+}]^3}{[Cr_2O_7^{2-}][H^+]^{14}}\right)$$

$$= +1.50 - \left(\frac{0.0592}{6}\right)\log\left(\frac{[0.10]^5}{[0.10][1.9 \times 10^{-3}]^{14}}\right)$$

$$= 1.16 \text{ V}$$

CHAPTER 19 Biochemistry: The Compounds of Life

OUTLINE

REVIEW

Chapter Overview

Biologically important molecules come in relatively few categories. **Amino acids** are the building blocks of peptides and proteins. **Monosaccharides** are the building blocks of sugars and carbohydrates. **Nucleotides** are made from a sugar, a phosphate, and an organic base. Nucleotides make up deoxyribonucleic acid (DNA) and ribonucleic acid (RNA). **Lipids** include various molecules that are not soluble in water.

The action of biological molecules is very dependent on the arrangement of their atoms in space. Molecules in which the atoms are connected in the same way, but are arranged differently in space are **stereoisomers**. A subcategory of stereoisomers is geometric isomers. **Geometric isomers** include *cis* isomers, where atoms (or groups of atoms) of the same type are on the same side of a double bond, and *trans* isomers, where atoms of the same type are on opposite sides of the double bond. Another type of stereoisomers are **enantiomers**, where the two molecules are nonsuperimposable images of each other. Molecules that can have an enantiomer are called **chiral**. If a molecule contains a carbon that is bonded to four different groups, it will be chiral. That carbon is called the **chiral center**.

Enantiomers exhibit optical activity. They will rotate a plane of polarized light. If the molecule rotates polarized light to the left, the molecule is called *levorotary* and (–) is added before the name. If the molecule rotates light to the right, the molecule is called *dextrorotary* and the name indicates this with a (+) prefix. The type of enantiomers can also be denoted based on the actual arrangement of atoms rather than its action on polarized light. In that case, a D or L is used as a prefix. Molecules produced and used by the human body are all L except for the sugars in DNA and RNA, which are D. An equal mixture of enantiomers, called a **racemic mixture**, is generally what is produced in the laboratory setting. A racemic mixture will not rotate polarized light.

Proteins are composed of amino acids. Each amino acid has an amine (NH_2) group and a carboxylic acid (COOH) group. Both groups are generally attached to the same carbon, called the **α-carbon**. Such amino acids are called **α-amino acids**. In addition to the amine and carboxylic acid, the α-carbon generally is bonded to a hydrogen. Most of the variation in amino acids is in what is connected to the carbon by its fourth bond. That can be as simple as another hydrogen (glycine) or very complex. For a general amino acid, this fourth bonded group is designated as "R." Three letter abbreviations are often used to identify the common amino acids. For example, glycine is abbreviated Gly.

The amine of the amino acid is a Brønsted base, while the carboxylic acid is a Brønsted acid. Therefore, either can be protonated or deprotonated depending on the pH of the solution. At physiological conditions (pH = 7), the amine is protonated (NH_3^+) and the carboxylic acid is deprotonated (CO_2^-). The resulting substance, with both a positive and a negative charge, is called a **zwitterion**.

The amino acids are connected to each other when the amine group of one amino acid reacts with the carboxylic acid of another amino acid, producing water and a **peptide bond**. Sequences of less than 20 amino acids are called *oligopeptides*. If there are more than 20 amino acids in a chain it is called a *polypeptide*. Polypeptides are also called **proteins**.

The sequence of amino acids, their identity and order, is called the **primary structure** of the protein. Their arrangement in space is called the **secondary structure**. The secondary structure is usually determined by hydrogen bonding between the amine and carboxylic acid groups. A common secondary structure is the **α-helix**, where the amino acids form a helix with all their R groups pointed out. A **β-pleated sheet** has a zigzag pattern with R groups both above and below. A **random coil** has no specific structure. The **tertiary structure** is a three-dimensional structure determined by interactions of the R groups of the amino acids. When the proteins combine with each other they can create a **quaternary structure**.

One type of protein is an **enzyme**, which is a biological catalyst. A substrate, or reactant, bonds to the active site of the enzyme, so that it can react more readily and form a product. Current theory explains that the enzyme changes shape so that the substrate can react more readily (lower activation energy). **Inhibitors** can diminish or destroy the effectiveness of enzymes. However, enzymes can substantially increase the number of molecules catalyzed per second, its **turnover number**.

Carbohydrates are composed of sugars. Simple sugars, like glucose and fructose, can exist in many forms, both open chains and as cyclic compounds. They also have many chiral centers and therefore, optical activity. Disaccharides are two sugars connected by a glycosidic link. The **glycosidic link** is made when alcohol groups (ROH) from two sugars combine to release water and create an ether (ROR) that connects the two sugars. Glycosidic links can connect several sugars and create polysaccharides. There are two types glycosidic links: α and β. The type of glycosidic link is determined by the carbon that is bonded to the oxygen in the ring as well as the oxygen in the link. In an β-glycosidic link, the link (and the second ring) is below this carbon. In a α-glycosidic link, this link (and the second ring) is above this carbon. Starch is a polysaccharide made of glucose connected by α-glycosidic links. Cellulose is also a polysaccharide of glucose, but connected by β-glycosidic links.

Photosynthesis produces sugars from carbon dioxide and water, using energy from the sun. **Glycolysis** is the breaking down of sugars to produce energy.

Lipids are generally made from **fatty acids**. These are long hydrocarbon chains ending in a carboxylic acid. In saturated fatty acids, all the carbons are connected by single bonds. In unsaturated fatty acids, some of the carbons may be connected by double bonds. Both types of fatty acids can react with glycerol, where the OH group of the glycerol reacts with the OH of the carboxylic acid, so that water is released and an ester link between the glycerol and acid is formed. When all three hydroxyl groups of the glycerol have reacted with fatty acids, a **triglyceride** has been made. **Fats** tend to be triglycerides of saturated fatty acids, while **oils** are triglycerides of unsaturated fatty acids. **Phospholipids** have a phosphate group replace one of the hydroxyl groups on glycerol while the other two react with fatty acids. This makes one end of the molecule ionic and hydrophilic (water-loving), while the fatty acid end is hydrophobic (water-fearing). Thus the hydrophobic ends orient toward the hydrophobic ends of other phospholipids and create the **lipid bilayer** characteristic of many cell membranes.

Each **nucleotide** is made of a phosphate (PO_4^{3-}) and a nitrogen base bonded to a sugar. In DNA, the sugar is deoxyribose; in RNA, the sugar is ribose. The nucleotide is characterized by the nitrogen base. There are four of these bases in DNA: thymine (T), cytosine (C), adenine (A), and guanine (G). In RNA, the nitrogen base uracil (U) replaces cytosine. There are two chains of nucleotides in DNA in the form of a double helix held together by the hydrogen bonds between the nitrogen bases. Thus a thymine is always paired with an adenine in the opposite chain, and a cytosine is always paired with a guanine. RNA is a single strand of nucleotides.

Proteins are synthesized at sites in the nuclei of cells called ribosomes. The information from the DNA is transferred to **messenger RNA (mRNA)** in a process called **transcription**. mRNA pairs with DNA in the same way that the complementary strand of DNA does. The mRNA carries the genetic information to the site of protein synthesis. The mRNA binds to **transfer RNA (tRNA)** with hydrogen bonding of the nitrogen bases in the same way as the double helix of DNA and the creation of RNA. The tRNA binds to only one type of amino acid and is used in the actual synthesis of proteins. The conversion of the genetic code into an amino acid sequence of a protein is called **translation**.

Other nucleotides are adenosine diphosphate (ADP) and adenosine triphosphate (ATP). When ATP reacts to become ADP it releases energy that can be used in other anabolic (requiring energy) processes. If the biological reaction produces energy (catabolic), the energy can be stored by using the reverse process when ATP is produced from ADP.

Worked Examples

Chirality

Example 1.
Do the following molecules have chiral centers? If so, what atoms are chiral centers?

a. 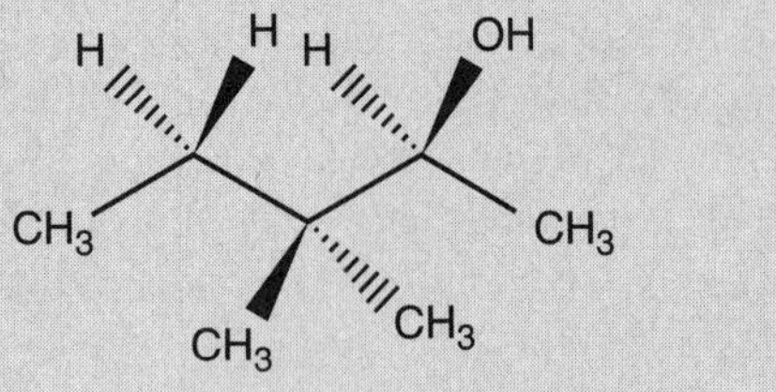 H H H OH CH3 CH3 CH3 CH3

b. 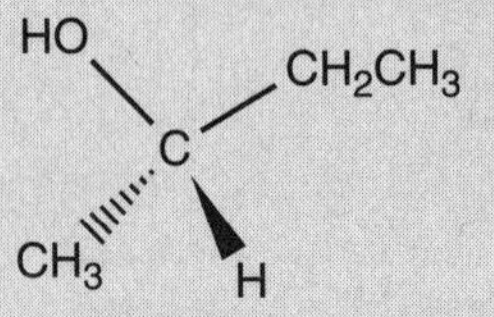 HO CH2CH3 C CH3 H

c. H CHO HO CH2CH3 C C CH3 H C C CH3 H H CH3

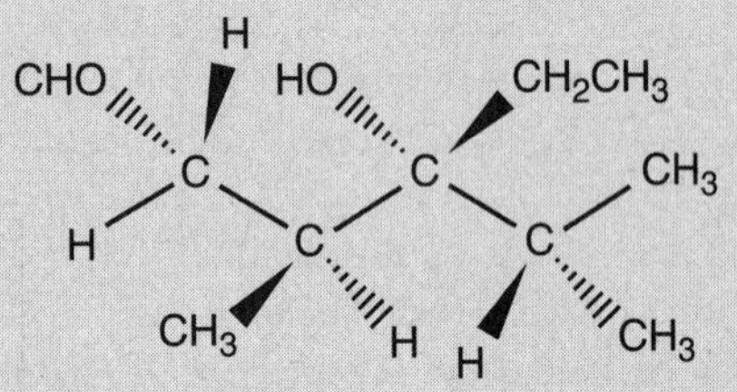

Collect and Organize We are given three molecules and are asked to determine which contain chiral centers.

Analyze A chiral center is an sp^3 carbon (four single bonds) that has a different group at each bond.

Solve

a. This molecule has one chiral center. It is the carbon on the right side of the chain, with a methyl (CH_3), hydroxyl (OH), and hydrogen (H) group. The fourth group has an isopropyl chain ($CH_2CH(CH_3)CH_3$).

b. This molecule also has one chiral center, which is the carbon in the center. The groups connected to this carbon are hydroxyl (OH), ethyl(CH_2CH_3), methyl (CH_3), and hydrogen (H).

c. This molecule has two chiral centers, the two carbons in the middle. The one on the left end is eliminated for its two hydrogen groups and the one on the right has two methyl (CH_3) groups.

Think About It Remember to consider not just the next atom in the bond, but all the bonds that follow it when determining whether the group is the same.

Proteins

Example 1.
Which of the following are amino acids?

a. 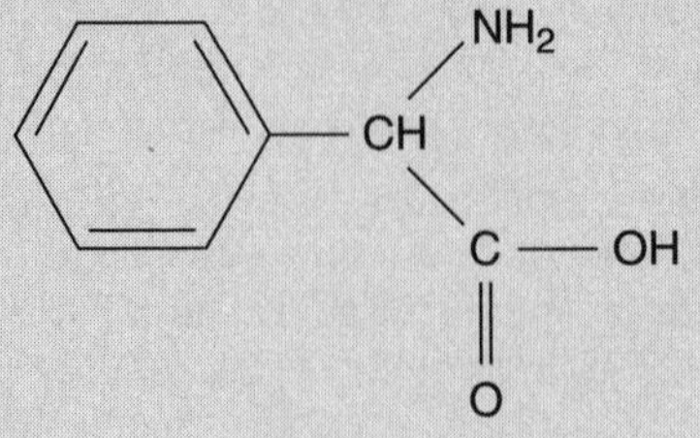 NH2 CH C OH O

b. H2N COOH CH CH CH3 CH3

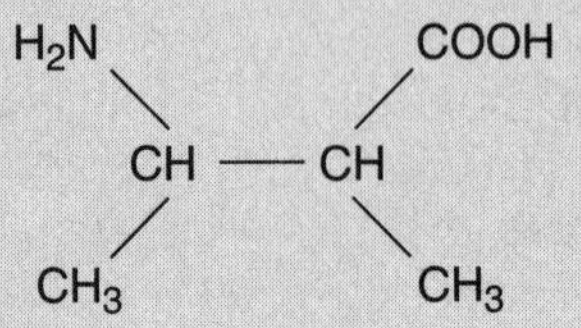

c. COOH CH3 CH NH2

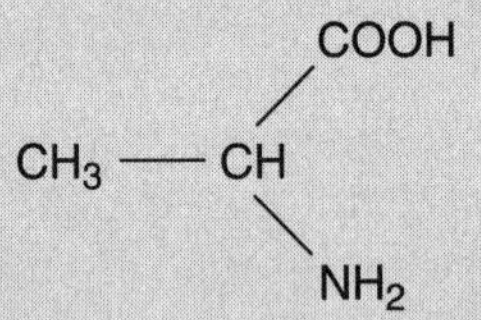

Collect and Organize We are given three molecular formulas and are asked to determine which are amino acids.

Analyze Amino acids are characterized by a carbon with single bonds to the following three groups: H, NH_2, and COOH. The fourth bond to that carbon can be any organic group.

Solve

a. This is an amino acid. The carboxylic acid is shown in its more drawn out form rather than as COOH.

b. This is not an amino acid. The COOH and NH_2 groups are on different carbons.

c. This is a classic amino acid. CH implies the carbon is bonded to the hydrogen as well as the COOH group and NH_2.

Think About It All but one of the amino acids (glycine) are chiral

Example 2.
What are the two products (other than water) of the reaction between these two amino acids, alanine and phenylalanine, respectively?

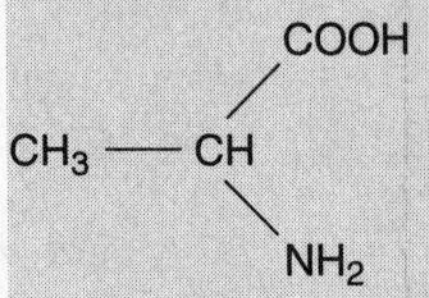

COOH CH3 CH NH2 and NH2 CH C OH O

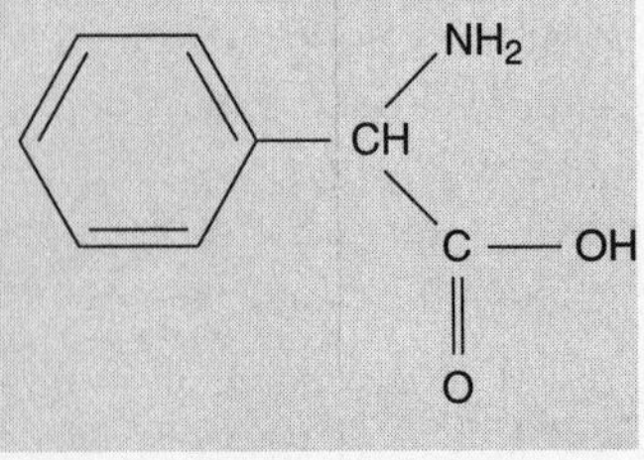

COLLECT AND ORGANIZE We are given two amino acids and are asked to determine the products of their reaction.

ANALYZE The amine group on one amino acid reacts with the hydroxyl of the carboxylic acid group of another amino acid in a condensation reaction that links the two amino acids together.

SOLVE One product is formed from the amine of alanine and the carboxylic acid group of phenylalanine.

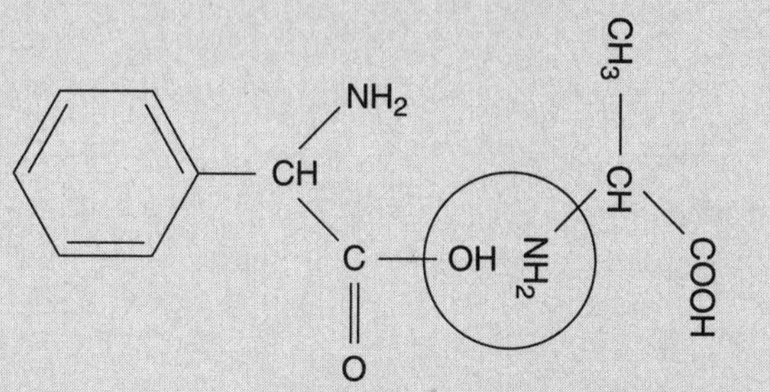

The groups circled react, the nitrogen and one hydrogen remain to give

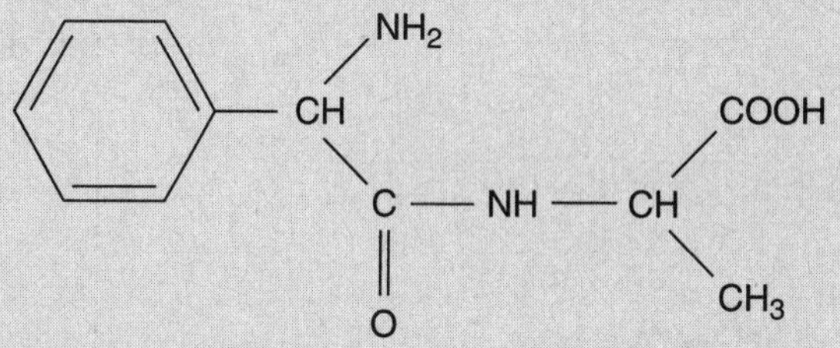

The other product reacts to the carboxylic acid of alanine with the amine of phenylalanine.

THINK ABOUT IT The two amino acids are linked through the nitrogen as a secondary amine. The link, consisting of a carbonyl and the secondary amine, is called a peptide bond.

EXAMPLE 3.

What are the two products of the reaction between the following peptide and amino acid?

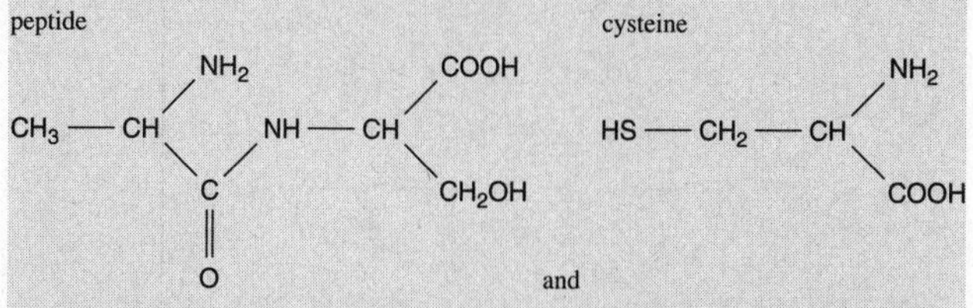

peptide and cysteine

COLLECT AND ORGANIZE We are given a peptide and an amino acid and are asked to determine the two products that result from their reaction.

ANALYZE Amino acids and peptides contain both amine and carboxylic acid groups. The reaction between the two can occur in two ways: by combination of the amino group of the peptide with carboxylic acid of the amino acid, or by combination of the amino group of the amino acid with the carboxylic acid of the peptide.

SOLVE One product is the amine of the peptide bonds with the carboxylic acid group of the cysteine.

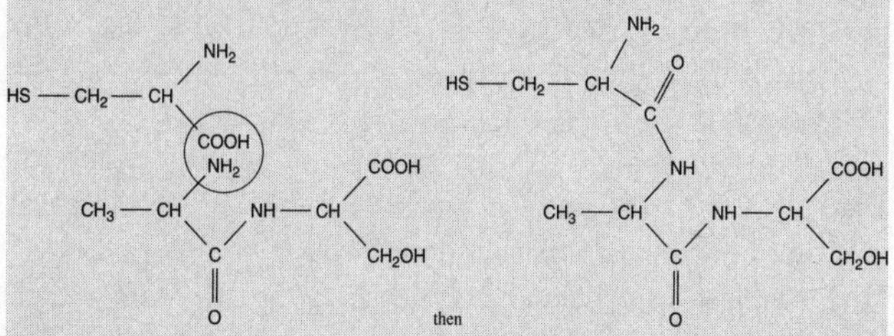

then

The other product combines the amine of cysteine with the COOH group of the peptide.

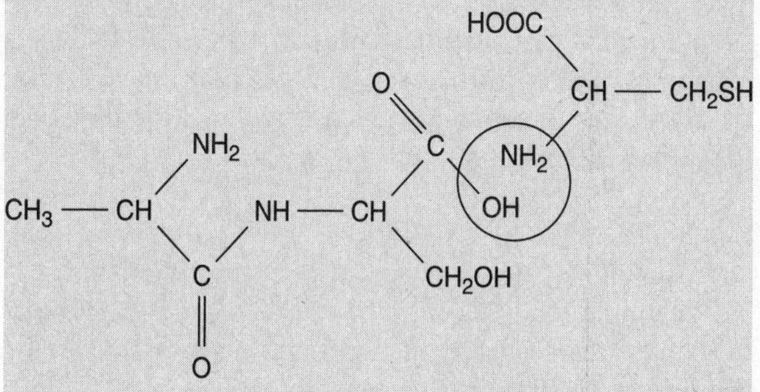

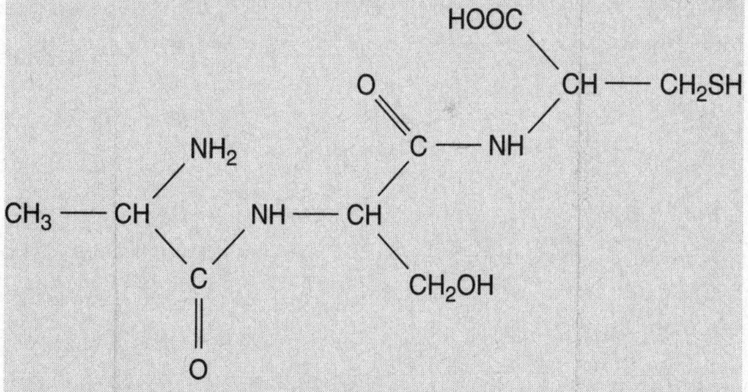

THINK ABOUT IT The two products each contain carboxylic acid groups and amino groups. Thus they can react with another amino acid (at either end!), creating another peptide bond and a three–amino-acid peptide.

CARBOHYDRATES

EXAMPLE 1.

What type of glycosidic linkages connect the disaccharides below?

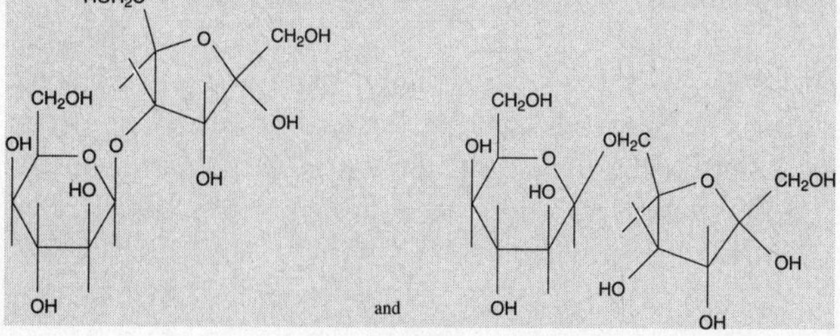

and

Collect and Organize We are given two disaccharides and asked to identify whether the glycosidic linkage connecting them is an α or β linkage.

Analyze The type of link is determined by the carbon next to the oxygen in the ring rather than the carbon it is connected to in the other ring. When this carbon of the ether link is in the lower ring, it is a β link. When this carbon is in the upper ring, it is an α link.

Solve In the left structure, the ether link connects a carbon in the upper ring that is also connected to two other carbons. However, the other side of the ether connects to a carbon that is connected to another oxygen in the ring. Since this carbon (the one with two ether links) is in the lower ring, this is a β linkage.

In the right structure, the carbon connected to two oxygens is in the left ring. If you consider how the right ring should sit in orientation to the left ring, you should be able to see that it should move up. Again, this is a β linkage.

Think About It Lewis structures can be drawn in many ways that do not always represent their orientation in space. To determine the type of glycosidic linkage, imagine that the oxygen that connects the two rings runs in a straight line and use the revised picture to determine whether it is an α or β linkage.

Lipids

Example 1.

Identify the following as saturated, monounsaturated, or polyunsaturated fatty acids. ("None of the above" is also a possible answer.)

a. $HO{-}C({=}O){-}CH{=}CH{-}CH_2{-}CH_2{-}CH{=}CH{-}CH_2{-}CH{=}CH{-}CH_2{-}H$

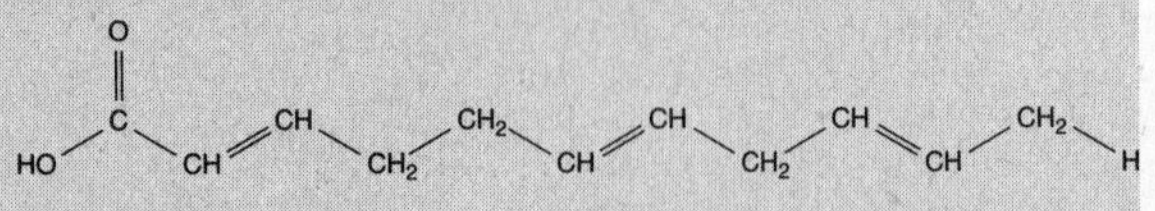

b. HOOC (line structure)

c. CH_3COOH

d. $CH_3{-}CH_2{-}CH_2{-}CH_2{-}CH_2{-}CH{=}CH{-}CH_2{-}CH_2{-}C({=}O){-}OH$

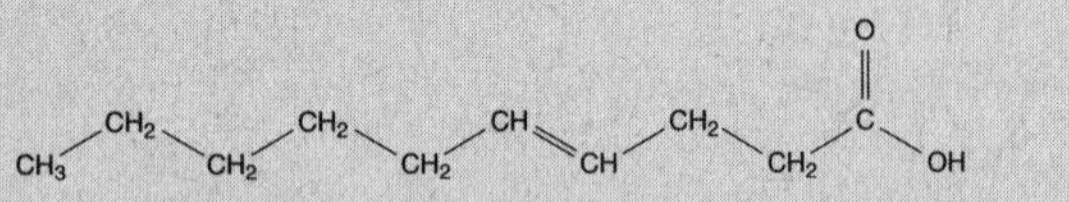

Collect and Organize We are given four molecular formulas and are asked to identify whether any are saturated, monounsaturated, or polyunsaturated fatty acids.

Analyze Fatty acids have a long (more than six) chain of carbons ending with a carboxylic acid group. They can be classified based on the number of carbon–carbon double bonds as saturated (none), monounsaturated (one), or polyunsaturated (more than one).

Solve

a. This is a polyunsaturated fatty acid. It is a fatty acid, since it has the carboxylic acid group on the left end and it contains a long (10 carbon) carbon chain. It is polyunsaturated, since it contains three double bonds between carbons.

b. This is a saturated fatty acid. It ends with a carboxylic acid group (on the left) and contains eight carbons beyond the COOH group. The carbons are not explicitly written in this structure but are implied where two line segments meet. If anything other than a CH_2 group (CH_3 at the end) was at that point, it would have been noted.

c. This is not a fatty acid. Although it has a carboxylic acid group, one carbon is not sufficient to give it hydrophobic (water-fearing) properties. Although there is no firm number that indicates a fatty acid, at least six carbons is usual.

d. This is a monounsaturated fatty acid. There is one (mono) carbon–carbon double bond. It has a long carbon chain ending (on the right) with a carboxylic acid group.

Think About It Because saturated fatty acids pack better, and therefore, have stronger dispersion forces, they are more likely to be solids. Unsaturated fatty acids are less organized and therefore, are more likely to be liquids. Unsaturated fatty acids are also more reactive, particularly at the site of the double bonds.

Example 2.

What is the product of the reaction between glycerol and three saturated dodecanoic (12-carbon) acid molecules?

Collect and Organize We are asked to predict the product of the reaction of one glycerol molecule and three dodecanoic acid molecules.

Analyze In lipids the fatty acids react with glycerol.

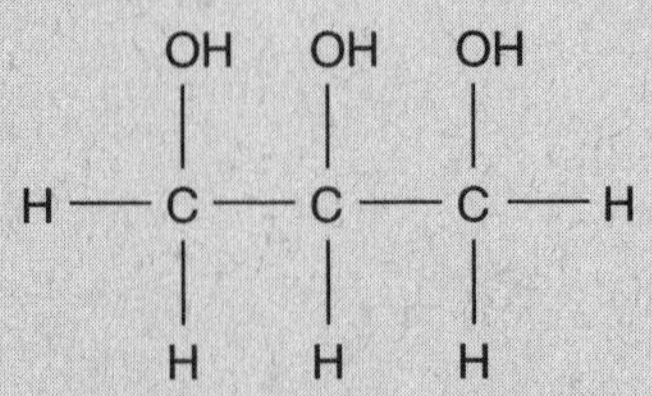

$$H{-}\underset{H}{\overset{OH}{C}}{-}\underset{H}{\overset{OH}{C}}{-}\underset{H}{\overset{OH}{C}}{-}H$$

Therefore, three fatty acids and one glycerol are needed to create a lipid.

Solve Each dodecanoic acid forms an ester with an alcohol group of the glycerol. The *dodec* root means that the acid has a total of 12 carbons. The last carbon is a COOH group (oic acid). The *an* means that this fatty acid is saturated, with no double bonds. (Otherwise it would have been *en*.) Therefore, the resulting structure is

$$CH_3-CH_2-CH_2-CH_2-CH_2-CH_2-CH_2-CH_2-CH_2-CH_2-CH_2-\overset{O}{\overset{\|}{C}}-O-CH_2$$
$$CH_3-CH_2-CH_2-CH_2-CH_2-CH_2-CH_2-CH_2-CH_2-CH_2-CH_2-\overset{O}{\overset{\|}{C}}-O-CH$$
$$CH_3-CH_2-CH_2-CH_2-CH_2-CH_2-CH_2-CH_2-CH_2-CH_2-CH_2-\overset{O}{\overset{\|}{C}}-O-CH_2$$

THINK ABOUT IT Since the link between the fatty acid and alcohol is an ester, the reaction can be considered an esterification.

NUCLEOTIDES AND NUCLEIC ACIDS

EXAMPLE 1.
What sequence is complementary to CCTAGTCATT? Write the new sequence as if it were base pairing with this sequence.

COLLECT AND ORGANIZE We are given one sequence of DNA and are asked to determine its complementary sequence.

ANALYZE The two chains of DNA are held together by hydrogen bonding between the nitrogen base part of the nucleotide. The interaction between bases is specific—the adenine base hydrogen-bonds to the thymine base and the cytosine base hydrogen-bonds to the guanine base. The complementary chain always consists of the opposite base. The first letter of the name of the base is used to represent that nucleotide segment.

SOLVE Replace each C with its complement G and C with G. Also replace T with A and A with T. Thus the complementary chain is GGATCAGTAA.

THINK ABOUT IT Base pairing of sequences to form the DNA double helix requires chains that have the correct sequences.

EXAMPLE 2.
What mRNA sequence forms from the DNA sequence CTTAAGCGTC? Write the mRNA sequence as if it were base-pairing with the DNA sequence.

COLLECT AND ORGANIZE From the given DNA sequence we are asked to determine the mRNA sequence that would form during the transcription process.

ANALYZE The system of copying a DNA sequence to an mRNA sequence is the same as the system of copying one DNA sequence to another DNA sequence, except that in RNA, uracil is substituted for thymine to base-pair with adenine.

SOLVE Replace C with G, T with A, A with U, and G with C. Therefore, the mRNA sequence is GAAUUCGCAG.

THINK ABOUT IT DNA is copied to mRNA through the process of transcription. During transcription, the DNA double helix "unzips," and one chain is copied.

Self-Test

KEY VOCABULARY COMPLETION QUESTIONS

__________ 1. Same formula, but the groups are oriented differently

__________ 2. Ether group connecting sugars

__________ 3. Molecule with both a positive and a negative charge

__________ 4. Substance containing unsaturated triglycerides

__________ 5. Made of two amino acids

__________ 6. Molecules that are nonsuperimposable mirror images

__________ 7. When it reacts it produces energy for biological reactions

__________ 8. Atom in amino acid that is bonded to NH_2 and to COOH

__________ 9. Contains the genetic code in a double-helix structure

__________ 10. Glucose units connected by α-glycosidic link

__________ 11. Long chain of amino acids

__________ 12. Takes a DNA sequence to ribosomes

__________ 13. Nitrogen base that hydrogen-bonds to thymine

__________ 14. Product when three fatty acids react with glycerol

__________ 15. When the same atom is on the same side of a double bond

__________ 16. A mixture of enantiomers that does not rotate polarized light

__________ 17. A type of carbon bonded to four different groups

__________ 18. Connection between amino acids

__________ 19. Sequence of amino acids in a protein

__________ 20. Made of a phosphate, an organic base, and a sugar

__________ 21. Biological catalyst

__________ 22. Secondary structure where all R-groups are oriented outward

__________ 23. Rate at which a substrate reacts when catalyzed by an enzyme

__________ 24. Biological substance that is not water soluble

__________ 25. Single unit of a carbohydrate

MULTIPLE-CHOICE QUESTIONS

1. Physical and chemical properties of enantiomers are
   a. very similar
   b. very different
   c. exactly the same
   d. dependent on how many chiral centers there are

2. Which of the following molecules has an enantiomer?
   a. CH_4
   b. $CH_3CHCHCH_2CH_3$
   c. CHFClBr
   d. CH_3COOH

3. Is the molecule

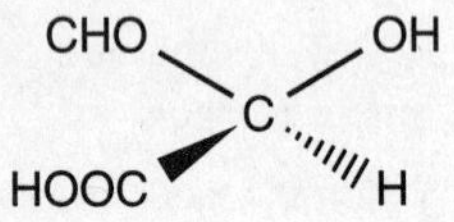

   CHO, OH, HOOC, H bonded to central C

   a. *cis*
   b. *trans*
   c. *chiral*
   d. *achiral*

4. Which of the following has an isomer?
   a. 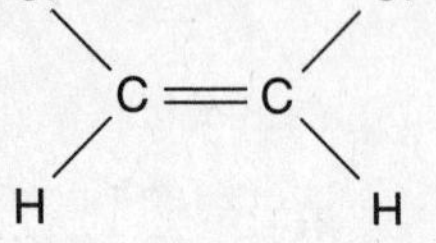 Cl(H)C=C(Cl)H
   b. 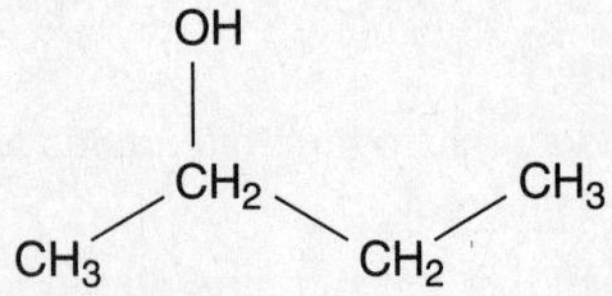 CH_3–CH_2(OH)–CH_2–CH_3
   c. 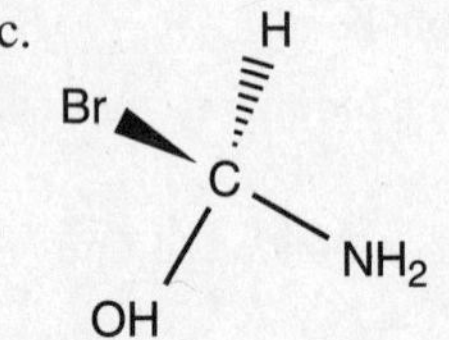

   d. all do

5. What is a product of the reaction between the amino acids of

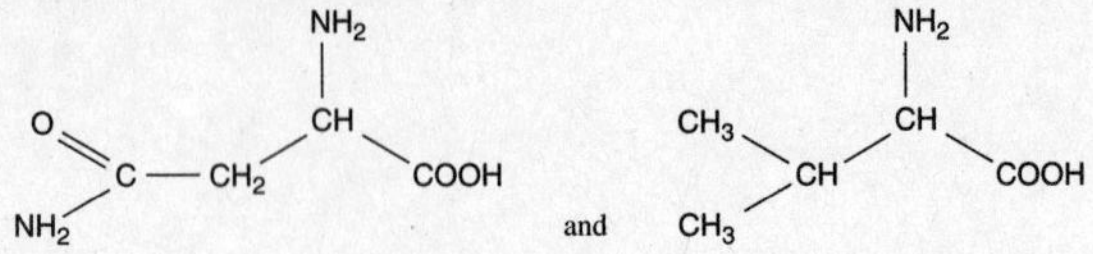

   H_2N–C(=O)–CH_2–CH(NH_2)–COOH and $(CH_3)_2$CH–CH(NH_2)–COOH

   a. 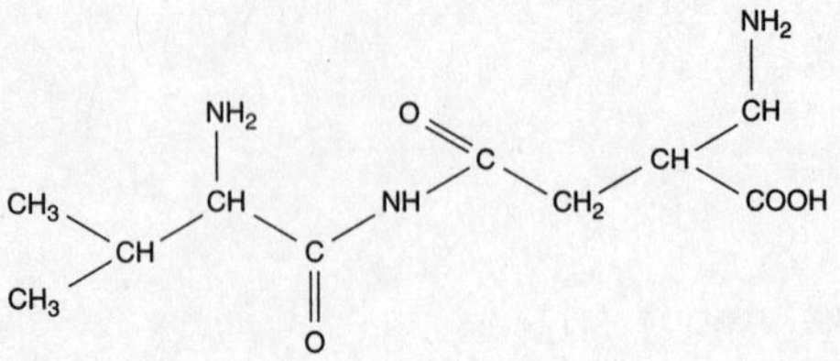 $(CH_3)_2$CH–CH(NH_2)–C(=O)–NH–C(=O)–CH_2–CH–CH(NH_2)–COOH
   b. 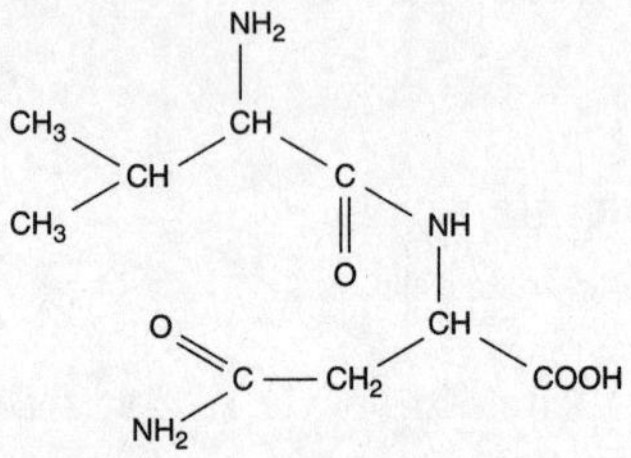 $(CH_3)_2$CH–CH(NH_2)–C(=O)–NH–CH(COOH)–CH_2–C(=O)–NH_2
   c. 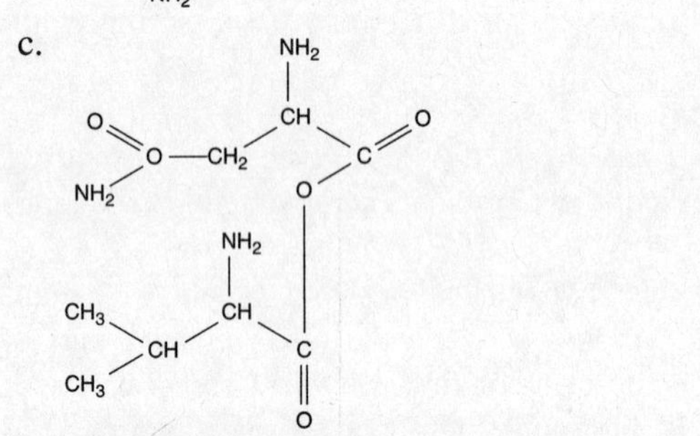

   d. none of the above

6. What is the complementary DNA sequence to CCGTAT?
   a. TTACGC
   b. AATGCG
   c. GGCAUA
   d. GGCATA

7. A tRNA sequence of AUGUC comes from a DNA strand of
   a. UACAG
   b. ATGTC
   c. TACAG
   d. GTATU

8. What kind of isomers are

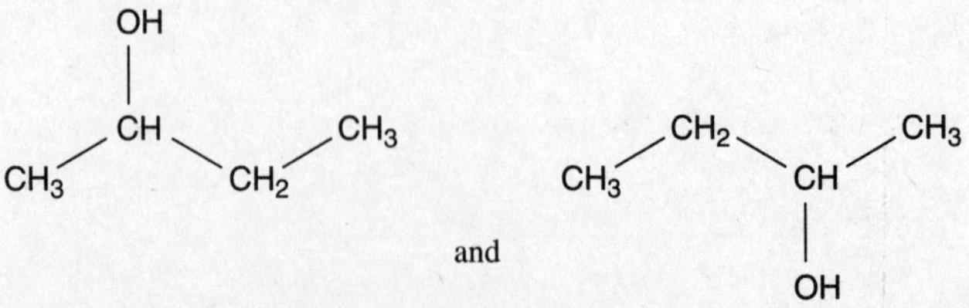

   CH_3–CH(OH)–CH_2–CH_3 and CH_3–CH_2–CH(OH)–CH_3

   a. enantiomers
   b. geometric
   c. structural
   d. none of the above

9. Which functional group is characteristic of a lipid?
   a. alcohol
   b. ketone
   c. ester
   d. ether

10. What functional group is typical of amino acids?
    a. primary amine
    b. alkene
    c. ester
    d. alcohol

11. The functional group that links together monosaccharides is
    a. alcohol
    b. ether
    c. amine
    d. hemiacetal
    e. carbonyl

12. At pH 7, an amino acid is a(n)
    a. anion
    b. cation
    c. neutral
    d. zwitterion

13. β-pleated sheets are an example of a protein's
    a. primary structure
    b. secondary structure
    c. tertiary structure
    d. quaternary structure

14. The tertiary structure of a protein is determined by interactions of
    a. the amine groups
    b. the carboxylic acid groups
    c. the side chain (R-groups)
    d. its interaction with other proteins

15. How many different primary structures can a tripeptide of three different amino acids have?
    a. one
    b. three
    c. six
    d. more than six

16. Which of the following type of molecules typically contain phosphate groups?
    a. amino acids
    b. sugars
    c. nucleotides
    d. alkanes

17. CH_2NH_2COOH is best classified as a(n)
    a. amino acid
    b. protein
    c. lipid
    d. nucleotide
    e. fatty acid

18. $CH_3(CH_2)_{16}COOH$ is best classified as a(n)
    a. amino acid
    b. protein
    c. fatty acid
    d. nucleotide
    e. triglyceride

19. Triglycerides are formed from
    a. glycerol and three fatty acids
    b. three glycerols and a fatty acid
    c. three monosaccharides and glycerol
    d. phosphate, sugar, and organic base

20. Enzymes are typically which class of compounds?
    a. proteins
    b. lipids
    c. nucleotides
    d. carbohydrates

21. A carbohydrate might contain which of the following?
    a. fatty acid
    b. amino acid
    c. monosaccharide
    d. organic base

22. Organic bases hydrogen bond because the hydrogen is attracted to the base's
    a. carbon
    b. oxygen
    c. nitrogen
    d. chlorine
    e. organic bases do not hydrogen bond

23. Oils tend to be liquid because they are made of
    a. unsaturated fatty acids
    b. saturated fatty acids
    c. monosaccharides
    d. polysaccharides

24. Enzymes catalyze biochemical reactions because
    a. they lower the activation energy of the reaction
    b. they have the right shape
    c. their shape changes to make the reaction more efficient
    d. all of the above

25. When a fatty acid reacts with glycerol, the two molecules are connected by an
    a. alcohol group
    b. ether group
    c. ester group
    d. amine group

Additional Practice Problems

1. What is the product (other than water) when these molecules react?

a.

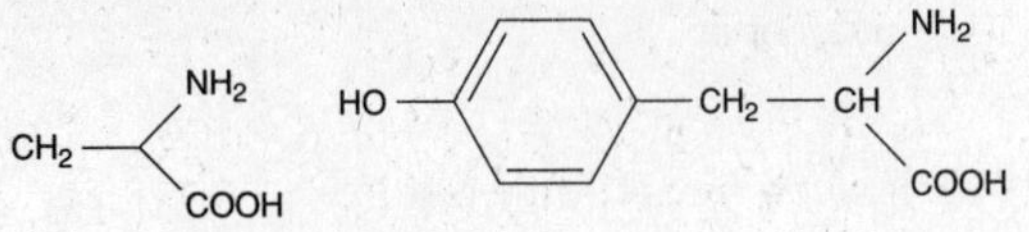

b.

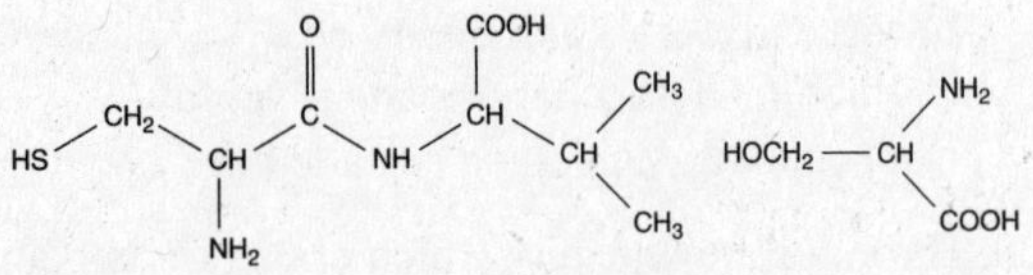

2. Find all the chiral centers on the following molecule.

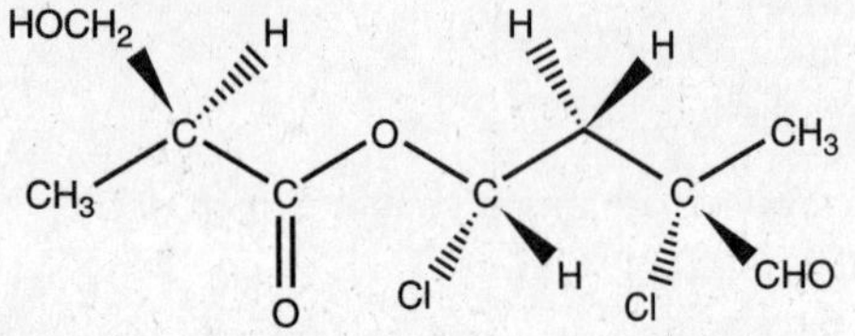

3. If tRNA has a nucleotide base sequence of CAUUGAG, what is the DNA sequence that originates it?

4. In the disaccharides below, are they connected by an α- or β-glycosidic link?

a.

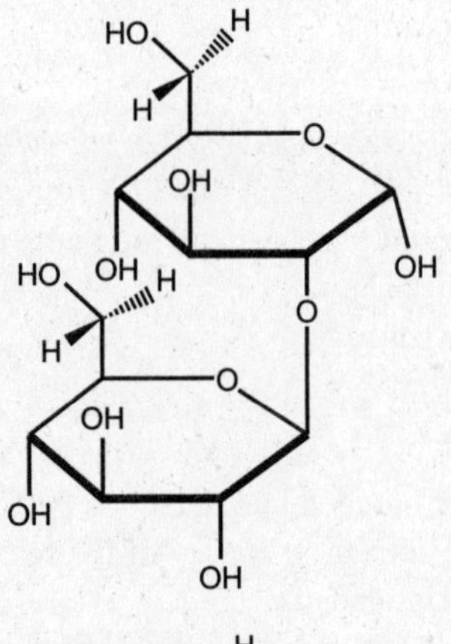

b.

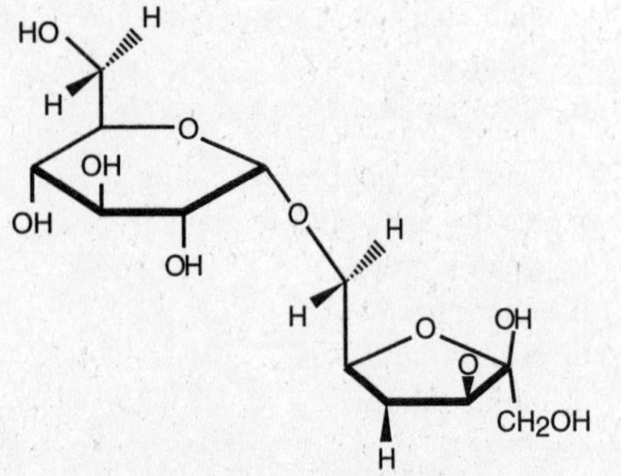

5. What is the complementary sequence to the DNA strand of AATCGGCCTT?

Self-Test Answer Key

Key Vocabulary Completion Answers

1. stereoisomer
2. glycosidic linkage
3. zwitterion
4. oil
5. dipeptide
6. enantiomers
7. adenosine triphosphate (ATP)
8. α-carbon
9. deoxyribonucleic acid (DNA)
10. polysaccharide, carbohydrate, or starch
11. protein, polypeptide
12. messenger RNA (mRNA)
13. adenine
14. triglyceride
15. *cis* isomer
16. racemic
17. chiral
18. peptide link or peptide bond
19. primary structure
20. nucleotide or nucleic acid
21. enzyme
22. α-helix
23. turnover number
24. lipid
25. monosaccharide

Multiple-Choice Answers

For further information about the topics these questions address, you should consult the text. The appropriate section of the text is listed in parentheses after the explanation of the answer.

1. a. Because the bonding is the same, only orientation of the bonds is different; the properties are very similar. There are a few differences, however, that allow enantiomers to be separated, and they often have different biological effects. (19.2)

2. c. An enantiomer has a chiral carbon, or a carbon with four different groups bonded to it. (19.2)

3. c. This is a chiral center, since the carbon has four different groups rather than a geometric isomer where groups are around a double bond or a ring. (19.2)

4. d. Enantiomers (a type of stereoisomer) occur when four different groups are bonded to a single carbon, as in choice "c." Choice "a" will have a geometric isomer and choice "b" will have structural isomers. Therefore, all these examples have isomers. (19.2)

5. b. When two amino acids react, the OH of the carboxylic acid and a hydrogen of the amine form water, and a peptide bond (carbonyl and secondary amine) is formed. Choice "a" looks like the peptide bond, but the wrong amine group reacted. (The amine on the same carbon as the COOH group reacts.) Choice "c" makes an ester rather than a peptide bond. (19.3)

6. d. In complementary DNA strands, A replaces T and T replaces A. Also G replaces C and C replaces G. (19.7)

7. b. tRNA comes from mRNA where base A pairs with U and C pairs with G. Therefore, the mRNA sequence is UACAG. It comes from a DNA where C pairs with G and A pairs with T and U with A. So the DNA strand is ATGTC. (19.7)

8. d. These are actually the same molecule. It helps to count the carbons. You start counting from the end closest to the functional group. The alcohol is on the second carbon for both structures when counted this way. (19.2)

9. c. Lipids are formed from the reaction of an alcohol (on glycerol) with a carboxylic acid (of a fatty acid) forming an ester. (19.6)

10. a. Amino acids are characterized by a primary amine and a carboxylic acid group bonded to the same carbon. (19.3)

11. b. To link two sugars, two alcohols combine to produce an ether link and a water molecule. (19.5)

12. d. The amine is protonated and therefore, has a positive charge, but the carboxylic acid is deprotonated and has a negative charge. (19.3)

13. b. β-pleated sheets are a secondary structure. (19.3)

14. c. Side-chain interactions create the tertiary structure. (19.3)

15. c. If the amino acids are called A, B, and C, then arrangements are ABC, ACB, BAC, BCA, CAB, and CBA. (19.3)

16. c. Nucleotides have a sugar, organic base and a phosphate. Amino acids have an amine and a carboxylic acid. The functional groups of sugars are primarily alcohols and ethers. Alkanes are simple hydrocarbons. (19.3, 19.5, and 19.7)

17. a. The NH_2 and COOH must both be connected to the carbon, which is typical of an amino acid. (19.3)

18. c. The "fatty" part is the long hydrocarbon chain, which ends in a carboxylic acid. (19.6)

19. a. The COOH of the fatty acid reacts with the OH of the glycerol. Glycerol has three OH groups. (19.6)

20. a. Enzymes are typically proteins. (19.4)

21. c. Carbohydrates are polysaccharides. (19.5)

22. c. Hydrogen bonds occur when hydrogen is bonded to an oxygen, nitrogen, or fluorine. Organic molecules are typically basic because of their nitrogen atoms. (19.7)

23. a. The double bonds in unsaturated hydrocarbon chains change the geometry of the molecules so that is does not stack well, so it does not easily become a solid. (19.6)

24. d. As the substrate fits into the geometric space in the enzyme, it changes shape so that it can react more efficiently, lowering the activation energy. (19.4)

25. c. The OH from glycerol reacts with the COOH of the acid removing water. Therefore, the carbonyl (CO) group is left and oxygen connects the two original molecules. This describes an ester. (19.6)

Additional Practice Problem Answers

1. a.

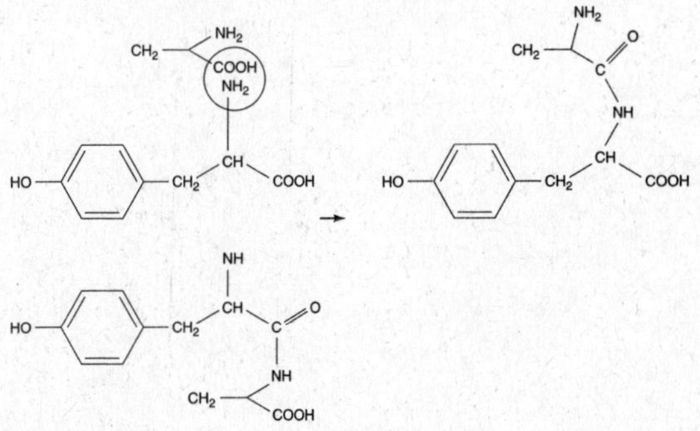

b.

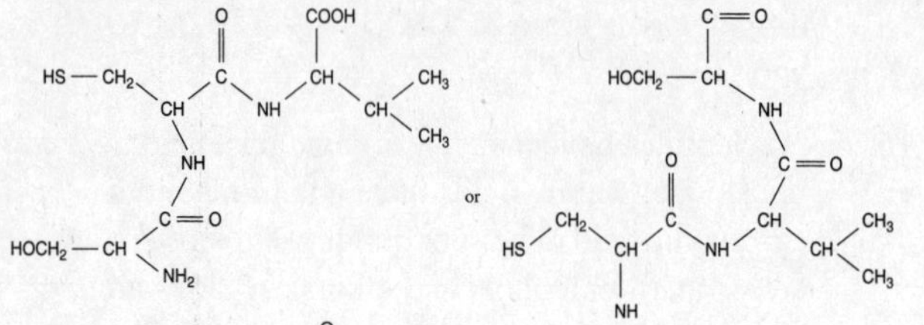

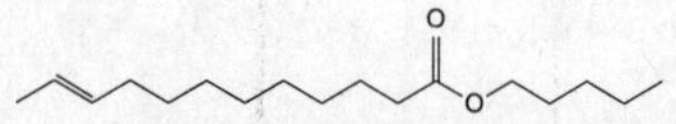

2. The chiral carbons are marked in the structure below with a star and numbered.

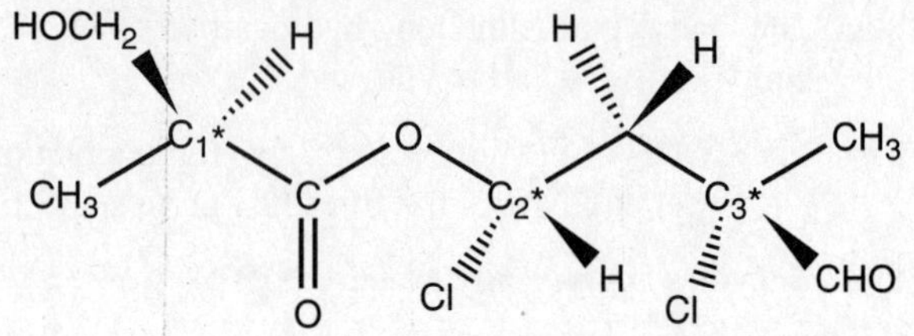

3. The mRNA is GUAACUG, and the DNA is CATTGAG.

4. a. The type of link is determined by the carbon next to the oxygen in the ring rather than the carbon it is connected to in the other ring. For this structure, that is in the lower ring. (The top monosaccharide is linked by a different carbon.) Because the ether link is above that ring, this is an α link.
   b. In this structure, the top ring has the relevant carbon. Therefore, this is an β glycosidic link.

5. The complement to A is T and the complement to C is G and the reverse. Therefore, the complementary chain to AATCGGCCTT is TTAGCCGGAA.

Chapter 20 Nuclear Chemistry

OUTLINE

REVIEW

Chapter Overview

Nuclear reactions differ from regular chemical reactions because the elements are changed into different elements; thus it is the nuclei not the bonds made by the outer-shell electrons that are changing. **Nuclear chemistry** is the study and application of these nuclear changes. To describe the details of this process, isotopic notation is used. For a nuclear reaction to be balanced, the sum of the mass numbers (superscripts) on the reactant side must equal the sum of the mass numbers on the product side. In addition, the sum of the charges of each nuclear particle (subscripts) must also be equal on each side for a balanced reaction. **Fusion** is the process whereby two small particles combine to make a larger particle. These are the type of reactions that are occurring in the sun. Fusion reactions produce energy if the atoms reacting are small (mass number < 56). **Fission** is the process of a large particle falling apart into smaller particles and is the type of reaction used in nuclear reactors. Fission reactions produce energy when the atom falling apart is large (mass number > 56). Nuclear reactions that produce products that induce further nuclear reactions are called **chain reactions**. The amount of material needed to sustain a chain reaction is called the **critical mass**.

Another type of nuclear reaction is **radioactive decay**. In this process an isotope spontaneously releases a small particle. These particles include: **alpha (*α*) particles**, which are equivalent to a helium nucleus ($^{4}_{2}He$); **beta (*β*) particles**, which are equivalent to an electron ($^{0}_{-1}e$); **positrons**, which have all the properties of an electron but the opposite charge ($^{0}_{1}e$); and **gamma (*γ*) rays**, which are high-energy electromagnetic radiation. Positrons are classified as **antimatter**. When a positron encounters its matter opposite (an electron), the two particles are **annihilated**, creating energy in the form of a gamma ray.

All radioactive decay occurs with first-order kinetics. Therefore, the rate is proportional to the number of reacting particles and the half-life ($t_{1/2}$) is constant. The rate of decay, or **activity**, is often expressed in units of **bequerels (Bq)**. One Bq is one decay event per second. Rate of decay can also be expressed in **curies (Ci)**, where $1 \text{ Ci} = 3.7 \times 10^{10}$ Bq. These rates can be measured with **scintillation counters** that measure the light produced when radiation strikes a particular type of material called a phosphor. **Geiger counters**, which measure the ionization caused by radioactivity, can also be used to measure the rate of radioactive decay.

The ionization caused by radioactive particles is hazardous, since it can break bonds, produce ions, and generally interfere with normal biological processes. The biological effects of ionizing radiation depend on how much energy is transferred to tissue. Therefore, the unit to measure biological effect, or **dose**, is the **gray (Gy)** which is 1 J/kg tissue. However, the type of particle is also important and accounted for with the **relative biological effect (RBE)**. A **sievert (Sv)** accounts for both, and is determined by the product of grays and the RBE. However, for the radiation to have an effect it must have **penetrating power**. For example, alpha particles have a low penetrating power and are stopped by as little as a sheet of paper; therefore, they rarely interact with tissue. However, gamma particles have a high penetrating power and will travel easily through the body.

There are several common types of radioactive decay. **Alpha (α) decay** occurs in elements where the atomic number (Z) is greater than 83. **Beta decay** occurs when an atom is neutron-rich. **Positron decay** or electron capture will occur when an atom is neutron-poor. **Electron capture** occurs when a nucleus captures one of its valence electrons. Atoms that do not undergo radioactive decay have a ratio of protons to neutrons in the **belt of stability**. This ratio is 1:1 for lighter atoms. However, as the atomic number increases, stable atoms will have more neutrons. Since the atomic weight is the average of all stable isotopes, atoms that have a mass number near this average are more likely to be stable. In addition, some atoms appear to be particularly stable. These atoms have a **magic number** of protons or neutrons. This is similar to the chemical effect of having a full shell of electrons, which is what makes the noble gases generally unreactive.

The energy that holds a nucleus together is called the **binding energy**. This can be calculated from the **mass defect**: the difference between the actual mass of the atom and the sum of the masses of all the protons, neutrons, and electrons in the atom. Using Einstein's $E = mc^2$ equation, this mass can be converted into energy units. The binding energy is significantly higher than energies typical of chemical reactions.

Worked Examples

Nuclear Binding Energies

Example 1.
What is the binding energy and the binding energy per nucleon for ^{118}Sn? The actual mass of the atom = 1.957827×10^{-22} g.

Collect and Organize We are asked to calculate the total binding energy and the binding energy per nucleon for ^{118}Sn. The binding energy holds protons and neutrons together as the nucleus.

Analyze The sum of the masses of the protons and neutrons in a nucleus is always greater than the actual mass of the nucleus. The difference is referred to as the mass defect and is converted into a binding energy given by: binding energy = BE = $(\Delta m)c^2$, where c is the speed of light and Δm is the mass defect described previously.

Solve ^{118}Sn has 50 protons, 68 neutrons, and 50 electrons. Since the mass given is the mass of the atom (rather than the mass of just the nucleus), the mass of electrons must be included in the calculation of the theoretical mass. Mass of protons = $50(1.67263 \times 10^{-24}$ g) = 8.3632×10^{-23} g; mass of neutrons = $68(1.67494 \times 10^{-24}$ g) = 1.139×10^{-22} g; mass of electrons = $50(9.100 \times 10^{-28}$ g) = 4.55×10^{-26} g. The sum = 1.975775×10^{-22} g. The mass defect = sum – actual mass = 1.975775×10^{-22} g – 1.957827×10^{-22} g = $1.7947975 \times 10^{-24}$ g = $1.7947975 \times 10^{-27}$ kg. Binding energy = BE = $(\Delta m)c^2 = (1.7947975 \times 10^{-27}$ g$)(2.998 \times 10^8$ m/s$)^2 = 1.613 \times 10^{-10}$ J. The use of four significant figures is based on the speed of light used. The binding energy per nucleon is BE/(protons + neutrons) = BE/nucleon = BE/118 = 1.613×10^{-10} J/118 = 1.367×10^{-12} J.

Think About It The loss in mass when protons and neutrons combine to form nuclei is converted into the energy needed to hold them together. It does not appear to be a lot of energy, but this is the binding energy of *one* nucleus. Energies of chemical reactants are usually given per mole. For comparison, 1.613×10^{-10} J/atom = 9.71×10^{13} J/mol.

Example 2.
What is the binding energy of ^{55}Mn? (Actual mass of nucleus = 9.120517×10^{-26} kg.)

Collect and Organize We are asked to calculate the binding energy of ^{55}Mn.

Analyze We will use the process outlined in the previous example to first determine the mass defect and then the binding energy.

Solve ^{55}Mn has 25 protons and 30 neutrons. Because the actual mass is that of the nucleus, the electrons don't need to be included. Mass of protons = $25(1.67263 \times 10^{-24}$ g) = 4.1816×10^{-23} g; mass of neutrons = $30(1.67494 \times 10^{-24}$ g) = 5.0248×10^{-23} g; the sum = 9.2064×10^{-23} g × (1 kg/1000 g) = 9.2064×10^{-26} kg. Recall that in doing addition and subtraction, each number must have the same units. Mass defect = $\Delta m = 9.2064 \times 10^{-26}$ kg – 9.120517×10^{-26} kg = 8.5883×10^{-28} kg. BE = $(8.5883 \times 10^{-28}$ kg$)(2.998 \times 10^8$ m/s$)^2$ = 7.719×10^{-11} J.

Think About It Although the binding energy of ^{55}Mn is less than the binding energy of ^{118}Sn, the best measure of how effectively the nucleons are held together is the ratio of binding energy to nucleons. Atoms with a higher binding energy per nucleon are more stable.

Example 3.
How much energy is created when 1000 atoms of ^{239}Pu decay in the following reaction?

$$^{239}Pu \rightarrow {}^{4}He + {}^{235}U$$

Actual masses: $^{239}Pu = 239.05218$ amu, $^{4}He = 4.0026033$ amu, $^{235}U = 235.04394$ amu, 1 amu = 1.66056×10^{-24} g.

COLLECT AND ORGANIZE We are asked to calculate the energy created during the nuclear decay of 1000 atoms of ^{239}Pu.

ANALYZE The equation used for binding energies can also be used to calculate the energy associated with nuclear reaction. Instead of a mass defect, the mass difference between the products and reactants is used for Δm.

SOLVE Mass difference = 239.05218 − (4.0026033 + 235.04394) = 0.0056367 amu. (With correct significant figures, the answer is 0.00564 amu. Because it was addition, decimal places are used to determine the significant figures of the answer. This answer has three significant figures. The entire number will be used for the rest of the calculation, but the final answer can't have more than three significant figures because of this first calculation.)

For use in the equation, atomic mass units must be converted to kilograms:

$$0.0056367 \text{ amu} \times \left(\frac{1.66056 \times 10^{-24}\text{ g}}{1 \text{ amu}}\right) \times \left(\frac{1 \text{ kg}}{1000 \text{ g}}\right)$$

$$= 9.360079 \times 10^{-30} \text{ kg}$$

An alternate way to obtain this answer is to convert each mass in atomic mass units to grams or kilograms before doing the subtraction.

$$E = \Delta mc^2 = (9.360079 \times 10^{-30} \text{ kg})(2.998 \times 10^{8} \text{ m/s})^2$$
$$= 8.4128 \times 10^{-13} \text{ J}$$

This value was calculated for one atom. Since the problem says there are 1000 atoms, energy generated = 1000 $(8.4128 \times 10^{-13}$ J$) = 8.41 \times 10^{-10}$ J. Since the mass difference only had three significant figures, the final answer has only three. We have assumed that 1000 atoms can be counted and that this value is exact (infinite significant figures).

THINK ABOUT IT The energy given off during either nuclear fusion or nuclear fission is the difference between the binding energies of the fuel and the fusion or fission products.

EXAMPLE 4.
How much energy is created from the annihilation of an electron (9.100×10^{-28} g) by a positron?

COLLECT AND ORGANIZE We are asked to calculate the energy created during the annihilation of an electron by a positron.

ANALYZE The equation used for binding energies can also be used to calculate the energy associated with the annihilation reaction. Instead of a mass defect, the mass difference between the products and reactants is used for *m*. Note that the mass of matter and antimatter are equal.

SOLVE

$$\Delta m = 2(9.100 \times 10^{-28} \text{ g}) = 1.820 \times 10^{-27} \text{ g} \times (1 \text{ kg}/1000 \text{ g})$$
$$= 1.820 \times 10^{-30} \text{ kg}$$

$$E = \Delta mc^2 = (1.820 \times 10^{-30})(2.998 \times 10^{8})^2 = 1.636 \times 10^{-13} \text{ J}$$

THINK ABOUT IT Since a positron is the antimatter of an electron, it will have the same mass as an electron. Since matter and antimatter annihilate each other, the mass of the product is zero. In other words, energy is the only product.

BELT OF STABILITY

EXAMPLE 1.
Write the nuclear decay reaction for the following isotopes:

a. ^{23}Al b. ^{241}Am c. ^{140}Cs d. ^{145}Eu e. ^{175}W

COLLECT AND ORGANIZE We are asked to predict the mode of decay for several unstable isotopes, and then write the nuclear reaction for that decay.

ANALYZE Large atoms (Z > 83) tend to undergo alpha decay; therefore, one of the products is ${}^{4}_{2}\alpha$. Isotopes that are neutron-rich (above the belt of stability) tend to undergo ${}^{0}_{-1}e$ decay. Isotopes that are neutron-poor (below the belt of stability) undergo positron (${}^{0}_{1}e$) emission or electron capture. Lighter elements (Z < 50) are more likely to undergo positron emission, whereas heavier elements are more likely to undergo electron capture.

SOLVE

a. The aluminum atom has an average atomic mass of about 27. Therefore, the ^{23}Al isotope is neutron-poor. It is also fairly light (Z < 50). Therefore, it will undergo positron emission.

$${}^{23}_{13}Al \rightarrow {}^{0}_{1}e + {}^{23}_{12}Mg$$

b. The americium isotope has Z = 95; consequently, it will undergo α decay.

$${}^{241}_{95}Am \rightarrow {}^{4}_{2}\alpha + {}^{237}_{93}Np$$

c. The cesium atom has an average atomic mass of about 133. Consequently, the cesium-140 isotope is neutron-rich. Therefore, it will undergo β decay.

$${}^{140}_{55}Cs \rightarrow {}^{0}_{-1}e + {}^{140}_{56}Ba$$

d. The europium atom has an average atomic mass of about 152. Therefore, europium-145 is neutron-poor and will decay by either positron emission or electron capture. As a heavier element, it is more likely to decay by electron capture than by positron emission.

$^{195}_{63}\text{Eu} + ^{0}_{-1}\text{e} \rightarrow ^{195}_{62}\text{Sm}$

e. Tungsten atoms have an average atomic mass of about 184. Consequently, the tungsten-179 isotope is neutron-poor. As a neutron-poor, heavier atom, it is likely to decay by electron capture.

$^{179}_{74}\text{W} + ^{0}_{-1}\text{e} \rightarrow ^{179}_{73}\text{Ta}$

THINK ABOUT IT The belt of stability shows the stable isotopes of each atom. The average atomic mass recorded on the periodic table is the average of these stable isotopes. Therefore, whether an isotope is above or below the belt of stability can be predicted using the periodic table.

RADIOCHEMICAL DATING

EXAMPLE 1.
What is the age of an object that has 31.5% of its initial ^{14}C activity? ($t_{1/2}$ = 5730 years)

COLLECT AND ORGANIZE We are asked to determine the age of an object based on the half-life of a radioactive element that it contains.

ANALYZE In order to determine the age of an object (t) we will use $t = (t_{1/2}/0.693)\ln(A_0/A_t)$, where t = time or the age of the object, $t_{1/2}$ = half-life of the isotope, A_t is the amount of the isotope at the time of measurement, and A_0 is the initial amount of the isotope.

SOLVE Since the remaining amount of ^{14}C is given as a percentage, the easiest way to interpret this into the equation is $A_0 = 1$ and $A_t = 0.315$. (*Note:* the 1 is exact; the 0.315 has three significant figures.)

$$t = \left(\frac{5730}{0.693}\right)\ln\left(\frac{1}{0.315}\right) = 9551.51$$

The units are years. There are three significant figures because of the value of 0.693 and $t_{1/2}$. Age = 9.55×10^3 yr.

THINK ABOUT IT Since radioactive decay occurs at a very well-defined rate, expressed as a half-life, it can be used to determine the age of objects. When the radioactive isotope is ^{14}C, the technique is referred to as radiocarbon dating and is a very powerful technique for dating animals and plants. The assumption is that the radioactive carbon-14 (^{14}C) was incorporated into the material only when it was alive. As the carbon-14 undergoes β decay to nitrogen-14, the ratio of carbon-14 to carbon-12 decreases. The amount of the decrease will depend on time according to the half-life of the isotope.

EXAMPLE 2.
^{244}Cf has a half-life of 17.9 years. How long will it take for 1.00 mg to be left from a 1.00-g original sample?

COLLECT AND ORGANIZE We are asked to determine the amount of time it will take for 1.00 g of a radioactive sample to decay to 1.00 mg given its half-life.

ANALYZE We can use the expression for radioactive decay to solve the problem:

$$t = \left(\frac{t_{1/2}}{0.693}\right)\ln\left(\frac{A_0}{A_t}\right)$$

SOLVE From the problem, $t_{1/2}$ = 17.9 yr, A_0 = 1.00 g, and A_t = 1.00 mg. The values for A must be in the same units. As there are 1000 mg in 1 g, A_0 is easily changed to 1000 mg.

$$t = \left(\frac{17.9}{0.693}\right)\ln\left(\frac{1000}{1}\right) = 25.8297(6.907755) = 178.425 \text{ yr}$$

(Remember time units must also match.) Note that each value only has three significant figures, so t = 178 yr.

THINK ABOUT IT Although californium does not occur naturally on Earth, the element and its decay products occur elsewhere in the universe. Their electromagnetic emissions are regularly observed in the spectra of supernovae.

EXAMPLE 3.
If after 348 days, 1.00 g of an isotope remains from a 5.00-g sample, what is the half-life of the isotope?

COLLECT AND ORGANIZE From a known initial amount radioactive isotope we are given the amount remaining after a specified amount of time and are asked to calculate its half-life.

ANALYZE We can use the expression for radioactive decay to solve the problem:

$$t = \left(\frac{t_{1/2}}{0.693}\right)\ln\left(\frac{A_0}{A_t}\right)$$

SOLVE From the problem, t = 348 days, A_0 = 5.00 g, and A_t = 1.00 g.

$$348 = \left(\frac{t_{1/2}}{0.693}\right)\ln\left(\frac{5}{1}\right) = 2.322t_{1/2}$$

$t_{1/2}$ = 149.84. The time unit is days. The answer should have three significant figures, so the final answer is 150 days.

THINK ABOUT IT We can determine the half-life of any isotope as long as we know how much of the initial material remains after a specified amount of time.

EXAMPLE 4.

^{210}Po has a half-life of 138.38 days. If 4.00 mg of ^{210}Po is allowed to decay for 30.0 days, how much would remain?

COLLECT AND ORGANIZE We are asked to calculate the amount of radioactive isotope that remains, given the initial amount, the time elapsed, and the isotope's half-life.

ANALYZE We can use the expression for radioactive decay to solve the problem.

$$t = \left(\frac{t_{1/2}}{0.693}\right)\ln\left(\frac{A_0}{A_t}\right)$$

SOLVE From the information given, t = 30.0 days, $t_{1/2}$ = 138.38 days, and A_0 = 4.00 mg.

$$30.0 = \left(\frac{138.38}{0.693}\right)\ln\left(\frac{4.00}{A_t}\right) = 199.68\ \ln\left(\frac{4.00}{A_t}\right)$$

$$e^{0.15024} = \frac{4.00}{A_t}$$

$$1.16211A_t = 4.00$$
$$A_t = 3.44\text{ mg.}$$

The three significant figures were pretty constant throughout the problem. Since the units on the original sample are milligrams, the same units are on the final answer.

THINK ABOUT IT We can easily calculate the amount of radioactive isotope that remains after any specified amount of time so long as we have the particular isotope's half-life and its initial mass.

Self-Test

KEY VOCABULARY COMPLETION QUESTIONS

________ 1. Region where the proton and neutron ratio indicates a stable nucleus

________ 2. Time required to reduce the quantity of isotopes by one-half

________ 3. Process that produces ${}^{0}_{1}e$

________ 4. Process where two nuclei combine to create a larger nucleus

________ 5. Unit of radioactivity equivalent to one decay event per second

________ 6. Difference between actual and theoretical mass

________ 7. Particle with a mass of 4 and a charge of +2

________ 8. Unit of dose that accounts for the type of particle

________ 9. Number of protons or neutrons that creates a very stable nucleus

________ 10. Nuclear reaction of a neutron-rich atom

________ 11. Device to measure radioactivity based on ionization

________ 12. Highest energy form of electromagnetic radiation

________ 13. Amount of material needed to sustain a chain reaction

________ 14. Order of nuclear decay reactions

________ 15. What happens when matter encounters its antimatter counterpart

________ 16. Nuclear reactions that induce more nuclear reactions

________ 17. Unit of radioactivity activity that expresses many decay events

________ 18. How far a radioactive particle will travel through matter

________ 19. Dose when one joule of energy is transferred to 1 kilogram of body tissue

________ 20. Energy holding a nucleus together

MULTIPLE-CHOICE QUESTIONS

1. Which particle has a mass of 1 amu and a charge of +1?
   a. n
   b. β
   c. p
   d. α

2. What is the appropriate symbol for a positron?
   a. ${}^{0}_{1}e$
   b. ${}^{1}_{0}e$
   c. ${}^{0}_{-1}e$
   d. ${}^{0}_{0}e$

3. What is the mass defect for a ^{232}Th atom (actual mass = 232.03807 amu)?
   a. 3.8531×10^{-22} g
   b. 0.3807 amu
   c. 3.234×10^{-24} g
   d. 3.8854×10^{-25} g

4. How many neutrons are in ^{11}B?
   a. 11
   b. 5
   c. 6
   d. 10

5. What is the binding energy of an atom with a mass defect of 9.96×10^{-27} g?
   a. 8.95×10^{-10} J
   b. 2.99×10^{-28} J
   c. 2.99×10^{-31} J
   d. 8.95×10^{-13} J

6. If ^{40}Ca captures a neutron, the isotope formed would be
   a. ^{40}Sc
   b. ^{40}K
   c. ^{41}Sc
   d. ^{41}Ca

7. Which isotopes produce energy when undergoing fusion?
   a. isotopes larger than ^{56}Fe
   b. isotopes smaller than ^{56}Fe
   c. isotopes with a high proton/neutron ratio
   d. isotopes with a low proton/neutron ratio

8. Which isotope of osmium is stable?
   a. ^{76}Os
   b. ^{200}Os
   c. ^{190}Os
   d. ^{152}Os

9. Which isotope is most likely to undergo β decay?
   a. ^{17}O
   b. ^{16}O
   c. ^{15}O
   d. ^{8}O

10. Which isotope is most likely to undergo α decay?
    a. ^{150}Sm
    b. ^{24}Al
    c. ^{222}Rn
    d. ^{56}Fe

11. What is a product of the β decay of ^{98}Mo?
    a. ^{98}Nb
    b. ^{98}Tc
    c. ^{99}Mo
    d. ^{97}Mo

12. What is a product of the positron emission of ^{11}C?
    a. ^{11}B
    b. ^{11}N
    c. ^{12}C
    d. ^{13}C

13. What is a product of the electron capture of ^{163}Ho?
    a. ^{163}Dy
    b. ^{163}Er
    c. ^{164}Ho
    d. ^{165}Ho

14. ^{207}Pb is formed as the α decay product of which isotope?
    a. ^{207}Bi
    b. ^{207}Tl
    c. ^{208}Pb
    d. ^{211}Po

15. Radon is particularly hazardous because
    a. it undergoes α decay
    b. it is a gas
    c. both of the above
    d. neither of the above

16. For a neutron-induced chain reaction to occur
    a. there must be enough of a target isotope to absorb the neutron
    b. the reaction must produce neutrons
    c. both of the above
    d. neither of the above

17. Neutrons are better at bombarding nuclei than α particles because
    a. neutrons are bigger
    b. neutrons are faster
    c. neutrons are neutral
    d. all of the above

18. If an isotope undergoes β decay at a rate of 6.8×10^8 decays/s, what is its radioactivity in curies?
    a. 6.8×10^8 Ci
    b. 1.4×10^{10} Ci
    c. 6.8×10^6 Ci
    d. 0.018 Ci

19. Why are γ rays considered the most dangerous form of radiation?
    a. γ rays cause the most biological damage
    b. γ rays are the most penetrating
    c. γ particles are very large
    d. all of the above

20. What is the advantage of sieverts over grays as a unit of radioactivity?
    a. sieverts take into account relative biological effectiveness
    b. sieverts take into account the energy of the radiation
    c. sieverts take into account the number of particles
    d. sieverts take into account the size of the person

21. How much of 1.0 g of A with $t_{1/2}$ is left after $2t_{1/2}$?
    a. 0.50 g
    b. 0.10 g
    c. 0.70 g
    d. 0.25 g

22. If the ^{14}C is 64.4% of its original value, how old is it?
    a. 5270 years
    b. 1.27×10^4 years
    c. 3600 years
    d. 1.86×10^4 years

23. What is the half-life of an isotope if 5.0 g remains of a 10.0-g sample after 20 days?
    a. 20 days
    b. 40 days
    c. 46 days
    d. 8.7 days

24. What is the half-life of an isotope if 10.0 g decays to 1.0 mg in 27.9 minutes?
    a. 8.41 min
    b. 2.10 min
    c. 371 min
    d. 92.7 min

25. How much energy is absorbed by a 70-kg man exposed to 100 mSv of β radiation?
    a. 2.7×10^{-10} J
    b. 7.0 J
    c. 14 J
    d. 1.4×10^{-3} J

Additional Practice Problems

1. Complete the following nuclear reactions.
   a. $^{16}O + {}^{3}He \rightarrow$ ____ + p
   b. ^{68}Zn + p → ____ + 2n
   c. $^{122}Te + \alpha \rightarrow$ ____ + 3n
   d. ^{201}Pb + e → ____

2. Predict the nuclear decay of the following isotopes and write the relevant reactions.
   a. ^{123}Xe b. ^{117}Cd c. ^{17}F d. ^{247}Bk e. ^{125}Sn

3. What is the dose, in sieverts, if a 75.0-kg person is exposed to 5.00 J of γ radiation with RBE = 1.0?

4. How old is an object if the ^{14}C activity has decreased to 3.55% of its original value?

5. What is the binding energy of ^{66}Zn (actual mass of atom = 65.926040 amu)?

Self-Test Answer Key

Key Vocabulary Completion Answers

1. belt of stability
2. half-life
3. positron decay
4. fusion
5. Bequerel
6. mass defect
7. alpha
8. sievert
9. magic number
10. beta decay
11. Geiger counter
12. gamma rays
13. critical mass
14. first
15. annihilation
16. chain reactions
17. Curie
18. penetrating power
19. Gray
20. binding energy

Multiple-Choice Answers

For further information about the topics these questions address, you should consult the text. The appropriate section of the text is listed in parentheses after the explanation of the answer.

1. c. A neutron has no charge, a β particle has no significant mass, and an α particle has a mass of 4 amu. (20.3)

2. a. A positron has the opposite charge as an electron ($e = -1$, so positron = +1) and the same mass. (20.3)

3. c. ^{232}Th has 90 protons, 90 electrons, and 142 neutrons. The sum of the masses of each particle is $90(1.67355 \times 10^{-24}$ g) + $90(9.100 \times 10^{-28}$ g) + $142(1.67495 \times 10^{-24}$ g) = 3.88544×10^{-22} g. To convert mass in atomic mass units to mass in grams = 232.03807 × (1.66056×10^{-24} g/1 amu) = 3.8531×10^{-22} g. The mass defect = sum – actual = 3.88544×10^{-22}g – 3.8531×10^{-22} g = 3.234×10^{-24} g. *Note:* Answer b is not correct, because p = 1.007825 amu and n = 1.008665 amu, not exactly 1 for either. (20.4)

4. c. The mass number is 11, which is the sum of protons and neutrons. Since boron has $Z = 5$, or five protons, there are six neutrons. (20.4 and 20.5)

5. d. $E = \Delta mc^2 = (9.96 \times 10^{-30}\text{ kg})(2.998 \times 10^8\text{ m/s})^2 = 8.95 \times 10^{-13}\text{ J}$ (20.4)

6. d. ${}^{40}_{20}\text{Ca} + {}^{1}_{0}n \rightarrow {}^{41}_{20}\text{Ca}$ (20.6)

7. b. Isotopes will produce energy by fusion if smaller than ^{56}Fe, and by fission if larger than ^{56}Fe. (20.5)

8. c. ^{190}Os is in the belt of stability; it is also near the average atomic mass of osmium. (20.6)

9. a. β decay occurs in neutron-rich isotopes. At low atomic numbers, the optimum proton/neutron ratio is 1. The most stable isotope of oxygen is ^{16}O, as evidenced by the atomic mass of almost exactly 16. ^{8}O would have no neutrons and is not a reasonable isotope of oxygen. (20.6)

10. c. Alpha decay occurs in elements with $Z > 83$. The only isotope that qualifies is ^{222}Rn with $Z = 86$. (20.6)

11. b. ${}^{98}_{42}\text{Mo} \rightarrow {}^{0}_{-1}\text{e} + {}^{98}_{43}\text{Tc}$ (20.6)

12. a. ${}^{11}_{6}\text{C} \rightarrow {}^{0}_{1}\text{e} + {}^{11}_{5}\text{B}$ (20.6)

13. a. ${}^{163}_{67}\text{Ho} + {}^{0}_{-1}\text{e} \rightarrow {}^{163}_{66}\text{Dy}$ (20.6)

14. d. ${}^{211}_{84}\text{Po} \rightarrow {}^{4}_{2}\alpha + {}^{207}_{82}\text{Pb}$ (20.6)

15. c. Alpha particles have a relative biological effectiveness of 20, which makes them more hazardous than β or γ radiation. Alpha particles are normally not a problem because they can be stopped by the skin. However, since radon is a gas, it can be breathed in. It undergoes α decay in the lungs, where it can do significant damage. (20.10)

16. c. A chain reaction requires that one reaction start another. Since neutrons start the reaction, more neutrons must be produced to sustain it. The target isotope is also a reactant and must be in sufficient quantities to be bombarded with the neutron. Therefore, both have to be true. (20.8)

17. c. The positive charge of the α particle is repelled by the nucleus. Since the neutron is uncharged, it is not repelled by the nucleus and is more likely to bombard it. (20.8)

18. d. $6.8 \times 10^8\text{ dps} \times \left(\dfrac{1\text{ Ci}}{3.70 \times 10^{10}\text{ dps}}\right) = 0.018\text{ Ci}$ (20.9)

19. b. Gamma rays are photons, with no mass. They have the same relative biological effect as β particles (RBE = 1). However, as electromagnetic radiation, γ rays are extremely penetrating. (20.9 and 20.10)

20. a. Grays take into account all the items listed except relative biological effectiveness (RBE). Sv = RBE × Gy. (20.9 and 20.10)

21. d. After one half-life, half the material will be left = 0.5 g. Half of that disintegrates with the passing of another half-life = 0.25 g. (20.2)

22. c. $t = \left(\dfrac{t_{1/2}}{0.693}\right)\ln\left(\dfrac{A_0}{A_t}\right) = \left(\dfrac{5370}{0.693}\right)\ln\left(\dfrac{1}{0.644}\right) = 3600\text{ years}$

(20.2 and 20.12)

23. a. 5.0 g is half of 10.0 g, so it has taken one half-life to reach that point, so 20 days.

24. b. $t = \left(\dfrac{t_{1/2}}{0.693}\right)\ln\left(\dfrac{A_0}{A_t}\right)$; $27.9\text{ min} = \left(\dfrac{t_{1/2}}{0.693}\right)\ln\left(\dfrac{10000\text{ mg}}{1\text{ mg}}\right)$

$27.9\text{ min} = \left(\dfrac{t_{1/2}}{0.693}\right)(9.210)$; $\dfrac{(27.9)(0.693)}{(9.210)} = t_{1/2}$;

$t_{1/2} = 2.10\text{ min.}$ (20.2)

25. b. For β radiation, Gy = Sv. 1 Gy = 1 J/kg.

$100\text{ mSv} \times \left(\dfrac{1\text{ Sv}}{1000\text{ mSv}}\right) \times \left(\dfrac{1\text{ Gy}}{1\text{ Sv}}\right) \times \left(\dfrac{1\text{ J/kg}}{1\text{ Gy}}\right) =$

$0.10\text{ J/kg} \times 70\text{ kg} = 7.0\text{ J.}$ (20.10)

Additional Practice Problem Answers

1. a. ${}^{16}_{8}\text{O} + {}^{3}_{2}\text{He} \rightarrow {}^{1}_{1}\text{p} + {}^{18}_{9}\text{F}$
   b. ${}^{68}_{30}\text{Zn} + {}^{1}_{1}\text{p} \rightarrow 2\,{}^{1}_{0}\text{n} + {}^{67}_{31}\text{Ga}$
   c. ${}^{122}_{52}\text{Te} + {}^{4}_{2}\alpha \rightarrow 3\,{}^{1}_{0}\text{n} + {}^{123}_{53}\text{Xe}$
   d. ${}^{201}_{82}\text{Pb} + {}^{0}_{-1}\text{e} \rightarrow {}^{201}_{81}\text{Tl}$

2. a. ^{123}Xe is neutron-poor and a large isotope, so it undergoes electron capture.

$${}^{123}_{54}\text{Xe} + {}^{0}_{-1}\text{e} \rightarrow {}^{123}_{53}\text{I}$$

   b. ^{117}Cd is neutron-rich, so it undergoes β decay.

$${}^{117}_{48}\text{Cd} \rightarrow {}^{0}_{-1}\text{e} + {}^{117}_{49}\text{In}$$

   c. ^{17}F is a light, neutron-poor isotope, so it decays by positron emission.

$${}^{17}_{9}\text{F} \rightarrow {}^{0}_{1}\text{e} + {}^{17}_{8}\text{O}$$

   d. ^{247}Bk is a heavy element, with $Z = 97$, so it undergoes α decay.

$${}^{247}_{97}\text{Bk} \rightarrow {}^{4}_{2}\alpha + {}^{243}_{95}\text{Am}$$

   e. ^{125}Sn is neutron-rich, so it undergoes β decay.

$${}^{125}_{50}\text{Sn} \rightarrow {}^{0}_{-1}\text{e} + {}^{125}_{51}\text{Sb}$$

3. Since 1 Gy = 1 J/kg and Sv = Gy × RBE,

$$\frac{5.00\text{ J}}{75.0\text{ kg}} = 0.0667\text{ J/kg} \times \left(\frac{1\text{ Gy}}{1\text{ J/kg}}\right) = 0.0667\text{ Gy}$$

The RBE of γ is 1, so Gy = Sv.

$$0.0667\text{ Gy} \times \left(\frac{1\text{ Sv}}{1\text{ Gy}}\right) \times \left(\frac{1000\text{ mSv}}{1\text{ Sv}}\right) = 66.7\text{ mSv}$$

4. The half-life of ^{14}C is 5730 years = $t_{1/2}$. If activity is 3.55% of the initial amount, $A_0 = 1$ and $A_t = 0.0355$.

$$t = \left(\frac{t_{1/2}}{0.693}\right)\ln\left(\frac{A_0}{A_t}\right)$$

$$t = \left(\frac{5370}{0.693}\right)\ln\left(\frac{1}{0.0355}\right) = 27601.75$$

$t = 2.76 \times 10^4$ years

5. ^{66}Zn has 30 protons, 36 neutrons, and 30 electrons. We have

$$30(1.67355 \times 10^{-24}) + 36(1.67495 \times 10^{-24}) + 30(9.100 \times 10^{-28}) = 1.10532 \times 10^{-22}\text{ g}$$

$$\text{actual mass} = 65.926040\text{ amu} \times \left(\frac{1.66056 \times 10^{-24}\text{ g}}{1\text{ amu}}\right)$$
$$= 1.094714 \times 10^{-22}\text{ g}$$

$$\text{mass defect} = 1.10532 \times 10^{-22} - 1.094714 \times 10^{-22}$$
$$= 1.057855 \times 10^{-24}\text{ g}$$
$$= 1.057855 \times 10^{-27}\text{ kg}$$

$$\text{energy} = (\Delta m)c^2$$

$$E = (1.057855 \times 10^{-27}\text{ kg})(2.998 \times 10^8\text{ m/s})^2$$

$$E = 9.508005 \times 10^{-11}\text{ J}$$

With correct significant figures, $E = 9.508 \times 10^{-11}$ J.

CHAPTER 21 Life and the Periodic Table

OUTLINE

REVIEW

Chapter Overview

Most elements can be found in the human body. Many of these are **essential elements**, required for the body to function. Of the **nonessential elements**, some are stimulatory, increasing growth, while others have no known function.

Of the essential elements, eleven are found in bulk, gram-sized quantities. These **bulk essential elements** include the main group elements of carbon, hydrogen, oxygen, nitrogen, sulfur, and phosphorus. These make up much of the proteins, fats, and water found in the body. The bulk essential metals are calcium, magnesium, sodium, and potassium. Calcium is used in the composite materials of bones and teeth. **Composite materials** are mixtures of substances of different composition and structure. In bones and teeth calcium appears primarily as hydroxyapatite, $Ca_5(PO_4)_3(OH)$. Magnesium is used in the transfer of phosphate groups in ATP. Sodium and potassium exist as ions, and are important in the generation of nerve impulses and maintaining proper osmotic pressure in cells. To do so, they must pass through the cell membrane. **Ion channels**, helical proteins that penetrate cell membranes, and **ion pumps**, that use the energy of ATP, aid in this process. Chlorine exists in the body as chloride ion, and provides a counter anion to the metal ions so that electrical neutrality can be maintained.

Many other elements exist in **trace** (mg) or **ultratrace** (μg) quantities. Many of these are components of metalloenzymes, where the metal complexes with a nitrogen or sulfur in the enzymes. If the metal bonds directly to a carbon, as it does in vitamin B_{12}, it is classified as an **organometallic**. Enzymes are catalysts for biochemical reactions. Some enzymes require a coenzyme partner to be effective. Our modern life exposes us to many elements that end up as trace elements and can replace elements with similar chemical and physical properties. This many be innocuous as when Rb^+ replaces K^+, beneficial as when F^- replaces the OH^- in hydroxyapatite, or harmful as when arsenic binds to the sulfur of some enzymes.

Many elements are used for medical purposes. Because they can easily be detected in very small quantities, radioactive isotopes are useful in many diagnostic tests. Positron emission tomography (PET) is used as an imaging technique. Gadolinium is used as a contrasting agent to make magnetic resonance imaging (MRI) more effective. Other elements are used to make medicines, such as lithium in the treatment of depression, gold in the treatment of rheumatoid arthritis, and platinum as a chemotherapy agent.

In the appropriate form and concentration, all elements are toxic. Some elements have a long history as poisons. These include lead, arsenic, and mercury. Other elements, such as cadmium, are a more modern concern as their use becomes common. Thus elements can be beneficial or harmful depending on how they are used.

Self-Test

Key Vocabulary Completion Questions

__________ 1. Time for the amount of a substance to decrease to half its original value

__________ 2. Elements that are required for life

__________ 3. Second-row transition metal that does not exist naturally

__________ 4. Calcium-containing material that is a major component of bones

__________ 5. A bulk element found in the body as an anion

__________ 6. Biological compound that catalyzes reactions

__________ 7. Metal found in hemoglobin

__________ 8. Elements found in milligram quantities in the body

__________ 9. Ultratrace element used by the thyroid gland to make thyroxine

__________ 10. Very toxic element that is a liquid at room temperature

__________ 11. ATP-powered method to move Na^+ and K^+ ions

__________ 12. Metallic cation found in chlorophyll

__________ 13. The effect of a nonessential element that leads to increased growth

__________ 14. Path for potassium ion

__________ 15. A mixture of substances with different compositions and structures

Multiple-Choice Questions

1. Which of the following is NOT a bulk essential element?
   a. carbon
   b. hydrogen
   c. oxygen
   d. phosphorus
   e. uranium

2. Which of the following is NOT a trace essential element?
   a. iron
   b. zinc
   c. radon
   d. selenium
   e. cobalt

3. Which of the following is NOT a way sodium ion migrates through cell membranes?
   a. ion channels
   b. diffusion
   c. dispersion forces
   d. ion pumps

4. Which of the following is NOT a factor in the toxicity of elements?
   a. form of the element
   b. identity of the element
   c. concentration of the element
   d. route of exposure
   e. all of the above *are* factors

5. The primary function of chloride ion in the body is
   a. as a part of enzymes
   b. to slow reactions
   c. as a structural building block
   d. to maintain electrical neutrality in cells

6. In chlorophyll, magnesium
   a. makes it green
   b. provides structure
   c. transfers an electron
   d. transfers a proton

7. When metals bond to proteins, the reaction is best classified as
   a. precipitation
   b. Lewis acid–base
   c. Brønsted acid–base
   d. redox

8. Technetium is
   a. an artificial element
   b. a low abundance, naturally occurring element
   c. a high abundance, naturally occurring element
   d. a naturally occurring element whose artificial isotope is used in medical diagnosis

9. Which of the following diagnostic techniques use radioactive isotopes?
   a. Magnetic resonance imaging (MRI)
   b. Positron emission tomography (PET)
   c. X-rays
   d. all of the above use radioactive isotopes

10. The half-life of isotopes used in medical diagnosis should be on the order of
    a. microseconds
    b. seconds
    c. hours
    d. years
    e. centuries

11. Which of the following metals is not a bulk essential element?
    a. magnesium
    b. sodium
    c. potassium
    d. thorium
    e. calcium

12. Which of the following elements is commonly found in bones and teeth?
    a. calcium
    b. nitrogen
    c. sodium
    d. chlorine
    e. iron

13. What type of radioactive decay is NOT suitable for medical imaging?
    a. alpha emission
    b. beta emission
    c. positron emission
    d. gamma emission

14. Which of the following metals is NOT typically found in enzymes?
    a. iron
    b. zinc
    c. nickel
    d. cobalt
    e. mercury

15. In hydroxyapatite ($Ca_5(PO_4)_3(OH)$), fluoride ion replaces
    a. calcium
    b. phosphate
    c. phosphorus
    d. hydroxide
    e. hydrogen

Self-Test Answer Key

Key Vocabulary Completion Answers

1. half-life
2. essential
3. technetium
4. hydroxyapatite
5. chloride
6. enzyme
7. iron
8. trace
9. iodine
10. mercury
11. ion pump
12. magnesium
13. stimulatory
14. ion channel
15. composite material

Multiple-Choice Answers

For further information about the topics these questions address, you should consult the text. The appropriate section of the text is listed in parentheses after the explanation of the answer.

1. e. Uranium has no known biological function. The bulk essential elements are C, H, O, N, S, P, Ca, Mg, Na, K, and Cl. (21.2)
2. c. Radon has no known biological function. (21.2 and 21.3)
3. c. Dispersion forces are intermolecular forces important in nonpolar molecules and not a mechanism of ion transfer through cell membranes. (20.2)
4. e. Form is very important; for example, methylmercury is much more toxic than mercury metal. Identity is important; for example, mercury is more toxic than iron. Concentration is important; for example, even iron is toxic at sufficiently high concentrations. Route of exposure is also important; for example, some materials are more toxic if absorbed through the lungs (breathing in) than through the stomach (swallowing). (21.5)
5. d. With transfer of sodium and potassium across cell membranes, chloride ion follows so that electrical neutrality is maintained. (21.2)
6. b. By bonding to magnesium, the rest of the structure is in the proper shape to transfer electrons and protons and interact with light. (21.2)
7. b. In most proteins, metals bond covalently with nitrogens or sulfurs. In a Lewis acid–base reaction, the nitrogen or sulfur provides both electrons needed for the bond. (21.3)
8. a. Tc is not found naturally. (21.4)
9. b. PET uses isotopes that undergo positron decay. The other techniques do not require radioactive isotopes. (21.4)
10. c. If the half-life is too short there is not time to run the tests. If the half-life is too long, the continual exposure to radiation could be harmful. (21.4)

11. d. Th has no known biological function. Bulk trace metals are Ca, Mg, Na, and K. (21.2)

12. a. Calcium is part of hydroxyapatite, a major component of bones. (21.2)

13. a. Alpha decay does not have sufficient penetrating ability to leave the body and be detected. It is also more likely to be harmful in the body. (21.4)

14. e. Mercury has no known biological function and is toxic even in very small quantities. (21.5)

15. d. Fluoride replaces hydroxide to make $Ca_5(PO_4)_3F$, which is less susceptible to reactions with acids. (21.2)